The Origin and Evolution of Cultures

EVOLUTION AND COGNITION
General Editor, Stephen Stich, Rutgers University

The Origin and Evolution of Cultures

Robert Boyd

Peter J. Richerson

OXFORD

UNIVERSITY PRESS

2005

OXFORD
UNIVERSITY PRESS

Oxford   New York
Auckland   Bangkok   Buenos Aires   Cape Town   Chennai
Dar es Salaam   Delhi   Hong Kong   Istanbul   Karachi   Kolkata
Kuala Lumpur   Madrid   Melbourne   Mexico City   Mumbai   Nairobi
Sao Paulo   Shanghai   Taipei   Tokyo   Toronto

Published by Oxford University Press, Inc.
198 Madison Avenue, New York, New York 10016
www.oup.com

Oxford is a registered trademark of Oxford University Press

Library of Congress Cataloging-in-Publication Data

Boyd, Robert.
The origin and evolution of cultures / Robert Boyd,
Peter J. Richerson.
p. cm.—(Evolution and cognition)
Includes bibliographical references and index.
ISBN-13 978-0-19-516524-1; 978-0-19-518145-6 (pbk.)
ISBN 0-19-516524-1; 0-19-518145-X (pbk.)
  1. Social evolution. 2. Culture—Origin. 3. Human evolution.
4. Sociobiology. I. Richerson, Peter J. II. Title. III. Series.

GN360.B69 2004
306—dc22   2004043408

9 8 7 6 5 4 3 2

Printed in the United States of America
on acid-free paper

ACKNOWLEDGMENTS

1: R. Boyd and P. J. Richerson. Social learning as an adaptation. In: *Lectures on Mathematics in the Life Sciences*, 20:1–26, 1989.

2: R. Boyd and P. J. Richerson. Why does culture increase human adaptability? *Ethology and Sociobiology*, 16:125–143, 1995.

3: R. Boyd and P. J. Richerson. Why culture is common, but cultural evolution is rare. *Proceedings of the British Academy*, 88:77–93, 1996.

4: P. J. Richerson and R. Boyd. Climate, culture, and the evolution of cognition. In: *The Evolution of Cognition*. Celia Heyes and Ludwig Huber, eds. Cambridge, MA: MIT Press, 2000.

5: R. Boyd and P. J. Richerson. Norms and bounded rationality. In: *The Adaptive Tool Box*, G. Gigerenzer and R. Selten, eds. Cambridge, MA: MIT press, 2000, pp. 281–296.

6: R. Boyd and P. J. Richerson. The evolution of ethnic markers. *Cultural Anthropology*, 2:65–79, 1987.

7: R. McElreath, R. Boyd, and P. J. Richerson. Shared norms can lead to the evolution of ethnic markers. *Current Anthropology*, 44:122–130, 2003.

8: R. Boyd and P. J. Richerson. The evolution reciprocity in sizable groups. *Journal of Theoretical Biology*, 132:337–356, 1988.

9: R. Boyd and P. J. Richerson. Punishment allows the evolution of co-operation (or anything else) in sizable groups. *Ethology and Sociobiology*, 13:171–195, 1992.

10: J. Henrich and R. Boyd. Why people punish defectors: Weak conformist transmission can stabilize costly enforcement of norms in cooperative dilemmas. *Journal of Theoretical Biology*, 208:79–89, 2001.

11: J. Soltis, R. Boyd, and P. J. Richerson. Can group functional behaviors evolve by cultural group selection? An empirical test. *Current Anthropology*, 36:473–494, 1995.

12: R. Boyd and P. J. Richerson. Group beneficial norms spread rapidly in a structured population. *Journal of Theoretical Biology*, 215:287–296, 2002.

13: R. Boyd, H. Gintis, S. Bowles, and P. J. Richerson. The evolution of altruistic punishment. *Proceedings of the National Academy of Sciences* (USA), 100:3531–3535, 2003.

14: P. J. Richerson, R. Boyd, and J. Henrich. The cultural evolution of human cooperation. In: *The Genetic and Cultural Evolution of Cooperation*, P. Hammerstein, ed. Cambridge, MA: MIT Press, 2003, pp. 357–388.

15: R. Boyd and P. J. Richerson. How microevolutionary processes give rise to history. In: *Evolution and History*, M. Niteki, ed. Chicago: University of Chicago Press, 1992.

16: R. Boyd, P. J. Richerson, M. Borgerhoff-Mulder, and W. H. Durham. Are cultural phylogenies possible? In: *Human by Nature, Between Biology and the Social Sciences*, P. Weingart, P. J. Richerson, S. D. Mitchell, and S. Maasen, eds. Mahwah, NJ: Lawrence Erlbaum Associates, 1997, pp. 355–386.

17: P. J. Richerson, R. Boyd, and R. L. Bettinger. Was agriculture impossible during the Pleistocene but mandatory during the Holocene? A climate change hypothesis. *American Antiquity*, 66:387–411, 2001.

18: R. Boyd and P. J. Richerson. Rationality, imitation, and tradition. In: *Nonlinear Dynamics & Evolutionary Economics*, R. Day and P. Chen, eds. New York: Oxford University Press, pp. 131–149, 1993.

19: P. J. Richerson and R. Boyd. Simple models of complex phenomena: The case of cultural evolution. In: *The Latest on the Best: Essays on Evolution and Optimality*, J. Dupre, ed. Cambridge, MA: MIT Press, 1987, pp. 27–52.

20: R. Boyd and P. J. Richerson. Memes: Universal acid or a better mouse trap? In: *Darwinizing Culture: The Status of Memetics as a Science*. R. Aunger, ed. Oxford: Oxford University Press, 2000, pp. 143–162.

CONTENTS

PART 3: HUMAN COOPERATION, RECIPROCITY, AND GROUP SELECTION   133

PART 4: ARCHAEOLOGY AND CULTURE HISTORY   283

PART 5: LINKS TO OTHER DISCIPLINES   375

The Origin and Evolution of Cultures

Introduction

Over the last 30 years, we have worked to develop a theory and supporting evidence to account for the evolution of the human capacity for culture and how this capacity leads to distinctive evolutionary patterns. Much of our early work is summarized in our book *Culture and the Evolutionary Process*, published in 1985. Since that time we have published numerous articles that expand the theory and discuss relevant data. We think that these articles fit together to tell a consistent story about how the capacity for culture evolved in the human lineage and why it has led to evolutionarily novel outcomes like large-scale cooperation. However, because this work is relevant to scholars in disciplines ranging from evolutionary biology to archaeology to economics, these essays are scattered among an equally wide range of journals. As a result, the overall story is not so easy to discern. So when Steve Stich suggested that we might bring a sampling of this work together in a single volume of his *Evolution and Cognition* series, we jumped at the chance.

Our research program can be summarized in five propositions:

1. *Culture is information that people acquire from others by teaching, imitation, and other forms of social learning.* On a scale unknown in any other species, people acquire skills, beliefs, and values from the people around them, and these strongly affect behavior. People living in human populations are heirs to a pool of socially transmitted information that affects how they make a living, how they communicate, and what they think is right and wrong. The information thus stored and transmitted varies from individual to individual and is a property of the population only in a statistical sense.

2. *Culture change should be modeled as a Darwinian evolutionary process.* Culture changes as some ideas and values or "cultural variants," become more common and others diminish. A theory of culture must account for the processes in the everyday lives of individuals that cause such changes. Some of these processes arise from human psychology because some ideas are more readily learned or remembered. Other processes are social and ecological. Some ideas make people richer, live longer, or migrate more often, and the resulting selective processes generate culture change. While making frequent use of ideas and mathematical tools from population biology in modeling such culture change, ultimately the theory must derive from the empirical facts of how culture is stored and transmitted.

3. *Culture is part of human biology.* The capacities that allow us to acquire culture are evolved components of human psychology, and the contents of cultures are deeply intertwined with many aspects of our biology. What we learn, what we feel, how we think, and how we remember are all shaped by the architecture of human minds and bodies shaped over the millennia by the ongoing action of organic evolution. As a result, much cultural variation can be understood in terms of human evolutionary history.

4. *Culture makes human evolution very different from the evolution of other organisms.* Humans, unlike any other living creature, have cumulative cultural adaptation. Humans learn things from others, improve those things, transmit them to the next generation, where they are improved again, and so on, leading to the rapid *cultural* evolution of superbly designed adaptations to particular environments. This ability has allowed human populations to become the most widespread and variable species on earth. At the same time, because cumulative cultural evolution makes available ideas that no individual could discover and technology that no individual could invent, it requires a degree of credulity. While individuals are not passive receptacles of their culture, they cannot vet every belief and value their culture makes available, and this opens the door to the spread of "maladaptive" ideas, ideas that would never evolve in a noncultural organism. Moreover, the fact that much culture is acquired from people other than parents means that such maladaptive ideas tend to accumulate.

5. *Genes and culture coevolve.* Because culture creates durable changes in human behavior, human genes evolve in a culturally constructed environment. This environment, in turn, generates selection on genes. The evolution of language is an example. We apparently have a complex innate system for hearing, speaking, and learning language. This capacity would likely be useless without complex languages to learn. Primitive languages presumably created a cultural world in which better innate language skills were favored by selection. Through repeated rounds of coevolution, complex languages and the costly apparatus necessary to operate them emerged. Such effects are

probably pervasive. The existence of complex technology depends upon great facility in observational learning, and complex social institutions depend on people being adept at learning the rules of social games. Our ape relations can learn only rudimentary bits of language and rudimentary technical and social skills. They have only rudimentary cultural traditions of any kind. Most of what *human* organic evolution has been about is the coevolution of capacities for culture and cultural traditions.

The first two propositions have to do with how culture works, and the last three have to do with how cultural evolution interacts with genetic evolution.

Both of us have a background in biology, and our first work was published during the heat of the sociobiology controversy, so you might think, as many do, that our work arose from an interest in culture and genetic evolution. However, the truth is that our entrée to the subject came from trying to understand how cultural evolution worked to generate human behavior, especially behavior affecting the environment. Our collaboration began in 1974 when we co-taught Environmental Studies 10, a survey of environmental studies for nonmajors at U. C. Davis. At that time Pete was an assistant professor in Environmental Studies and Rob was a finishing graduate student in the Ecology Graduate Group. ES 10 was typically organized around a series of environmental problems—the population explosion, resource depletion, air and water pollution, and so on. We had the idea of organizing it around the principle that individual, goal-seeking behavior sometimes led to outcomes that were bad for everyone and in this way bringing together ideas from ecology and economics. However, we also wanted to discuss human impacts on the environment in ancient and contemporary nonindustrial societies, so this meant going beyond economics. We knew that one of the then-dominant schools in anthropology, cultural ecology, held that much cultural variation could be understood as adaptations to local environments, so this didn't seem like it would be much of a problem. Such is the way of young men.

When we actually sat down to learn what the social sciences had to say about culture, and how cultures adjusted to their environment, we were frustrated and disappointed. Cultural ecologists provided lots of interesting empirical examples of how behavioral variation could be understood as adaptations to environmental differences. However, there was little discussion, and no consensus, about *how* such adaptation occurred. To make matters worse, prominent authors like Marvin Harris explained some behaviors in terms of their function at the group level (the male supremacist complex in the Amazon conserved game) and others in terms of individual advantage (Indians do not eat cows because they are more useful for traction). Since environmental problems often arise because the interests of individuals and groups conflict, we found this more than disconcerting. Other social scientists, symbolic anthropologists, social anthropologists, and many sociologists, refused to explain culture in terms of individual decisions and characteristics as a matter of principle. (A distinguished sociologist once astounded us with the claim that it had been *proven* that it was impossible to do so.)

Of course, the rational actor model that predominates in economics and political science provides a very clear picture of how aggregate behavior arises from individual choices. Human actors are assumed to come equipped with preferences that describe how they rank outcomes and beliefs that express what they think is the connection between their actions and outcomes. Behavior emerges as people rationally choose the actions that produce the best mix of outcomes. Variation between groups of people arises because different groups face different conditions. The problem is that rational actor theorists do not offer an account of where the preferences and beliefs come from. Scholars working in such traditions usually don't deny that culture is real and important but maintain that worrying about how it gives rise to preferences and beliefs is just not part of their job description.

Darwinian students of human behavior proposed to rectify the lack of a theory of preferences and beliefs with evolutionary theory. Organisms should prefer to maximize their genetic fitness, or rather prefer and believe things that would have led to fitness maximization in the past. This is a strong theory, and certainly part of the answer. Darwinians, like economists, do not usually deny that culture plays a role in the formation of preferences and beliefs. But, like economists, they seldom enter terms representing culture into their models or collect much data about cultural variation.

This benign neglect of culture is usually accompanied by a largely unarticulated prejudice against cultural explanations. Confronted with differences in marriage systems, inheritance rules, or economic organization, such scholars prefer almost any economic or ecological explanation, no matter how far-fetched, over explanations that invoke cultural history. From table talk we gather that one reason is that those students of human behavior who aspire to "hard" scientific explanations are reacting to the "soft" methods of the historians, anthropologists, and sociologists who frequently propose cultural explanations. Blaming the messengers, if such is the case, seems to us unwise.

We think the way to make cultural explanations "hard" enough to enter into principled debates is to use Darwinian methods to analyze cultural evolution. Think of culture as a pool of information, mainly stored in the brains of a population of people. This information gets transmitted from one brain to another by various social learning processes. We define culture as follows: *Culture is information capable of affecting individuals' behavior that they acquire from other members of their species by teaching, imitation, and other forms of social transmission.* By "information," we mean any individual attribute that is acquired or modified by social learning and affects behavior. Most culture is mental states, but not all. Think of the blacksmith's proverbial muscular arms or the model's waif-like figure—essential parts of their crafts. We often use everyday words like *idea, knowledge, belief, value, skill,* and *attitude* to describe this information, but we do not mean that such socially acquired information is always consciously available, or that it corresponds to folk-psychological categories. People in culturally distinct groups behave differently mostly because they have acquired different beliefs, preferences, and skills, and these differences persist through time because the people of one generation acquire their beliefs and attitudes from those around them.

To understand how cultures change, we set up an accounting system that describes how cultural variants are distributed in the population and how various processes, some psychological, others social and ecological, cause some variants to spread and others to decline. The processes that cause such cultural change arise in the everyday lives of individuals as people acquire and use cultural information. Some values are more appealing and thus more likely to spread from one individual to another. These will tend to persist while less attractive alternatives tend to disappear. Some skills are easy to learn accurately while others are likely to be transformed during social learning. Some beliefs cause people to be more likely to be imitated, because the people who hold those beliefs are more likely to survive or more likely to achieve social prominence. Such beliefs will tend to spread while beliefs that lead to early death or social notoriety will disappear. We want to explain how these processes, repeated generation after generation, account for observed patterns of cultural variation.

We find it hard to recollect the exact pathway that brought us to this way of thinking. For sure, we were influenced by Donald T. Campbell's famous 1965 essay, and by an early (1973) article of Luca Cavalli-Sforza and Marc Feldman. The general idea was somehow in the air in the early 1970s as F. T. Cloak, Eugene Ruyle, Richard Dawkins, Bill Durham, and Ron Pulliam and Christopher Dunford published work espousing a similar approach to culture. Somewhat later (in 1978), we were fortunate to sit in on a class taught by Cavalli-Sforza and Feldman that was very helpful, especially in adapting models from population genetics to model the population dynamics of cultural variants. We recall thinking that applying the evolutionary biologists' concepts and methods to the study of culture was a rather obvious thing to do. We were more than pleasantly surprised that our predecessors had left so much relatively easy and interesting work undone. As Geoff Hodgson and Robert Richards have discovered, a properly evolutionary social science formed in the late nineteenth and early twentieth centuries before dying an untimely death.

As we were first thinking these thoughts, what came to be called the sociobiology controversy burst into full bloom. The mid-1960s saw the birth of the modern theory of the evolution of animal behavior. Bill Hamilton's seminal articles on inclusive fitness and George Williams's book *Adaptation and Natural Selection* were the foundations. The next decade saw an avalanche of important ideas on the evolution of sex ratio, animal conflicts, parental investment, and reciprocity, setting off a revolution in our understanding of animal societies, a revolution still going on today. By the mid-1970s a number of people, including Dick Alexander, Ed Wilson, Nap Chagnon, Bill Irons, and Don Symons, began applying these ideas to understand human behavior. Humans are evolved creatures, and quite plausibly our societies were shaped by the same evolutionary forces that shaped the societies of other animals. Moreover, the new theory of animal behavior—especially kin selection, parental investment, and optimal foraging theory—seemed to fit the data on human societies fairly well. The reaction from much of the social sciences was, to put it mildly, pretty negative.

The causes of this reaction are complex, as Ullica Segerstråle has shown. The association of biological ideas with racist, eugenicist ideas during the early part of the last century surely played an important role. Another big problem was that many social scientists mistakenly thought about these problems in terms of nature versus nurture. On this view, biology is about nature; culture is about nurture. Some things, like whether you have sickle-cell anemia, are determined by genes—what we call nature. Other things, like whether you speak English or Chinese, are determined by the environment—nurture. Evolution shapes genetically determined behaviors, but not learned behaviors. Social scientists knew that culture played an overwhelmingly important role in shaping human behavior, and since culture is learned, evolutionary theory has little to contribute to understanding human behavior.

The problem was that this argument cut no ice with anybody who knew much about evolutionary biology. Although the nature-nurture way of thinking is common, biologists know that it is deeply mistaken. Traits *do* vary in how sensitive they are to environmental differences, and it is sensible to ask whether differences in traits are mainly due to genetic differences or differences in the environment. However, the answer you get to this question tells you *nothing* about whether the traits in question are adaptations shaped by natural selection. The reason is that *every bit* of the behavior (or physiology or morphology, for that matter) of every single organism living on the face of the earth results from the interaction of genetic information stored in the developing organism and the properties of its environment, and if we want to know why the organism develops one way in one environment and a different way in a different environment, we have to find out how natural selection has shaped the developmental process of the organism. This logic applies to any trait, learned or not. Moreover, biologists have been quite successful in applying adaptationist reasoning to explain learned behavior.

Because it was framed in terms of nature versus nurture, the evolutionary social science community by and large rejected the idea that culture makes any *fundamental* difference in the way that evolutionary thinking should be applied to humans. The genes underlying the psychological machinery that gives rise to human behavior were shaped by natural selection, so, at least in ancestral environments, the machinery *must have* led to fitness-enhancing behavior. If it goes wrong in modern environments, it is not culture that is the culprit, but the fact that our evolved, formerly adaptive psychology "misfires" these days.

We think that both sides in this debate got it wrong. Culture completely changes the way that human evolution works, but not because culture is learned. Rather, the capital fact is that human-style social learning creates a novel evolutionary trade-off. Social learning allows human *populations* to accumulate reservoirs of adaptive information over many generations, leading to the cumulative cultural evolution of highly adaptive behaviors and technology. Because this process is much faster than genetic evolution, it allows human populations to evolve (culturally) adaptations to local environments—kayaks in the arctic and blowguns in the Amazon—an ability that was a masterful adaptation to the chaotic, rapidly changing world of the Pleistocene epoch. However, the same psychological mechanisms that create this benefit *necessarily* come with a built-in cost. To get the benefits of social learning, humans have to be credulous, for the most part accepting the ways that they observe in their society as sensible and proper, but such credulity opens human minds to the spread of maladaptive beliefs. The problem is one of information costs. The advantage of culture is that individuals don't have to invent everything for themselves. We get wondrous adaptations like kayaks and blowguns on the cheap. The trouble is that a greed for such easy adaptive traditions easily leads to perpetuating maladaptions that somehow arise. Even though the capacities that give rise to culture and shape its content must be (or at least have been) adaptive on average, the behavior observed in any particular society at any particular time may reflect *evolved* maladaptations. Empirical evidence for the predicted maladaptations is not hard to find.

Much of our work has been directed at understanding the evolution of the psychological capacities that both permit and shape human culture (see part I). Most evolutionary thinkers approach this problem by first asking how evolution should have shaped the psychology of a group-living, foraging hominid. Then, having answered that question, they ask how the evolved psychology will shape human culture. The implicit evolutionary scenario seems to be that Pleistocene hominids were just extra-smart chimpanzees, clever social animals in whom *social* learning played a negligible role until the evolution of our brain was more or less complete. *Then* we took up culture, whose evolution is completely controlled by the preexisting evolved mind. *First*, we got human nature by genetic evolution; *then*, culture happened as an evolutionary by-product.

This way of thinking neglects the feedback between the nature of human psychology and the kind of social information that this psychology should be designed to process. For us to take bitter medicine, our psychology has to have evolved both to learn socially and to let social learning override aversive stimuli from time to time. As we discuss in chapters 1 and 2, social learning can be adaptive because the behavior of other individuals is a rich source of information about which behaviors are adaptive and which are not. We all know that plagiarism is often easier than the hard work of writing something oneself, and imitating the behavior of others can be adaptive for the same reason. The trick is that once social learning becomes important, the nature of the behavior that is available to imitate is itself strongly affected by the psychology of social learning. Suppose, for example, that everyone relied completely on imitation. Then, even if we somehow started with highly adapted traditions, behavior would gradually become dysfunctional as the environment changed and errors crept into the traditions. To understand the evolution of the psychology that underlies social learning, one must take this sort of feedback into account. We want to know how evolving psychology shapes the social information available to individuals and how selection shapes psychology in an environment with direct information from personal experience *and* the potential to use the behavior of others at a lower cost but perhaps greater risk of error. The research reported in these chapters suggests that this kind of reasoning leads to conclusions quite different from those of other evolutionary theories of human behavior. Under the right conditions, selection can favor a psychology that causes *most* people to adopt behaviors "just" because the people around them are using those behaviors. Weak psychological forces that derive from people occasionally tweaking their traditions in adaptive directions are sufficient to maintain the tradition in an adapted state so long as the environment is not changing too rapidly and the cultural analog of mutation is not too disruptive.

If the only processes shaping culture arose from our innate evolved psychology, then culture would be a strictly proximate cause of human behavior.

However, not all of the processes shaping culture arise from our innate psychology. From the beginning of our work, we have emphasized that culture leads to the spread of maladaptive cultural variants (see Richerson and Boyd 1976, 1978). Culture is not always, or even typically, transmitted from parents to offspring. Instead, cultural variants are acquired from all kinds of people. This is a good thing because sampling a wider range of models increases the chance of acquiring useful information. However, acquiring adaptive information from others also opens a portal into people's brains through which maladaptive ideas can enter—ideas whose content makes them more likely to spread, but do not increase the genetic fitness of their bearers. Such ideas can spread because they are not transmitted as genes are. For example, in the modern world, beliefs that increase the chance of becoming an educated professional can spread even if they limit reproductive success because educated professionals have high status and thus may likely be emulated. Professionals who are childless can succeed culturally as long as they have an important influence on the beliefs and goals of their students, employees, or subordinates. The spread of such maladaptive ideas is a *predictable* by-product of cultural transmission.

Selection acting on culture is an ultimate cause of human behavior just like natural selection acting on genes. In several of the chapters in part III we argue that much cultural variation exists at the group level. Different human groups have different norms and values, and the cultural transmission of these traits can cause such differences to persist for long periods. The norms and values that predominate in a group plausibly affect the probability that the group is successful, whether it survives, and whether it expands. For illustration, suppose that groups with norms that promote patriotism are more likely to survive. This selective process leads to the spread of patriotism. Of course, this process may be opposed by an evolved innate psychology that biases social learning, making us more prone to imitate, remember, and invent nepotistic beliefs than patriotic beliefs. The long-run evolutionary outcome would then depend on the balance of these two processes. Again, for illustration, let us suppose that the net effect of these opposing processes causes patriotic beliefs to predominate. Then, the population behaves patriotically *because* such behavior promotes group survival, in exactly the same way that the sickle-cell gene is common in malarial areas *because* it promotes individual survival. Human culture participates in ultimate causation.

This way of thinking about cultural evolution leads to a picture of a powerful adaptive system necessarily accompanied by exotic side effects. Some of our evolutionist friends take a dim view of this notion, seeing it as giving aid and comfort to those who would deny the relevance of evolution to human affairs. We prefer to think that the population-based theories of cultural evolution strengthen Darwin's grasp on the human species by giving us for the first time a tentative picture of the engine that powered the furious pace of change in the human species over the last few hundred thousand years. Compare us to our ape cousins. They still live in the same tropical forests in the same small social groups and eat the same fruits, nuts, and bits of meat as our common ancestors did. By the late Pleistocene epoch (say 20,000 years ago), human foragers already occupied a much wider geographical and ecological range than any other vertebrate species, using a remarkable range of subsistence systems and social arrangements. Over the last ten millennia we have exploded to become the earth's dominant organism by dint of deploying ever more sophisticated technology and social systems. The human species is a spectacular evolutionary anomaly, and we ought to expect that the evolutionary system behind it is anomalous as well. Our quest is for the evolutionary motors that drove our divergence from our ancestors, and we believe that the best place to hunt is among the anomalies of cultural evolution. This does not mean that gene-based evolutionary reasoning is worthless. On the contrary, human sociobiologists and their successors have explained a lot about human behavior, even though most work ignores the novelties introduced by cultural adaptation. However, there is still much to explain, and we think that the Darwinian, population-based properties of culture are essential components of a satisfactory theory of human behavior.

REFERENCES

Campbell, D. T. 1965. Variation and selective retention in sociocultural evolution. In: *Social change in developing areas*, H. R. Barringer, G. I. Blanksten, & R. W. Mack, eds. (pp. 190–49).

Cavalli-Sforza, L., & M. Feldman. 1973. Models for cultural inheritance, I: Group mean and within group variation. *Theoretical Population Biology* 4:42–55.

Cloak, F. T. 1975. Is a cultural ethology possible? *Human Ecology* 3:161–182.

Dawkins, R. 1976. *The selfish gene*. Oxford: Oxford University Press.

Durham, W. H. 1976. The adaptive significance of cultural behavior. *Human Ecology* 4(2):89–121.

Hamilton, W. D. 1964a. The genetical evolution of social behaviour. I. *Journal of Theoretical Biology* 7:1–16.

Hamilton, W. D. 1964b. The genetical evolution of social behaviour. II. *Journal of Theoretical Biology* 7:17–32.

Hodgson, G. M. 2004. *The evolution of institutional economics: Agency, structure and Darwinsm in American institutionalism*. London: Routledge.

Pulliam, H. R., & C. Dunford. 1980. *Programmed to learn*. New York: Columbia University Press.

Richards, R. J. 1987. *Darwin and the emergence of evolutionary theories of mind and behavior*. Chicago: University of Chicago Press.

Richerson, P. J., & R. Boyd. 1976. A simple dual inheritance model of the conflict between social and biological evolution. *Zygon* 11:254–262.

Richerson, P. J., & R. Boyd. 1978. A dual inheritance model of the human evolutionary process. *Journal of Social and Biological Structures* 1:127–154.

Ruyle, E. E. 1973. Genetic and cultural pools: Some suggestions for a unified theory of biocultural evolution. *Human Ecology* 1:201–215.

Segerstråle, U. 2000. *Defenders of the truth: The battle for science in the sociobiology debate and beyond*. Oxford: Oxford University Press.

Williams, G. C. 1966. *Adaptation and natural selection*. Princeton: Princeton University Press.

PART 1

The Evolution of Social Learning

The human species presents evolutionists with a vexing puzzle. Complex, cumulatively evolving culture is rare in nature. Simple traditions are widespread, and in a few species—whales, dolphins, primates, and birds—traditions are fairly complex. However, even the most complex traditions in other animals are manifestly simpler than those in human cultures. Our capacities to imitate and teach support exceedingly complex and variable technological, social, and symbolic systems like art and language, a capability that is qualitatively different from that possessed by any other species. If another species has a language with thousands of words, a toolkit with hundreds of intricate items, and societies composed of a few thousand unrelated individuals, we would know of it by now. This fact raises the obvious questions: Why now? And why only us? True, some fancy adaptations like the elephant's trunk are unique, but really good tricks like the camera eye tend to have evolved repeatedly among the world's millions of species. Given that fancy culture has made humans extraordinarily successful, why isn't it much more common? And why didn't it arise with the dawn of complex animals hundreds of millions of years ago?

The chapters in this part address these questions. In chapter 1, we construct a very simple model of the evolution of social learning. We imagine a population in which individuals can learn for themselves but can also imitate someone of the previous generation (their mothers, for example). These organisms live in a spatially variable world, and their adaptive task is to combine their own experience and the vicarious experience acquired from their mother to guess how they should behave. There are two types of environments: wet and dry. In the dry environment, the best subsistence strategy is, say, hunting and gathering. In the wet environment, farming is the best subsistence strategy. The information available to individuals is noisy. On average, individual learning gets the right answer, but sometimes it leads to errors. Even in the dry environment, a run of rainy years might lead one who depends on individual experience to believe that the environment is really wet and hence to mistakenly adopt farming instead of hunting and gathering. Individuals can evaluate the quality of their individual experience and use this rule: if individual experience is sufficiently accurate, rely on it; otherwise, imitate. Some individuals move about on the landscape and may find themselves in a wet environment, whereas their mother came from a dry one or vice versa. Thus, depending upon a mother's traditional wisdom has the advantage of evading errors due to noisy individual learning. So long as a mother's lineage has not recently switched environments, both natural selection and individual learning will have tended to make her ideas about the nature of the environment accurate on average. On the other hand, if migration has recently removed the mother's lineage to the other environmental state, her received wisdom may well be wrong.

Though very simple and stylized, this model captures one much-noted structural feature of the cultural system, namely, that it is a system for the inheritance of acquired variation (often called "Lamarckian inheritance"; ironically, this process was as much a part of Darwin's ideas as Lamarck's). The results of the model are quite intuitive. If there is little migration between different environment types, the optimal thing to do is rely on individual experience only when it is highly accurate and, as a result, imitate most of the time. The effect of occasional individual learners is sufficient to keep most traditions adapted to the local environment. In the opposite limit, when individuals move so much that each generation is placed at random with respect to their mom's environment, imitation information is useless, and the adaptive strategy is to depend only upon individual learning—personal experience. In between these limits, some weighted average of personal and vicarious experience should determine an individual's choice: more individual experience in the mix when migration is relatively frequent and individual learning not so error-prone, more tradition when migration is relatively infrequent and individual learning relatively error-prone. Given that all environments are spatially heterogeneous and all animals, and plants for that matter, migrate, this model suggests that culture should be common, if not ubiquitous. It certainly does not solve the puzzle of human uniqueness; it makes that puzzle more difficult.

These results do not depend too much on the details of the model—various models have very similar properties. The spatial model can easily be modified to reflect temporal variations with similar results (Boyd and Richerson, 1988). Other interactions between individuals' psychology and culture lead to similar effects. We have studied a variety of biased social learning effects in which individuals do not learn new variants for themselves but rather preferentially copy existing ones using a number of biasing rules (Boyd and Richerson, 1985). The models can also be modified to take account of social learning within, as well as between, generations. The take-home message is that a cultural system of inheritance is an evolutionarily flexible system that natural selection could tune to cope with many patterns of environmental variability. These models support Darwin's intuition that imitation and other forms of social learning should be common, but they give us no clue about why our species' unusually hypertrophied cultural system evolved.

However, one assumption *is* crucial. In 1989, Alan Rogers published a model with very different qualitative properties. Here, the population consisted of two innate types: learners and imitators. Learners learn individually and imitators copy someone at random. Rogers showed that the evolution of imitation in such a population behaves curiously. Social learning tends to be favored; under many conditions, a fair frequency of imitators exists at evolutionary equilibrium. However, when the system equilibrates, imitators and learners have exactly the same fitness, and since learners always have the same fitness, this mixture of imitators and learners has the same fitness as an all-learner population before imitators began evolving in it. Social learning evolves, but it is not adaptive because the population at equilibrium copes no better with a variable environment than a population that doesn't imitate at all. In contrast, in the models introduced in chapter 1, the mean fitness of the population is higher at equilibrium—imitation *does* increase the population's ability to adapt. In chapter 2 we show that the key difference is the effect of imitation on individual learning. In Rogers's model, the only benefit to imitation is that it allows individuals to avoid the costs of learning; imitators are scroungers who profit from the costly learning efforts of others. In the model presented in chapter 1, the possibility of imitation *increases* the efficiency of learning by allowing learners to be selective. We show in chapter 2 that the ability to accumulate improvements in many small steps can have the same effect.

Nevertheless, Rogers's model does illustrate an important feature of the relationship between individual and social learning. In a cultural population, effortful individual attempts to learn or to bias imitation tend to improve the average quality of cultural traditions to the benefit of everyone. Selection at the individual level will tend to produce less individual learning and bias than would be optimal from the point of view of the population because of the altruistic effect of social learning on future members of the population. Kameda and Nakanishi (2002) have shown experimentally that some human subjects produce information while others free ride on the efforts of others. Intellectual property protections are a modern method of trying to adjust incentives to individuals to gain a more optimal level of creative work than in a society in which inventors are parasitized by imitators. Henrich and Gil-White (2001) argue that human prestige systems evolved to compensate those who seem to have the best ideas to imitate. If, as we argue in part III, humans are subject to cultural group selection, many institutions and even (via coevolution) innate predispositions may arise to increase the individual effort devoted to information updating beyond that favored by individual advantage alone.

Chapter 3 tackles the uniqueness of human culture. The models of chapter 1 and 2 suggest that the capacities that give rise to culture can readily evolve. Given that culture has made humans so successful, shouldn't many more animals have evolved this world-beating adaptation? In one sense, many do. Simple systems of social transmission are quite common (Heyes and Galef, 1996). What seems unique about human social learning is our ability to accumulate adaptive information over many generations, building complex artifacts and institutions composed of many small innovations. Even something as simple as a good stone-tipped spear reflects cumulative innovations applied to the shaft, the hafting, and the stone point. No nonhuman tradition yet described approaches such a spear in complexity. Humans can maintain complex traditions because we are more accurate imitators than any other animal yet tested. Accurate imitation plausibly depends upon costly cognitive structures such as a theory of mind. As we show in chapter 3, the evolution of such structures faces a major hurdle. Complex cultural traditions are the product of a *population* of minds. Many people and the passage of time are necessary for a complex tradition to evolve. In the absence of such a population, the costly structures necessary for accurate imitation are useless. The rare individual who happens to have the costly structure, perhaps only in rudimentary form, will be born into a world with no complex traditions to learn and hence no use for the capacity to imitate accurately.

If correct, this model suggests that the capacities that permit accurate imitation must have been favored initially for some other purpose. For example, a theory of mind may have been favored because it allows better manipulation of the social world. Then, this capacity gave rise to more accurate imitation, and the cultural evolution of complex adaptive traditions as a side effect. This argument provides one explanation for the rarity of cumulative cultural tradition: humans were the first species to chance on some devious path around this constraint, and then we have preempted most of the niches requiring culture, inhibiting the evolution of any competitors.

Chapter 4 provides a different explanation for the rarity of culture, one based on recent discoveries about the nature of Pleistocene climates. Evolutionists divide explanations of the large-scale, long-term patterns of evolution into those internal to the evolutionary process itself and those external to it, such as changes in climate. The argument in chapter 3 is a typical internal explanation. Evolution always favored a capacity for complex culture, but it took life a long time to find its way around constraints and evolve complex, cumulative traditions. Such internal explanations are implicit in many accounts of our origins. Such accounts flatter our species because they assume that an intelligent culture-bearing species is superior to the common run of animals. Considering the possibility of external causes is a useful antidote to the implicit acceptance of internal explanations, especially as they may be the product of anthropocentrism.

The correlation of brain size with climate variation favors an external explanation for the timing of the evolution of culture in humans and other animals. Terrestrial vertebrates have been around for some 350 million years. Dinosaurs and their allies were not simple animals, but they did have small brains. The mammals that coexisted with dinosaurs also had small brains. Brain tissue is quite expensive. All else equal, selection will favor the stupidest possible creatures. Perhaps dinosaurs and ancient mammals lived in a world that did not require much brainpower. For the last 65 million years, the average size of mammalian brains has gradually increased. The rate of increase has jumped during the last couple of million years. Brain size increase in mammals has an interesting parallel in the cooling and drying of climates over the last 65 million years, culminating in the sharp average cooling and drying and the onset of cycles of glacial advance and retreat that became more pronounced about 2.5 million years ago. If in addition to cooling and drying, this world has become more variable, we'd have an explanation for why brain size has increased in so many lineages.

The long-known advances and retreats of glaciers take tens of thousands of years and thus are far too slow to require much brainpower to cope with. However, ice core data published in the early 1990s began to paint a picture of hugely variable glacial environments, much more variable than we have experienced on the long march to our present civilizations during the last 11,500 years. Much of this variation is on time scales ranging from a millennium to the limits of resolution of the data (a few years; chapter 17 includes more recent references, see also Helmke, Schultz, and Bauch [2002]). These are just the time scales of variation that the models suggest should favor a cultural system that can mix and match the conservatism of faithful transmission with flexibility of individual learning to generate rapidly evolving traditions adapted to rapidly changing environments. Variability on short time scales probably also favors individual behavioral flexibility. If this argument is correct, we can interpret brain size as a rough bioindicator of the amount of fine-scale environmental variability in space and time. Ancient mammals were dull because they lived in a dull, little-varying world, whereas modern mammals are sharp because they live in a world alive with rapid change. The field of paleoclimatology is currently advancing rapidly, and consequently our ability to formulate and test such conjectures is increasing.

Chapter 5 introduces two forms of biased transmission, conformity and success-based, that can produce both adaptive and maladaptive evolution-ary outcomes. These biases can be thought of as adaptive rules of thumb for acquiring adaptive information. If information is costly to acquire, evolution will favor fast, frugal heuristics for solving adaptive problems. (The Dahlem Conference book from which this chapter is drawn covers this general topic in considerable detail.) Imitating mom in the face of the costs of learning for one's self in the style of the model of chapter 1 is a trick to finesse information costs. Conforming to the majority is an inexpensive rule to apply, compared, say, to doing experiments on the alternative behaviors one might adopt. Many adaptive forces will tend to make adaptive behaviors common, so adopting the commonest is generally not a bad guess. Similarly, if other people's adaptive success is in any way public knowledge, imitating the successful is a good rule to follow.

These quick-and-dirty rules of thumb have interesting evolutionary side effects. In part III we discuss how conformity reduces within-group cultural variation, making group-level selection a more plausible process than group selection on genes is usually thought to be. Imitating the successful can also lead to a form of rapid group selection. In part II, we will see how this process leads to symbolically marked group boundaries. The other interesting evolutionary feature of these rules is that under some conditions they can give rise to maladaptive behavior. Consider a moral norm that is maintained by a combination of conformity and success-based bias. Some such norms, for example, the mutilation of genitalia and high rates of female infanticide, are probably quite maladaptive. Yet if people conform and if those who violate the norms are punished in some way, those who attempt to abandon such practices in favor of more adaptive ones will become a stigmatized minority. In this way, normally adaptive learning mechanisms can perpetuate dysfunctional behavior under the right circumstances. Perhaps one reason why complex human-style culture is so rare is that these complexities impose a burden that is worth meeting only when the adaptive advantages of culture outweigh this cost.

We do not tout this family of models and our interpretations of them as any more than a first attempt at explaining why social learning evolves, especially how our own extraordinary system of complex culture has evolved. We do hope to have demonstrated how we can think in a more rigorous way about the Big Questions of human life using simple models of cultural evolution as a tool. Cultural evolution is rooted in the psychology of individuals, but it also creates population-level consequences. Keeping these two balls in the air is a job for mathematics; unaided reasoning is completely untrustworthy in such domains.

REFERENCES

Boyd, R., & P. J. Richerson. 1985. *Culture and the evolutionary process.* Chicago: University of Chicago Press.
Boyd, R., & P. Richerson. 1988. An evolutionary model of social learning: The effects of spatial and temporal variation. In: *Social Learning: Psychological and biological approaches*, T. Zentall & B. G. Galef, eds. Hillsdale, NJ: Lawrence Erlbaum.
Helmke, J. P., M. Schultz, & H. A. Bauch. 2002. Sediment-color record from the Northeast Atlantic reveals patterns of millennial-scale climate variability during the past 500,000 years. *Quaternary Research* 57:49–57.
Henrich, J., & F. J. Gil-White. 2001. The evolution of prestige—Freely conferred deference as a mechanism for enhancing the benefits of cultural transmission. *Evolution and Human Behavior* 22(3):165–196.
Heyes, C. M., & B. G. Galef. 1996. *Social learning in animals: The roots of culture.* San Diego: Academic Press.
Kameda, T., & D. Nakanishi. 2002. Cost-benefit analysis of social/cultural learning in a nonstationary uncertain environment: An evolutionary simulation and an experiment with human subjects. *Evolution and Human Behavior* 23:373–393.
Rogers, A. R. 1989. Does biology constrain culture? *American Anthropologist* 90: 819–831.

1   Social Learning as
an Adaptation

Learning is widespread in the animal kingdom. While the mechanisms of learning range from relatively simple conditioning in invertebrates to elaborate cognitive mechanisms in mammals, most animals use some form of learning to acquire behavior that is adaptive in the local habitat. Despite this fact, the great bulk of evolutionary theory assumes that organisms adapt to variable environments through genetic mechanisms alone. The neglect of learning may result from the difficulty of understanding the evolution of learned behaviors. Learning entails an evolutionary trade-off. The advantages of learning are obvious; it allows the same individual to behave appropriately in different environments. For example, by sampling novel foods and learning to avoid noxious food types, a cosmopolitan species like the Norway rat can acquire an appropriate diet in a wide range of environments. However, learning also has disadvantages. First, the learning process itself may be costly. By sampling novel foods, the rat may accidentally poison itself, a risk that could be avoided by an animal with rigid, genetically specified food preferences. Second, because learned behavior is based on imperfect information about the environment, it can lead to errors. For example, the rat may fail to sample or mistakenly reject a nutritious food item. To understand variation in learned behavior among species, one must understand how this evolutionary trade-off is resolved.

Recently, several authors have used statistical decision theory to show why the learning rules of different species vary (McNamara and Houston, 1980; Staddon, 1983; Stephens and Krebs, 1988). One can think of individual organisms as having to "choose" among alternative behaviors to maximize their fitness in the local environment. They have some genetically inherited "prior" information about the state of local environment, some data from their experience, and usually the opportunity to gather more data at some cost in terms of fitness. Decision theory is useful because it tells us the best way to make decisions with imperfect information. Assuming that natural selection has shaped the learning rules of different species so that they are adaptive, decision theory should help us to understand why different animals learn differently. In the same way that mechanics helps us understand the comparative morphology of skeletons, decision theory may help us understand comparative behavior of animals.

We are interested in understanding the adaptive function of one particular form of learning, social learning. By social learning, we mean the acquisition of behavior by observation or teaching from other conspecifics. Social learning has been implicated in the acquisition of behavior in a variety of taxa. Many songbirds acquire their song by copying the song of other adult birds (Marler and Tamura, 1964). Rats seem to acquire food preferences both from taste cues in their mothers' milk, and from the smell of other rats' pelage (Galef, 1976). There is circumstantial evidence that individuals of several different primate species may acquire complex new behaviors by social learning (Kawai, 1965; McGrew and Tutin, 1978; Hauser, 1988). Finally, social learning plays an essential role in human adaptation (Boyd and Richerson, 1985). For reviews of the literature on social learning in nonhuman animals, see Galef (1976, 1988).

In this essay we present several simple mathematical models, of social learning. Our aim is to use these models to explain social learning as an adaptation in the same way that decision theoretic models have been used to explain other forms of learning. The decision theoretic models alone are not sufficient to understand the conditions under which social learning is adaptive. Instead, decision theoretic models must be generalized to allow for the fact that behaviors acquired by social learning are transmitted from individual to individual. Thus, to understand social learning, we need models that keep track of the processes that change the frequency of alternative behaviors in a population through time. Consider a young rat learning food preferences. To predict whether it acquires a preference for some food, say cilantro, by social learning, we need to know whether its mother's diet includes cilantro. Its mother's diet will depend on both her experience and her own mother's diet. More generally, to understand why a preference for cilantro among a population of rats is becoming more common (or more rare), we must know its frequency among rats of previous generations, and how this generation's individual learning experiences changed the frequency of the preference between the time that they acquired their initial food preferences by social learning and the time that they serve as models for members of the next generation. Because behavioral variants are transmitted from individual to individual, and thus from generation to generation, understanding social learning requires understanding the dynamic processes that act to change the frequency of different socially learned behaviors in a population of organisms through time. We must link models of individual learning to models of social learning to determine the evolutionary dynamics of behavioral variants in a population.

We will use these models to address two questions about the adaptive function of social learning:

1. *Under what circumstances should natural selection favor increased reliance upon social learning at the expense of individual learning?* We will begin by analyzing

a model in which a population of organisms acquires behavior by a combination of individual and social learning in a uniform and constant environment. This model indicates that, on average, in constant environments reliance on social learning always leads to higher fitness than reliance on individual learning. We will then add environmental variability to the model. Under these conditions, there is an optimal mix of social and individual learning. The relative importance of social learning in the optimal mix is increased when environments are predictable and when individual learning is error-prone.

2. *Given that naive individuals experience the behavior of a number of experienced individuals, and that this behavior varies, how should social learning be structured?* Here we will consider a model in which naive individuals are exposed to a finite sample of the behavior of members of the previous generation. We will refer to this set of observed and potentially imitated individuals as "models." Naive individuals will be exposed to different combinations of behavior that they can imitate. The analysis suggests that in a variable environment, selection favors individuals who are predisposed to acquire the most common behavior among their models. It also suggests that selection favors individuals whose propensity to rely on individual learning increases as the variability among their set of models increases.

A Model of Individual and Social Learning

We begin by addressing this question: when does social learning allow a more effective tracking of the environment than individual learning? To answer this question, we want to construct a model that embodies the following assumptions about the interaction of social and individual learning:

1. A population of organisms is potentially confronted with a variable environment in which different behaviors are favored by selection in different habitats.
2. Individuals in the population can acquire their behavior by some mixture of social learning and individual learning, where:
3. Social learning involves the faithful copying of the behavior of a single other individual in the population, and:
4. Individual learning occasionally leads to errors.
5. All individuals pay any fitness costs associated with individual learning whether they ultimately acquire a behavior by social learning or by individual learning.

Given these assumptions, we want to determine the conditions under which selection will favor individuals who rely significantly on social rather than individual learning. Consider a population that occupies an environment that can be in one of two distinct states: habitat 1 or habitat 2. Each individual in the population will acquire one of two alternative behaviors, also labeled 1 and 2. As shown in Table 1.1, each individual has a "baseline" fitness W; individuals who acquire the behavior that is best in their environment achieve an increase in fitness, D. Thus, individuals that acquire behavior 1 have higher fitness in habitat 1

Table 1.1. Fitness associated with two behaviors

	Behavior 1	Behavior 2
Habitat 1	$W+D$	W
Habitat 2	W	$W+D$

than individuals that acquire behavior 2. Similarly, behavior 2 yields higher fitness in habitat 2 than does behavior 1. Once an individual has acquired one of the two behaviors, it does not change. Nor does the environment change, so that an individual experiences only one of the two environmental states during its lifetime.

The adaptive problem that faces each individual is to determine which of the two habitats it is in. Individuals in the model have two sources of information available to help them solve this problem.

Each individual obtains evidence from its own experience: any observations, learning trials, or other nonsocial information that can help determine the state of the environment. We assume the result of each individual's experience can be quantified in terms of a single normally distributed random variable, x. If the environment is in state 1, the mean value of x is M; if it is in state 2, the mean value of x is $-M$. In other words, the true state of the environment is either M or $-M$. Individuals acquire an imperfect estimate of the state of the environment, x, from personal experience. The standard deviation of the distribution of x, S, is an inverse measure of the quality of the evidence available to the members of the population. The larger S is, the poorer the individual's estimate of the state of the environment. If $S \ll |M|$, then most individuals' experiences will clearly indicate the state of the environment. If $S \gg |M|$, the results of gathering direct evidence will not be very informative.

Assume that the population is structured into nonoverlapping cohorts. Individuals in one cohort can observe the behavior of individuals from the previous cohort who have already acquired either behavior 1 or behavior 2. Individuals in one cohort act as models for individuals in the next cohort.

We imagine that individuals in the population use these sources of information to decide between the two alternatives in the following way: if the outcome of direct observation, x, is greater than a threshold value d $(d \geq 0)$, the individual acquires behavior 1; if x is less than $-d$, then it acquires behavior 2. This is our attempt to capture the essence of the processes of individual learning. Finally, if $-d \leq x \leq d$, then the individual imitates the behavior of a single individual chosen at random from the population, its model. This, in turn, is our attempt to capture the essence of social learning. The order in which the two kinds of learning occur is not crucial; the model applies equally well to a situation in which individuals begin by imitating others and then adopt a new behavior only if confronted with decisive personal experience.

The parameter d serves two functions. First, as shown in figure 1.1, it is analogous to a confidence interval. The larger the value of d that characterizes the population, the more decisive the evidence must be before it will affect the individual's decision. Second, the value of d simultaneously determines the

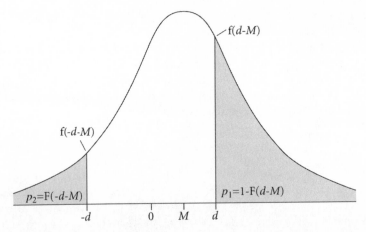

Figure 1.1. Illustrates the definition of p_1 and p_2 and their relationship to the parameter d. $F(x)$ is the cumulative normal distribution, and $f(x)$ is the normal density function.

relative importance of social learning and individual learning. We assume that when individuals are in doubt on the basis of their own experience, they utilize behaviors acquired by imitation. Let p_1 be the probability that $x > d$, and let p_2 be the probability that $x < -d$. If d is large, then individuals attend to their own experience only if it provides compelling evidence about the state of the environment (i.e., p_1, $p_2 \approx 0$). For the most part, they imitate another individual. If d is small, behavior is mainly determined by an individual's experience, and social learning has little importance (i.e., $p_1 + p_2 \approx 1$).

Effects of Learning on the Distribution of Behavior in the Population

To predict the likelihood that an individual will acquire a particular behavior by social learning, we must know what behavior characterizes the individual's model. Suppose that a fraction q_t of individuals in cohort t acquired behavior 1. A fraction p_1 of the naive individuals in cohort t will acquire behavior 1 based on their own experience, and a fraction $q_t(1 - p_1 - p_2)$ acquire alternative 1 by imitation. Thus, in cohort t the frequency of individuals acquiring behavior 1, q_t', is

$$q_t' = q_t(1 - p_1 - p_2) + p_1 \tag{1}$$

Now suppose that these individuals then serve as models for individuals in the cohort $t + 1$. Then the frequency of behavior 1 among the models for cohort $t + 1$, q_{t+1}, is approximately

$$q_{t+1} = q_t \tag{2}$$

We say "approximately" because we have ignored the effect of natural selection. In environment 1, differential mortality will increase the frequency of behavior 1. Here we are assuming that the effect of learning on the relative frequencies of the two behaviors is so much greater than the effect of selection that selection can be safely ignored.

Suppose that this process is repeated many times. That is, members of a cohort acquire their behavior by a combination of social and individual learning and then serve as models for the next cohort, and this process is repeated for many successive cohorts. Eventually the fraction of each cohort acquiring behavior 1 will stabilize at the equilibrium value

$$\hat{q} = \frac{1}{1 + p_2/p_1} \tag{3}$$

Thus, the fraction of individuals acquiring behavior 1 at equilibrium depends only on the ratio of the probability that an individual will choose alternative 2 based on its own experience (p_2) to the probability that it will choose alternative 1 based on its own experience (p_1). If $p_2/p_1 > 1$, then the equilibrium frequency of individuals choosing alternative 1 is less than half; if $p_2/p_1 < 1$, $\hat{q} > \frac{1}{2}$. The fraction choosing alternative 1 at equilibrium does not depend (directly) on the relative importance of social learning versus individual learning in determining the behavior of individuals (i.e., on the magnitude of $1 - p_1 - p_2$). However, from equation 1 we know that the rate at which the population converges to the equilibrium value depends crucially on the amount of social learning. If there is little individual learning, p_1 and p_2 will be very small, and social learning will ensure that the population remains very similar from one generation to the next. Thus, as individual learning becomes less important in determining individual behavior, the population will converge more slowly to equilibrium. This property is crucial to our understanding of the evolution of mixed systems of social and individual learning in variable environments, as we will see.

The Evolution of Social Learning

We now consider the evolution of social learning. The relative importance of individual learning and social learning in determining phenotype is given by the parameter d. If d is affected by heritable genetic variation, then it will evolve under the influence of natural selection. We will model the evolution of d using the evolutionarily stable strategies (ESS) approach. That is, we assume that an individual's learning rule is affected by a genetic locus at which two alleles, a common allele, H, and a very rare allele, h, are segregating. Most individuals in a population are characterized by the genotype HH, which results in them having a learning rule characterized by the parameter value d; however, there are a few rare mutant Hh individuals whose learning rule is characterized by a slightly different parameter value, $d + \delta$. We assume that the hh genotype is so rare that it can be neglected. We then determine the conditions under which the rare allele can invade. The ESS value of d is that value which prevents any rare alleles from invading. When the ESS value of d is very large, we will say that social learning is adaptive, since when d is large, most individuals will depend on social learning.

As a first step in understanding the evolution of social learning, we calculate the ESS value of d, assuming that the environment is entirely in state 1. In this case, the expected fitness of an individual whose learning rule is characterized by the parameter d' in a population in which most individuals have a learning rule characterized by parameter d (where d' may or may not equal d) is given by:

$$E\{w(d')\} = W + D\{\hat{q}(d)[1 - p_1(d') - p_2(d')] + p_1(d')\} \tag{4}$$

where $\hat{q}(d)$ is the frequency of behavior 1 at the equilibrium value given in equation (3), assuming that most individuals in the population are characterized by learning parameter d. The rare allele, h, can invade the population if Hh individuals whose learning rule is characterized by learning parameter $d + \delta$ have a higher expected fitness than HH individuals whose learning rule is characterized by learning parameter d, that is, if $E\{w(d + \delta)\} > E\{w(d)\}$. Since δ is small, $E\{w(d + \delta)\} \approx E\{w(d)\} + \delta(\partial E\{w(d)\}/\partial d)$, this condition can be rewritten in the following form:

$$\delta\left[\frac{\partial p_1}{\partial d} p_2(d) - \frac{\partial p_2}{\partial d} p_1(d)\right] < 0 \tag{5}$$

Suppose that the invading allele increases d, so that $\delta > 0$. It follows from the definitions of d, p_1 and p_2 that a given change in d causes a larger absolute decrease in p_1 than in p_2, or $\partial p_1/\partial d < \partial p_2/\partial d < 0$. Thus, inequality (5) says that the rare allele can invade whenever the percent decrease in the probability of acquiring the wrong behavior by individual learning exceeds the percent decrease in the probability of getting the right behavior by individual learning. It can be shown that this expression is satisfied for all values of d. This means that the ESS value of d is as large as possible.

We draw two lessons from this simple result. First, some social learning is always better than relying completely on the results of experience. (That is, the expected fitness of an individual using a learning rule characterized by $d = 0$ is always less than the expected fitness of individuals using a learning rule characterized by any positive value of d.) Second, in a population characterized by the ESS value of d, individuals may virtually ignore the evidence presented by direct experience and depend entirely on social learning, even when the only cost associated with learning is the occasional error.

It is important to notice that this result was derived assuming that every individual in every cohort experienced habitat 1. This assumption of an invariant environment is crucial because, as we have seen, the equilibrium frequency of the superior variant does not depend on the amount of individual relative to social learning, but the rate of approach to that frequency does. In a variable environment, the expected fitness of individuals in the population likely will depend on the rate at which the population can respond to changes as well as the eventual equilibrium.

Social Learning in Variable Environments

To introduce environmental variation into the model, suppose that half of each cohort experiences environment 1 and the other half of each cohort experiences state 2. (The assumption that the habitats are the same size greatly simplifies the mathematical argument without altering the essential aspects of the problem.) Let p_{jk} be the probability an individual's choice is based on direct experience and that it results in behavior k given that the state of the environment is j. Because of the symmetry of the model, the following is true:

$$p_{11} = p_{22}$$
$$p_{12} = p_{21} \tag{6}$$

Variable environments are interesting in an evolutionary context only if events in one environment affect the other. Migration, a flow of behavioral variants from one environment into the other, will likely influence evolution in spatially variable environments. To model this effect, we suppose that there is a probability $1 - m$ that each model to whom a given individual is exposed experienced the same environment that the given individual will experience, and therefore a probability m that the model experienced the other environmental state. Thus, m measures the effective rate of migration of individuals from one habitat to the other. We assume throughout that $0 \leq m \leq \frac{1}{2}$. Let $q_{t,j}$ be the fraction of individuals that acquire behavior 1 within the subpopulation of individuals that experience environmental state j in cohort t. Then the frequency of behavior 1 in environment j after learning but before migration will be:

$$q'_{t,j} = q_{t,j}(1 - p_{j1} - p_{j2}) + p_{j1} \tag{7}$$

and the frequency of models exhibiting alternative t in habitat j during cohort $t+1$ is

$$q_{t+1,1} = (1 - m)q'_{t,1} + mq'_{t,2}$$
$$q_{t+1,2} = (1 - m)q'_{t,2} + mq'_{t,1} \tag{8}$$

Once again let us suppose that this process is repeated until a stable equilibrium is reached. Due to the assumed symmetry of the model, we know that any equilibrium at which both behaviors are present must satisfy

$$\hat{q}_1 = 1 - \hat{q}_2 \tag{9}$$

where $\hat{q}_1$ is the fraction of individuals acquiring behavior 1 in environment 1, and $\hat{q}_2$ is the fraction of individuals acquiring behavior 1 in environment 2. Using this fact one can show that

$$\hat{q}_1 = \frac{(1 - 2m)p_{11} + m}{(1 - 2m)(p_{11} + p_{12}) + 2m} \tag{10}$$

Notice that when $m = 0$, equation (10) reduces to the equilibrium derived in the model without any environmental variation. Also notice that if individuals are equally likely to imitate models drawn from both environments (i.e., $m = \frac{1}{2}$), then $\hat{q}_1 = \frac{1}{2}$. For intermediate values of m, q_1 falls between these two extreme values.

These properties make sense. In a uniform environment the behavior that results in higher fitness will increase in frequency according to the simplified model of the previous section; individuals should depend entirely on social learning and not take a chance on trial and error learning. When $m = 0$, there is no contact between individuals who experience the different environments, and the correct behavior in each environment becomes overwhelmingly common. Individual learning cannot do better than a perfected tradition, and it will frequently lead to errors. Within-cohort environmental variation, represented now by the movement of individuals among groups exposed to different environments, causes individuals to be exposed to some immigrant models who are likely to have acquired the behavior favored by individual learning in the other environment. Therefore, the movement of models among groups in a spatially variable environment causes social learning to be a less reliable method of acquiring one's behavior than it is in a homogeneous environment. When $m = \frac{1}{2}$ the frequency of the superior behavior is increased in each environment by the effects of individuals' experience, but the mixing of models from the two environments exactly erases the gains, and the individuals in the next cohort must start from scratch. In this case, social learning is useless.

The most interesting cases are the ones at intermediate values of m, where both social and individual learning are likely important. We will now compute the ESS amount of social learning in a variable environment for $0 < m < \frac{1}{2}$. The expected fitness of individuals using a learning rule characterized by the learning parameter d' is given by

$$E\{w(d')\} = W + D[\hat{q}_1(d)(1 - p_{11}(d') - p_{12}(d')) + p_{11}(d')] \qquad (11)$$

where $\hat{q}_1(d)$ is the equilibrium frequency of trait 1 in habitat 1, assuming that virtually all of the population is characterized by learning parameter d. To determine the ESS value of d, the confidence-interval-like parameter that determines the relative importance of social and individual learning, we once again determine which value of d can resist invasion by modifying alleles. A population in which d predominates can resist invaders that increase d whenever:

$$(1 - 2m)\left[\frac{\partial p_{11}}{\partial d}p_{12}(d) - \frac{\partial p_{12}}{\partial d}p_{11}(d)\right] + m\left[\frac{\partial p_{11}}{\partial d} - \frac{\partial p_{12}}{\partial d}\right] < 0 \qquad (12)$$

Consider how varying d affects the sign of the left-hand side of expression (12). We know from the models of a constant environment that the first term on the left-hand side of (12) is always positive (see equation 5). It is clear from the definition of p_{11} and p_{12} (see fig. 1.1) that the second term equals zero when $d = 0$, and is negative for all larger values of d. This means that when $d = 0$, the left-hand side of (12) will be positive and alleles that increase d can invade. Next notice that as d becomes large, both p_{11} and p_{12} approach zero, and therefore for large enough values of d, the left-hand side of (12) is negative, and alleles that decrease d can invade. Taken together, these facts mean that expected fitness is maximized for some amount of social learning intermediate between zero and one as long as $\frac{1}{2} > m > 0$. While we have not been able to solve (12) analytically, it is easy to solve numerically. The results, shown in figures 1.2 and 1.3, suggest that under a wide combination of migration rates and quality of individual experience, it is optimal to employ a mixture of social and individual learning. There is a broad region with combinations of modest migration rates and moderate to low information quality where social learning should be rather more important than individual learning in determining individual behavior. In figure

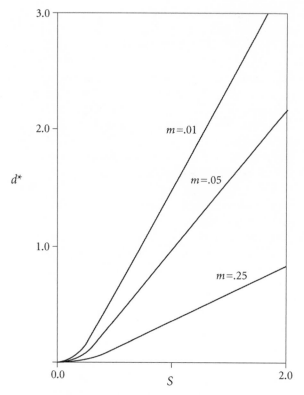

Figure 1.2. Plots the evolutionary equilibrium value of d, d^*, as a function of the quality of information available for individual learning, S, and for three levels of environmental heterogeneity, measured by m.

1.2, the ESS value of d, d^*, is plotted as a function of S, the measure of the quality of the information available to individuals, and the probability that naive individuals are exposed to models who learned from the wrong environment (m). There are two things to notice about these results: first, as individual experience becomes less reliable (i.e., S becomes large), the optimal amount of social learning is increased. Second, as the environment becomes less predictable (i.e., m increases), the optimal amount of social learning decreases. In figure 1.3, we plot the probability that individuals rely on social learning ($L^* = 1 - p_{11}(d^*) - p_{12}(d^*)$), given that d equals its optimal value.

This model suggests that the adaptiveness of social learning relative to individual learning depends on two factors: the accuracy of individual learning and the chance that an individual's social models experienced the same environment that the individual experiences. A substantial dependence upon social learning seems to be most adaptive when individual learning is inaccurate and there is not too much migration among habitats. The occasional use of individually acquired compelling evidence, coupled with faithful copying in the absence of such evidence, is sufficient to keep the locally adaptive behavior common.

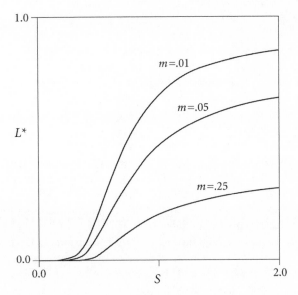

Figure 1.3. Plots the fraction of the population acquiring behavior by social learning when d is at its equilibrium value, $L^* = (1 - p_1(d^*) - p_2(d^*))$, as a function of S and m.

Increasing the importance of individual learning would entail more errors and would reduce the frequency of the adaptive behavior. In contrast, when there is extensive migration among habitats, relatively rare instances of individual learning would not be sufficient to maintain a high frequency of the locally adaptive behavior. Under such conditions, individuals must rely on individual learning if they are to have any chance of acquiring locally adaptive behavior.

Similar results derived using different models suggest that these conclusions are robust. We have analyzed the same dichotomous model in a temporally fluctuating environment (Boyd and Richerson, 1988). Assuming a Markov model of environmental change, we showed that the ESS reliance on social learning has the same qualitative properties as the model analyzed here. Elsewhere (Boyd and Richerson, 1985, ch. 4) we have analyzed a model that embodies the same qualitative assumptions about the nature of social learning and individual learning but in which behaviors are formalized as quantitative characters. These models have the same qualitative conditions for the evolution of social learning that result from this model. Finally, we have also extended the analysis of these models to allow for the genetic transmission of behavioral predispositions in addition to the genes that affect learning (Boyd and Richerson, 1983, 1985, ch. 4).

Social Learning with More than One Model

One can think of social learning as using the behavior of others as a source of information about the environment. Adaptive processes such as individual learning

will often cause the more common behavior to also be the most adaptive behavior, and, therefore, copying the behavior of a randomly chosen individual can
be adaptive under the right circumstances. In many species, however, naive
individuals may be able to observe the behavior of a number of experienced
conspecifics. That is, each naive individual often has a set of models. When this is
the case, one can think of such sets of models as samples of the behavior in the
population. Then if there is behavioral variation in a population, different individuals will be exposed to different samples of that behavior. Since different
samples of behavior lead to different inferences about the commonness of one or
the other behaviors in the population, it seems plausible that naive individuals
exposed to different samples of behavior might differ in the extent to which they
rely on social learning versus individual learning.

To address this question, we have modified the model so that individuals are
exposed to the behavior of n models. An individual's models may differ in their
behavior, and the naive individual can confront the problem of deciding which
variant to adopt. There is also an opportunity afforded by a large set of models.
Since individual learning will tend to increase the frequency of adaptive behaviors in a local habitat, there may well be information in the model "sample" as
to what behaviors are adaptive, especially as the size of the sample of the previous generation increases. Selection might structure social learning to use this
information. We want to determine the evolutionarily stable solutions to this
problem.

Begin by considering an individual exposed to i models using behavior 1 and
$n - i$ models using behavior 2. Once again assume that the individual observes
the variable x that indicates the state of the environment and then adopts each
behavior with the probabilities given in table 1.2. As before, the value of d_i
determines the minimum quality of information necessary before the individual
will rely on individual learning. It is indexed by i to indicate that individuals may
have different thresholds depending on the number of models who use one behavior or the other. We further assume that $d_i = d_{n-i}$. This assumption formalizes
the idea that it is the *number* of models who use a given behavior that governs the
usefulness of information acquired by social learning, not which trait they use.
The value of A_i determines the conditional probability that the individual will
acquire behavior 1 given that it is going to rely on social learning. To represent the
idea that there is no innate predisposition to adopt either trait in the absence of
information about the environment, we assume that $A_i = 1 - A_{n-i}$.

As before, suppose that there are two habitats linked by migration, one in
which behavior 1 is favored and one in which behavior 2 is favored. Let the

Table 1.2. Probability of acquiring behavior

Event	Behavior 1	Behavior 2
$d_i < x$	1	0
$-d_i \leq x \leq d_i$	A_i	$(1 - A_i)$
$x < -d_i$	0	1

frequency of the behavior 1 among models in environment j be $q_{t,j}$. Further suppose that models are sampled at random from the population. With these assumptions, the frequency of behavior 1 in environment j after individual and social learning, $q_{t,j}$, is

$$q'_{t,j} = \sum_{i=1}^{n} \binom{n}{i} q_{t,j}^i (1 - q_{t,j})^{n-i}$$
$$\times \{A_i(1 - p_{j1}(d_i) - p_{j2}(d_i)) + p_{j1}(d_i)\} \qquad (13)$$

The frequencies in each habitat after migration are given by equation (8).

The next step is to determine the equilibrium frequency of behavior 1 in each habitat. Because equation (13) is quite complex, we have not been able to derive an analytical expression for these equilibrium frequencies. However, it follows from the symmetry of the model that there is a stable symmetric equilibrium such that the favored behavior is common in each habitat, that is, $\hat{q}_1 = 1 - \hat{q}_2 > \frac{1}{2}$. We will refer to this as the symmetric equilibrium. Depending on the values of A_i and d_i, there may also be other stable internal equilibria at which one behavior is common in both habitats.

To determine an evolutionarily stable pattern of social learning, assume that most of the population has a learning rule characterized by the sets of parameters $d = \{d_0, \ldots, d_n\}$ and $A = \{A_0, \ldots, A_n\}$, and that the population has reached the resulting symmetric equilibrium. Then an individual with a different learning rule characterized by the sets of parameters $d' = \{d'_0, \ldots, d'_n\}$ and $A' = \{A'_0, \ldots, A'_n\}$, has expected fitness given by

$$E\{w(d', A')\} = W + D \sum_{i=0}^{n} \binom{n}{i} \hat{q}_1^i (1 - \hat{q}_1)^{n-i}$$
$$\times \{A'_i(1 - p_{11}(d'_i) - p_{12}(d'_i)) + p_{11}(d'_i)\} \qquad (14)$$

where $\hat{q}_1$ is the frequency of the favored behavior in each habitat at the symmetric equilibrium resulting from A and d. Then using the fact that $A_i = 1 - A_{n-i}$ and $d_i = d_{n-i}$, we can show that alleles that lead to a small increase in A_i can invade if

$$\hat{q}_1^i (1 - \hat{q}_1)^{n-i} - \hat{q}_1^{n-i}(1 - \hat{q}_1)^i > 0 \qquad (15)$$

which is always satisfied for $i > n/2$. Thus, the ESS values of A_i, A_i^*, are given by

$$A_i^* = \begin{cases} 1 & i > n/2 \\ \frac{1}{2} & i = n/2 \\ 0 & i < n/2 \end{cases} \qquad (16)$$

Given that an individual is going to rely on social learning, he should always adopt the more common behavior exhibited by his models. At the symmetric equilibrium the favored behavior is more common in each habitat. Thus, if individual experience is not determinative, the best thing to do is copy the behavior that is most common among models as it is more likely to be the locally favored behavior.

To determine the ESS value of d_i, d_i^*, assume that the set of A_i are at their ESS values given by (16). Then alleles which lead to a small increase in d_i can invade if

$$\frac{\partial p_{12}}{\partial d_i}\hat{q}_1^i(1-\hat{q}_1)^{n-i} - \frac{\partial p_{11}}{\partial d_i}\hat{q}_1^{n-i}(1-\hat{q}_1)^i > 0 \tag{17}$$

Substituting the definitions of p_{11} and p_{12} and simplifying yields the following expression for the ESS value of d_i:

$$d_i^* = (S/M)(n/2 - i)\{\ln \hat{q}_1 - \ln (1-\hat{q}_1)\} \tag{18}$$

This expression says that when an equal number of models use each behavior $(i = n/2)$, individuals should ignore their models and rely completely on individual learning. As the number of models exhibiting one behavior increases, d_i^* also increases linearly, and therefore the relative importance of individual learning declines. This effect becomes stronger as the frequency of the favored behavior in each habitat increases and as the size of the set of models increases. When nearly everyone in a given habitat uses the optimal trait and the set of models gives clear indication which behavior is more common in the local habitat, then you should adopt the alternative behavior only if the evidence from your own experience is very strong. On the other hand, if both behaviors are almost equally common in both habitats, the fact that one behavior is common among your models gives little information about which behavior is favored locally (especially if the number of models is small), and individuals should mainly rely on their own experience.

Discussion

The models presented in this chapter lead to three qualitative conclusions about the evolution of social learning. First, the adaptiveness of social learning depends on a trade-off. Increasing the importance of social learning increases fitness because it allows a reduction in the error rate of individual learning. However, increasing the importance of social learning also decreases the ability of the population to track a variable environment. A heavy dependence on social learning relative to individual learning seems to be most adaptive when individual learning is error-prone and environments are predictable. Second, the models suggest that when individuals do depend on social learning in a variable environment, they should not imitate randomly chosen individuals. Rather, they should tend to imitate the more common behavior among their models. This result follows from the fact that the behaviors favored by selection in a particular environment will tend to be more common in that environment. Finally, the models presented here suggest that selection will favor a pattern of social learning in which individuals exposed to more variable sets of models rely more heavily on individual learning. Given that models are numerous and sampled at random from the population, a predominance of one behavior among the models indicates that the behavior is more common in the population from which the models were drawn and, therefore, likely to be adaptive. An even mix of behavior among models indicates little about which behavior is common, especially if the number of models is small. Therefore, it may make sense to depend heavily on individual learning.

The models presented in this chapter can be thought of as a generalization of statistical decision theory. Within the context of that body of theory, decision makers seek to choose the best decision from among a set of possibilities, given specified information about the relationship between alternative decisions and outcomes. While this information may be imperfect, its statistical properties are specified, and they are independent of the decisions made by others. Given these assumptions, it is possible to specify the best decision procedures by considering each decision maker in isolation. Social learning involves decision makers who use the behavior of others as part of the information on which they base their decisions. The behavior of others depends on the decisions those individuals made, and therefore their decision rules. To specify the best rules for social learning, one must determine how a given decision rule affects the distribution of observed behavior in a population of decision makers. The models presented here provide one simple example of how this might be done in the context of the evolution of social learning.

The models presented here are very general and should apply to many situations in which animals could get information about the environment by observing conspecifics. The apparent rarity, or at least lack of sophistication, of social learning in species besides humans (Galef, 1988) is a considerable puzzle given our results. The adaptive properties of social learning present an array of fascinating theoretical and empirical problems.

REFERENCES

Boyd, R., & P. J. Richerson. 1983. The cultural transmission of acquired variation: Effect on genetic fitness. *Journal of Theoretical Biology* 100:567–596.

Boyd, R., & P. J. Richerson. 1985. *Culture and the evolutionary process.* Chicago: University of Chicago Press.

Boyd, R., & P. J. Richerson. 1988. The evolution of cultural transmission: The effects of spatial and temporal variation. In: *Social learning: A biopsychological approach,* T. Zentall & B. G. Galef, eds. (pp. 29–48). Hillsdale, NJ: Lawrence Erlbaum.

Galef, B. G. 1976. Social transmission of acquired behavior: A discussion of tradition and social learning in vertebrates. In: *Advances in the Study of Behavior,* vol. 6, J. S. Rosenblatt, R. A. Hinde, E. Shaw, & C. Beer, eds. (pp. 77–100). New York: Academic Press.

Galef, B. G. 1988. Imitation in animals: History, definition, and interpretation of data from the psychological laboratory. In: *Social learning: A biopsychological approach,* T. Zentall & B. G. Galef, eds. (pp. 1–28). Hillsdale, NJ: Lawrence Erbaum.

Hauser, M. 1988. Invention and social tranmission: New data from wild vervet monkeys. In: *Machiavellian intelligence: Social expertise and the evolution of intellect in monkeys, apes, and hen,* R. W. Byrne & A. Whitten, eds. (pp. 327–344). Oxford: Clarendon Press.

Kawai, M. 1965. Newly acquired pre-cultural behavior of the natural troop of Japanese monkeys on Koshima Island. *Primates* 6:1–30.

McNamara, J., & A. Houston. 1980. The application of statistical decision theory to animal behavior. *Journal of Theoretical Biology* 85:673–690.

Marler, P., & M. Tamura. 1964. Culturally transmitted patterns of vocal behavior in sparrows. *Science* 146:1483–1486.

McGrew, W. C., & C. E. G. Tutin. 1978. Evidence for a social custom in wild chimpanzees? *Man* 234:234–251.

Staddon, J. 1983. *Adaptive behavior and learning.* Cambridge: Cambridge University Press.

Stephens, D., & J. Krebs. 1987. *Foraging theory.* Princeton, NJ: Princeton University Press.

2 Why Does Culture Increase Human Adaptability?

Culture has made the human species a spectacular ecological success. Since the first appearance of tools and other evidences of culture in the archaeological record, the human species has expanded its range from part of Africa to the entire world, increased in numbers by many orders of magnitude, exterminated competitors and prey species, and radically altered the earth's biota.

It is not clear, however, *why* culture improves human adaptability. There has been a lot written about this topic, often in the introductions to articles and books on other topics, but very little careful analysis. In previous work, we (e.g., Boyd and Richerson, 1985) suggested that social learning allows us to avoid the costs of individual learning. Learning is costly, and without social learning everybody would have to learn everything for themselves. Teaching, imitation, and other forms of social learning, we argued, allow us to acquire a vast store of useful knowledge without incurring the costs of discovering and testing this knowledge ourselves. Recently, however, Alan Rogers (1989) has shown that this argument is, at best, incomplete and, at worst, plain wrong. Using a mathematical model of the evolution of social learning, he showed that the fact that social learning allows individual organisms to avoid the costs of learning does not increase the ability of that species of organisms to adapt. In fact, in the long run, social learning has no effect at all on the evolving organism's average fitness.

Here we have two goals: first, we argue that Rogers's result is robust, not an artifact of the specific form of his model. To do this, we analyze two models that incorporate Rogers's fundamental assumption that social learning allows individuals to avoid the costs of individual learning, but incorporate quite different assumptions about how social learning works and how the environment varies. Because these models also show that social learning does not increase the average fitness, we conclude that Rogers's result is robust. Culture will not increase the ability of a population to adapt if its only benefit is to allow individuals to avoid learning costs. We then analyze two models of the evolution of social learning that incorporate different assumptions about the evolutionary benefit of social learning. They assume that social learning increases the fitness of individuals who do *not* imitate by reducing the cost or increasing the accuracy of individual learning. In these models, culture does increase the average fitness of populations.

Why Avoiding Learning Costs Does Not Increase Average Fitness

Rogers's Model

Rogers's conclusions are based on a mathematical model of the evolution of imitation in a very simple hypothetical organism. These animals live in an environment that can be in one of two states; let's call them wet and dry. The environment has a constant probability of switching from wet to dry each generation, and the same probability of switching from dry to wet, which means that over the long run the environment is equally likely to be in each state. The probability of switching is a measure of the predictability of the environment. When this probability is high, knowing the state of the environment in one generation tells little about the state of the environment in the next generation. In contrast, when the probability of switching is low, the environment in the next generation is likely to be the same as the environment this generation. There are two behaviors available to the organism: one best in wet conditions and the other in dry conditions. There also are two genotypes—learners and imitators. Learners figure out whether the current environment is wet or dry and always adopt the appropriate behavior. However, the learning process is costly in that it reduces learners' chances of survival or reproduction. Imitators simply pick a random individual from the population and copy it. Copying does not have any direct effect on survival or reproduction. Rogers then used some simple but clever mathematics to determine which genotype wins in the long run.

The answer is surprising. The long-run outcome of evolution is always a mixture of learners and imitators in which both types have the same fitness as learners in a population in which there are no imitators. In other words, natural selection favors culture, but culture provides no benefit to the species. The organisms are no better off than they were without any imitation.

To understand the logic of this result, think about the fitness of learners and imitators as the frequency of imitators changes. As shown in figure 2.1, when imitators are rare, they have higher fitness than learners. They are nearly certain to acquire the best behavior because the population is composed of almost all learners, and learners always acquire the right behavior. But imitators don't suffer the cost of learning, so their fitness must be higher than learners. Thus, new mutations that give rise to copying will always be able to invade a population of

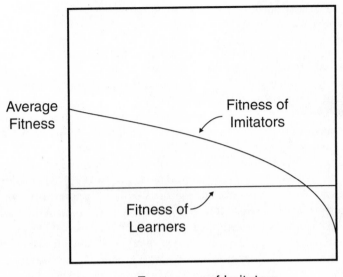

Figure 2.1. The average fitness of learners and imitators as a function of the frequency of imitators in the population. The frequency of learners is one minus the frequency of imitators. This figure is redrawn from Rogers (1989).

learners. On the other hand, when learners are rare, they have higher fitness than imitators. When there are very few learners, most of the imitators copy imitators who themselves copied imitators and so on. Because the environment changes periodically, this means that when learners are rare, imitators, in effect, choose behavior at random. In contrast, learners still acquire the best behavior. Thus, rare learners will be able to invade a population of imitators any time that the benefits of learning are sufficient to compensate for its costs. Because both types can increase when they are rare, the population will always be a mix of the two types. But only mixtures in which the two types have the same fitness can be stable long-run outcomes. Since the fitness of the learners is constant, it follows that the evolutionarily stable mix of learners and imitators has the same fitness as a population composed only of learners.

Two Extensions of Rogers's Model

One might think that this paradoxical result is an artifact. After all, the model *is* very simple. Perhaps if we add just a little realism, the paradox would go away. But such is not the case. We show that as long as the only benefit of imitation is the avoidance of learning costs, then changing rules of cultural transmission, the nature of environmental variability, and the number of traits leaves Rogers's basic result unchanged.

Spatially Varying Environment, More than Two Behaviors, Learning Errors

Rogers's model assumes that the environment varies in time but not space, that there are only two behaviors, and that learners always acquire the correct behavior. Each of these assumptions can be changed without changing the qualitative result.

Consider a model in which organisms live in an environment that consists of a large number of discrete islands, each with a different environment in which a different behavior is favored by natural selection. The populations on different islands are linked by migration of individuals from each island to all other islands. Thus, in this model the rate of migration measures the predictability of the environment. If migration rates are high, individuals' environments are unlikely to be similar to their parents'. If migration rates are low, most individuals live in environments just like the one their parents lived in. Learners engage in costly learning trials that usually allow them to acquire the locally optimal behavior but also sometimes lead to errors. As shown in Appendix 1, this model yields the same qualitative result as Rogers's model. Imitation evolves but does not benefit the population in the long run.

Imitators Can Detect Learners

Unlike the simple organisms in Rogers's model, humans do not blindly imitate a randomly chosen individual. Rather, they often evaluate the behavior of many individuals and choose the one that seems best, a process we have labeled *biased transmission* (Boyd and Richerson, 1985). Once a beneficial innovation arises, biased transmission allows it to spread through a population without further individual learning. Thus, it seems plausible that if Rogers's model were extended to allow biased transmission, the average fitness of the population might increase. However, a little analysis shows that this intuition is wrong.

Consider a model in which there are learners and imitators. As before, learners always acquire the currently favored behavior but at some cost. After learners learn, each imitator surveys the behavior of n individuals living in his social group. Imitators query each potential model to find out whether he acquired behavior by copying or by learning. If there is even a single learner in their group, imitators copy the learner and thereby acquire the behavior that is best in the current environment. If there are no learners, imitators copy a randomly chosen individual. This model allows imitators a great deal more information than Rogers's model: they can imitate n others rather than one, and they don't copy at random. However, as is shown in Appendix 2, the qualitative result is exactly the same—both types are present, and their long-run average fitness is the same as a pure population of learners.

Why Rogers's Result Is Robust

As Rogers argued in his original article, his result is robust because it reveals a basic evolutionary property of social learning: the advantage that imitators get from avoiding learning costs cannot increase fitness of a population because the frequency of imitators will increase until this advantage is exactly balanced by the disadvantage that imitators often acquire the wrong behavior. The fundamental logic underlying Rogers's result can be represented graphically as in figure 2.1, which plots the expected fitness of learners and imitators as a function of the fraction of imitators in the population. The fitness of imitators declines as the frequency of imitators increases because the more imitators there are, the more poorly the population tracks the changing environment, the lower the frequency of adaptive behavior, and, therefore, the dumber it is to copy. Moreover, there always have to be some learners in the population, because a population consisting only of imitators behaves at random. Thus, the expected fitness of imitators and learners has to be the same at equilibrium. But the fitness of learners isn't affected by the number of imitators. Thus, at equilibrium the average fitness of the population is the same as that of a population without culture.

How Culture Can Increase Average Fitness

Thinking about the problem this way points to its solution. Social learning would improve the average fitness of a population if it increased the fitness of *learners* as well as imitators. Consider figure 2.2. Here, we assume that the average fitness of learners increases as the frequency of imitators increases, and the paradox disappears—learners and imitators still have the same fitness at equilibrium, but

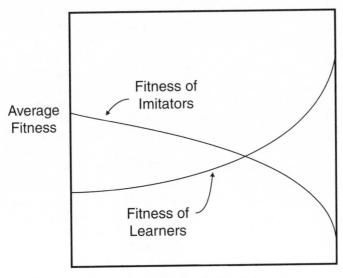

Frequency of Imitators

Figure 2.2. If increasing the frequency of imitators reduces the cost or increases the accuracy of individual learning, then the average fitness of the population can be increased by imitation.

now that fitness is higher than for a population composed entirely of learners. Thus, to improve the average fitness of the population, imitation must make individual learning cheaper or more accurate.

Of course, this formal possibility would be of little importance if there were no plausible means by which increasing the amount of imitation would cause individual learning to be more efficient. However, we suggest that there are at least two ways that imitation can benefit learners.

Imitation Allows Selective Learning

Imitation can increase the average fitness of learners by allowing individuals to learn more selectively. Learning opportunities often vary. Sometimes it may be easy to determine the best behavior while other times it may be very difficult. Without imitation, an organism must rely on learning even when it is difficult and error-prone. In contrast, an imitating organism can learn when learning is cheap and accurate and imitate when it is costly or inaccurate. The following model shows that imitation plus selective learning can increase average fitness in a population even when most individuals imitate.

As before, consider a population that lives in an environment that switches between two states, and assume that there are two behaviors, one best in each environmental state. However, now suppose that all individuals attempt to discover the best behavior in the current environment. Each individual experiments with both behaviors and then compares the results. The results of such experiments vary for many reasons, and, thus, the behavior that is best during any particular trial may be inferior over the long run. To avoid errors, individuals adopt a particular behavior only if it appears *sufficiently* better than its alternative. The larger the observed difference in the payoffs between the two behaviors, the more likely that the behavior with the higher payoff actually is best. By insisting on a large difference in observed payoff, individuals can reduce the chance that they will mistakenly adopt the inferior behavior. Of course, being selective will also cause more trials to be indecisive, and, in that case, they imitate a randomly chosen individual. Thus, there is a tradeoff: You can increase the accuracy of learning, but only by also increasing the probability that learning will be indecisive, and you will have to rely on imitation. The exact nature of the trade-off depends on the probability distribution of the outcome of learning trials. In Appendix 3, we analyze a model in which the observed difference in payoffs is a normal random variable. For one set of parameters ($\mu = 0.5$, $\sigma = 1$), the relationship between imitation and the accuracy of learning has the form shown in figure 2.3. If the individual adopts a behavior any time that it yields a higher payoff during the learning trial, it will acquire the wrong behavior around 30 percent of the time. If it requires a larger difference in payoffs, then it can reduce the chance of such errors, but sometimes it will have to imitate. If it is sufficiently picky, it will almost never err, but it will also almost always acquire its behavior by imitation.

To model the evolution of social learning, we assume that an individual's position on this continuum is a genetically heritable trait. Suppose that most individuals use a learning rule that causes them to imitate x percent of the time—we

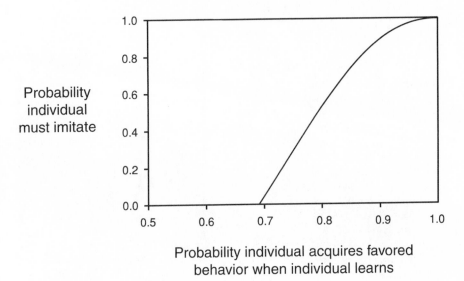

Probability
individual
must imitate

Probability individual acquires favored
behavior when individual learns

Figure 2.3. The trade-off between imitation and learning, assuming that the outcomes of
learning trials are normally distributed, with mean equal to 0.5 and variance equal to 1.0.

call these "common-type individuals." There are also a few rare "mutant" individuals who imitate slightly more often. Compared to the common type,
mutants are less likely to make learning errors. Thus, when mutants learn, they
have higher fitness than the common-type individuals when they learn. When
mutants imitate, they have the same fitness as the common type. However, mutants must imitate more often, and imitators always have lower fitness than
learners. To see why, think of each imitator as being connected to a learner by a
chain of imitation. If the learner at the end of the chain learned in the current
environment, then the imitator has the same chance of acquiring the favored
behavior as does a learner. If the learner at the end of the chain learned in a
different environment, the imitator will have a lower chance of acquiring the best
behavior. Thus, the mutant type will have higher fitness if the advantage of
making fewer learning errors is sufficient to offset the disadvantage of imitating
more.

This evolutionary trade-off depends on how much the common type imitates. When the common type rarely imitates, the fitnesses of individuals who
imitate and individuals who learn will be similar because most imitators will imitate somebody who learned, and, therefore, the fact that mutants make fewer
learning errors will allow them to invade. However, as the amount of imitation
increases, the fitness of imitating individuals relative to those who learn declines
because increased imitation lengthens the chain connecting each imitator to a
learner. Eventually an equilibrium is reached at which the common type can
resist invasion by mutants that change the rate of imitation. We refer to the
fraction of time that the common type imitates at equilibrium as the "evolutionary equilibrium amount of imitation."

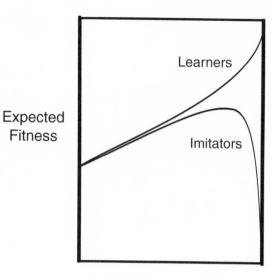

Figure 2.4. Individuals either learn or imitate according to the outcome of their learning trial. As individuals become more selective, the frequency of imitating individuals increases. This figure plots the expected fitness of individuals who imitate and those who learn as a function of the frequency of imitating individuals, assuming the outcome of learning experiments is normally distributed with mean 0.5 and variance 1.

The average fitness of a population at the evolutionary equilibrium is greater than the average fitness of individuals who do not imitate as long as the probability that the environment changes is less than half (see Appendix 3 for a formal proof). You can get an intuitive feel for why by considering figure 2.4, which plots the average fitness of imitating and learning individuals as a function of the fraction of common-type individuals who imitate. The fitness of learning individuals increases as the amount of imitation increases because learners make fewer errors. The fitness of imitating individuals also increases at first because they are imitating learners who make fewer errors. If imitation is common enough, fitness eventually declines because the population fails to track the changing environment. The first effect is apparently sufficient to lead to a net increase in average fitness at evolutionary equilibrium.

It is important to understand that this increase in average fitness is only a side effect of selection at the individual level. The evolutionary equilibrium amount of imitation does not maximize the average fitness of the population. Selection at the individual level favors more imitation than is optimal for the population because it ignores the effect on the population as a whole of increased imitation, and after a certain point this effect is deleterious.

Imitation Allows Cumulative Improvement

Imitation may increase the average fitness of learners by allowing learned improvements to accumulate from one generation to the next. So far we have considered only two alternative behaviors. Thus, learning is an either/or proposition. Many kinds of behaviors admit successive improvements toward some optimum. Individuals start with some initial "guess" about the best behavior and then invest time and effort at improving their performance. For a given amount of time and effort, the better an individual's initial guess, the better on average its final performance. Now, imagine that the environment varies, so that different behaviors are optimal in different environments. Organisms who cannot imitate must start with whatever initial guess is provided by their genotype. They can then learn and improve their behavior. However, when they die, these improvements die with them, and their offspring must begin again at the genetically given initial guess. In contrast, an imitator can acquire its parents' behavior after their behavior has been improved by learning. Therefore, it will start its search closer to optimal behavior, and for a given amount of searching, it will achieve a better adult phenotype. Thus, if the learning cost per unit improvement is smaller for small improvements than for big ones, imitation makes learning more efficient and therefore increases the average fitness of the population.

The following simple model illustrates this idea (a more realistic model with the same properties is analyzed in Boyd and Richerson, 1985, ch. 4). Consider an organism that lives in an environment that can be in a continuum of states. For example, suppose that the population density of prey species varies. In each generation there is a chance that the environment switches to a new state (more or less prey), but also some chance that it remains unchanged. There is also a continuum of behaviors, such as the amount of effort devoted to foraging versus hunting. We measure the environmental state in terms of the optimal behavior in that environment and assume that an individual's fitness decreases as the difference between the environmental state and its behavior value increases.

All individuals modify their behavior by learning. Each individual begins with an initial guess about the state of the environment and then experimentally modifies this behavior. In doing so, individuals reduce the difference between their behavior and the optimum behavior in the current environment. Learning is costly— individuals who devote more time and effort to experimenting suffer greater learning costs but move closer to the current optimum. There are two genotypes. Learners use a fixed, genetically inherited norm of reaction as their initial guess about the environment, and they always acquire the optimum behavior. Imitators acquire their initial guess by imitating the behavior of a randomly chosen member of the previous generation. They invest much less in learning than do learners and, as a result, improve on their initial behavior only a small amount. However, as long as the environment does not change, the population of imitators will converge slowly toward the optimum as each generation moves toward the optimum. Thus, imitators may start their learning nearer to the optimum than do learners.

Imitators have higher fitness at evolutionary equilibrium in this model as long as (1) the environment does not change too often compared to the rate at which the population of imitators converges toward the optimum, and (2) learners suffer substantially greater learning costs than do imitators. If the environment changes slowly enough, the gradual cumulative improvement achieved by imitators will be sufficient to ensure that their behavior is near the current optimum most of the time. Of course, imitators will never track the environment as accurately as learners, but if the small improvements realized by imitators are cheaper than the large improvements of learners, imitators will have higher average fitness. Because only imitators are present at such an equilibrium, imitation increases average fitness.

Discussion

Culture increases average fitness if it makes the learning processes that generate new knowledge less costly or more accurate. Culture may do this in at least two ways: first, social learning allows individual learning to be selective. Individuals can learn opportunistically when it is likely to be more accurate or less costly and imitate when conditions are less favorable. Second, social learning allows learned improvements to accumulate from one generation to the next. When learning in small steps is less costly per unit improvement in fitness than learning in large steps, the cumulative learning over many generations can increase average fitness.

These results help us understand the importance of the evolution of true imitation. There are a number of examples of social traditions in other animals. For example, some populations of chimpanzees in West Africa regularly use stone tools to crack open tough nuts, while other nearby populations never use stones to crack nuts. The stones and nuts are available to both populations, and the environments are otherwise very similar (Boesch et al., 1994). Students of social learning in nonhuman animals (e.g., Galef, 1988; Visalberghi and Fragaszy, 1990) distinguish two classes of processes that could maintain such cultural differences between different populations: *social enhancement* occurs when the activity of older animals increases the chance that younger animals will learn the behavior on their own. Young individuals do not acquire the behavior by observing older individuals. Social facilitation could cause tool use to persist in some populations but not others, as in the following scenario: in populations in which chimpanzees use tools to crack nuts, young chimpanzees spend a lot of time in proximity to both nuts and hammer stones. Nuts are a greatly desired food, and young chimpanzees find eating nutmeats highly reinforcing. Young chimpanzees experiment with the hammers and anvils until they master the skill of opening the nuts. In populations in which chimpanzees do not use stones to open nuts, young chimpanzees never spend enough time in proximity to both nuts and hammer stones to acquire the skill. *Imitation* occurs when younger animals observe the behavior of older animals and learn how to perform the behavior by watching them. In this case, the tradition is preserved because young chimpanzees actually imitate the behavior of older chimpanzees.

Students of animal social learning have distinguished between social enhancement and imitation because the necessary psychological mechanisms are quite different. Our results suggest that this distinction is also of evolutionary importance because selective social learning and cumulative culture change are possible only when there is imitation. Social enhancement can preserve variation only in behavior that organisms can learn on their own, albeit in favorable circumstances, but it does not allow individuals to avoid learning when information is poor or costly. Even more important, only imitation allows cumulative cultural change. Suppose that on her own in especially favorable circumstances an early hominid learned to strike rocks together to make useful flakes. Her companions, who spent time near her, would be exposed to the same kinds of conditions, and some of them might learn to makes flakes too, entirely on their own. This behavior could be preserved by social enhancement because groups in which tools were used would spend more time in proximity to the appropriate stones. However, that would be as far as it would go. Even if an especially talented individual found a way to improve the flakes, this innovation would not spread to other members of the group because each individual learns the behavior anew. With imitation, on the other hand, innovations can persist as long as younger individuals are able to acquire the modified behavior by observational learning. As a result, imitation can lead to the cumulative evolution of behaviors that no single individual could invent on his own.

Recent reviews (Galef, 1992; Tomasello, 1990; Visalberghi and Fragaszy, 1990) suggest that all known cases of animal social traditions can be explained as the result of social enhancement. If this is correct, our results explain why animal cultures seem to play such a small role in the lives of such species. It also suggests that understanding the evolution of the psychological mechanisms that allow imitation is of key importance for understanding human evolution.

APPENDIX 1: Spatially Varying Environment, More than Two Variants, Learning Errors

Consider an organism that lives in a spatially varying environment in which there are a large number of islands. A different behavior is favored on each island so that the fitness of behavior i on island j is

$$W_i = \begin{cases} W_0 + D & \text{in environment } i \\ W_0 - D & \text{in environment } j \end{cases} \tag{A1.1}$$

There are two genotypes:

Learners $=$ Discover locally optimal behavior with probability $1 - e$.
Imitators $=$ Imitate a randomly chosen individual from the previous generation.

After learning and imitating, a fraction m of the individuals on each island emigrate and are replaced by individuals drawn from all other islands at random. Because the number of behaviors is large, the frequency of the favored behavior among immigrants is approximately zero.

After migration, selection occurs. We assume that selection is weak so that the frequency of innovators and imitators is the same on all islands. Then let

$q =$ frequency of imitators on the focal island.
$p =$ frequency of the locally favored behavior among imitators.

The probability that an imitator encounters a single individual who has the locally optimal behavior is $(1 - q)(1 - e) + qp$, and thus the frequency of the locally optimal trait among imitators after imitation, p', is

$$p' = (1 - q)(1 - e) + qp \tag{A1.2}$$

And after migration the frequency of the favored behavior among imitators, p'', is

$$p'' = (1 - m)[(1 - q)(1 - e) + pq] \tag{A1.3}$$

Thus, there is a unique stable equilibrium frequency of the locally favored variant, $\tilde{p}$

$$\tilde{p} = \frac{(1 - m)(1 - e)(1 - q)}{1 - q(1 - m)} \tag{A1.4}$$

The average fitness of learners is $W_L = W_0 + D(1 - 2e) - C$, where C is the cost of individual learning. The average fitness of imitators, $W_I = W_0 + D(2\tilde{p} - 1)$. If imitators are rare ($q \approx 0$), then the equilibrium frequency of the favored variant among rare imitators is approximately $(1 - e)(1 - m)$, the same frequency as among learners, and since imitators incur no learning cost, they increase in frequency. If imitators are common ($q \approx 1$), then the equilibrium frequency of the favored variant is zero, and therefore imitators have lower fitness than do learners as long as learning pays $[D(1 - 2e) > C]$. Since $\tilde{p}$ is a monotonically decreasing function of q, there is a unique stable equilibrium value of q at which imitators have the same fitness as learners.

APPENDIX 2: Imitators Can Identify Learners

Consider an organism that lives in an environment that can be in one of two states. Each generation there is a probability γ that the environment switches from one state to the other. There are two behaviors with fitnesses as given in table A2.1:

Table A2.1. Fitness in environments 1 & 2

	Environment 1	Environment 2
Behavior 1	$W_0 + D$	$W_0 - D$
Behavior 2	$W_0 - D$	$W_0 + D$

There are two genotypes:

Learners = Always acquired the best behavior in the current environment but at a cost C.

Imitators = Observe n individuals after learning. If there is a learner among these individuals, imitators acquire the best behavior in the current environment. Otherwise they copy a random individual from within the group.

And let q equal the frequency of imitators, and p the frequency of the currently favored behavior among imitators. Assume that selection is sufficiently weak so that the effect of selection on cultural evolution can be ignored (i.e., on dynamics of p), and genetic evolution (the dynamics of q) responds to the stationary distribution of p.

Then the frequency of the currently favored behavior after learning and imitation is

$$p' = \begin{cases} 1 - q^n + q^n p & \text{if no environmental change} \\ 1 - q^n + q^n(1 - p) & \text{if environment changes} \end{cases} \tag{A2.1}$$

Suppose at some time t the probability density for p is $f_t(p)$ with mean P_t. Then the mean of $f_{t+1}(p)$ given by

$$P_{t+1} = \int [(1 - \gamma)(1 - q^n + q^n p) + \gamma(1 - q^n + q^n(1 - p))] f_t(p) dp \qquad \text{(A2.2)}$$

where γ is the probability that the environment switches states. Integrating and simplifying yields the following recursion for P_t:

$$P_{t+1} = 1 - q^n + q^n[(1 - 2\gamma)P_t + \gamma] \qquad \text{(A2.3)}$$

Thus, the equilibrium value of mean frequency of the favored behavior is:

$$P = \frac{1 - q^n + q^n \gamma}{1 - q^n(1 - 2\gamma)} \qquad \text{(A2.4)}$$

The average fitness of learners is $W_L = W_0 + D - C$, which is independent of changes in the environment. The average fitness of imitators once P_t has reached its equilibrium value is $W_I = W_0 - D(2P - 1)$. The frequency of imitators will increase whenever $W_I > W_L$. Substituting the expression for P given in equation A2.4 and solving for q yields the following inequality:

$$q < q^* = \left(\frac{C/D}{2\gamma(1 - C/D) + C/D} \right)^{1/n} \qquad \text{(A2.5)}$$

Thus, q^* is a unique stable equilibrium value for the frequency of imitators, and at this frequency the average fitness of imitators and learners is equal.

APPENDIX 3: Selective Learning

Consider an organism that lives in an environment that can be in one of two states. Each generation there is a probability γ that the environment switches from one state to the other. There are two behaviors with fitnesses as given in the table A2.1.

Each individual performs a learning trial in which it estimates the payoff of each behavior in the current environment. The difference between the payoff of the currently favored behavior and that of the alternative behavior observed by each individual is an independent, normally distributed, random variable, x, with mean equal to m, and variance equal to 1. The mean, m, is positive because, on average, the currently favored behavior yields a higher payoff in the current environment. All individuals use the learning rule:

Outcome of Learning Trial	Decision
$x > d$	Adopt favored behavior
$d \geq x \geq -d$	Imitate
$-d > x$	Adopt other behavior

The threshold parameter d determines how selectively individuals learn. Individuals regard trials that yield positive outcomes greater than d as decisive evidence that the environment is in the state that is currently favored, and trials in which x is less than $-d$ as decisive evidence that the environment is in the other state. When a trial produces an outcome in between d and $-d$, it is indecisive and individuals imitate.

The value of d is a genetically heritable trait. At any time there are two genotypes present in the population. Most of the population has $d = d^*$, but there are a very few rare mutants who have $d = d^* + \delta d$. We seek to determine the values of d^* that can resist invasion by mutants with slightly smaller or slightly larger values of d. Such continuous ESS solutions often yield the same outcome as genetically more realistic models.

Let v be the frequency of the favored behavior in the population. Assume that selection is sufficiently weak so that the effect of selection on cultural evolution can be ignored (i.e., on dynamics of v) and genetic evolution responds to the stationary distribution of v. Finally, let $p_1(d) = \Pr(x > d)$, $p_2(d) = \Pr(x < -d)$, and $L(d) = 1 - p_1(d) - p_2(d)$; $p_1(d)$ is the probability of correctly choosing the currently favored behavior, $p_2(d)$ is the probability of mistakenly choosing the other behavior, and $L(d)$ is the probability of imitating. Then the frequency of the favored variant in the next generation, v', is:

$$v' = \begin{cases} vL(d^*) + p_1(d^*) & \text{if no change in environment} \\ (1 - v)L(d^*) + p_2(d^*) & \text{if environment changes} \end{cases} \tag{A3.1}$$

Suppose at some time t the probability density for v is $f_t(v)$ with mean V_t. Then the mean of $f_{t+1}(v)$ is given by

$$V_{t+1} = \int [(1 - \gamma)(vL + p_1) + \gamma((1 - v)L + p_2)] f_t(v) dv \tag{A3.2}$$

Integrating and simplifying yield the following recursion for V_t:

$$V_{t+1} = (1 - 2\gamma)(V_t L + p_1) + \gamma \tag{A3.3}$$

Thus, the equilibrium value of the mean frequency of the favored behavior is:

$$V = \frac{(1 - 2\gamma)p_1 + \gamma}{(1 - 2\gamma)(p_1 + p_2) + 2\gamma} \tag{A3.4}$$

The fitness of the common genotype averaged over the stationary distribution of v is:

$$\overline{W}(d^*) = W_0 + D[VL(d^*) + p_1(d^*)] - D[(1 - V)L(d^*) + p_2(d^*)] \tag{A3.5}$$

and the fitness of the mutant type is

$$\overline{W}(d^* + \delta d) = W_0 + D[VL(d^* + \delta d) + p_1(d^* + \delta d)] \\ - D[(1 - V)L(d^* + \delta d) + p_2(d^* + \delta d)] \tag{A3.6}$$

Thus, because δd is small, the difference in fitness between the mutant and common types, δW, is

$$\delta W = D\left((2V - 1)\left(\frac{\partial L}{\partial d}\right)_{d^*} \delta d + \left(\frac{\partial p_1}{\partial d}\right)_{d^*} \delta d - \left(\frac{\partial p_2}{\partial d}\right)_{d^*} \delta d \right) \tag{A3.7}$$

Setting $\delta W = 0$, substituting the expression for V given in A3.4, and simplifying yield the following necessary condition for the ESS:

$$0 = (1 - 2\gamma)\left(\left(\frac{\partial p_1}{\partial d}\right)_{d^*} p_2(d^*) - \left(\frac{\partial p_2}{\partial d}\right)_{d^*} p_1(d^*)\right) + \gamma\left(\left(\frac{\partial p_1}{\partial d}\right)_{d^*} - \left(\frac{\partial p_2}{\partial d}\right)_{d^*}\right) \tag{A3.8}$$

Given that x is normal with a known mean and variance, this equation can be solved numerically for the value of d^*.

We now prove that the average fitness of a population at the ESS value of d, d^*, is greater than the average fitness of a population with no imitation (i.e., $d = 0$) whenever $m > 0$ and $\gamma < 1/2$. It follows from A3.8 that when $\gamma = 1/2$ then $d^* = 0$ and, therefore, that $\overline{W}(d^*) - \overline{W}(0) = 0$. Next, we show that $\overline{W}(d^*) - \overline{W}(0)$ is a monotonically decreasing function of γ as long as m is positive. Compute

$$\frac{\partial}{\partial \gamma}(\overline{W}(d^*) - \overline{W}(0)) = 2\frac{\partial V}{\partial \gamma}L(d^*) + \frac{\partial d^*}{\partial \gamma}\left\{(2V-1)\left(\frac{\partial L}{\partial d}\right)_{d^*} + \left(\frac{\partial p_1}{\partial d}\right)_{d^*} - \left(\frac{\partial p_2}{\partial d}\right)_{d^*}\right\}$$

(A3.9)

But the ESS condition (A3.7) guarantees that the term in braces on the right-hand side of A3.9 is zero. Thus,

$$\frac{\partial}{\partial \gamma}(\overline{W}(d^*) - \overline{W}(0)) \propto \frac{\partial V}{\partial \gamma} = p_2(d^*) - p_1(d^*)$$

$$+ (1 - 2\gamma)\frac{\partial d^*}{\partial \gamma}\left\{\gamma\left(\frac{\partial p_1}{\partial d} - \frac{\partial p_2}{\partial d}\right)_{d^*}\right.$$

$$\left. + (1 - 2\gamma)\left(\frac{\partial p_1}{\partial d}p_2 - \frac{\partial p_2}{\partial d}p_1\right)_{d^*}\right\}$$

(A3.10)

Once again the ESS condition guarantees that the term in braces on the right-hand side of A3.10 is zero, and since $p_1(d^*) > p_2(d^*)$ for $m > 0$, it follows that the average fitness of an ESS population is greater than the fitness of a population with no imitation as long as $\gamma < \frac{1}{2}$.

APPENDIX 4: Cumulative Learning

Consider an organism that lives in an environment that can be in a continuum of states. Each generation, there is a probability γ that the environment switches from its current state to a new state drawn at random from a probability distribution with mean equal to zero and variance equal to H. There is a probability $1 - \gamma$ that the environment will remain unchanged. There is also a continuum of behaviors. In each environment, fitness is a gaussian function of behavior so that there is a unique optimum behavior θ_t. We choose to measure the state of the environment as the optimal behavior in that environment. All individuals modify their behavior by learning so that the difference between their behavior and optimum behavior in the current environment is reduced. There are two genotypes:

Learners = Acquire the optimal behavior. Learning costs reduce fitness by a factor e^{-C_L}.

Imitators = Imitate a randomly chosen individual from the previous generation, and then adjust their behavior a small fraction, a ($a \ll 1$) by learning. Learning costs reduce fitness by a factor e^{-C_I}.

Suppose most individuals in a population are imitators, but that there are a small number of rare learners. Because they always acquire the optimal behavior, the expected fitness of learners is simply:

$$W_L = \exp(1 - C_L)$$

(A4.1)

and the expected fitness of copiers is:

$$W_I = \exp[-(1-a)^2(Z_t - \theta_t)^2 - C_I]$$

(A4.2)

where Z_t is the behavior of imitators during period t, which will change from period to period according to the following recursion.

$$Z_{t+1} = a\theta_t + (1-a)Z_t$$

(A4.3)

Thus, the behavior of imitators will converge toward the current optimum at a rate a. When the environment changes, it will converge toward a different value. Assume that selection is weak enough that changes in gene frequency respond to the stationary distribution of Z_t. Thus, imitation is evolutionarily stable if

$$-(1 - a)^2 E\{(Z_t - \theta_t)^2\} - C_I > - C_L \tag{A4.4}$$

where the expectation is taken with respect to the joint stationary distribution of θ_t and Z_t.

$$E\{(Z_t - \theta_t)^2\} = E\{Z_t^2\} - 2E\{Z_t\theta_t\} + E\{\theta_t^2\} \tag{A4.5}$$

To compute $E\{Z_t\theta_t\}$ multiply both sides of A4.3 by θ_{t+1}.

$$\theta_{t+1}Z_{t+1} = a\theta_t\theta_{t+1} + (1 - a)Z_t\theta_{t+1} \tag{A4.6}$$

Taking the expectation of both sides yields:

$$E\{\theta_{t+1}Z_{t+1}\} = a[(1 - \gamma)V + \gamma 0] + (1 - a)[(1 - \gamma)E\{\theta_tZ_t\} + \gamma 0] \tag{A4.7}$$

The moments of the stationary distribution are constant, and thus setting $E\{Z_{t+1}\theta_{t+1}\} = E\{Z_t\theta_t\}$ and solving yields:

$$E\{Z_t\theta_t\} = \frac{a(1 - \gamma)V}{1 - (1 - a)(1 - \gamma)} \tag{A4.8}$$

To compute $E\{Z_t^2\}$ square both sides of A4.3.

$$Z_{t+1}^2 = a^2\theta_t^2 + 2a(1 - a)Z_t\theta_t + (1 - a)^2Z_t^2 \tag{A4.9}$$

Again taking the expectation of both sides, setting $E\{Z_{t+1}^2\} = E\{Z_t^2\}$, and substituting the expression for $E\{Z_t\theta_t\}$ yields:

$$E\{Z_t^2\} = \frac{a[1 + (1 - a)(1 - \gamma)]}{(2 - a)[1 - (1 - a)(1 - \gamma)]} \tag{A4.10}$$

Substituting the expressions for $E\{Z_t\theta_t\}$ and $E\{Z_t^2\}$ into A4.5 and simplifying yields:

$$E\{(Z_t - \theta_t)^2\} = \frac{2\gamma V}{(2 - a)[1 - (1 - a)(1 - \gamma)]} \tag{A4.11}$$

Substituting this expression into A4.4, ignoring terms of order a^2, and simplifying yield the following condition for imitation to be an ESS.

$$\left(\frac{\delta}{1 - \delta}\right)a > \gamma \tag{A4.12}$$

where $\delta = \frac{C_L - C_I}{V}$ is the fitness advantage of imitators due to lower cost learning measured in units of V the average log fitness increase of learners due to learning. Because learning would not be favored by selection for learners if $V < C_L$, we know that $\delta < 1$. Recall that a is the rate at which imitators converge toward the current optimum. Thus, the ESS condition, A4.12, says that the rate of environmental change must be less than the rate at which imitators converge toward the current optimum as modified by the term in parentheses. This term is greater than one when the learning cost advantage of imitators is a large fraction of the total benefit of learning and less than one when the learning cost advantage of imitators is relatively small.

NOTE

We thank Alan Rogers for many useful discussions of these ideas and his careful reading of this manuscript. RB also thanks Dorothy Cheney and Robert Seyfarth for providing electricity and other facilities at Baboon Camp, where the first draft of this chapter was written.

REFERENCES

Boesch, C., P. Marchesi, N. Marchesi, B. Fruth, & F. Joulian. 1994. Is nut cracking in wild chimpanzees a cultural behavior? *Journal of Human Evolution* 26:325–328.

Boyd, R., & P. J. Richerson. *Culture and the evolutionary process*. 1985. Chicago: University of Chicago Press.

Boyd, R., & P. J. Richerson. 1988. An evolutionary model of social learning: The effects of spatial and temporal variation. In: *Social Learning: Psychological and Biological Approaches*. T. Zentall & B. G. Galef, eds. (pp. 29–48). Hillsdale, NJ: Lawrence Erlbaum.

Boyd, R., & P. J. Richerson. 1989. Social learning as an adaptation. *Lectures on Mathematics in the Life Sciences* 20:1–26.

Galef, B. G. 1988. Imitation in animals: History, definitions, and interpretations from the psychological laboratory. In: *Social Learning: Psychological and Biological Approaches*. T. Zentall & B. G. Galef, eds. (pp. 1–28). Hillsdale, NJ: Lawrence Erlbaum.

Galef, B. G. 1992. The question of animal culture. *Human Nature* 3:157–178.

Rogers, A. R. 1989. Does biology constrain culture? *American Anthropologist* 90: 819–831.

Tomasello, M. 1990. Cultural transmission in the tool use and communicatory signaling of chimpanzees. In: *"Language" and Intelligence in Monkeys and Apes. Comparative and Developmental Perspectives*, S. T. Parker & K. R. Gibson, eds. (pp. 273–311). Cambridge: Cambridge University Press.

Visalberghi, E., & D. M. Fragaszy. 1990. Do monkeys ape? In: *"Language" and Intelligence in Monkeys and Apes. Comparative and Developmental Perspectives*, S. T. Parker & K. R. Gibson, eds. (pp. 247–273). Cambridge: Cambridge University Press.

3 Why Culture Is Common, but Cultural Evolution Is Rare

Cultural variation is common in nature. In creatures as diverse as rats, pigeons, chimpanzees, and octopuses, behavior is acquired through social learning. As a result, the presence of a particular behavior in a population makes it more likely that individuals in the next generation will acquire the same behavior, which, in turn, results in persistent differences between populations that are not due to genetic or environmental differences.

In sharp contrast, cumulative cultural evolution is rare. Most culture in nonhuman animals involves behaviors that individuals can, and do, learn on their own. There are only a few well-documented cases in which cultural change accumulates over many generations leading to the evolution of behaviors that no individual could invent—the only well-documented examples are song dialects in birds, perhaps some behaviors in chimpanzees, and, of course, many aspects of human behavior.

We believe that this situation presents an important evolutionary puzzle. The ability to accumulate socially learned behaviors over many generations has allowed humans to develop subtle, powerful technologies and to assemble complex institutions that permit us to live in larger, and more complex, societies than any other mammal species. These accumulated cultural traditions allow us to exploit a far wider range of habitats than any other animal, so that even with only hunting and gathering technology, humans became the most widespread mammal on earth. The fact that simple forms of cultural variation exist in a wide variety of organisms suggests that intelligence and social life alone are not sufficient to allow cumulative cultural evolution. Cumulative cultural change seems to require some special, derived, probably psychological, capacity. Thus, we have the puzzle, if cultural traditions are such a potent means of adaptation, why is this capacity rare?

In this chapter we suggest one possible answer to this question. We begin by reviewing the literature on animal social learning. We then analyze two models of the evolution of the psychological capacities that allow cumulative cultural evolution. The results of these models suggest a possible reason why such capacities are rare.

Culture in Other Animals

There has been much debate about whether other animals have culture. Some authors define culture in human terms. That is, the investigator essays human cultural behavior and extracts a number of "essential" features. For example Tomasello, Kruger and Ratner (1993) argue that culture is learned by all group members, faithfully transmitted, and subject to cumulative change. Then to be cultural, the behavior of other animals must exhibit these features. Moreover, a heavy burden of proof is placed on those who would claim culture for other animals—if there is any other plausible interpretation, it is preferable. Others (McGrew, 1992; Boesch, 1993) argue that a double standard is being applied. If the behavioral variation observed among chimpanzee populations were instead observed among human populations, they argue, anthropologists would regard it as cultural.

Such debates make little sense from an evolutionary perspective. The psychological capacities that underpin human culture must have homologies in the brains of other primates and perhaps other mammals as well. Moreover, the functional significance of social transmission in humans could well be related to its functional significance in other species. The study of the evolution of human culture must be based on categories that allow human cultural behavior to be compared to potentially homologous, functionally related behavior of other organisms. At the same time, such categories should be able to distinguish between human behavior and the behavior of other organisms because it is quite plausible that human culture is different in important ways from related behavior in other species.

Here we define cultural variation as differences among individuals that exist because they have acquired different behavior as a result of some form of social learning. Cultural variation is contrasted with genetic variation, differences among individuals that exist because they have inherited different genes from their parents, and environmental variation, differences among individuals due to the fact that they have experienced different environments. Cultural variation is often lumped together with environmental variation. However, as we have argued at length elsewhere (Boyd and Richerson, 1985), this is an error. Because cultural variation is transmitted from individual to individual, it is subject to population dynamic processes analogous to those that effect genetic variation and quite unlike the processes that govern other environmental effects. Combining cultural and environmental effects into a single category conceals these important differences.

There is much evidence that cultural variation, defined this way, is very common in nature. In a review of social transmission of foraging behavior, Levebre and Palameta (1988) give 97 examples of cultural variation in foraging behavior in animals as diverse as baboons, sparrows, lizards, and fish. Song dialects are socially transmitted in many species of songbirds. Three decades of study shows that chimpanzees have cultural variation in subsistence techniques, tool use, and social behavior (Wrangham, McGrew, DeWaal, and Heltne, 1994; McGrew, 1992).

There is little evidence, however, of cumulative cultural evolution in other species. With a few exceptions, social learning leads to the spread of behaviors that individuals could have learned on their own. For example, food preferences are socially transmitted in rats. Young rats acquire a preference for a food when they smell the food on the pelage of other rats (Galef, 1988). This process can cause the preference for a new food to spread within a population. It can also lead to behavioral differences among populations living in the same environment because current foraging behavior depends on a history of social learning. However, it does not lead to the cumulative evolution of new, complex behaviors that no individual rat could learn on its own.

In contrast, human cultures do accumulate changes over many generations, resulting in culturally transmitted behaviors that no single human individual could invent on his own. Even in the simplest hunting and gathering societies, people depend on such complex, evolved knowledge and technology. To live in the arid Kalahari, the !Kung San need to know what plants are edible, how to find them during different seasons, how to find water, how to track and find game, how to make bows and arrow poison, and many other skills. The fact that the !Kung can acquire the knowledge, tools, and skills necessary to survive the rigors of the Kalahari is not so surprising—many other species can do the same. What is amazing is that the same brain that allows the !Kung to survive in the Kalahari also permits the Inuit to acquire the very different knowledge, tools, and skills necessary to live on the tundra and ice north of the Arctic circle, and the Aché the knowledge, tools, and skills necessary to live in the tropical forests of Paraguay. No other animal occupies a comparable range of habitats or utilizes a comparable range of subsistence techniques and social structures. Two kinds of evidence indicate that such differences result from cumulative cultural evolution of complex traditions. First, such gradual change is documented in both the historical and archaeological records. Second, cumulative change leads to a branching pattern of descent with modification in which more closely related populations share more derived characters than distantly related populations. Although the possibility of horizontal transmission among cultural lineages makes reconstructing such cultural phylogenies difficult for "cultures" (Boyd, Richerson, Borgerhoff Mulder, and Durham, 1997), patterns of cultural descent can be reconstructed for particular cultural components, such as languages or technologies.

Circumstantial evidence suggests that the ability to acquire novel behaviors by observation is essential for cumulative cultural change. Students of animal social learning distinguish *observational learning* or *true imitation*, which occurs when younger animals observe the behavior of older animals and learn how to perform a novel behavior by watching them, from a number of other mechanisms of social transmission that also lead to behavioral continuity without observational learning (Galef, 1988; Visalberghi and Fragaszy, 1990; Whiten and Ham, 1992). One such mechanism, *local enhancement*, occurs when the activity of older animals increases the chance that younger animals will learn the behavior on their own. If younger, naive individuals are attracted to the locations in the environment where older, experienced individuals are active, they will tend to learn the same behaviors as the older individuals. Young individuals do not acquire the information necessary to perform the behavior by observing older individuals. Instead, the activity of others causes them to be more likely to acquire this information through interaction with the environment. Imagine a young monkey acquiring its food preferences as it follows its mother around. Even if the young monkey never pays any attention to what its mother eats, she will lead it to locations where some foods are common and others rare, and the young monkey may learn to eat much the same foods as mom.

Local enhancement and observational learning are similar in that they both can lead to persistent behavioral differences among populations, but only observational learning allows cumulative cultural change (Tomasello et al., 1993). To see why, consider the cultural transmission of stone tool use. Suppose that on their own in especially favorable circumstances, an occasional early hominid learned to strike rocks together to make useful flakes. Their companions, who spent time near them, would be exposed to the same kinds of conditions and some of them might learn to make flakes too, entirely on their own. This behavior could be preserved by local enhancement because groups in which tools were used would spend more time in proximity to the appropriate stones. However, that would be as far as it would go. Even if an especially talented individual found a way to improve the flakes, this innovation would not spread to other members of the group because each individual learned the behavior anew. Local enhancement is limited by the learning capabilities of individuals and the fact that each new learner must start from scratch. With observational learning, on the other hand, innovations can persist as long as younger individuals are able to acquire the modified behavior by observational learning. To the extent that observers can use the behavior of models as a starting point, observational learning can lead to the cumulative evolution of behaviors that no single individual could invent on her own.

Most students of animal social learning believe that observational learning is limited to humans and, perhaps, chimpanzees and some bird species. Several lines of evidence suggest that observational learning is not responsible for cultural traditions in other animals. First, many of the behaviors, like potato washing in Japanese macaques, are relatively simple and could be learned independently by individuals in each generation. Second, new behaviors like potato washing often take a long time to spread through the group, a pace more consistent with the idea that each individual had to learn the behavior on her own. Finally, extensive laboratory experiments capable of distinguishing observational learning from other forms of social transmission like local enhancement have usually failed to demonstrate observational learning (Galef, 1988; Whiten and Ham, 1992; Tomasello et al., 1993; Visalberghi, 1993), except in humans and songbirds. (In many songbirds, song traditions are transmitted by imitation, but little or nothing else is.) The fact that observational learning appears limited to humans seems to confirm that observational learning is necessary for cumulative cultural change. However, one must be cautious here because most students of animal social learning refuse to invoke observational learning unless all other possible explanations have been excluded. Thus, there actually may be many cases of observational learning that are interpreted as social enhancement or some putatively simpler mechanism. A few well-controlled laboratory studies do apparently show some true imitation in nonhuman animals (Heyes, 1993; Dawson and Foss, 1965), and striking anecdotes suggest that observational learning may occur in organisms as diverse as parrots (Pepperberg, 1988) and orangutans (Russon and Galdikas, 1993).

Adaptation by cumulative cultural evolution is apparently not a by-product of intelligence and social life. Cebus monkeys are among the world's cleverest creatures. In nature, they use tools and perform many complex behaviors, and in captivity, they can be taught extremely demanding tasks. Cebus monkeys live in social groups and have ample opportunity to observe the behavior of other individuals of their own species. Yet good laboratory evidence suggests that cebus monkeys make no use of observational learning. This suggests that observational learning is not simply a by-product of intelligence and opportunity to observe conspecifics. Rather, observational learning seems to require special psychological mechanisms (Bandura, 1986). This conclusion suggests, in turn, that the psychological mechanisms that enable humans to learn by observation are adaptations that have been shaped by natural selection because culture is beneficial. Of course, this need not be the case. Observational learning could be a by-product of some other adaptation that is unique to humans, such as bipedalism, dependence on complex vocal communication, or the capacity for deception. However, given the great importance of culture in human affairs, it is reasonable to think about the possible adaptive advantages of culture. In what follows we consider two mathematical models of the evolution of the capacity for observational learning based on this assumption.

Models of the Evolution of Social Learning

The maintenance of cultural variation involves two different processes (figure 3.1). First, there must be some kind of *transmission* of information from one brain to another. Consider, for example, the maintenance of the use of a particular kind of tool. Individuals have information stored in their brain that allows them to manufacture and use the tool. For use of the tool to persist through time, observing tool use and manufacture must cause individuals in the next "generation" to acquire information that allows them to manufacture and use the same tool. (We put generation in quotes because the same model can be used to represent culture change occurring on much shorter time scales. See Boyd and Richerson, 1985: 68–69.) As we have seen, this transmission may occur because individuals can learn how to make and use tools by observation, or because observation stimulates them to learn on their own how to make and use the tool, for example by local enhancement. Second, individuals must preserve the information that

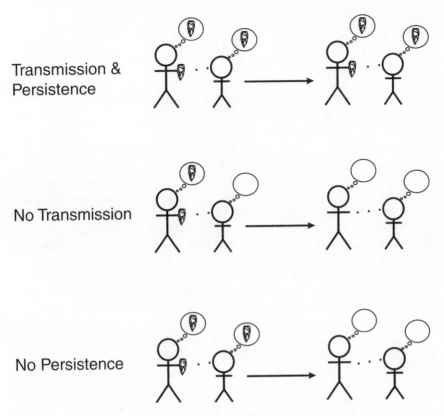

Transmission &
Persistence

No Transmission

No Persistence

Figure 3.1. The maintenance of cultural transmission requires both the accurate transmission of mental representations from experienced to inexperienced individuals and the persistence of those representations through the lives of individuals until such time that they act as models for others.

allows them to make and use the tool until such time that they serve as models for the next generation of individuals. Such *persistence* may fail to occur for two different reasons: individuals may forget how to make or use the tool, or they may, as a result of interacting with the environment, modify the information stored in their brains so that they make or use the tool in a significantly different way. Without both transmission and persistence, there can be no culturally transmitted variation.

Our previous work on the evolution of culture (Boyd and Richerson 1985, 1988, 1989, 1995) has focused on the evolution of persistence. All of the models analyzed in these studies assume that transmission occurs and consider the evolution of genes that affect the extent to which behavior acquired by imitation is modified by individual learning. They differ in how the trait is modelled (discrete vs. continuous), how environmental variation is modelled, whether individuals are sensitive to the number of models who exhibit a particular cultural variant, and a number of other features. This work leads to the robust conclusion that natural selection will favor individuals who do not modify culturally acquired behavior when individual learning is costly or error-prone, and environments are variable, but not too variable. Thus, natural selection can favor persistence. (See Rogers, 1989, for a related model.)

In several articles, Feldman and his co-workers (Cavalli-Sforza and Feldman, 1983a, 1983b; Aoki and Feldman, 1987) have considered the evolution of genes that affect transmission. In these models it is assumed that there is a beneficial trait that can be acquired only by cultural transmission, not by individual learning. They further allow for the possibility that successful transmission requires new behavior both on the part of the individual acquiring the behavior and in the individual modelling the behavior. Thus, there are two different genetic loci, one affecting the behavior of the transmitter and a second affecting the behavior of the receiver. For transmission to evolve, there must be substitutions at both loci. These models are very relevant to the evolution of communication systems. However, they cannot address the questions posed here because the culturally transmitted trait cannot be acquired or modified by individual learning.

Here we consider two models of the evolution of psychological capacities that allow the transmission of behavior that can be acquired or modified through individual learning. Each model is designed to answer the same basic question: what are the conditions under which selection can favor a costly psychological capacity that allows individuals to acquire behavior by imitation? The primary difference between the models is the nature of the culturally transmitted behavior. In the first model, the behavior is discrete—individuals are either skilled or unskilled, and the skill can be acquired either by social or individual learning. In the second model, there is a continuum of behaviors subject to stabilizing selection. Only the continuous trait model allows true cumulative cultural change leading to behaviors that individuals cannot learn on their own. However, the discrete model allows us to investigate the effects of several factors that are difficult to include in the continuous character model. As we will see, both models tell a similar story about why there is a selective barrier to the evolution of the capacity for observational learning and why capacities that allow local enhancement and related mechanisms do not face a similar barrier.

Discrete Character Model

Consider an organism that lives in a temporally variable environment that can be in an infinite number of states. In each state, individuals can acquire a skill that increases fitness, so that unskilled individuals have fitness W_0, and skilled individuals have fitness $W_0 + D$. Each generation there is a probability γ that the environment switches from its current state to a different state. When this occurs, the old skill is no longer useful in the new environment.

There are two genotypes with different learning rules. *Individual learners* acquire the skill appropriate to the current environment with probability δ at a cost C_I. *Social learners* observe n randomly selected members of the previous generation. If there is a skilled individual among the n, an imitator acquires the skill at cost C_S. Otherwise they acquire the skill with probability δ at a cost C_I. The ability to acquire the skill by social learning reduces the fitness of an individual by an amount K. Thus, parameters C_I and C_S give the variable costs of individual and social learning, respectively, and K gives the fixed cost associated with the capacity for social learning.

It is shown in the appendix that social learning can increase when rare and is the only ESS when the following condition holds:

$$(1 - (1 - \delta)^n)(1 - \gamma)[D(1 - \delta) + C_I - C_S] > K \tag{1}$$

When expression (1) is true, social learning has higher fitness than individual learning no matter what the mix of the two types in the population. The term in square brackets gives the fitness benefit of acquiring the skill through social rather than individual learning—$C_I - C_S$ is the advantage that results from the fact that social learning may reduce the cost of acquiring the trait, and $D(1 - \delta)$ is the advantage that results from being more likely to acquire the skill. Sensibly, the latter term implies that the fitness advantage of social learning increases as the likelihood that individuals will learn the trait on their own, δ, decreases. The less likely it is that individual learners will acquire the skill, the bigger the relative advantage that accrues to social learning. The fitness benefit is discounted by the two factors on the left-hand side of expression (1). The term $1 - \gamma$ expresses the fact that social learning is beneficial only if the environment has not changed, and term $1 - (1 - \delta)^n$ gives the probability that at least one of the n individuals from the previous generation will have acquired the behavior when social learning is rare. Notice that this latter term decreases as the probability of learning the trait decreases. Thus, the net advantage of social learning is highest at intermediate values of δ, when there is a good chance that individuals will learn the skill on their own, but also a good chance that they won't.

When (1) is not satisfied, there is a range of conditions in which social learning cannot increase when rare, but is an ESS once it becomes common. In this analysis we are limited to the case $n = 1$ because when $n > 1$ the dynamics of the cultural traits are nonlinear, and such systems are difficult to analyze in autocorrelated random environments. With this assumption, social learning is an ESS when:

$$\frac{\delta(1 - \gamma)(D(1 - \delta) + C_I - C_S)}{\gamma + (1 - \gamma)\delta} > K \tag{2}$$

To compare this expression with (1), notice that when $n = 1$, $1 - (1 - \delta)^n = \delta$, and, thus, the benefit of social learning when it is common is the benefit when rare divided by the term $\gamma + (1 - \gamma)\delta$. When individual learners are likely to acquire the skill (so that δ is large), the conditions for social learning to increase when rare (1) and to persist when common (2) will be similar. However, when individual learners are unlikely to acquire the skill ($\delta \ll 1$) and the rate of environmental change is slow ($\gamma \ll 1$), social learning will be able to persist when common under a much wider range of conditions than it can increase when it is rare. When social learning is rare, most of the population will be individual learners who have little chance of acquiring the skill. As a consequence, social learning will provide little benefit because there will be few skilled individuals to observe. When social learning is common, the population will slowly accumulate the skill over many generations. If the environment does not change too often, the social learning population will spend most of the time with the skill at high frequency, and thus the cost of the capacity for social learning need be only less than the net benefit of acquiring the skill by individual learning.

Continuous Character Model

Consider an organism that is characterized by a single quantitative character subject to stabilizing selection. During generation t the optimum value of the quantitative character is θ_t. Each generation there is a probability γ that the environment changes. If the environment does not change then $\theta_{t+1} = \theta_t$. If it does change, then θ_{t+1} is a normal random variable with mean Θ, and variance H. Notice that this assumption implies that Θ is the long-run optimum trait value.

Each individual acquires its trait value through a combination of genetic transmission, imitation, and individual learning. The adult trait value, x, is given by:

$$x = (1 - a)[(1 - i)\Theta + iy] + a\theta_t \tag{3}$$

The term $(1 - i)\Theta + iy$ represents a "norm of reaction," which forms the basis for subsequent individual learning. It is acquired as the result of a combination of a genetically acquired norm of reaction at the long-run optimum, Θ, and the observed trait value, y, of a randomly selected member of the previous generation. The parameter i governs the relative importance of genetic inheritance and imitation in determining the norm of reaction. When $i = 0$, the norm of reaction is completely determined by an innate, genetically inherited value. As i increases, the observed trait value of another individual has greater influence on the trait until, when $i = 1$, the norm of reaction is completely determined by observational learning. Because observational learning is assumed to require special-purpose cognitive machinery, individuals incur a fitness cost proportional to the importance of observational learning in determining their norm of reaction, iC. Thus, C measures the incremental cost of the capacity for observational learning. Individuals adjust their adult behavior from the norm of reaction toward the current optimum a fraction a. To capture the idea that cumulative change is possible, we assume that a is small, so that the repeated action of learning and social transmission can lead to fitness increases that could not be attained by individual learning.

With these assumptions it is shown in the appendix that a population in which most individuals do not imitate can be invaded by rare individuals who imitate a little bit only if

$$(1 - \gamma)aH > C \tag{4}$$

The parameter H is a measure of how far the population is from the optimum in fitness units, on average, immediately after an environmental change. Since a population without imitation always starts from the same norm of reaction, Θ, the term aH is a measure of the average fitness improvement due to individual learning in a single generation. Thus, (4) says that imitation can evolve only when the benefit of imitating what individuals can learn on their own is sufficient to compensate for the costs of the capacity to imitate.

In contrast, the condition for social learning to be maintained once it is common is much more easily satisfied. It is shown in the appendix that a population in which $i = 1$ can resist invasion by rare alleles that reduce the reliance on imitation whenever:

$$\frac{(1-\gamma)aH}{\gamma + (1-\gamma)a} > C \tag{5}$$

If the rate at which the population adapts by individual learning, a, is greater than the rate at which the environment changes, γ, then a population in which social learning is common spends most of its time with the mean behavior near the optimum. Thus, (5) says that imitation is evolutionarily stable as long as the cost of the capacity is less than a substantial fraction of the total improvement in fitness due to many generations of social learning.

Discussion

Both of these models tell a similar story about the evolution of capacities that allow social learning. When social learning is rare, the only useful behavior that is present in the population, and thus the only behavior that can be acquired by social learning, is behavior that individuals can learn on their own. In contrast, when social learning is common, the population accumulates adaptive behavior over many generations, and, as long as the environment does not change faster than adaptive behavior accumulates, social learning allows individuals to acquire behaviors that are much *more* adaptive than they could acquire on their own.

This result provides a potential explanation for why cultural variation is so common in nature but cumulative cultural evolution so rare. Capacities that increase the chance that individuals will learn behaviors that they could learn on their own will be favored as long as they are relatively cheap. On the other hand, even though the benefits of cumulative cultural evolution are potentially substantial, selection cannot favor a capacity for observational learning when rare. Thus, unless observational learning substantially reduces the cost of individual learning, it will not increase because there is an "adaptive valley" that must be crossed before benefits of cumulative cultural change are realized. This argument suggests, in turn, that it is likely that the capacities that allow the initial evolution of observational learning must evolve as a side effect of some other adaptive change. For example, it has been argued that observational learning requires that individuals have what psychologists and philosophers call a "theory of mind" (Cheney and Seyfarth, 1990; Tomasello et al., 1993). That is, imitators must be able to understand that others have different beliefs and goals from theirs. Lacking such a theory, typical animals cannot make a connection between the acts of other animals and their own goal states and thus can't interpret the acts of other animals as acts they might usefully perform. A theory of mind may have initially evolved to allow individuals to better predict the behavior of other members of their social group. Once it had evolved for that reason, it could be elaborated because it allowed observational learning and cumulative cultural evolution.

APPENDIX 1: Analysis of Discrete Character Model

Individual learners always have the same fitness:

$$W_t = W_0 + \delta D - C_l \tag{A1.1}$$

The expected fitness of social learners depends on the frequency of social learners in the previous generation, q, the frequency of skilled individuals among social learners, p, and whether the environment has changed during the previous generation.

$$
\begin{aligned}
W_S = {} & \gamma(W_0 + \delta D - C_l) \\
& + (1 - \gamma)(W_0 + \pi(D - C_S) + (1 - \pi)(\delta D - C_l))
\end{aligned} \tag{A1.2}
$$

where π is the probability that at least one of the n individuals in the sample of models has acquired the skill favored in the previous environment, and can be calculated as:

$$\pi = \sum_{i=0}^{n} \binom{n}{i} q^i (1 - q)^{n-i} [1 - (1 - p)^i (1 - \delta)^{n-i}] \tag{A1.3}$$

To understand this expression, assume that there are i social learners among the n models observed by a given, naive social learner. The probability that all i of the social learners are not skilled is $(1 - p)^i$, and the probability that the remaining $n - i$ individual learners are not skilled, is $(1 - \delta)^{n-i}$, and therefore, the probability that there is at least one skilled individual among the n given that there are i social learners, is $1 - (1 - p)^i (1 - \delta)^{n-i}$. Then to calculate π, take the expectation over all values of i.

Thus, social learners will have higher fitness in a particular generation if

$$W_S - W_l = \pi(1 - \gamma)(D(1 - \delta) + C_l - C_S) - K > 0 \tag{A1.4}$$

We consider two special cases. Case 1: $q \approx 0$, $\pi \approx 1 - (1 - \delta)^n$. When social learners are rare, they will observe only individual learners, and thus the probability of observing at least one skilled individual does not depend on q or p. Thus, social learning will increase when rare as in this expression

$$(1 - (1 - \delta)^n)(1 - \gamma)(D(1 - \delta) + C_l - C_S) - K > 0 \tag{A1.5}$$

Immediately after an environmental change, the frequency of skilled individuals among social learners is δ and then increases monotonically until the next environmental change. Thus, the expected value of π is greater than $(1 - (1 - \delta)^n)$, and if social learning can increase when rare, it will continue to increase until it reaches fixation.

Case 2: $n = 1$, $\pi = 1 - q(1 - p) - (1 - q)(1 - \delta)$. Assume that selection is sufficiently weak so that the effect of selection on cultural evolution can be ignored (i.e., on dynamics of p), and genetic evolution (the dynamics of q) responds to the stationary distribution of p.

Then the frequency of the currently favored behavior after learning and imitation is:

$$p' = \begin{cases} \delta & \text{if environment changes} \\ (qp + (1 - q)\delta)(1 - \delta) + \delta & \text{if environment does not change} \end{cases} \tag{A1.6}$$

Suppose at some time t the probability density for p is $f_t(p)$ with mean P_t. Then the mean of $f_{t+1}(p)$ is given by:

$$P_{t+1} = \int [(1 - \gamma)((qp + (1 - q)\delta)(1 - \delta) + \delta) + \gamma\delta] f_t(p) dp \tag{A1.7}$$

Integrating yields the following recursion for P_t:

$$P_{t+1} = \gamma\delta + (1-\gamma)[(qP_t + (1-q)\delta)(1-\delta) + \delta] \tag{A1.8}$$

Thus, the equilibrium value of mean frequency of the favored behavior is:

$$P = \frac{\delta + (1-\gamma)(1-q)\delta(1-\delta)}{1 - (1-\gamma)(1-\delta)q} \tag{A1.9}$$

Assume that selection is weak enough that the dynamics of q respond to the stationary distribution of p. Then, since the expression for W_s is linear in p when $n=1$, we can substitute P for p.

$$\pi = \frac{\delta}{1 - (1-\gamma)(1-\delta)q} \tag{A1.10}$$

Notice that $\pi > \delta$, which implies that social learners are more likely on average to acquire the skill. Substituting A1.10 into A1.4 yields the following condition for social learning to increase in frequency:

$$\frac{(1-\gamma)(D(1-\delta) + C_l - C_S)\delta}{1 - (1-\gamma)(1-\delta)q} > K \tag{A1.11}$$

APPENDIX 2: Analysis of Continuous Character Model

Since we are free to determine the scale of measurement of trait values, we can, without loss of generality, set $\Theta = 0$. Then the mean value of x in the population during generation t, X_t is:

$$X_t = (1-a)iX_{t-1} + a\theta_t \tag{A2.1}$$

The logarithm of the fitness of an individual with adult trait value x is proportional to:

$$\ln(W) \propto -(x - \theta_t)^2 - C(i) \tag{A2.2}$$

Thus, the expected fitness of an individual whose behavioral acquisition is governed by the parameter i is:

$$E\{\ln(W)\} \propto -(1-a)^2 E\{(iX_{t-1} - \theta_t)^2\} - C(i) \tag{A2.3}$$

Consider the competition between two genotypes. The common type has development characterized by parameter i and the rare type by $i+\delta$, where δ is very small. If one assumes that changes in i have no effect on the variance of the trait among the invading type individuals, the expected fitness of the invading type is approximately proportional to:

$$E\{\ln(W)\} \propto -(1-a)^2[(i^2 + 2i\delta)E\{X_{t-1}^2\} - 2(i + \delta)E\{X_{t-1}\theta_t\} + \theta_t^2]$$
$$- C(i) - \frac{\partial C}{\partial i}\delta \tag{A2.4}$$

Combining expression A2.3 and A2.4 shows that the invading type will increase in frequency if:

$$-(1-a)^2[2i\delta E\{X_{t-1}^2\} - 2\delta E\{X_{t-1}\theta_t\}] - \frac{\partial C}{\partial i}\delta > 0 \tag{A2.5}$$

To calculate $E\{X_{t-1}\theta_t\}$ first notice the following:

$$\theta_t = \begin{cases} \theta_{i-t} & \text{with probability } 1-\gamma \\ \varepsilon & \text{with probability } \gamma \end{cases} \tag{A2.6}$$

where ε is an independent normal random variable with mean zero and variance H. Thus, it follows that:

$$E\{\theta_t X_{t-1}\} = (1 - \gamma)E\{\theta_{t-1} X_{t-1}\} + \gamma E\{X_{t-1}\varepsilon\} \qquad (A2.7)$$

Multiplying both sides of A2.1 by θ_t and taking the expectation with respect to the joint stationary distributions yield:

$$E\{\theta_t X_t\} = (1 - a)iE\{\theta_t X_{t-1}\} + aH \qquad (A2.8)$$

Combining A2.7 and A2.8 yields the following expression for $E\{X_{t-1}\theta_t\}$:

$$E\{X_{t-1}\theta_t\} = \frac{(1 - \gamma)aH}{1 - i(1 - \gamma)(1 - a)} \qquad (A2.9)$$

To calculate $E\{X_{t-1}^2\}$ square both sides of A2.1, take the expectation, and using A2.9 solve:

$$E\{X_{t-1}^2\} = \frac{a^2 - 2i(1 - a)E\{X_{t-1}\theta_t\}}{1 - i^2(1 - a)^2} \qquad (A2.10)$$

Substituting A2.9 and A2.10 into A2.5 and simplifying yield expressions (4) and (5) in the text.

REFERENCES

Aoki, K., & M. W. Feldman. 1987. Toward a theory for the evolution of cultural communication: Coevolution of signal transmission and reception. *Proceedings of the National Academy of Sciences, U.S.A.* 84:7164–7168.

Bandura, A. 1986. *Social foundations of thought and action: A social cognitive theory.* Englewood Cliffs, NJ: Prentice Hall.

Boesch, C. 1993. Aspects of transmission of tool-use in wild chimpanzees. In: *Tools, Language, and Cognition in Human Evolution*, K. R. Gibson, & T. Ingold, eds. (pp. 171–189). Cambridge University Press, Cambridge.

Boyd, R., & P. J. Richerson. 1985. *Culture and the evolutionary process.* University of Chicago Press, Chicago.

Boyd, R., & P. J. Richerson. 1988. An evolutionary model of social learning: the effects of spatial and temporal variation. In: *Social Learning, Psychological and Biological Perspectives*, T. Zentall, & B. G. Galef, eds. (pp. 29–48). Hillsdale, NJ: Lawrence Erlbaum.

Boyd, R., & P. J. Richerson. 1989. Social learning as an adaptation. *Lectures on Mathematics in the Life Sciences* 20:1–26.

Boyd, R., & P. J. Richerson. 1995. Why does culture increase human adaptability? *Ethology and Sociobiology* 16:125–141.

Boyd, R., P. J. Richerson, M. Borgerhoff Mulder, & W. H. Durham. 1997. Are cultural phylogenics possible? In: *Human Nature, Between Biology and the Social Sciences*, P. Weingart, P. J. Richerson, S. Mitchell, & S. Maasen, eds. pp. 355–386.

Cavalli-Sforza, L. L., & M. W. Feldman. 1983a. Cultural versus genetic adaptation. *Proceedings of the National Academy of Sciences, U.S.A.* 80: 4993–4996.

Cavalli-Sforza, L. L., & M. W. Feldman. 1983b. Paradox of the evolution of communication and of social interactivity. *Proceedings of the National Academy of Sciences, U.S.A.* 80:2017–2021.

Cheney, D., & R. Seyfarth. 1990. *How monkeys see the world.* Chicago: University of Chicago Press.

Dawson, B. V., & B. M. Foss, 1965. Observational learning in budgerigars. *Animal Behavior* 13:470–474.

Galef, B. G. 1988. Imitation in animals: History, definitions, and interpretation of data from the psychological laboratory. In: *Social Learning: Psychological and Biological Perspectives*, T. Zentall, & B. G. Galef, eds. (pp. 355–386). Hillsdale, NJ: Lawrence Erlbaum.

Heyes, C. M. 1993. Imitation, culture and cognition. *Animal Behavior* 46:999–1010.

Levebre, L., & B. Palameta. 1988. Mechanisms, ecology, and population diffusion of socially-learned, food-finding behavior in feral pigeons. In: *Social Learning: Psychological and Biological Perspectives*, T. Zentall, & B. G. Galef, eds. (pp. 141–165). Hillsdale, NJ: Lawrence Erlbaum.

McGrew, W. 1992. *Chimpanzee material culture*. Cambridge: Cambridge University Press.

Pepperberg, I. 1988. The importance of social interaction and observation in the acquisition of communicative competence: Possible parallels between avian and human learning. In: *Social Learning: Psychological and Biological Perspectives*, T. Zentall, & B.G. Galef, eds. (pp. 3–29). Hillsdale, NJ: Lawrence Erlbaum.

Rogers, A. R. 1989. Does biology constrain culture? *American Anthropologist* 90: 819–831.

Russon, A. E., & B. Galdikas. 1993. Imitation in free-ranging rehabilitant orangutans (Pongo pygmacus). *Journal of Comparative Psychology* 107:147–161.

Tomasello, M., A. C. Kruger, & H. H. Ratner. 1993. Cultural learning. *Behavioral and Brain Sciences* 16:495–552.

Visalberghi, E. 1993. Capuchin monkeys: A window into tool use in apes and humans. In: *Tools, Language, and Cognition in Human Evolution*, K. R. Gibson, & T. Ingold, eds. (pp. 138–150). Cambridge: Cambridge University Press.

Visalberghi, E., & D. M. Fragazy. 1990. Do monkeys ape? In: *Language and Intelligence in Monkeys and Apes*, S. Parker, & K. Gibson, eds. (pp. 247–273). Cambridge: Cambridge University Press.

Whiten, A., & R. Ham. 1992. On the nature and evolution of imitation in the animal kingdom: A reappraisal of a century of research. In: *Advances in the Study of Behavior, Vol. 21*, P. J. B. Slater, J. S. Rosenblatt, C. Beer, & M. Milkinski, eds. (pp. 239–283). New York: Academic Press.

Wrangham, R. W., W. C. McGrew, F. B. M. DeWaal, & P. G. Heltne. 1994. *Chimpanzee cultures*. Cambridge, MA: Harvard University Press.

4   Climate, Culture, and the Evolution of Cognition

What are the causes of the evolution of complex cognition? Discussions of the evolution of cognition sometimes seem to assume that more complex cognition is a fundamental advance over less complex cognition, as evidenced by a broad trend toward larger brains in evolutionary history. Evolutionary biologists are suspicious of such explanations because they picture natural selection as a process leading to adaptation to local environments, not to progressive trends. Cognitive adaptations will have costs, and more complex cognition will evolve only when its local utility outweighs them.

In this chapter, we argue that Cenozoic trends in cognitive complexity represent adaptations to an increasingly variable environment. The main support for this hypothesis is a correlation between environmental deterioration and brain size increase in many mammalian lineages.

We would also like to understand the sorts of cognitive mechanisms that were favored in building more complex cognitions. The problem is difficult because little data exist on the adaptive trade-offs and synergies between different cognitive strategies for adapting to variable environments. Animals might use information-rich, innate decision-making abilities, individual learning, social learning, and, at least in humans, complex culture, alone or in various combinations, to create sophisticated cognitive systems.

We begin with a discussion of the correlated trends in environmental deterioration and brain size evolution and then turn to the problem of what sorts of cognitive strategies might have served as the impetus for brain enlargement.

Plio-Pleistocene Climate Deterioration

The deterioration of climates during the last few million years should have dramatically increased selection for traits increasing animals' abilities to cope with more variable environments. These traits include more complex cognition. Using a variety of indirect measures of past temperature, rainfall, ice volume, and the like, mostly from cores of ocean sediments, lake sediments, and ice caps, paleoclimatologists have constructed a stunning picture of climate deterioration over the last 14 million years (Lamb, 1977; Schneider and Londer, 1984; Dawson, 1992; Partridge et al., 1995). The Earth's mean temperature has dropped several degrees and the amplitudes of fluctuations in rainfall and temperature have increased. For reasons as yet ill understood, glaciers wax and wane in concert with changes in ocean circulation, carbon dioxide, methane, and dust content of the atmosphere and changes in average precipitation and the distribution of precipitation. The resulting pattern of fluctuation in climate is very complex. As the deterioration has proceeded, different cyclical patterns of glacial advance and retreat involving all these variables have dominated the pattern. A 21,700-year cycle dominated the early part of the period, a 41,000-year cycle between about 3 and 1 million years ago, and a 95,800-year cycle the last million years.

This cyclic variation is very slow with respect to the generation time of animals and is not likely to have directly driven the evolution of adaptations for phenotypic flexibility. However, increased variance on the time scales of the major glacial advances and retreats also seems to be correlated with great variance at much shorter time scales. For the last 120,000 years, quite high-resolution data are available from ice cores taken from the deep ice sheets of Greenland and Antarctica. Resolution of events lasting only a little more than a decade is possible in ice 90,000 years old, improving to monthly after 3,000 years ago. During the last glacial period, ice core data show that the climate was highly variable on time scales of centuries to millennia (GRIP, 1993; Lehman, 1993; Ditlevsen, Svensmark, and Johnson, 1996). Even when the climate was in the grip of the ice, there were brief spikelike ameliorations of about a thousand years duration in which the climate temporarily reached near interglacial warmth. The intense variability of the last glacial period carries right down to the limits of the nearly 10-year resolution of the ice core data. Sharp excursions lasting a century or less occur in estimated temperatures, atmospheric dust, and greenhouse gases. Comparison of the rapid variation during this period with older climates is not yet possible. However, an internal comparison is possible. The Holocene (the last relatively warm, ice-free 10,000 years) has been a period of very stable climate, at least by the standards of the last glacial epoch. At the decadal scale, the last glacial climates were much more variable than climates in the Holocene. Holocene weather extremes have had quite significant effects on organisms (Lamb, 1977). It is hard to imagine the impact of the much greater variation that was probably characteristic of most if not all of the Pleistocene epoch. Floods, droughts, windstorms, and the like, which we experience once a century, might have occurred once a decade. Tropical organisms did not escape the impact of climate variation; temperature and especially rainfall were highly variable at low latitudes (Broecker, 1996). During most periods in the Pleistocene, plants and animals must generally have lived under conditions of rapid, chaotic, and ongoing reorganization of ecological communities as species' ranges adjusted to the noisy variation in climate. Thus, since the late Miocene epoch, organisms have had to cope with increasing variability in many environmental parameters at time scales on which strategies for phenotypic flexibility would be highly adaptive.

Brain Size Evolution in the Neogene

Mammals show clear signs of responding to climate deterioration by developing more complex cognition. Jerison's (1973) classic study of the evolution of brain size documents major trends toward increasing brain size in many mammalian lineages that persist up through the Pleistocene. The time trends are complex. There is a progressive increase in average encephalization (brain size relative to body size) throughout the Cenozoic era. However, many relatively small-brained mammals persist even in orders where some species have evolved large brains. The *diversity* of brain size increases toward the present. Mammals continue to evolve under strong selective pressure to minimize brain size (see section on cognitive economics), and those that can effectively cope with climatic deterioration by range changes or noncognitive adaptations do so. Other lineages evolve the means to exploit the temporal and spatial variability of the environment by using behavioral flexibility. The latter, we suppose, pay for the cost of encephalization by exploiting the ephemeral niches that less flexible, smaller brained species leave underexploited.

Humans anchor the tail of the distribution of brain sizes in mammals; we are the largest brained member of the largest brained mammalian order. This fact supports a Darwinian hypothesis. Large gaps between species are hard to account for by the processes of organic evolution. That we are part of a larger trend suggests that a general selective process such as we propose really is operating. Nevertheless, there is some evidence that human culture is more than just a more sophisticated form of typical animal cognitive strategies. More on this vexing issue follows.

The largest increase in encephalization per unit time by far is the shift from Miocene and Pliocene species to modern ones, coinciding with the Pleistocene climate deterioration. In the last 2.5 million years, encephalization increases were somewhat larger than during the steps from Archaic to Paleogene and Paleogene to Neogene, each of which represents tens of millions of years of evolution.

General Purpose versus Special Purpose Mechanisms

To understand how evolution might have shaped cognitive adaptations to variable environments, we need to know something about the elementary properties of mental machinery. Psychologists interested in the evolution of cognition have generated two classes of hypotheses about the nature of minds. A long-standing idea is that cognitively sophisticated mammals and birds have evolved powerful

and relatively general-purpose mental strategies that culminate in human intelligence and culture. These flexible general-purpose strategies replace more rigidly
innate ones as cognitive sophistication increases. For example, Donald Campbell
(1965, 1975) emphasizes the general similarities of all knowledge-acquiring
processes ranging from organic evolution to modern science. He argues that even
a quite fallible cognitive apparatus could nevertheless obtain workable mental
representations of a complex variable environment by trial and error methods,
much as natural selection shapes random mutations into organic adaptations.
Bitterman's (2000) empirical argument that simple and complex cognitions use
rather similar learning strategies is a kindred proposal. Jerison (1973) argues that
the main region of enlargement of bird and mammal brains in the Cenozoic era
has been the forebrain, whose structures serve rather general coordinating
functions. He believes that it is possible to speak of intelligence abstracted from
the particular cognition of each species, which he characterizes as the ability to
construct perceptual maps of the world and use them to guide behavior adaptively. Edelman's (1987) theory of neuronal group selection is based on the
argument that developmental processes cannot specify the fine details of the
development of complex brains and hence that a lot of environmental feedback
is necessary just to form the basic categories that complex cognition needs to
work. This argument is consistent with the observation that animals with more
complex cognition require longer juvenile periods with lots of "play" to provide
the somatic selection of the fine details of synaptic structure. In Edelman's
argument, a large measure of phenotypic flexibility comes as a result of the
developmental constraints on the organization of complex brains by innate programming. If cognition is to be complex, it must be built using structures that are
underdetermined at birth.

Against general-purpose hypotheses, there has long been the suspicion that
animal intelligence can be understood only in relationship to the habitat in which
the species lives (Hinde, 1970:659–663). Natural selection is a mechanism for
adapting the individuals of a species to particular environmental challenges. It
will favor brains and behaviors specialized for the niche of the species. There is
no reason to think that it will favor some general capacity that we can oper-
ationalize as intelligence across species. A recent school of evolutionary psychologists has applied this logic to the human case (Barkow, Cosmides, and
Tooby, 1992; Pinker, 1997; Shettleworth, 2000). The brain, they argue, even the
human brain, is not a general problem-solving device but a collection of modules
directed at solving the particular challenges posed by the environments in which
the human species evolved. General problem-solving devices are hopelessly
clumsy. To work at all, a mental problem-solving device must make a number of
assumptions about the structure of its world, assumptions that are likely to hold
only locally. Jack of all trades, master of none. Human brains, for example, are
adapted to life in small-scale hunting and gathering societies of the Pleistocene.
They will guide behavior within such societies with considerable precision but
behave unpredictably in other situations. These authors are quite suspicious of
the idea that culture alone forms the basis for human behavioral flexibility. As
Tooby and Cosmides (1992) put it, what some take to be cultural traditions
transmitted to relatively passive imitators in each new generation could actually

be partly, or even mainly, "evoked culture," innate information that leads to similar behavior in parents and offspring simply because they live in similar environments. In this model, human cognition is complex because we have many content-rich, special-purpose, innate algorithms, however much we also depend upon transmitted culture.

This debate should not be trivialized by erecting straw protagonists. On the one hand, it is not sensible for defenders of cognitive generalism to ignore that the brain is a complex organ with many specialized parts, without which no mental computations would be possible. No doubt, much of any animal's mental apparatus is keyed to solve niche-specific problems, as is abundantly clear from brain comparative anatomy (Krubitzer, 1995) and from performance on learning tasks (Garcia and Koelling, 1966; Poli, 1986). Learning devices can be only *relatively* general; all of them must depend upon an array of innate processing devices to interpret raw sense data and evaluate whether they should be treated as significant (an actual or potential reinforcer). The more general a learning rule is, the weaker it is liable to be.

On the other hand, one function of all brains is to deal with the unforeseeable. The dimensionality of the environment is very large even for narrow specialists, and even larger for weedy, succeeds-everywhere species like humans. Being preprogrammed to respond adaptively to a large variety of environmental contingencies may be costly or impossible. If efficient learning heuristics exist that obviate the need for large amounts of innate information, they will be favored by selection.

When the situation is sufficiently novel, like most of the situations that rats and pigeons face in Skinner Boxes, every species is forced to rely upon what is, in effect, a very general learning capability. An extreme version of the special-purpose modules hypothesis would predict that animals should behave completely randomly in environments as novel as they usually face in the laboratory. The fact that adaptive behavior emerges at all in such circumstances is a clear disproof of such an extreme position. Likewise, humans cannot be too tightly specialized for living in small hunting and gathering societies under Pleistocene conditions. We are highly successful in the Holocene epoch using far different social and subsistence systems.

A Role for Social Learning in Variable Environments

Our own hypothesis is that culture plays a large role in the evolution of human cognitive complexity. The case for a role for social learning in other animals is weaker and more controversial, but well worth entertaining. Social learning and culture furnish a menu of heuristics for adapting to temporally and spatially variable environments. Learning devices will be favored only when environments are variable in time or space in difficult to predict ways. Social learning is a device for multiplying the power of individual learning. Systems of phenotypic adaptation have costs. In the case of learning, an individual will have to expend time and energy, incur some risks in trials that may be associated with costly errors, and support the neurological machinery necessary to learn. Social learning can economize on the trial and error part of learning. If kids learn from mom, they can avoid repeating her mistakes. "Copy mom" is a simple heuristic that may save one a lot of effort and be almost as effective as learning for oneself, provided the environment in one's own generation is pretty much like mom's. Suppose the ability to somehow copy mom is combined with a simple check of the current environment that warns one if the environment has changed significantly. If it has, one learns for oneself. This strategy allows social learners to avoid some learning costs but rely on learning when necessary.

We have constructed a series of mathematical models designed to test the cogency of these ideas (Boyd and Richerson, 1985, 1989, 1995, 1996; see also Cavalli-Sforza and Feldman, 1973; Pulliam and Dunford, 1980). The formal theory supports the story. When information is costly to obtain and when there is some statistical resemblance between models' and learners' environments, social learning is potentially adaptive. Selection will favor individual learners who add social learning to their repertoire so long as copying is fairly accurate and the extra overhead cost of the capacity to copy is not too high. In some circumstances, the models suggest that social learning will be quite important relative to individual learning. It can be a great advantage compared to a system that relies on genes only to transmit information and individual learning to adapt to the variation. Selection will also favor heuristics that bias social learning in adaptive directions. When the behavior of models is variable, individuals who try to choose the best model by using simple heuristics like "copy dominants" or "go with the majority," or by using complex cognitive analyses, are more likely to do well than those who blindly copy. Contrarily, if it is easy for individuals to learn the right thing to do by themselves, or if environments vary little, then social learning is of no utility.

A basic advantage common to many of the model systems that we have studied is that a system linking an ability to make adaptive decisions to an ability to copy speeds up the evolutionary process. Both natural selection and the biasing decisions that individuals make act on socially learned variation. The faster rate of evolution tracks a variable environment more faithfully, providing a fitness return to social learning.

Our models of cultural evolution are much like the learning model Bitterman describes (2000). In fact, one of our most basic models adds social learning to a model of individual learning virtually identical to his in order to investigate the inheritance-of-acquired-variation feature of social learning. Such models are simple and meant to be quite general. We expect that they will apply, at least approximately, to most examples of social learning in nature.

Social learning strategies could represent a component of general-purpose learning systems. Social learning is potentially an adaptive supplement to a weak, relatively general-purpose learning rule. (We accept the argument that the more general a learning rule is, the weaker it has to be.) However, we have modeled several different kinds of rules for social learning. These would qualify as different modules in Shettleworth's terms (2000). The same rule, with different inputs and different parameter settings, can be implemented as a component of many narrowly specialized modules. Psychological evidence suggests that human culture involves numerous subsystems and variants that use a variety of patterns of transmission and a variety of biasing heuristics (Boyd and Richerson, 1985). Although all nonhuman social learning systems are, as far as we know, much simpler than human culture, they probably obey a similar evolutionary logic and vary adaptively from species to species (Chou and Richerson, 1992; Laland, Richerson, and Boyd, 1996).

In no system of social learning have fitness effects yet been estimated; the adaptivness of simple social learning warrants skepticism. Rogers (1989, see also Boyd and Richerson, 1995) constructed a plausible model in which two genotypes were possible: individual learners and social learners. In his model, the social learning genotype can invade because social learners save on the cost of learning for themselves. However, at the equilibrium frequency of social learners, the fitness of the two types is equal. Social learners are parasites on the learning efforts of individual learners. Social learning raises the average fitness of individuals only if individual learners also benefit from social learning. The well-studied system of social learning of food preference in rats is plausibly an example of adaptive social learning (Galef, 1996), but the parasitic hypothesis is not yet ruled out. Lefebvre's (2000) data indicating a positive correlation of individual and social learning suggest an adaptive combination of social and individual learning, although his data on scrounging in aviaries show that pigeons are perfectly willing to parasitize the efforts of others. We will be surprised if no cases of social learning corresponding to Rogers's model ever turn up.

The complex cognition of humans is one of the great scientific puzzles. Our conquest of *the* ultimate cognitive niche seems to explain our extraordinary success as a species (Tooby and Devore, 1987). Why then has the human cognitive niche remained empty for all but a tiny slice of the history of life on earth, finally to be filled by a single lineage? Human culture, but not the social learning of most other animals, involves the use of imitation, teaching, and language to transmit complex adaptations subject to progressive improvement. In the human system, socially learned constructs can be far more sophisticated than even the most inspired individual could possibly hope to invent. Is complex culture the essence of our complex cognition or merely a subsidiary part?

The Problem of Cognitive Economics

To understand how selection for complex cognition proceeds, we need to know the costs, benefits, trade-offs, and synergies involved in using elementary cognitive strategies in compound architectures to adapt efficiently to variable environments. In our models we have merely assumed costs, accuracies, and other psychological properties of learning and social learning. We here sketch the kinds of knowledge necessary to incorporate cognitive principles directly into evolutionary models.

Learning and decision making require larger sensory and nervous systems in proportion to their sophistication, and large nervous systems are costly (Eisenberg, 1981:235–236). Martin (1981) reports that mammalian brains vary over about a 25-fold range, controlling for body size. Aiello and Wheeler (1995) report that human brains account for 16 percent of our basal metabolism. Average mammals have to allocate only about 3 percent of basal metabolism to their brains, and many marsupials get by with less than 1 percent. These differences are large enough to generate significant evolutionary trade-offs. In addition to metabolic requirements, there are other significant costs of big brains, such as increased difficulty at birth, greater vulnerability to head trauma, increased potential for developmental snafus, and the time and trouble necessary to fill these large brains with usable information. On the cost side, selection will favor as small a nervous system as possible.

If our hypothesis is correct, animals with complex cognition foot the cost of a large brain by adapting more swiftly and accurately to variable environments. Exactly how do they do it? Given just three generic forms of adaptation to variable environments—innate information, individual learning, and social learning—and two kinds of mental devices—more general-purpose and less general-purpose—the possible architectures for minds are quite numerous. What sorts of trade-offs will govern the nature of structures that selection might favor? What is the overhead cost of having a large repertoire of innate special-purpose rules? Innate rules will consume genes and brain tissue with algorithms that may be rarely called upon. The gene-to-mind translation during development may be difficult for complex innate rules. If so, acquiring information from the environment using learning or social learning may be favored. Are there situations where a (relative) jack-of-all-trades learning rule can outcompete a bevy of specialized rules? What is the penalty paid in efficiency for a measure of generality in learning? Are there efficient heuristics that minds can use to gain a measure of generality without paying the full cost of a general-purpose learning device? Relatively general-purpose heuristics might work well enough over a wide enough range of environmental variation to be almost as good as several sophisticated special-purpose algorithms, each costing as much brain tissue as the general heuristic (see Gigerenzer and Goldstein, 1996, on simple but powerful heuristics).

Hypothesis building here is complicated because we cannot assume that individual learning, social learning, and innate knowledge are simply competing processes. For example, more powerful or more general learning algorithms may generally require more innate information (Tooby and Cosmides, 1992). More sophisticated associative learning will typically require more sense data to make finer discriminations of stimuli. Sophisticated sense systems depend upon powerful, specialized, innate algorithms to make useful information from a mass of raw data from the sensory transducers (Spelke, 1990; Shettleworth, 2000). Hypothesis building is also complicated because we have no rules describing the efficiency of a compound system of some more and some less specialized modules. For example, a central general-purpose associative learning device might be the most efficient processor for such sophisticated sensory data because redundant implementation of the same learning algorithm in many modules might be costly. Intense modularity in parts of the mind may favor general-purpose, shared, central devices in other parts. Bitterman's (2000) data are consistent with a central associative learning processor that is similar by homology across most of the animal kingdom. However, his data are also consistent with several or many encapsulated special-purpose associative learning devices that have converged on a relatively few efficient association algorithms.

Shettleworth's (2000) argument for modularity by analogy with perception has appeal. If the cost of implementing an association algorithm is small relative to the cost of sending sensory data large distances across the brain, selection will favor association algorithms in many modules. However, the modularity of perception is surely driven in part by the fact that the different sense organs must transduce very different physical data. Bitterman's (2000) data show that, once reduced to a more abstract form, many kinds of sense data can be operated on by the same learning algorithm, which might be implemented centrally or modularly. The same sorts of issues will govern the incorporation of social learning into an evolving cognitive system.

There may be evolutionary complications to consider. For example, seldom-used special-purpose rules (or the extreme seldom-used ranges of frequently exercised rules) will be subject to very weak selection. More general-purpose structures have the advantage that they will be used frequently and hence be well adapted to the prevailing range of environmental uncertainty. If they work to any approximation outside this range, selection can readily act to improve them. Narrowly special-purpose algorithms could have the disadvantage that they can be "caught out" by a sudden environmental change, exhibiting no even marginally useful variation for selection to seize upon, whereas more general-purpose individual and social learning strategies can expose variation to selection in such cases (Laland et al., 1996). On the other hand, we might imagine that there is a reservoir of variation in outmoded special-purpose algorithms, on which selection has lost its purchase, that furnishes the necessary variation in suddenly changed circumstances.

The high dimensionality of the variation of Pleistocene environments puts a sharp point on the innate information versus learning/social learning modes of phenotypic flexibility. Mightn't the need for enough information to cope with such complex change by largely innate means exhaust the capacity of the genome to store and express it? Recall Edelman's (1987) neuronal group selection hypothesis in this context. Immelman (1975) suggests that animals use imprinting to identify their parents and acquire a concept of their species because it is not feasible to store a picture of the species in the genes or to move the information from genes to the brain during development. It may be more economical to use the visual system to acquire the picture after birth or hatching by using the simple heuristic that the first living thing one sees is mom and a member of one's own species. In a highly uncertain world, wouldn't selection favor a repertoire of heuristics designed to learn as rapidly and efficiently as possible?

As far as we understand, psychologists are not yet in a position to give us the engineering principles of mind design the way that students of biological mechanics now can for muscle and bone. If these principles turn out to favor complex, mixed designs with synergistic, nonlinear relationships between parts, the mind design problem will be quite formidable. We want to avoid asking silly questions analogous to "which is more important to the function of a modern PC, the hardware or the software?" However, in our present state of ignorance, we do run the risk of asking just such questions!

With due care, perhaps we can make a little progress. In this chapter, we use a method frequently used by evolutionary biologists, dubbed "strategic modeling" by Tooby and Devore (1987). In strategic modeling, we begin with the tasks that the environment sets for an organism and attempt to deduce how natural selection should have shaped the species' adaptation to its niche. Often, evolutionary biologists frame hypotheses in terms of mathematical models of alternative adaptations that predict, for instance, what foraging or mate choice strategy organisms with a given general biology should pursue in a particular environment. This is just the sort of modeling we have undertaken in our studies of social learning and culture. We ask: how should organisms cope with different kinds of spatially and temporally variable environments?

Social Learning versus Individual Learning versus Innate Programming?

Increases in brain size could signal adaptation to variable environments via individual learning, social learning, or more sophisticated innate programming. Our mathematical models suggest that the three systems work together. Most likely increases in brain size to support more sophisticated learning or social learning will also require at least some more innate programming. There is likely an optimal balance of innate and acquired information dictated by the structure of environmental variability. Given the tight cost/benefit constraints imposed on brains, at the margin we would expect to find a trade-off between social learning, individual learning, and innate programming. For example, those species that exploit the most variable niches should emphasize individual learning, whereas those that live in more highly autocorrelated environments should devote more of their nervous systems to social learning.

Lefebvre (2000) reviews studies designed to test the hypothesis that social and opportunistic species should be able to learn socially more easily than the more conservative species, and the conservative species should be better individual learners. Surprisingly, the prediction fails. Species that are good social learners are also good individual learners. One explanation for these results is that the synergy between these systems is strong. Perhaps the information-evaluating neural circuits used in social and individual learning are partly or largely shared. Once animals become social, the potential for social learning arises. The two learning systems may share the overhead of maintaining the memory storage system and much of the machinery for evaluating the results of experience. If so, the benefits in quality or rate of information gained may be large relative to the cost of small bits of specialized nervous tissue devoted separately to each capacity. If members of the social group tend to be kin, investments in individual learning may also be favored because sharing the results by social learning will increase inclusive fitness. On the other hand, Lefebvre notes that not all learning abilities are positively correlated. Further, the correlation may be due to some quite simple factor, such as low neophobia, not a more cognitively sophisticated adaptation.

The hypothesis that the brain tissue trade-off between social and individual learning is small resonates with what we know of the mechanisms of social learning in most species. Galef (1988, 1996), Laland et al. (1996), and Heyes and

Dawson (1990) argue that the most common forms of social learning result from very simple mechanisms that piggyback on individual learning. In social species, naive animals follow more experienced parents, nestmates, or flock members as they traverse the environment. The experienced animals select highly nonrandom paths through the environment. They thus expose naive individuals to a highly selected set of stimuli that then lead to acquisition of behaviors by ordinary mechanisms of reinforcement. Social experience acts, essentially, to speed up and make less random the individual learning process, requiring little additional, specialized, mental capacity. Social learning, by making individual learning more accurate without requiring much new neural machinery, tips the selective balance between the high cost of brain tissue and advantages of flexibility in favor of more flexibility. As the quality of information stored on a mental map increases, it makes sense to enlarge the scale of maps to take advantage of that fact. Eventually, diminishing returns to map accuracy will limit brain size.

Once again, we must take a skeptical view of this adaptive hypothesis until experimental and field investigations produce better data on the adaptive consequences of social learning. Aside from Rogers's parasitic scenario, the simplicity of social learning in most species and its close relationship to individual learning invite the hypothesis that most social learning is a by-product of individual learning that is not sufficiently important to be shaped by natural selection. Human imitation, by contrast, is so complex as to suggest that it must have arisen under the influence of selection.

Eisenberg's (1981, ch. 23) review of a large set of data on the encephalization of living mammals suggests that high encephalization is associated with extended association with parents, late sexual maturity, extreme iteroparity, and long potential life span. These life cycle attributes all seem to favor social learning (but also any other form of time-consuming skill acquisition). We would not expect this trend if individual and social learning were a small component of encephalization relative to innate, information-rich modules. Under the latter hypothesis animals with a minimal opportunity to take advantage of parental experience and parental protection while learning for themselves ought to be able to adapt to variable environments with a rich repertoire of innate algorithms. Eisenberg's data suggest that large brains are not normally favored in the absence of social learning or social facilitation of individual learning. The study of any species that run counter to Eisenberg's correlation might prove very rewarding. Large-brained species with a small period of juvenile dependence should have a complex cognition built disproportionately of innate information. Similarly, small-brained social species with prolonged juvenile dependence or other social contact may depend relatively heavily on simple learning and social learning strategies. Lefebvre and Palameta (1988) provide a long list of animals in which social learning has been more or less convincingly documented. Recently, Dugatkin (1996) and Laland and Williams (1997) have demonstrated social learning in guppies. Even marginally social species may come under selection for behaviors that enhance social learning, as in the well-known case of mother housecats who bring partially disabled prey to their kittens for practice of killing behavior (Caro and Hauser, 1992).

Some examples of nonhuman social learning are clearly specialized, such as birdsong imitation, but the question is open for other examples. Aspects of the social learning system in other cases do show signs of adaptive specialization, illustrating the idea that learning and social learning systems are only general purpose *relative* to a completely innate system. For example, Terkel (1996) and Chou (1989, personal communication) obtained evidence from laboratory studies of black rats that the main mode of social learning is from mother to pups. This is quite unlike the situation in the case of Norway rats, where Galef (1988, 1996) and coworkers have shown quite conclusively that mothers have no special influence on pups. In the black rat, socially learned behaviors seem to be fixed after a juvenile learning period, whereas Norway rats continually update their diet preferences (the best-studied trait) based upon individually acquired and social cues. Black rats seem to be adapted to a more slowly changing environment than Norway rats. Terkel studied a rat population that has adapted to open pinecones in an exotic pine plantation in Israel, a novel and short-lived niche by most standards, but one that will persist for many rat generations. Norway rats are the classic rats of garbage dumps, where the sorts of foods available change weekly.

Human versus Other Animals' Culture

The human species' position at the large-brained tail of the distribution of late Cenozoic encephalization suggests the hypothesis that our system of social learning is merely a hypertrophied version of a common mammalian system based substantially on the synergy between individual learning and simple systems of social learning. However, two lines of evidence suggest that there is more to the story.

First, human cultural traditions are often very complex. Subsistence systems, artistic productions, languages, and the like are so complex that they must be built over many generations by the incremental, marginal modifications of many innovators (Basalla, 1988). We are utterly dependent on learning such complex traditions to function normally.

Second, this difference between humans and other animals in the complexity of socially learned behaviors is mirrored in a major difference in mode of social learning. As we saw, the bulk of animal social learning seems to be dependent mostly on the same techniques used in individual learning, supplemented at the margin by a bit of teaching and imitation. Experimental psychologists have devoted much effort to trying to settle the question of whether nonhuman animals can learn by "true imitation" or not (Galef, 1988). True imitation is learning a behavior by seeing it done. True imitation is presumably more complex cognitively than merely using conspecifics' behavior as a source of cues to stimuli that it might be interesting to experience. Although there are some rather good experiments indicating some capacity for true imitation in several socially learning species (Heyes, 1996; Moore, 1996; Zentall, 1996), head-to-head comparisons of children's and chimpanzee's abilities to imitate show that children begin to exceed chimpanzees' capabilities at about three years of age (Whiten and Custance, 1996; Tomasello, 1996, 2000). The lesson to date from comparative studies of social learning suggests that simple mechanisms of social learning are much more common and more important than imitation, even in our close relatives and other highly encephalized species.

Why Is Complex Culture Rare?

One hypothesis is that an intrinsic evolutionary impediment exists, hampering the evolution of a capacity for complex traditions. We show elsewhere that, under some sensible cognitive-economic assumptions, a capacity for complex cumulative culture cannot be favored by selection when rare (Boyd and Richerson, 1996). The mathematical result is quite intuitive. Suppose that to acquire a complex tradition efficiently, imitation is required. Suppose that efficient imitation requires considerable costly, or complex, cognitive machinery, such as a theory of mind/imitation module (Cheney and Seyfarth, 1990:277–230; Tomasello, 2000). If so, there will be a coevolutionary failure of capacity for complex traditions to evolve. The capacity would be a great fitness advantage, but only if there are cultural traditions to take advantage of. But, obviously, there cannot be complex traditions without the cognitive machinery necessary to support them. A rare individual who has a mutation coding for an enlarged capacity to imitate will find no complex traditions to learn and will be handicapped by an investment in nervous tissue that cannot function. The hypothesis depends upon a certain lumpiness in the evolution of the mind. If even a small amount of imitation requires an expensive or complex bit of mental machinery, or if the initial step in the evolution of complex traits does not result in particularly useful traditions, then there will be no smooth evolutionary path from simple social learning to complex culture.

If such an impediment to the evolution of complex traditions existed, evolution must have traveled a roundabout path to get the frequency of the imitation capacity high enough to begin to bring it under positive selection for its tradition-supporting function. Some suggest that primate intelligence was originally an adaptation to manage a complex social life (Humphrey, 1976; Byrne and Whiten, 1988; Kummer, Daston, Gigerenzer, and Silk, 1997; Dunbar, 1992, 2000). Perhaps in our lineage the complexities of managing the sexual division of labor, or some similar social problem, favored the evolution of the capacity to develop a sophisticated theory of mind. Such a capacity might incidentally make efficient imitation possible, launching the evolution of elementary complex traditions. Once elementary complex traditions exist, the threshold is crossed. As the evolving traditions become too complex to imitate easily, they will begin to drive the evolution of still more sophisticated imitation. This sort of stickiness in evolutionary processes is presumably what gives evolution its commonly contingent, historical character (Boyd and Richerson, 1992).

Conclusion

The evolution of complex cognition is a complex problem. It is not entirely clear what selective regimes favor complex cognition. The geologically recent increase in the encephalization of many mammalian lineages suggests that complex cognition is an adaptation to a common, widespread, complex feature of the environment. The most obvious candidate for this selective factor is the deterioration of the earth's climate since the late Miocene epoch, culminating in the exceedingly noisy Pleistocene glacial climates.

In principle, complex cognition can accomplish a system of phenotypic flexibility by using information-rich innate rules or by using more open individual and social learning. Presumably, the three forms of phenotypic flexibility are partly competing, partly mutually supporting mechanisms that selection tunes to the patterns of environmental variation in particular species' niches. Because of the cost of brain tissue, the tuning of cognitive capacities will take place in the face of a strong tendency to minimize brain size. However, using strategic modeling to infer the optimal structure for complex cognitive systems from evolutionary first principles is handicapped by the very scanty information on trade-offs and constraints that govern various cognitive information-processing strategies. For example, we do not understand how expensive it is to encode complex innate information-rich computational algorithms relative to coping with variable environments with relatively simple, but still relatively efficient, learning heuristics. Psychologists and neurobiologists might usefully concentrate on such questions.

Human cognition raises the ante for strategic modeling because of its apparently unique complexity and yet great adaptive utility. We can get modest but real leverage on the problem by investigating other species with cognitive complexity approaching ours, which in addition to great apes may include other monkeys, some cetaceans, parrots, and corvids (Moore, 1996; Heinrich, 2000; Clayton, Griffiths, and Dickinson, 2000). Our interpretation of the evidence is that human cognition mainly evolved to acquire and manage cumulative cultural traditions. This capacity probably cannot be favored when rare, even in circumstances where it would be quite successful if it did evolve. Thus, its evolution likely required, as a preadaptation, the advanced cognition achieved by many mammalian lineages in the last few million years. *In addition*, it required an adaptive breakthrough, such as the acquisition of a capacity for imitation as a by-product of the evolution of a theory of mind capacity for social purposes.

REFERENCES

Aiello, L. C., & P. Wheeler. 1995. The expensive tissue hypothesis: The brain and the digestive system in human and primate evolution. *Current Anthropology* 36:199–221.

Barkow, J. H., L. Cosmides, & J. Tooby. 1992. *The adapted mind: Evolutionary psychology and the generation of culture*. Oxford: Oxford University Press.

Basalla, G. 1988. *The evolution of technology*. Cambridge: Cambridge University Press.

Bitterman, M. E. 2000. Cognitive evolution: A psychological perspective. In: *The Evolution of Cognition*, C. Heyes, & L. Huber, eds. (pp. 61–80). Cambridge MA: MIT Press.

Boyd, R., & P. J. Richerson. 1985. *Culture and the evolutionary process*. Chicago: University of Chicago Press.

Boyd, R., & P. J. Richerson. 1989. Social learning as an adaptation. *Lectures on Mathematics in the Life Sciences* 20:1–26.

Boyd, R., & P. J. Richerson. 1992. How microevolutionary processes give rise to history. In: *History and evolution*, M. H. Nitecki, & D. V. Nitecki, eds. (pp. 179–209). Albany: State University of New York Press.

Boyd, R., & P. J. Richerson. 1995. Why does culture increase human adaptability? *Ethology and Sociobiology* 16:125–143.

Boyd, R., & P. J. Richerson. 1996. Why culture is common but cultural evolution is rare. *Proceedings of the British Academy* 88:77–93.

Broecker, W. S. 1996. Glacial climate in the tropics. *Science* 272:1902–1903.

Byrne, R. W., & A. Whiten. 1988. *Machiavellian intelligence: Social expertise and the evolution of intellect in monkeys, apes, and humans.* New York: Oxford University Press.

Campbell, D. T. 1965. Variation and selective retention in sociocultural evolution. In: *Social change in developing areas: A reinterpretation of evolutionary theory*, H. R. Barringer, G. I. Blanksten, & R. W. Mack, eds. (pp. 19–49). Cambridge, MA: Schenkman.

Campbell, D. T. 1975. On the conflicts between biological and social evolution and between psychology and moral tradition. *American Psychologist* 30: 1103–1126.

Caro, T., & M. Hauser. 1992. Is there teaching in nonhuman animals? *Quarterly Review of Biology* 67:151–174.

Cavalli-Sforza, L. L., & M. W. Feldman. 1973. Models for cultural inheritance. I. Group mean and within group variation. *Theoretical Population Biology* 4:42–55.

Cheney, D. L., & R. M. Seyfarth. 1990. How monkeys see the world: Inside the mind of another species. Chicago: University of Chicago Press.

Chou, L-S. 1989. *Social transmission of food selection by rats.* Ph.D. Dissertation, University of California, Davis.

Chou, L., & P. J. Richerson. 1992. Multiple models in social transmission among Norway rats, *Rattus norvegicus*. *Animal Behaviour* 44:337–344.

Clayton, N. S., D. P. Griffiths, & A. Dickinson. 2000. Declarative and episodic-like memory in animals: Personal musings of a scrub jay. In: *The Evolution of Cognition*, C. Heyes, & L. Huber, eds. (pp. 273–288). Cambridge MA: MIT Press.

Dawson, A. G. 1992. *Ice age earth: Late quaternary geology and climate.* London: Routledge.

Ditlevsen, P. D., H. Svensmark, & S. Johnsen. 1996. Contrasting atmospheric and climate dynamics of the last-glacial and Holocene periods. *Nature* 379:810–812.

Dugatkin, L. A. 1996. Copying and mate choice. In: *Social learning in animals: The roots of culture*, C. M. Heyes, & B. G. Galef, Jr., eds. (pp. 85–105). San Diego: Academic Press.

Dunbar, R. I. M. 1992. Neocortex size as a constraint on group size in primates. *Journal of Human Evolution* 20:469–493.

Dunbar, R. 2000. Causal reasoning, mental rehearsal, and the evolution of primate cognition. In: *The Evolution of Cognition*, C. Heyes, & L. Huber, eds. (pp. 202–220). Cambridge MA: MIT Press.

Edelman, G. M. 1987. *Neural Darwinism: The theory of neuronal group selection.* New York: Basic Books.

Eisenberg, J. F. 1981. *The mammalian radiations: An analysis of trends in evolution, adaptation, and behavior.* Chicago: University of Chicago Press.

Galef, Jr., B. G. 1988. Imitation in animals: History, definition, and interpretation of data from the psychological laboratory. In: *Social learning: Psychological and biological perspectives*, T. R. Zentall, & B. G. Galef, Jr., eds. (pp. 3–28). Hillsdale, NJ: Lawrence Erlbaum.

Galef, Jr., B. G. 1996. Social enhancement of food preferences in Norway rats: A brief review. In: *Social learning in animals: The roots of culture*, C. M. Heyes, & B. G. Galef, Jr., eds. (pp. 49–64). San Diego: Academic Press.

Garcia, J., & R. Koelling. 1966. Relation of cue to consequence in avoidance learning. *Psychonomic Science* 4:123–124.

Gigerenzer, G., & D. G. Goldstein. 1996. Reasoning the fast and frugal way: Models of bounded rationality. *Psychological Review* 103:650–669.

GRIP (Greenland Ice-core Project Members). 1993. Climate instability during the last interglacial period recorded in the GRIP ice core. *Nature* 364:203–207.

Heinrich, B. 2000. Testing insight in ravens. In: *The Evolution of Cognition*, C. Heyes, & L. Huber, eds. (pp. 289–306). Cambridge MA: MIT Press.

Heyes, C. M. 1996. Introduction: Identifying and defining imitation. In: *Social learning in animals: The roots of culture*, C. M. Heyes, & B. G. Galef, Jr., eds. (pp. 211–220). San Diego: Academic Press.

Heyes, C. M., & G. R. Dawson. 1990. A demonstration of observational learning using a bidirectional control. *Quarterly Journal of Experimental Psychology* 42B: 59–71.

Hinde, R. A. 1970. *Animal behavior: A synthesis of ethology and comparative psychology.* New York: McGraw Hill.

Humphrey, N. K. 1976. The social function of intellect. In: *Growing points in ethology*, P. P. G. Bateson, & R. A. Hinde, eds. (pp. 303–317). Cambridge: Cambridge University Press.

Immelman, K. 1975. Ecological significance of imprinting and early learning. *Annual Review of Ecology and Systematics* 6:15–37.

Jerison, H. J. 1973. *Evolution of the brain and intelligence.* New York: Academic Press.

Krubitzer, L. 1995. The organization of neocortex in mammals: Are species differences really so different? *Trends in Neurosciences* 18:408–417.

Kummer, H., L. Daston, G. Gigerenzer, & J. B. Silk. 1997. The social intelligence hypothesis. In: *Human by nature: Between biology and the social sciences*, P. Weingart, S. D. Mitchell, P. J. Richerson, & S. Maasen, eds. (pp. 159–179). Mahwah, NJ: Lawrence Erlbaum.

Laland, K. N., P. J. Richerson, & R. Boyd. 1996. Developing a theory of animal social learning. In: *Social learning in animals: The roots of culture*, C. M. Heyes, & B. G. Galef, Jr., eds. (pp. 129–154). San Diego: Academic Press.

Laland, K. N. & K. Williams. 1997. Shoaling generates social learning of foraging information in guppies. *Animal Behaviour* 53:1161–1169.

Lamb, H. H. 1977. *Climatic history and the future.* Princeton, NJ: Princeton University Press.

Levebre, L. 2000. Feeding innovations and their cultural transmission in bird populations. In: *The Evolution of Cognition*, C. Heyes, & L. Huber, eds. (pp. 311–328). Cambridge MA: MIT Press.

Lefebvre, L., & B. Palameta. 1988. Mechanisms, ecology and population diffusion of socially learned, food-finding behavior in feral pigeons. In: *Social learning: Psychological and biological perspectives*, T. R. Zentall, & B. G. Galef, Jr., eds. (pp. 141–164). Hillsdale, NJ: Lawrence Erlbaum.

Lehman, S. 1993. Climate change: Ice sheets, wayward winds and sea change. *Nature* 365:108–109.

Martin, R. D. 1981. Relative brain size and basal metabolic rate in terrestrial vertebrates. *Nature* 293:57–60.

Moore, B. R. 1996. The evolution of imitative learning. In: *Social learning in animals: The roots of culture*, C. M. Heyes, & B. G. Galef, Jr., eds. (pp. 245–265). San Diego: Academic Press.

Partridge, T. C., G. C. Bond, C. J. H. Hartnady, P. B. deMenocal, & W. F. Ruddiman. 1995. Climatic effects of late Neogene tectonism and vulcanism. In: *Paleoclimate and evolution with emphasis on human origins*, E. S. Vrba, G. H. Denton, T. C. Partridge, & L. H. Burckle, eds. (pp. 8–23). New Haven: Yale University Press.

Pinker, S. 1997. *How the mind works*. New York: Norton.

Poli, M. D. 1986. Species-specific differences in animal learning. In: *Intelligence and evolutionary biology*, H. Jerison, & I. Jerison, eds. (pp. 277–297). Berlin: Springer.

Pulliam, H. R., & C. Dunford. 1980. *Programmed to learn: An essay on the evolution of culture*. New York: Columbia University Press.

Rogers, A. R. 1989. Does biology constrain culture? *American Anthropologist* 90: 819–831.

Schneider, S. H., & R. Londer. 1984. *The coevolution of climate and life*. San Francisco: Sierra Club Books.

Shettleworth, S. 2000. Modularity and the evolution of cognition. In: *The Evolution of Cognition*, C. Heyes, & L. Huber, eds. (pp. 43–60). Cambridge MA: MIT Press.

Spelke, E. S. 1990. Principles of object perception. *Cognitive Science* 14:29–56.

Terkel, J. 1996. Cultural transmission of feeding behavior in the black rat (*Rattus rattus*). In: *Social learning in animals: The roots of culture*, C. M. Heyes, & B. G. Galef, Jr., eds. (pp. 17–47). San Diego: Academic Press.

Tomasello, M. 1996. Do apes ape? In: *Social learning in animals: The roots of culture*, C. M. Heyes, & B. G. Galef, Jr., eds. (pp. 319–346). New York: Academic Press.

Tomasello, M. S. 2000. Two hypotheses about primate cognition. In: *The Evolution of Cognition*. C. Heyes, & L. Huber, eds. (pp. 165–184). Cambridge MA: MIT Press.

Tooby, J., & L. Cosmides. 1992. Psychological foundations of culture. In: *The adapted mind: Evolutionary psychology and the generation of culture*, J. H. Barkow, L. Cosmides, & J. Tooby, eds. (pp. 19–136). Oxford: Oxford University Press.

Tooby, J., & I. DeVore. 1987. The reconstruction of hominid behavioral evolution through strategic modeling. In: *The evolution of human behavior: Primate models*, W. G. Kinzey, ed. (pp. 183–237). Albany: State University of New York Press.

Whiten, A., & D. Custance. 1996. Studies of imitation in chimpanzees and children. In: *Social learning in animals: The roots of culture*, C. M. Heyes, & B. G. Galef, Jr., eds. (pp. 291–318). San Diego: Academic Press.

Zentall, T. 1996. An anlysis of imitative learning in animals. In: *Social learning in animals: The roots of culture*, C. M. Heyes, & B. G. Galef, Jr., eds. (pp. 221–243). San Diego: Academic Press.

5   Norms and Bounded Rationality

Do Norms Help People Make Good Decisions without Much Thought?

Many anthropologists believe that people follow the social norms of their society without much thought. According to this view, human behavior is mainly the result of social norms and rarely the result of considered decisions. In recent years, there has been increased interest within anthropology in how individuals and groups struggle to modify and reinterpret norms to further their own interests. However, we think it is fair to say that most anthropologists still believe that culture plays a powerful role in shaping how people think and what they do.

Many anthropologists also believe that social norms lead to adaptive behavior; by following norms, people can behave sensibly without having to understand why they do what they do. For example, throughout the New World, people who rely on corn as a staple food process the grain by soaking it in a strong base (such as calcium hydroxide) to produce foods like hominy and masa (Katz, Hediger, and Valleroy, 1974). This alkali process is complicated, requires hard work, and substantially reduces the caloric content of corn. However, it also increases the amount of available lysine, the amino acid in which corn is most deficient. Katz et al. argue that alkali processing plays a crucial role in preventing protein deficiency disease in regions where the majority of calories are derived from corn. Traditional peoples had no understanding of the nutritional value of alkali processing; rather, it was a norm: we Maya eat masa because that is what we do. Nonetheless, by following the norm, traditional people were able to solve an important and difficult nutritional problem. The work of cultural ecologists, such as Marvin Harris (1979), provides many other examples of this kind, although few are as well worked out. Other varieties of functionalism (for a discussion, see Turner and Maryanski, 1979) also hold that social norms evolve to adapt to the local environment. While nowadays anthropologists are explicitly critical of functionalism, cryptic functionalism still pervades much thinking in anthropology (Edgerton, 1992).

Norms may also lead to sensible behavior by proscribing choices that people find tempting in the short run but are damaging in the long run. Moral systems around the world have proscriptions against drunkenness, laziness, gluttony, and other failures of self-control. There is evidence that such proscriptions can increase individual well-being. For example, Jensen and Ericson (1979) show that Mormon youths in Tucson are less likely to be involved in "victimless crimes," such as drinking and marijuana use, than members of a nonreligious control group. Moreover, these differences seem to have consequences. McEvoy and Land (1981) report that age-adjusted mortalities for Mormons in Missouri are approximately 20 percent lower than those for control populations, and the differences were biggest for lung cancer, pneumonia/influenza, and violent death, sources of mortality that should be reduced if the abstentious Mormon norms are being observed. Apparently, living in a group in which there are norms against alcohol use makes it easier for young Mormons to do what is in their own long-term interest.

What Are Norms, and Why Do People Follow Them?

Examples like these present a series of interesting questions to economists, psychologists, and others who start with individuals as the basic building blocks of social theory. First, what are norms? How can we incorporate the notion that there are shared social rules into models that assume that people are goal-oriented decision makers? Second, why should people follow norms? Norms will change behavior only if they prescribe behavior that differs from what people would do in the absence of norms. Finally, why should norms be sensible? If individuals cannot (or do not) determine what is sensible, why should norms prescribe sensible behavior? It seems more plausible that they will simply represent random noise or even superstitious nonsense.

A recent efflorescence of interest in norms among rational choice theorists provides one cogent answer to the first two questions. Norms are the result of shared notions of appropriate behavior and the willingness of individuals to reward appropriate behavior and punish inappropriate behavior (for a review, see McAdams, 1997). Thus, it is a norm for men to remove their hats when they enter a Christian church because they will suffer the disapproval of others if they do not. In contrast, it is not a norm for men to remove their hats in an overheated country and western bar, even if everyone does so. By this notion, people obey norms because they are rewarded by others if they do and punished if they do not. As long as the rewards and punishments are sufficiently large, norms can stabilize a vast range of different behaviors. Norms can require property to be passed to the oldest son or to the youngest; they can specify that horsemeat is a delicacy or deem it unfit for human consumption.

There is no consensus in this literature about why people choose to punish norm violators and reward norm followers. There have been a number of proposals: Binmore (1998) argues that social life is an infinite game and that norms are game theoretic equilibria of the kind envisioned in the folk theorem. Norm violators are punished, and so are people who fail to punish norm violators, people who fail to punish them, and so on ad infinitum. McAdams (1997) suggests that all people desire the esteem of others, and because esteem can be "produced" at very low cost, it is easy to punish norm violators by withholding esteem. Bowles and Gintis (1999) and Richerson and Boyd (1998) argue that group selection acting over the long history of human evolution created a social environment in which natural selection favored genes leading to a reciprocal psychology. Here we will simply assume that the problem of why people choose to enforce norms has somehow been solved.

How Do Norms Solve Problems That People Cannot Solve on Their Own?

Virtually all of the recent literature on norms focuses on how norms help people solve public goods and coordination problems (e.g., Ostrom, 1991; Ellickson, 1994). It does not explain why norms should be adaptive. If people do not understand why alkali treatment of corn is a good thing, why should they require their neighbors to eat masa and hominy and be offended if they do not? Nor does the recent literature on norms explain why norms should commonly help people with problems of self-control. If people cannot resist the temptations of alcohol, why should they insist that their neighbors do so? We sketch possible answers to these questions.

Occasional Learning plus Conformism Leads to Adaptive Norms

In this section we show how a small amount of individual learning, when coupled with cultural transmission and a tendency to conform to the behavior of others, can lead to adaptive norms, even though most people simply do what everyone else is doing.

Why It May Be Sensible for Most People to Imitate

It is easy to see why people may choose to imitate others when it is costly or difficult to determine the best behavior—copying is easier than invention, and plagiarism is easier than creation. However, as these examples illustrate, it is not clear that by saving such costs, imitation makes everybody better off; this is why we have patents and rules against plagiarism. We have analyzed a series of mathematical models which indicate that when decisions are difficult, everyone can be better off if most people imitate the decisions of others under the right circumstances (Boyd and Richerson, 1985, 1988, 1989, 1995, 1996). The following simple model illustrates our reasoning.

Consider a population that lives in an environment that switches between two states with a constant probability. Further assume that there are two behaviors, one best in each environmental state. All individuals attempt to discover the best behavior in the current environment. First, each individual experiments with both behaviors and then compares the results. The results of such experiments vary for many reasons, and so the behavior that is best during any particular trial may be inferior over the long run. To represent this idea mathematically, assume that the observed difference in payoffs is a normally distributed random variable, X (figure 5.1). Second, each individual can observe the behavior of an individual from the previous generation who has already made the decision.

We assume that individuals combine sources of information by adopting a particular behavior if its payoff appears *sufficiently* better than its alternative; otherwise, they imitate. The larger the observed difference in the payoffs between the two behaviors, the more likely it is that the behavior with the higher payoff actually is best. By insisting on a large difference in observed payoff, individuals can reduce the chance that they will mistakenly adopt the inferior behavior. Of course, being discriminating will also cause more trials to be indecisive, and, then, they must imitate. Thus, there is a trade-off. Individuals can increase the accuracy of learning but only by also increasing the probability that learning will be indecisive and having to rely on imitation.

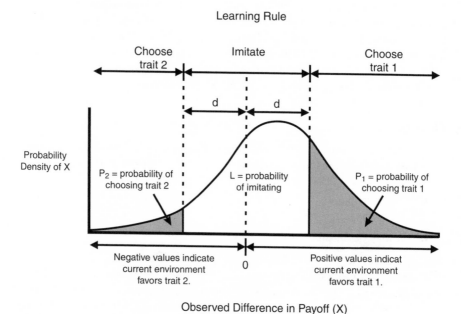

Figure 5.1. A graphical representation of the model of individual and social learning. Each individual observes an independent, normally distributed environmental cue, X. A positive value of X indicates that the environment is in state 1; a negative value indicates that the environment is in state 2. If the value of X is larger than the threshold value, d, an individual adopts trait 1. This occurs with probability, p_1. If the value of the environmental cue is smaller than $-d$, the individual adopts trait 2, which occurs with probability, p_2. Otherwise, the individual imitates. Thus, the larger the standard deviation of the cue compared to its mean value, the greater is the predictive value of the cue.

The optimal decision rule depends on what the rest of the population is doing. Assume that most individuals use a learning rule that causes them to imitate x percent of the time—call these "common-type" individuals. There are also a few rare invaders who imitate slightly more often. Compared to the common type, invaders are less likely to make learning errors. Thus, when invaders learn, they have a higher payoff than the common-type individuals when they learn. When invaders imitate, they have the same payoff as the common-type individuals. However, invaders must imitate more often and those who must imitate always have lower fitness than those whose personal information is above the learning threshold. To see why, think of each imitator as being connected to a learner by a chain of imitation. If the learner at the end of the chain learned in the current environment, then the imitator has the same chance of acquiring the favored behavior as does a learner. If the learner at the end of the chain learned in a different environment, the imitator will acquire the wrong trait. Thus, the invading type will achieve a higher payoff if the advantage of making fewer learning errors is sufficient to offset the disadvantage of imitating more.

This trade-off depends on how much the common type imitates. When the common type rarely imitates, the payoff of individuals who imitate and individuals who learn will be similar because most imitators will imitate somebody who learned, and the fact that mutants make fewer learning errors will allow them to invade. However, as the amount of imitation increases, the payoff of imitating individuals relative to those who learn declines because increased imitation lengthens the chain connecting each imitator to a learner. Eventually the population reaches an equilibrium at which the common type can resist invasion by mutants that change the rate of imitation. Figure 5.2 plots the probability that individuals imitate (denoted as L in figure 5.1) at evolutionary equilibrium as a function of the quality of the information available to individuals for three different rates of environmental change (for details of the calculation, see Boyd and Richerson, 1988). Notice that when it is difficult for individuals to determine the best behavior and when environments change infrequently, more than 90 percent of a population at equilibrium simply copies the behavior of others.

As long as environments are not completely unpredictable, the average payoff at the evolutionary equilibrium is greater than the average payoff of individuals who do not imitate (Boyd and Richerson, 1995). The reason is simple: imitation allows the population to learn when the information is good and imitate when it is bad. Figure 5.3 plots the average payoff of imitating and learning individuals as a function of the fraction of individuals who imitate. The payoff of learning individuals increases as the amount of imitation increases because individuals are demanding better evidence before relying on their individual experience and therefore are making fewer learning errors. The payoff of individuals who imitate because their evidence does not happen to meet rising standards *also* increases at first because they are directly or indirectly imitating learners who make fewer errors. If imitation is too common, the payoff to imitation declines because the too-discriminating population does too little learning to track the changing environment. The first effect is sufficient to lead to a net increase in average payoff at evolutionary equilibrium.

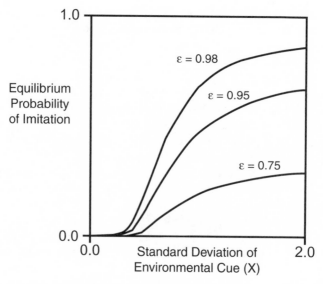

Figure 5.2. Probability that individuals imitate (L) at evolutionary equilibrium as a function of the quality of the environmental cue for three different rates of environmental change. The mean of the environmental cue (X) is 1.0, so as the standard deviation of X increases, the extent to which the cue predicts the environmental state decreases. Thus, the results plotted here indicate that as the predictive quality of the cue decreases, the probability of imitation at evolutionary equilibrium increases. The parameter ε is the probability that the environment remains unchanged from one time period to the next. Thus, as the rate of environmental change decreases, the probability of imitation at evolutionary equilibrium increases. See Boyd and Richerson (1988) for details.

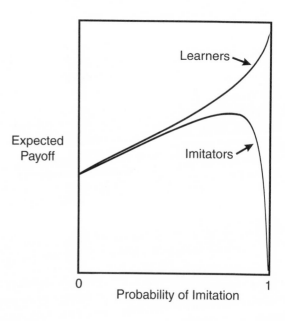

Figure 5.3. Individuals either learn or imitate according to the outcome of their learning trial. As individuals become more selective, the frequency of imitating individuals increases. This figure plots the expected fitness of individuals who imitate and those who learn as a function of the probability that an individual randomly chosen from the population imitates (assuming the outcome of learning experiments is normally distributed with mean 0.5 and variance 1).

We believe that the lessons of this model are robust. It formalizes three basic assumptions:

1. The environment varies.
2. Cues about the environment are imperfect, so individuals make errors.
3. Imitation increases the accuracy (or reduces the cost) of learning.

We have analyzed several models that incorporate these assumptions but differ in other features. All of these models lead to the same qualitative conclusion: when learning is difficult and environments do not change too fast, most individuals imitate at evolutionary equilibrium. At that equilibrium, an optimally imitating population is better off, on average, than a population that does not imitate.

Adding Conformism

So far we have shown only that it may be best for most people to copy others rather than try to figure things out for themselves. Recall that for something to be a norm, there has to be a conformist element. People must agree on the appropriate behavior and disapprove of others who do not behave appropriately. We now show that individuals who respond to such disapproval by conforming to the social norm are *more* likely to acquire the best behavior. We will also show that as the tendency to conform increases, so does the equilibrium amount of imitation.

To allow the possibility for conformist pressure, we add the following assumption to the model described. When an individual imitates, she may be disproportionately likely to acquire the more common variant. Let q be the fraction of the population using trait 1. As before, individuals collect information about the best behavior in the current environment, and then if the information is not decisive, they imitate. However, now the probability (Prob) that an imitating individual acquires trait 1 is:

$$\text{Prob}(1) = q + \Delta q(1 - q)(2q - 1) \tag{1}$$

Thus, Δ represents the extent to which individuals respond to the blandishments of others. When $\Delta = 0$, individuals ignore conformist pressures, and the model is the same as the one described in the previous section. When $\Delta > 0$, social pressure (or merely a desire to be like others) induces individuals to adopt the more common of the two behaviors. When $\Delta \approx 1$, individuals almost always adopt the same behavior as the majority.

We now determine the equilibrium values of Δ and L (the probability of relying on imitation) in the same way that we determined the equilibrium amount of imitation. Assume that most of the population is characterized by one pair of values of Δ and L. Then, consider whether that population can be invaded by individuals using slightly different values of Δ and L. The evolutionary equilibrium is the combination of values of Δ and L that cannot be invaded in this way.

This analysis leads to two robust results. First, all conditions that lead a substantial fraction of the population to rely on imitation also lead to very strong conformity. Consider, for example, figure 5.4, which plots the equilibrium values

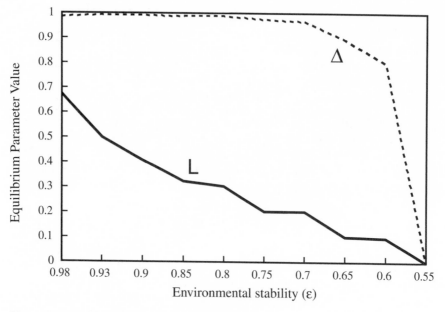

Figure 5.4. Equilibrium values of L and Δ for different rates of environmental variation. At evolutionary equilibrium, the strength of conformist transmission is high for a wide range of rates of environmental change. However, reliance on social learning (L as proportion ranging from 0 to 1.0) decreases rapidly over the same range of environmental stability. When there is no conformist effect (Δ is constrained to be zero), the evolutionary equilibrium value of L is lower than when Δ is free to evolve to its equilibrium value. Since the conformist effect causes the population to track the environment more effectively, it makes social learning more useful. For more details on this calculation, see Henrich and Boyd (1998).

of Δ and L as a function of the rate of environmental change. Notice that as long as the environment is not completely unpredictable, the equilibrium value of Δ is near its maximum value—when people imitate, they virtually always do what the majority of the population is doing. As detailed in Henrich and Boyd (1998), the equilibrium values of Δ and L are equally insensitive to other parameters in the model. Second, as conformism increases, so does the fraction of the population that relies on imitation. Figure 5.4 shows that the equilibrium value of L, when both L and Δ are allowed to evolve, is larger than the equilibrium value of L in a model in which Δ is constrained to be zero. Thus, a tendency to conform increases the number of people who follow social norms and decreases the numbers who think for themselves.

These results are easy to understand. Just after the environment switches, most people acquire the wrong behavior. Then, the combination of occasional learning and imitation causes the best behavior to become gradually more common in the population until an equilibrium is reached at which most of the people are characterized by the better behavior. For rates of environmental change that favor substantial reliance on imitation, the best behavior is more common than the alternative averaged over this entire cycle. Thus, individuals with a conformist tendency to adopt the most common behavior when in doubt are more likely to acquire the best behavior. Conformism continues to increase until it becomes so strong that it prevents the population from responding adaptively after an environmental shift. Optimal conformism leads to increased imitation because on average conformism causes imitators to be more likely to acquire the best behavior in the current environment.

Imitation of Successful Neighbors Leads to the Spread of Beneficial Norms

There is a large literature that indicates that people often have time-inconsistent preferences and, as a result, they often make choices in the short run that they know are not in their long-term interest. It is plausible that social norms help people solve these problems by creating short-term incentives to do the right thing. I may not be able to resist a drink when the costs are all in the distant future but will make a different decision if I suffer immediate social disapproval. It is also easy to see why such norms persist once they are established. If everyone agrees that self-control is proper behavior and punish people who disagree, then the norm will persist. The problem is that the same mechanism can stabilize any norm. People could just as easily agree that excessive drinking is proper behavior and punish teetotalers. If it is true that norms often promote self-control, then we need an explanation of why such norms are likely to arise and spread. In this section, we sketch one such mechanism.

Suppose people modify their beliefs by imitating the successful. If they sometimes imitate people from neighboring groups with different norms, then under the right circumstances norms that solve self-control problems will spread from one group to another because their enforcement makes people more successful and therefore more likely to be imitated.

Consider a model in which a population is subdivided into n social groups (numbered $d = 1, \ldots, n$). There are two alternative behaviors: individuals can be self-indulgent or abstentious. Self-indulgent individuals succumb to the temptations of strong drink, while abstentious individuals restrain themselves. Abstentious individuals are better off in the long run. They make more money, live longer, are healthier, and so on, and everyone agrees that the short-term pleasures of the bottle are not sufficient to compensate for the long-term costs that result. Nonetheless, because individuals do not have time-consistent preferences, everyone succumbs to the temptations of the table and drinks to excess.

Next, assume that there are two social norms governing consumption behavior. People can be *puritanical* or *tolerant*. Puritans believe that alcohol consumption is wrong and disapprove of those who drink. Tolerant people believe everyone should make their own consumption decisions. Each type disapproves of the other: puritans believe that no one should tolerate excess, and the tolerant think that others should be tolerant as well. These norms affect the costs and benefits of the two behaviors. When puritans are present, the people who drink suffer social disapproval, and because this cost is incurred *immediately*, it can cause people to choose not to drink when they otherwise might. Thus, as the proportion of the population who hold puritanical beliefs increases, the proportion of people who drink decreases, and people are better off in the long run. To formalize these ideas, let p_d be the frequency of the puritanical norm in group d. Then W_1, the average payoff of puritanical individuals in group d, is given by:

$$W_1(p_d) = W_0 - s(1 - p_d) + g\,p_d \qquad (2)$$

W_2, the average payoff of tolerant individuals in group d, is given by:

$$W_2(p_d) = W_0 - sp_d + (g + \delta)p_d \qquad (3)$$

W_0 is the baseline payoff of drinkers in a completely tolerant group. Individuals of each type suffer disapproval and a reduction in welfare when the other type is present in their social group. These social effects on welfare are represented by the terms proportional to s in equations 2 and 3. However, the welfare of all individuals is increased by the fraction of puritanical individuals because everybody is less likely to drink when puritans are present to shame them. These effects are represented by the terms proportional to g and $g + \delta$. The parameter δ captures the idea that puritans may have a different effect on each other than they do on the tolerant: perhaps bigger because they are more sensitive to the opinions of their own kind; perhaps smaller because they are already avoiding strong drink.

Next, the following process governs the evolution of these norms within a group. During each time period, each individual encounters another person, compares his welfare to the person he encounters, and then, with probability proportional to the difference between their payoffs during the last time period, adopts that person's norm. In particular, suppose that an individual with norm i from group f encounters an individual with norm j from group d. After the encounter, the probability that an individual switches to j is:

$$\text{Prob}(j\,|i,\,j) = \tfrac{1}{2}\{1 + \beta[W_j(p_f) - W_i(p_d)]\} \qquad (4)$$

When the parameter β equals zero, payoffs do not affect imitation—people imitate at random. When $\beta > 0$, people are more likely to imitate high payoff individuals. Notice that since an individual's payoff depends on the composition of his group, there will be a tendency for ideas to spread from groups in which beneficial norms are common to groups in which less beneficial norms are common.

Let m_{df} be the probability that an individual from group f encounters an individual from group d. Δp_f, the change in p_f during one time period, is given by:

$$\Delta p_f = \beta p_f[W_1(p_f) - \overline{W}(p_f)]$$
$$+ \sum_{d \neq f} m_{df}\beta\{p_d[W_1(p_d) - \overline{W}(p_d)] - p_f[W_1(p_f) - \overline{W}(p_f)]\}$$
$$+ \sum_{d \neq f} m_{df}(p_d - p_f)\{1 + \beta[\overline{W}(p_d) - \overline{W}(p_f)]\} \qquad (5)$$

To make sense of this expression, first assume that people only encounter individuals from their own social group.

$$\Delta p_f = \beta p_f[W_1(p_f) - \overline{W}(p_f)] \qquad (6)$$

This is the ordinary replicator dynamic equation. This equation simplifies to have the form:

$$\Delta p_f = \alpha p_f (1 - p_f)(p - \tilde{p})$$

(7)

where $\alpha = \beta(2s - \delta)$ and $\tilde{p} = s/(2s - \delta)$. Thus, when each social group is isolated and the effects of social sanctions are large compared to the effects of drinking ($2s > \delta$), there are two stable evolutionary equilibria: groups consisting of all puritans or all tolerant individuals. If the presence of puritans benefits other puritans more than it benefits the tolerant ($\delta > 0$), then the all-puritan equilibrium has a larger basin of attraction. If puritans benefit the tolerant more, then the all-tolerant equilibrium has a larger basin of attraction.

When there is contact between different groups, the last two terms in equation 5 affect the change in frequency of norms within social groups. The third term is of most interest here. If $\beta = 0$, this term is proportional to the difference in the frequency of puritanism between the groups and simply represents passive diffusion. If, however, $\beta > 0$, there is a greater flow of norms from groups with high average payoff to groups with lower average payoff. This differential flow arises because people imitate the successful and norms affect the average welfare of group members. Can this effect lead to the systematic spread of beneficial norms?

For the beneficial puritanical norm to spread, two things must occur. First, such a norm must increase to substantial frequency in one group. Second, it must spread to other groups. Here we address only the second question. To keep things simple, we further assume that social groups are arranged in a ring and that individuals have contact only with members of two neighboring groups. Now, suppose that a random shock causes the puritan norm to become common in a single group. Will this norm spread? To answer this question, we have simulated this model for a range of parameter values. Representative results are shown in figure 5.5 that plots the ranges of parameters over which the beneficial norm spreads. The vertical axis gives the ratio of m (the probability that individuals interact with others outside of their group) to α (rate of change due to imitation within groups), and the horizontal axis plots $\tilde{p}$ (the unstable equilibrium that separates the domains of attraction of puritanical and tolerant equilibria in isolated groups). The shaded areas give the combinations of m/α and $\tilde{p}$, which lead to the spread of the puritanical norm to all groups, given that it was initially common in a single group for two values of g.

First, notice that the beneficial norm spreads most easily when the level of interaction between groups is intermediate. If there is too much mixing, the puritanical norm cannot persist in the initial population. It is swamped by the flow of norms from its two tolerant neighbors. If there is too little mixing, the puritanical norm remains common in the initial population but cannot spread because there is not enough interaction between neighbors for the beneficial effects of the norm to cause it to spread.

Second, to understand the effect of g, consider the case in which $g = 0$. Even when the norm produces no benefit to individuals as it becomes common, it can still spread if the puritanical norm has a larger basin of attraction in an isolated population ($\delta < 0$). In this case, the costly disapproval of harmless pastimes can

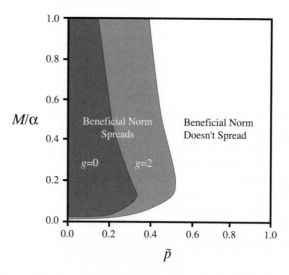

Figure 5.5. Plots parameter combinations that lead to the spread of the group beneficial norm between groups. The vertical axis gives the ratio of m, the probability that individuals interact with others outside of their group to α, the rate of change due to imitation within groups. The horizontal axis plots $\tilde{p}$, the unstable equilibrium that separates the domains of attraction of all puritanical and tolerant equilibria in isolated groups. The shaded areas give the combinations of m/α and $\tilde{p}$ that lead to the spread of the puritanical norm given that it has become common in a single group for two values of g, the extent to which individual behavior is affected by norms. Notice that the beneficial norm spreads when the level of interaction between groups is intermediate. If there is too much mixing, the puritanical norm cannot persist in the initial population. If there is too little mixing, it can persist in the initial population but cannot spread.

seriously handicap the tolerant when puritans are only moderately common. To understand why, consider a focal group at the boundary between the spreading front of puritan groups and the existing population of tolerant groups. The focal group, in which both norms are present, is bounded on one side by a group in which puritan norms are common and on the other side by a group in which tolerant norms are common. Since groups on both sides of the boundary have the same average payoff, the flow of norms will tend to move the focal group toward an even balance of the two norms. If the domain of attraction of the puritanical norm includes 0.5 and if there is enough mixing, then mixing with neighboring groups can be enough to tip the focal group into the basin of attraction of the puritanical norm. This is true even though the differential success owes only to puritans avoiding the costs imposed on the tolerant by puritans. To see why increasing g increases the range of values of $\tilde{p}$ that allow the beneficial norm to spread, consider again a focal group on the boundary between the regions in which the puritanical norm is common and uncommon. When $g > 0$, individuals in the focal group are more likely to imitate someone from the neighboring group where the puritanical norm is common than the other neighboring group where tolerant individuals are common because individuals from the former group are more successful. Therefore, the flow of norms will tend to move the focal group toward a frequency of puritans greater than 0.5.

It is interesting to note that the rate at which this process of equilibrium selection goes on seems to be roughly comparable to the rate at which traits spread within a single group under the influence of the same learning process. Game theorists have considered a number of mechanisms of equilibrium selection that arise because of random fluctuations in outcomes due to sampling variation and finite numbers of players (e.g., Samuelson, 1997). These processes also tend to pick out the equilibrium with the largest domain of attraction. However, unless the number of individuals in the population is very small, the rate at which this occurs is very slow. In contrast, in the simulations we performed, the group beneficial trait spread from one population to the next at a rate roughly half the rate at which the same imitate-the-successful mechanism led to the spread of a trait within an isolated group. Of course, we have not accounted for the rate at which the beneficial norm becomes common in an initial group. This requires random processes. However, only the group, not the whole population, needs be small, and the group must be small only for a short period of time for random processes to give rise to an initial "group mutation," which can then spread relatively rapidly to the population as a whole.

Conclusion: Are Norms Usually Sensible?

We have shown that it is possible for norms to guide people toward sensible behavior that they would not choose if left to their own devices. Norms could be sensible, just as functionalists in anthropology have claimed. However, the fact that they could be sensible does not mean that they *are* sensible. There are some well-studied examples, like the alkali treatment of corn, and there are many other plausible examples of culturally transmitted norms that seem to embody adaptive wisdom. However, as documented in Robert Edgerton's book, *Sick Societies* (1992), there are also many examples of norms that are not obviously adaptive, and, in fact, some seem spectacularly maladaptive. Such cases might result from the pathological spread of norms that merely handicap the tolerant without doing anyone any good (and perhaps harm puritans as well?). Or they might result from antiquated norms that persist in a frequency above a large basin of attraction for tolerance, having lost their original fitness-enhancing effect due to social or environmental change. More careful quantitative research on the costs and benefits of alternative norms would clearly be useful.

We believe that it is also important to focus more attention on the processes by which norms are shaped and transmitted. Anthropologists and other social scientists have paid scant attention to estimating the magnitude of evolutionary processes affecting culture change in the field or lab, although several research programs demonstrate that such estimates are perfectly practical (Aunger, 1994;

Insko et al., 1983; Labov, 1980; Rogers, 1983; Rosenthal and Zimmerman, 1978; Soltis, Boyd, and Richerson, 1995). What happens to a Maya who does not utilize the normative form of alkali treatment of corn in her traditional society? What are the nutritional effects? The social effects? From whom do people learn how to process corn? How does this affect which variants of the process are transmitted and which are not? Only by answering such questions will we learn why societies have the norms they have and when norms are adaptive.

REFERENCES

Aunger, R. 1994. Are food avoidances maladaptive in the Ituri Forest of Zaire? *Journal of Anthropological Research* 50:277–310.

Binmore, K. G. 1998. *Just playing: Game theory and the social contract.* Cambridge, MA: MIT Press.

Bowles, S., & H. Gintis. 1999. *The evolution of strong reciprocity.* Santa Fe Institute Working Paper 98-08-073E. Santa Fe, NM.

Boyd, R., & P. J. Richerson. 1985. *Culture and the evolutionary process.* Chicago: University of Chicago Press.

Boyd, R., & P. J. Richerson. 1988. An evolutionary model of social learning: The effects of spatial and temporal variation. In: *Social learning: Psychological and biological perspectives*, T. Zentall, & B. G. Galef, eds. (pp. 29–48). Hillsdale, NJ: Lawrence Erlbaum.

Boyd, R., & P. J. Richerson. 1989. Social learning as an adaptation. *Lectures in Mathematics and the Life Sciences* 20:1–26.

Boyd, R., & P. J. Richerson. 1995. Why does culture increase human adaptability? *Ethology and Sociobiology* 16:125–143.

Boyd, R., & P. J. Richerson. 1996. Why culture is common, but cultural evolution is rare. *Proceedings of the British Academy* 88:77–93.

Edgerton, R. 1992. *Sick societies: Challenging the myth of primitive harmony.* New York: Free Press.

Ellickson, R. C. 1994. *Order without law: How neighbors settle disputes.* Cambridge, MA: Harvard University Press.

Harris, M. 1979. *Cultural materialism: The struggle for a science of culture.* New York: Random House.

Henrich, J., & R. Boyd. 1998. The evolution of conformist transmission and the emergence of between-group differences. *Evolution and Human Behavior* 19: 215–242.

Insko, C. A., R. Gilmore, S. Drenan, A. Lipsitz, D. Moehle, & J. Thibaut. 1983. Trade versus expropriation in open groups: A comparison of two types of social power. *Journal of Personality and Social Psychology* 44:977–999.

Jensen, G. F., & M. L. Erickson. 1979. The religious factor and delinquency: Another look at the hellfire hypothesis. In: *The religious dimension: New directions in quantitative research*, R. Wenthow, ed. (pp. 157–177). New York: Academic.

Katz, S. H., M. L. Hediger, & L. A. Valleroy. 1974. Traditional maize processing techniques in the New World. *Science* 184:765–773.

Labov, W. 1980. *Locating language in time and space.* Philadelphia: University of Pennsylvania Press.

McAdams, R. H. 1997. The origin, development, and regulation of norms. *Michigan Law Review* 96:338–443.

McEvoy, L., & G. Land. 1981. Life-style and death patterns of Missouri RLDS Church members. *American Journal of Public Health* 71:1350–1357.

Ostrom, E. 1991. *Governing the commons: The evolution of institutions for collective action.* Cambridge: Cambridge University Press.

Richerson, P. J., & R. Boyd. 1998. The evolution of human ultra-sociality. In: *Indoctrinability, ideology, and warfare: Evolutionary perspectives,* I. Eibl-Eibisfeldt, & F. Salter, eds. (pp. 71–96). New York: Berghahn.

Rogers, E. M. 1983. *Diffusion of innovations.* 3d ed. New York: Free Press.

Rosenthal, T., & B. Zimmerman. 1978. *Social learning and cognition.* New York: Academic.

Samuelson, L. 1997. *Evolutionary games and equilibrium selection.* Cambridge, MA: MIT Press.

Soltis, J., R. Boyd, & P. J. Richerson. 1995. Can group-functional behaviors evolve by cultural group selection? An empirical test. *Current Anthropology* 36:473–494.

Turner, J. H., & A. Maryanski. 1979. *Functionalism.* Menlo Park, CA: Benjamin/ Cummings.

PART 2

Ethnic Groups and Markers

Human populations are richly subdivided into groups marked by seemingly arbitrary symbolic traits, including distinctive styles of dress, cuisine, or dialect. Such symbolically marked groups often have distinctive moral codes and norms of behavior, and sometimes exhibit economic specialization. Ethnic groups provide the most obvious example of such groups, but the phenomenon includes groups based on class, region, religion, gender, and profession. Ethnic groups, present in all historical periods, often split and merge through time, yet many have substantial historical continuity. Nowadays ethnic groups can have millions of members, but even in simple hunting and gathering societies, symbolically marked groups are much larger than the residential band, typically linking roughly one thousand people.

The evidence is fairly clear that the symbolic marking is not simply a by-product of a common cultural heritage. If cultural boundaries were impermeable, like species boundaries, then this fact would explain the association between symbolic markers and other traits. However, group boundaries are highly permeable. The movement of people and ideas between groups attenuates group differences. Thus, the persistence of existing boundaries and the birth of new ones indicate that other social processes resist the homogenizing effects of migration and the strategic adoption of ethnic identities. Moreover, since groups are typically fairly large, such processes likely produce symbolic marking as an unintended by-product of human choices made for some other reason.

The following two chapters explore the idea that symbolically marked groups arise and are maintained because dress, dialect, and other markers allow people to identify in-group members. In chapter 6, we analyze a model that assumes that identifying in-group members is useful because it allows selective imitation. Rapid cultural adaptation makes the local population a valuable source of information about what is adaptive in the local environment. Individuals are well advised to imitate locals and avoid learning from immigrants who bring ideas adapted to other environments. In chapter 7, we (along with Richard McElreath) study a model in which markers allow selective social interaction. Rapid cultural adaptation can preserve differences in moral norms between groups. It's best to interact with people who share beliefs about what is right and wrong, what is fair, and what is valuable. Thus, once there are reliable symbolic markers, selection will favor the psychological propensity to imitate and interact selectively with individuals who share those markers.

These models have several interesting and, at least to us, less-than-obvious properties. First, the same nonrandom interaction that makes markers useful also creates and maintains variation in symbolic marker traits as an *unintended* by-product. Nonrandom interaction acts to increase correlation between arbitrary markers and locally adaptive behaviors. This, in turn, makes markers more useful, setting up a positive feedback process that can amplify small differences in markers between groups. Second, this process is not sufficient by itself to generate group markers. There must be some initial, perhaps weak correlation between symbolic expression and group membership, and there has to be some kind of population structure so that groups are at least partly isolated from each other. Otherwise, the positive feedback process cannot get started. Third, once groups have become sharply marked, the feedback process is sufficient by itself to maintain group marking even if groups are perfectly mixed and there is no population structure other than that caused by the markers. The models also make a number of interesting predictions about the spatial and temporal patterns of symbolic expression.

If the processes captured in these models are important in creating ethnic groups, then ethnic groups should have arisen as soon as cumulative cultural evolution became important in the human lineage. Rapid cumulative cultural evolution is an engine for generating important differences between groups, both in subsistence technology and other kinds of local ecological adaptation, and differences in moral norms and other determinants of social behavior and institutions. Thus, this picture of ethnicity predicts that symbolic markers should appear in the archaeological record around the same time as the signs of cumulative cultural adaptation, which we take to be increased variation in space and highly refined cultural adaptations.

One very important and widespread component of ethnicity is *not* obviously an entailment of these models of ethnicity, namely, ethnocentrism. Quite commonly, but as Brewer and Campbell (1976) showed, by no means universally, people derogate members of other ethnic groups. In many times and places, these feelings led to interethnic conflict. We can think of two reasons why the kinds of ethnic groups that arise in these models might be associated with in-group favoritism and out-group bias. The first is that, on our view, groups of people who share distinctive moral norms, particularly norms that govern social interaction, quite likely become ethnically marked. This suggests that ethnocentric judgments easily arise because "we the people" behave properly, while those "others" behave improperly, doing disgusting, immoral things, and showing no remorse for it, either. Second, as we explain in part 3 on group cooperation, we expect group selection to work at levels of population structure at which there is lots of cultural variation affecting group success. Quite plausibly, the differences in subsistence and moral systems assumed in these models would give rise to group selection at the level of ethnic groups, particularly group selection driven by differential imitation of successful groups. This in turn implies that ethnic groups should be one locus of economic, political, and military cooperation. Of course, cooperation within groups creates competition between groups for the resources that people want, and resulting norms lead to in-group cooperation.

REFERENCE

Brewer, M. B., & D. T. Campbell. 1976. *Ethnocentrism and intergroup attitudes: East African evidence.* Beverly Hills: Sage Publications.

6 The Evolution of
Ethnic Markers

Much of the debate about human sociobiology has been framed as a binary opposition. Sociobiologists argue that evolutionary theory is useful for understanding humans because much of our behavior is currently adaptive, or was adaptive under food-foraging conditions. To be sure, they aver, culture occasionally causes human behavior to drift away from the fitness-maximizing optimum, but in the long run behaviors that have important effects on Darwinian fitness should tend to be adaptive. Critics of this view argue that the existence of culture has allowed the human species to transcend ordinary evolutionary imperatives. Culturally transmitted behavior must not be so maladaptive as to lead to the extinction of the social group, but as long as this rather weak constraint is satisfied, it is argued, people are free to elaborate their culture more or less as they please.

We believe that this dichotomy is false. Culture is neither autonomous and free to vary independently of genetic fitness, nor is it simply a prisoner of genetic constraints. Our rejection of this dichotomy is based on what we call the "dual inheritance" theory of the interaction of genes and culture (Boyd and Richerson, 1985). The essential feature of this theory is that, like genes, culture should be viewed as a system of inheritance. People acquire beliefs, attitudes, and values from others by social learning and then transmit them to others. Human behavior results from the interaction of genetically and culturally inherited information. In the theoretical models we have constructed to represent this interaction, two results stand out: (1) The cultural system of inheritance has many properties that make it quite different from the genetic system. For example, an individual can observe the behavior of a number of peers and choose the "best" behavior. Such properties may often enhance genetic fitness because they allow modes of adaptation not available to noncultural species. (2) These same properties can lead to the evolution of many cultural traits that are costly divergences from those that would increase genetic fitness. Culture is an evolutionarily active part of a system that, jointly with genes and environment, can account for much of human behavioral variation.

Here we will illustrate this general argument in the context of a particular problem, the evolution of markers of group membership. One of the most striking and unusual features of the human species is that it is subdivided into ethnic groups. Barth (1969) identified what we take to be the critical feature of ethnicity: people identify themselves, and are identified by others, as members of an ethnic group based on a set of culturally transmitted characters. Some of these traits, such as language, dress style, ritual, and cuisine, appear to be arbitrary symbolic "markers" of ethnic affiliation, while others are more directly functional cultural traits such as basic moral values and standards of excellence. Membership in a particular ethnic group can have important effects on an individual's economic behavior and political and social interactions.

The interpretation of ethnic markers is controversial. Sahlins (1976) has argued that one must choose between functional explanations and nonfunctional cultural explanations of symbolic marker characters. We will show that this dichotomy oversimplifies the relationship between genetic and cultural evolution. Ethnicity provides a good example of how functional organic adaptation and symbolic cultural processes are thoroughly intertwined in human evolution. Our argument is based on an evolutionary model embodying two mechanisms that cause a population occupying a variable environment to be subdivided on the basis of ethnic markers. These mechanisms result from a pattern of enculturation in which individuals are disproportionately influenced by two kinds of people: those who are similar to themselves and those who are successful. Even though these two mechanisms cause groups to become differentiated based on arbitrary symbolic markers in a way that could not be predicted from fitness maximization alone, they will be favored by natural selection because they allow more accurate adaptation to variable environments.

This application of dual inheritance theory emphasizes the fitness-enhancing properties of culture. We have chosen this emphasis for two reasons. First, it is interesting to try to understand why a cultural system of inheritance arose in the hominid lineage and how that process shaped the way that culture is transmitted. Most likely, the organic capacities that allow culture to be stored and transmitted arose through the action of natural selection. In the context of this example, we are interested in why selection favored mechanisms of cultural transmission that give rise to ethnic groups. Second, the reasons why culture is adaptive are both subtle and interesting. *Even when culture is highly adaptive, it has its own evolutionary properties and can lead to patterns of behavior that could not be understood in the absence of knowledge of how cultural processes operate.* To understand why ethnic markers allow more accurate adaptation to variable environments, one must understand how the cultural processes that give rise to ethnic differentiation operate. We have discussed the properties of cultural inheritance that lead to genetically maladaptive behavior elsewhere (Boyd and Richerson, 1985). Knauft (1987) also gives an intriguing empirical example of how the differences between genetic and cultural inheritance can give rise to behavior that is genetically maladaptive.

Models of Cultural Evolution

We define culture as information—skills, attitudes, beliefs, values—capable of affecting individuals' behavior, which they acquire from others by teaching, imitation, and other forms of social learning. A particular member of a set of attitudes, beliefs, and values will be referred to as a cultural variant. (See Boyd and Richerson, 1985, ch. 3 for an extended discussion of this definition.) We have adopted this definition because it focuses attention on the means by which cultural traditions are perpetuated. Culture is acquired by individuals by teaching, imitation, and other forms of social learning from other individuals, stored in individual brains, and transmitted by teaching and imitation to others.

Recently, there has been a fair amount of interest in applying concepts drawn from evolutionary biology to the problem of cultural evolution (e.g., Campbell, 1975; Cavalli-Sforza and Feldman, 1981; Boyd and Richerson, 1985). Despite the fact that cultural and genetic evolution differ in important ways, this methodological borrowing has been fruitful because genes and culture both have population-level properties. That is, individual behavior depends in part on the cultural variation in the population from which individuals acquire cultural variants. At the same time, which cultural variants are available in the population to be acquired depends on what happened to individuals with different variants in the population in the past. For example, in every generation some individuals will invent or learn new behaviors, modifying the variants they originally imitated and transmitting the new variants to others in the process of enculturation. Cultural evolution can be viewed as a complex of sampling and modifying processes that operate iteratively on a population of variable culture-bearing individuals. That there is a very general analogy between genes and culture is a commonplace observation; what is new is the reworking of methods of analysis developed by evolutionary biologists to build a useful theory from the old analogy.

Simple mathematical models are among the most important tools that biologists use to study population-level processes. The tradition of their use began in evolutionary biology with Wright, Fisher, and Haldane in the first part of the last century and is continued today by people like John Maynard Smith, W. D. Hamilton, and many others. The goal of such models is to isolate the population-level consequences of a limited set of processes by stripping away all of the confusing detail due to other processes. For example, kin-selection models address the question: when can selection favor behaviors that reduce the fitness of the individual performing them, given that they increase the fitness of other individuals affected by the behavior? In such models virtually all the actual behavioral and ecological detail is suppressed, so that exactly the same mathematical model is applied, for example, to coalition behavior among macaques and communal nesting in scrub jays. The intent of the model is to give insight into kin selection as a generic evolutionary process, not to account for the details of particular examples of the process. Evolutionary biologists construct many such simple models, each isolating one or a few processes. That a particular process is neglected in a model is not to say that it is unimportant, only that we desire to focus on something else for the moment. This sort of theorizing is sometimes stigmatized as "reductionistic." A more apt characterization would be "modular." Real evolutionary phenomena are complex; except for deliberately controlled experiments, we expect to link several such models together to achieve a satisfactory explanation of real events.

Nevertheless, the study of the simple modules in isolation is useful because it has proven difficult to deduce the population-level consequences of individual processes using verbal reasoning alone. Population processes involve the interaction of phenomena occurring at two different levels of organization and two distinct time scales. The individual and population levels of organization interact through the sampling processes inherent in reproduction or socialization. The day-to-day ecological time scale, on which processes of change act (e.g., selective mortality), interacts with the long-run evolutionary time scale on which adaptations of particular kinds are or are not produced. Even the simplest examples of evolutionary processes are thus rather complex. Mathematics makes it relatively easy to consistently and systematically trace the implications of a given set of assumptions, especially when the processes modelled are probabilistic or quantitative. Simple but formal models are a useful mental prosthesis to reduce the handicap of a certain kind of a cognitive limitation. It is important to realize that such models serve a rather narrow function, the testing of explanations for logical consistency. While they are tremendously useful in this role, they are only a supplement to other theoretical and empirical tools in the social and biological sciences, not a replacement for them.

A Model of the Evolution of Ethnic Markers

The existence of ethnic groups and similarly marked social units suggests two evolutionary questions: (1) What are the processes that would cause a human population to split into two groups distinguished by cultural marker traits? (2) Could such processes give rise to cultural variation that is biologically adaptive in the sense of increasing reproductive success?

Motivating the Model

Let us approach these two questions by turning the second one around: how should natural selection have shaped the processes by which individuals acquire culture? At the very least, this way of viewing the problem ought to be appropriate for considering the origin of organic capacities that make culture possible. Consider an ancient human population that has recently expanded into a new habitat. Some individuals in the new habitat will have adopted beliefs and values that are appropriate in the new habitat, but many will share the values and beliefs of individuals in the old habitat. This lag in cultural adaptation could result from at least two factors: (1) innovation is slow and the occupation of the new habitat is recent; (2) there is an exchange of individuals between habitats, so that some individuals in the new habitat acquired their beliefs and values in the old habitat. If either of these two factors obtain, many individuals will carry variants that are appropriate in the old habitat, but not in the new one. Assuming that natural selection plays a strong role shaping cultural capacities, it will structure the acquisition of culture so that individuals in each habitat have the best chance of acquiring the set of beliefs and values that are appropriate there.

If one set of beliefs or values has easily observable advantages relative to the others, then there is an easy answer: individuals should adopt the beliefs and values that maximize reproductive success. It seems likely, however, that people commonly must choose among variant beliefs where it is quite difficult to determine which belief is most advantageous, even though the beliefs, in fact, differ in utility. Behavioral decision theorists (Nisbett and Ross, 1980) and students of social learning (Rosenthal and Zimmerman, 1978) argue from empirical evidence that the complexity and number of real decisions forces people to use simple rules of thumb. Chief among these is a heavy reliance on imitation to acquire most of their behavior.

Studies of the diffusion of innovations (summarized in Rogers with Shoemaker, 1971) suggest that people often use two simple rules to increase the likelihood that they acquire locally adaptive beliefs by imitation. The chance that individual A will adopt an innovation modeled by individual B often seems to depend upon (1) how *successful* B is, and (2) the *similarity* of A to B. When it is difficult to evaluate whether an innovation is sensible, imitating the successful seems like a good general rule; if the innovation is beneficial, people who use it will be more successful, on the average, than those who do not. It also seems sensible to condition adoption on similarity. If a model is very different from oneself, the model's success might not indicate that the innovation would be useful in one's own circumstances. In the interests of simplicity, we will model a situation in which success and similarity are the only adoption rules people use. As in the case of kin-selection models, the model is meant to yield insight into the operation of this particular pair of decision rules as their effects are integrated over individuals and time to produce evolutionary results. Since many other important processes are left out, the model is meant to apply partially and qualitatively to a great many cases, but to be a complete quantitative description of none.

How cultural populations will evolve under the influences of these two processes depends a great deal on what people use as indicators of success and similarity. Because our focus here is on the problem of the origin of a capacity for culture under the influence of natural selection, we will assume that the index of success is a correlate of genetic fitness and that the index of similarity is a conspicuous symbolic character, like dialect, acquired from primary socializers such as parents. As far as the formal model is concerned, any standard of success or similarity can be substituted. If these assumptions are relaxed, the model may still be appropriate to understanding how ethnic groups form, but not to the problem of how such a capacity evolved in the first place. Ethnicity might be a costly by-product of some other advantage associated with ability to recognize success and similarity. The narrow interpretation we give here is not meant to prejudge these empirical issues. (See Boyd and Richerson, 1985, ch. 8, for a model in which the standard of success is explicitly cultural and in which it departs very sharply from what selection on genes would favor.)

Is the evolution of ethnic markers possibly an adaptive result of using these two rules in cases where more direct decisions are too costly to use? It is fairly obvious that if most people adopt beliefs or values modeled by successful people, beliefs or values that lead to success will spread. It seemed possible to us that coupling a propensity to imitate the successful with a propensity to adopt the beliefs and values of those who are similar to oneself might cause groups occupying different habitats to become culturally isolated from each other because the cultural markers used to judge similarity would diverge in the two populations. To check the cogency of this intuition, we analyzed the following model.

Formalizing the Intuition

Real environments and real means of exploiting them are complex. However, we think that the cogency of intuitions can be evaluated using quite simple models. Accordingly, we imagine that there are two ecological "niches" that differ according to the optimal value of an "adaptive" character. For example, suppose that there are two habitats—one moist, one dry. The adaptive character could be a belief that affects the extent to which a person relies on stock raising as opposed to cultivation. This belief might be the extent to which an individual believes that cattle ownership is an intrinsic measure of a person's worth as a human being. In the dry habitat the most successful subsistence strategy might be pure pastoralism, and thus the optimal value of the adaptive character is a heavy valuation of cattle. In the moist habitat the most successful strategy might involve mostly horticulture, and a lesser valuation of cattle might lead to a more successful subsistence strategy.

To represent these assumptions mathematically, we suppose that each individual's subsistence strategy can be characterized by a single number labeled A. This can be thought of as an index of the extent to which individuals' beliefs lead them to depend on stock raising. The habitats are labeled 1 and 2, and the optimal values of A are θ_1 and θ_2. The more that an individual's adaptive character deviates from the optimum in his or her habitat, the lower on average will be his or her success (and genetic fitness). More mathematical detail is given in the appendix in Boyd and Richerson (1987). In terms of the example, θ_1 might be the value of A that corresponds to mostly pastoralism, and θ_2 mostly horticulture. In what follows we will sometimes refer to the adaptive character as the amount of pastoralism, in order to make the presentation less abstract. The reader should keep in mind, though, that the adaptive trait is not meant to refer to any specific situation. Rather, it is meant to formalize the idea that different beliefs and values are more or less adaptive in different environments.

We assume, further, that each individual is characterized by an arbitrary neutral "marker" character. For example, the marker trait might be an index of dialect, such as the extent to which people pronounce r's. It is arbitrary and neutral in the sense that many dialect variants with no direct effect on adaptive success are possible, although, as we shall see, there may be very strong indirect

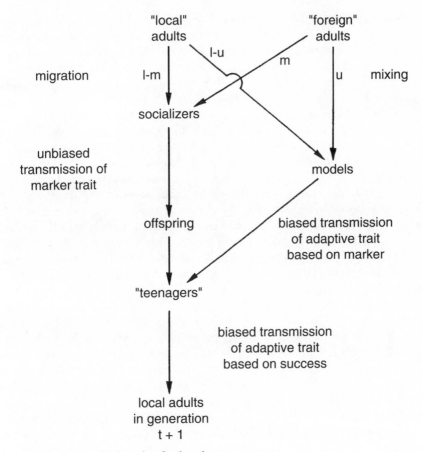

Figure 6.1. Assumed life cycle of cultural transmission.

effects of marker traits upon fitness. Once again, we will assume that the marker trait can be described by a single number, labeled M. Thus, in the context of the model, each individual's culturally acquired beliefs can be described by a pair of numbers, A and M.

We assume that these two cultural traits are transmitted according to the life cycle shown in figure 6.1. This life cycle is meant to reflect the fact that children, adolescents, and young adults have different patterns of enculturation. Individuals acquire their *marker* trait (e.g., their dialect) at an early age from a set of primary socialization agents ("socializers" for short). They acquire their *adaptive* trait at a later age by observing the behavior of a much wider range of individuals whom we will refer to as "models." Socializers need not be biological kin—*the key assumption is only that the amount of mixing between habitats is much greater for models than for socializers* ($u \gg m$). As we shall see, this condition allows the differentiation of marker traits, hence a sense of ethnic distinctiveness, to build in the local environment. We further assume that dialect is acquired through a process of faithful copying. That is, on average people acquire the dialect of the community in which they were raised. We formalize this idea by assuming that each naive individual has the opportunity to observe the behavior of n socializers. Naive individuals then adopt a weighted average of the dialect of n socializers as their own dialect. The fact that socializers may have different weights is meant to represent the idea that some individuals may be more important in transmission than others due to kinship, social status, or some other factor.

In later phases of the life cycle, the adaptive trait is not acquired through faithful copying. Rather, the acquisition of the adaptive trait is biased by two processes. When individuals initially acquire their adaptive trait from models as teenagers, they are predisposed to imitate individuals who have similar marker traits (i.e., have similar dialects). This idea is represented mathematically by assuming that the basic influence of a model (due to social role and the like) is reduced as the difference between the individual's and the model's marker trait increases (in absolute value). Subsequently, individuals modify both their adaptive and marker traits by imitating the successful individuals among their local young adult peers. We represent this idea mathematically by assuming that individuals select one peer to imitate and weight this peer's modifying influence in proportion to his or her success.

Our goal is to study how these transmission and choice processes might change the distribution of culturally transmitted variation in a population through time. In particular, we want to know whether different values of the marker trait will come to predominate in the two habitats and whether this difference ensures that more people acquire the locally adaptive trait. The first step is to describe the nature of the cultural variation in the population at some point in time. To do this, we use the joint distribution of the two traits in the population. This distribution simply specifies the fraction of the population that is characterized by each pair of values, A and M. The shape of such a distribution can be summarized by five numbers. The two means give the "position" of the distribution. For example, $\overline{A}$ tells us the degree to which, on average, the population relies on stock raising. The two variances describe the spread of the distribution. For example, a large variance of A would mean a wide range of subsistence techniques in use in the population. The covariance tells us the extent to which the two traits are correlated. A nonzero covariance means that individuals who rely largely upon pastoralism tend to have a similar dialect, different from the dialect most commonly used by horticulturalists.

The next step is to see how the distributions of A and M in the two populations change through a single generation. To do this, we must determine how events in the lives of individuals change the distribution of cultural variants in the population. First, we assume that when the generation begins, the means and variances that describe the distribution of cultural variants in the population are at initial values. Then we construct submodels to represent individual movement from population to population and the two forms of biased imitation. The effects of each individual's behavior on the properties of the population are very small, but aggregated over all individuals they may cause an appreciable change by the beginning of the next generation. *It is this part of the model that does the important work of linking individual- and population-level processes.* In what follows we provide a qualitative description of the most important effects of each process.

Faithful copying leaves the mean value of the marker trait, M, in each habitat unchanged. This result follows from the assumption that naive individuals faithfully copy the marker of their socializers, who are in turn an unbiased sample of the previous generation.

Mixing of individuals between the environments creates covariance between the adaptive character and the marker character in the populations of models, even if there was no association before mixing in either habitat. To see why, suppose that in habitat 1 people have beliefs that cause them to depend more on pastoralism than do people in habitat 2. This means that the value of $\overline{A}$, the mean value of the adaptive trait, is larger in habitat 1 than in habitat 2. Now suppose that the values of $\overline{M}$, the mean values of the marker trait, in the two habitats are different—for example, individuals in habitat 1 might be more likely to pronounce their *r*'s. Then a model drawn from habitat 1 will be more likely to have large values of A and M, while a model from habitat 2 will tend to have small values. Thus, models who practice pastoralism will tend to pronounce their *r*'s and those who practice horticulture will tend not to, even if there was no association between the two traits in either habitat before mixing. Mixing also moves the mean values of A and M in the two habitats toward each other. If no other processes affect the means, the populations in both habitats will eventually be characterized by the same values of A and M, even though the habitats are quite different.

Biased transmission based on similarity causes the mean value of the adaptive trait among individuals who have just acquired their adaptive trait to be closer to the mean in their habitat before mixing than the mean adaptive trait among their models. By imitating the adaptive trait of people who are like themselves with regard to the marker trait, naive individuals reduce the chance that they will imitate a model drawn from the other habitat. Thus, this form of biased imitation has the effect of reducing the amount of mixing. The strength of this effect depends on the difference between the mean marker trait in the two habitats. If the dialects are not very different, biased imitation based on similarity will have little effect. If the dialects are quite different, the result will be to substantially reduce the effect of mixing.

Biased transmission based on success moves the mean value of the adaptive trait toward the optimum in both habitats and causes the mean values of the marker traits in the two habitats to diverge from each other. Suppose that in habitat 1 individuals who rely mostly on pastoralism are more successful on the average than individuals who rely mostly on horticulture. Then individuals whose beliefs cause them to rely more on pastoralism will be more likely to be imitated, and such beliefs will spread. The same process will cause the mean values of the marker traits to diverge because of the covariance between the marker trait and the adaptive trait that is induced by mixing. Suppose that individuals who rely on pastoralism tend to pronounce their *r*'s. Then the practice of imitating successful people will cause the pronunciation of *r*'s to spread because successful people will tend to pronounce their *r*'s.

The analysis presented so far tells us only what will happen to the distribution of cultural variants in the two habitats over the course of one generation. Normally such changes will be quite small, there will be competing effects, and the direction of change will be dependent on several interacting factors. Our goal is to find out what will happen to the population over the long run. To accomplish this goal, we use various techniques to iterate the equations that describe the change over one generation. *These techniques allow us to accomplish the second difficult step of evolutionary reasoning, the connection of short-time-scale ecological processes with their eventual evolutionary results.* Assuming that the amount of mixing of primary socializers is small enough that it may be neglected, two important results emerge from such an analysis. Starting with a single, nearly uniform population that comes to occupy two habitats: (1) The mean value of the adaptive trait in each habitat approaches the optimum, and (2) the mean values of the marker trait in the populations become quite different.

These general properties are illustrated by the numerical simulation of the model shown in figure 6.2.

These qualitative results make sense in the light of the processes described previously. The mean value of the adaptive trait is affected by two forces—mixing causes the mean in the two habitats to approach each other, while biased cultural transmission based on success causes the means to approach the optimum in each habitat. The impact of mixing depends on the difference in the mean marker traits, both because increasing this difference increases the covariance created by mixing and because it makes biased transmission based on similarity more effective in causing people to imitate models with more adaptive variants. Thus, increasing the difference in the mean marker traits will cause the mean adaptive trait in each habitat to move toward the optimum. This in turn

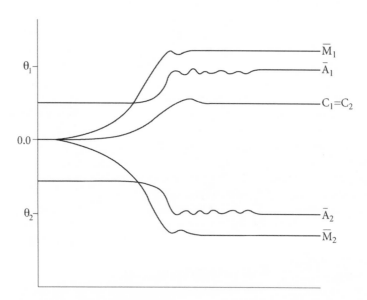

Figure 6.2. Representative trajectory of the mean value of the adaptive character, the marker character, and the covariance in the two habitats.

will cause the mean marker traits to diverge. This positive feedback cycle will come to a halt only when the mean adaptive trait stops changing, which occurs when the adaptive trait in each habitat is at the optimum.

These results suggest that subdivision of a population into culturally semi-isolated groups based on arbitrary symbolic traits such as dialect can result from using the success and similarity choice rules. The same analysis also indicates that the tendency to imitate similar individuals can be genetically adaptive. Consider an individual who does not use similarity as a criterion in weighting potential models for the adaptive trait. On average, such an individual will acquire a value of the adaptive trait that is farther from the optimum in his or her habitat than an individual who does use similarity. If, as we have assumed, the criteria by which success is judged are correlated with reproductive success, then individuals who use similarity to bias their enculturation will have higher fitness than those who do not. If one further assumes that the nature of the imitative process is affected by heritable genetic variation, then natural selection will give rise to a cultural transmission system that is biased in favor of imitating culturally similar individuals.

Testing the Model

The modeling exercise sketched here tells us only that the posited forms of biased cultural transmission *could* give rise to ethnically differentiated populations. How can we find out whether the posited mechanisms have anything to do with the actual formation of ethnic groups in the real world? Such an extremely simple model cannot be expected to produce precise numerical predictions that can be sharply tested in the fashion of physics. However, empirical data can be brought to bear on the veracity of the model in two different ways: First, one can investigate whether the basic individual-level processes assumed in the model are reasonable. Are ethnic markers typically acquired at an early age compared to other cultural traits? Do people use success and similarity as criteria for imitation? If the assumed processes do not capture at least part of the way that ethnicity structures cultural transmission, then the model is unlikely to be useful. Second, we can examine the model for qualitative predictions, and use comparative or historical data to test them. In this section we present three predictions that can be used in this way. Again, we can expect only qualitative predictions from the model, and only a statistical pattern of confirmation, given noisy data from a complex real world. Nevertheless, if these three predictions were to fit a significant number of empirical cases, our confidence that similarity and success rules play a substantial role in the evolution of arbitrary marker traits would increase.

1. *If two neighboring ethnic groups are of unequal size, the smaller of the groups will have more extreme values of ethnic marker traits and a higher covariance between ethnic markers and any adaptive specialty that characterizes the ethnic group.* The results shown in figure 6.2 are based on the assumption that populations living in each habitat are the same size. When the populations are different sizes,

the mean value of the marker trait diverges farther from its original state in the smaller population than in the larger one, and the covariance between marker trait and adaptive trait at equilibrium is larger. Both of these effects result from the smaller group's experiencing a greater amount of mixing than the larger group.

2. *If two groups come into contact, the larger the initial difference between them with respect to ethnic markers, the more likely they will come to adopt different ecological specialties.* It may be quite difficult for groups that are too similar to diverge so as to optimally adapt to two different environments. The rate at which two groups diverge with respect to both the marker character and the adaptive character depends critically on the initial difference between the two populations with regard to the marker character. If the mean value of the marker character is very similar in both populations, then they will diverge very slowly because the covariance created by mixing will be very small. A small covariance will, in turn, slow down the rate at which both the adaptive and marker traits in the two populations diverge, which in turn keeps the covariance small. In contrast, when the populations are initially quite different, the divergence of both adaptive and marker characters will be rapid. In the context of the model, the ultimate equilibrium states are the same. However, in the real world, in which many processes affect the spread and persistence of ethnic groups, we would expect effects that occur rapidly to be much more important. Thus, we would expect that when two ethnic groups come into contact, the chance that they will come to adopt different ecological specialties will be increased if they are initially more distinct. We would also predict that the chance that a group entering an area can displace an existing population from an ecological or economic specialization will increase if the entering group is ethnically distinct. If the groups are sufficiently similar, techniques sufficiently superior to be the basis for displacement will rapidly reach similar frequency in both groups due to the effects of migration.

3. *When ethnic groups occupy large, contiguous territories, the greatest amount of ethnic differentiation should occur at the boundary between groups.* In many cases of interest, ethnic groups occupy contiguous spatial territories. Individuals who are close to the boundary between two groups will have more encounters with members of another ethnic group than individuals more distant from the boundary. To model this situation, imagine a number of populations arranged along a transect in space, and at some point along the transect the optimal value of the adaptive character changes abruptly. For example, an abrupt change in altitude might lead to a change in rainfall regime. Finally, suppose that there is migration or mixing of individuals along the transect so that mixing is more likely among neighboring populations than among distant ones. With these assumptions, the model predicts that the degree of differentiation with respect to the marker trait is greatest at the boundary between the two environments, where mixing creates the greatest covariance between the adaptive trait and the marker trait.

We are not aware of the existence of data to test these predictions, but the information required is of a type that anthropologists are well equipped to obtain.

Discussion

The model presented here suggests that the modes of cultural transmission that give rise to ethnically subdivided populations are adaptive because they allow populations to more accurately track a heterogeneous environment. Similar processes may favor the development of symbolically marked caste, class, occupational, and professional subgroups within complex societies. The process of imitating people like oneself sets up a self-reinforcing process that causes subpopulations occupying different habitats, or pursuing different economic strategies in the same environment, to become culturally isolated. Thus, the mean value of the adaptive trait in each habitat converges to the optimum. A population using only transmission based on success would adapt much less quickly to a variable habitat.

It is noteworthy that this mode of adaptation is closed to animals that lack a capacity for culture. Such differences between genetic and cultural evolution ought to be reflected in basic differences between the natural history of humans and other animals. It is interesting that the human species occupies a much broader range of habitats than any other mammalian species. Consider the primates: if all baboons are classified as a single species, then it is the primate species with the widest geographical range, a substantial fraction of sub-Saharan Africa. Our closest relatives, chimpanzees and gorillas, are restricted to the tropical forests of Africa. In contrast, even with only hunting and gathering technology, humans occupied virtually every terrestrial habitat.

Most contemporary theories of speciation hold that a population must occupy more than one ecological niche in order for speciation to occur (Templeton, 1981). Once a portion of a population has adapted *genetically* to a particular niche, selection will favor mechanisms that prevent mating with individuals living in some other niche, because the offspring that result from such matings will be inferior in both niches. Whether multiple niches are sufficient, or some additional factor such as an isolating barrier is necessary, is not completely clear. The data from other primate species suggest, however, that typical primate species occupy much smaller ranges than the human species, presumably because reproductive barriers were favored by selection as successful primates extended their ranges to sufficiently different habitats.

Unlike other mammals, humans acquire massive amounts of adaptive information culturally. Perhaps it is not coincidental that symbol-using humans of the late Pleistocene epoch became very widely distributed for a biological species. The processes modeled here, by allowing the protection of culturally transmitted adaptations to local conditions without genetic isolation, can be considered a cultural substitute for speciation. Undoubtedly many aspects of cultural transmission allow adaptation to a wide range of habitats. However, it does seem plausible that the fact that the human species is divided into distinct groups that are culturally isolated from each other may play a role in allowing humans to be culturally polymorphic and thus to occupy such a wide range of ecological niches. This intuition is reinforced by studies like those of Fredrik Barth, which suggest that contemporary ethnic groups often occupy different ecological niches.

This interpretation illustrates, in the context of a rather simple model, how adaptive modes of cultural transmission lead to outcomes that could not be predicted without taking cultural processes explicitly into account. Even if one assumes that the criteria by which success is judged are coincident with reproductive success, only the properties of cultural transmission allow populations to adapt rapidly to a variable environment. An adaptive outcome—the differentiation of local groups with regard to marker traits—can be understood only in terms of cultural processes. We believe that this argument ought to be very interesting to cultural anthropologists. *We have not had to leave the confines of adaptationist assumptions to show how the properties of culture play a fundamental role in human evolution.*

However, once the use of such rules as success and similarity arise, selection on genes underlying the capacity for culture may not be able to prevent the violation of adaptationist assumptions. For example, processes closely related to those modeled here can lead to the "runaway" evolution of marker and preference traits, which have no adaptive or functional explanation (Boyd and Richerson, 1985, ch. 8). It is easy to imagine that the adaptive uses of cultural markers are common enough so that selection on genes maintains a cognitive capacity to use them despite the runaway process carrying some to maladaptive extremes. We are convinced that complexities of this sort are a pervasive feature of the coevolutionary process that links genes and culture. If this idea is correct, any attempt to reduce the problems of human evolution to binary choices between sociobiological and cultural explanations is bound to fail. The real puzzle is to determine how the genetic and cultural systems interact in a unified evolutionary process.

NOTE

We thank Bruce Knauft, Robert Paul, and Joan Silk for thoughtful comments on the first draft of this chapter.

REFERENCES

Barth, F. 1969. *Introduction. Ethnic groups and boundaries.* F. Barth, ed. Boston: Little Brown.

Boyd, R., & P. J. Richerson. 1985. *Culture and the evolutionary process.* Chicago: University of Chicago Press.

Boyd, R., & P. J. Richerson. 1987. The evolution of ethnic markers. *Cultural Anthropology* 2:65–79.

Campbell, D. T. 1975. On the conflicts between biological and social evolution and between psychology and moral tradition. *American Psychologist* 30:1103–1126.

Cavalli-Sforza, L. L., & M. W. Feldman. 1981. *Cultural transmission and evolution.* Princeton, NJ: Princeton University Press.

Knauft, B. M. 1987. Divergence between cultural success and reproductive fitness in preindustrial cities. *Cultural Anthropology* 2:94–114.

Nisbett, R., & L. Ross. 1980. *Human inference: Strategies and shortcomings of social judgment*. Englewood Cliffs, NJ: Prentice-Hall.

Rogers, E. M., with F. F. Shoemaker. 1971. *The communication of innovations*: A cross-cultural approach. New York: Free Press.

Rosenthal, T., & B. Zimmerman. 1978. *Social learning and cognition*. New York: Academic Press.

Sahlins, M. 1976. *Culture and practical reason*. Chicago: University of Chicago Press.

Templeton, A. 1981. Mechanisms of speciation—A population genetic approach. *Annual Review of Ecology and Systematics* 12:23–48.

7   Shared Norms and the Evolution of Ethnic Markers

With Richard McElreath

Unlike other primates, human populations are often divided into ethnic groups that have self-ascribed membership and are marked by seemingly arbitrary traits such as distinctive styles of dress or speech (Barth, 1969, 1981). The modern understanding that ethnic identities are flexible and ethnic boundaries porous makes the origin and existence of such groups problematic because the movement of people and ideas between groups will tend to attenuate group differences. Thus, the persistence of existing boundaries and the birth of new ones suggests that there must be social processes that resist the homogenizing effects of migration and the strategic adoption of ethnic identities.

One recurring intuition in the social sciences is that, since ethnic markers signal ethnic group membership and ethnic groups are often loci of cooperation, markers persist because they allow people to direct altruistic behavior selectively toward coethnics (Van den Berghe, 1981; Nettle and Dunbar, 1997). On closer analysis, however, this argument turns out not to be cogent. Altruism can evolve only if some cue allows altruists to interact with each other preferentially so that they receive a disproportionate share of the benefits of altruism. One such cue is kinship (Hamilton, 1964), and another is previous behavior (Trivers, 1971; Axelrod, 1984). Another idea is that selection might favor altruists who carried an external, visible marker that would allow them to limit their cooperation to others who exhibited the marker. However, evolutionary theorists argue that this mechanism is unlikely to be important (Hamilton, 1964; Grafen, 1990). Nonaltruists with the marker do best because they get the benefit without paying the cost. Thus, if any process breaks up the association between the cooperator strategies and the markers, such individuals will rapidly proliferate and altruists will disappear.

Here we argue that markers function to allow individuals to interact with others who share their social norms. We present a simple mathematical model showing that marked groups can arise and persist if three empirically plausible conditions are satisfied: (1) Social behavior in groups is regulated by norms in such a way that interactions between individuals who share beliefs about how people should behave yield higher payoffs than interactions among people with discordant beliefs. (2) People preferentially interact with people with whom they share easily observable traits like dress style or dialect. (3) People imitate successful people, with the result that behaviors that lead to higher payoffs tend to spread. We also show that the preference to interact with people with markers like one's own may be favored by natural selection under plausible conditions. We conclude by outlining several qualitative, empirically testable predictions of our model.

A Simple Model of the Evolution of Ethnic Markers

Consider a population divided into a number of large groups. In each time period, each individual interacts with another individual from the same group. People's behavior in these interactions depends on culturally acquired beliefs. We will refer to this culturally transmitted belief as the *behavioral trait*. There are two alternative beliefs, labeled 1 and 0. Individuals' payoffs from the social interaction depend on their own behavior and the behavior of their partners in the way given in table 7.1. This simple coordination game is meant to capture the intuition that many real social interactions go well if people have the same beliefs about proper behavior. It is likely that human societies face many problems of this kind. An example familiar to many of us is the one of problems in cross-cultural communication that result from different expectations about interactions and codes for communicating (Gumperz, 1982). The parameter δ measures the strength of this effect.

We also assume that it is difficult to determine another individual's beliefs about proper behavior before an interaction occurs. Given the large number of norms and the fact that some of them will be used only a few times in one's lifetime (Nave, 2000), people cannot always reliably predict the behavior of everyone they must interact with or even predict their own behavior, since many such norms are unconsciously held. Much the same argument can be made for rules enforced by third-party punishment. A stranger who moves to a new village

Table 7.1. Payoffs in the coordination game

Player 1's behavior	Player 2's behavior	
	1	0
1	$1+\delta$	1
0	1	$1+\delta$

Note: Payoffs shown for player 1: δ is assumed to be positive.

cannot guess ahead of time all of the social rules that regulate behavior in his new home. People may be able to tell him some of the things that he needs to know, but it is still likely that he will make many costly social blunders, perhaps even run afoul of basic moral principles (field anthropologists should be familiar with this sort of problem). As long as people are sometimes ignorant in these ways, people with uncommon behaviors will be at a disadvantage, and the model targets these situations, not the entire scope of interaction.

Of course, people have many traits, such as dialect, clothing style, and cuisine, that *can* be observed, and often these traits are the basis of assortative social interaction. To formalize this idea, we assume that there is also a readily observable *marker trait*. This trait also has variants, labeled 0 and 1, and we assume that individuals tend to interact with others who have the same variant of marker trait. The strength of this propensity is given by the parameter e. When $e = 1$, individuals interact at random; when $e = 0$, they always interact with someone with the same marker trait.

There is much evidence that people who do well in life are more likely to be imitated (Henrich and Gil-White, 2001). To incorporate this process, we assume that the probability that an individual with behavior i and marker j will be imitated is proportional to $W_{ij}/\overline{W}$, where $\overline{W}$ is the average payoff in the group. This means that combinations of behavior and marker that lead to higher than average payoffs will be more likely to be imitated (see Gintis, 2000, for derivation).

With these assumptions it is possible to derive expressions that describe how imitation and social interaction change the frequency of the behavior and marker traits in each group. The change in the fraction of the people with marker 1 within a group, p_1, is

$$\Delta p_1 = \delta U\{(p_1 - p_0)(1 - (1 - e))R^2\} \tag{1}$$

where $R\{= D/(UV)^{1/2}\}$ is the correlation of behavior and marker, U and V are the variances of behavior and marker, and D is the covariance between marker and behavior. If $R = 1$, everyone who has marker 1 also has behavior 1; if $R = -1$, then everyone who has marker 1 has behavior 0, and if $R = 0$, the traits are randomly associated. Equation 1 says that if more individuals use behavior 1 than behavior 0, it increases; if fewer individuals use it, it decreases. The rate at which this occurs depends on whether the marker allows individuals to interact preferentially with people who have the same behavior. When R^0 is near 1, most individuals with a given behavior have the same marker, and if e is small, they almost always interact with individuals with the same behavior as themselves, and thus there is little advantage in having the common behavior. When R^2 is near zero, most interactions occur at random and individuals with the most common behavior have an advantage.

The change in frequency of the marker 1, q_1, is approximately given by equation (2):

$$\Delta q_1 \approx 2\delta D(p_1 - p_0)\left(1 - \frac{e}{2}\right) \tag{2}$$

This expression is valid when the covariance between marker and behavior is small—when individuals' markers predict little about their behavior. When D is positive, marker 1 is associated with behavior 1, and if behavior 1 increases, so does marker 1. The complete expression for the change in q_1 shows that this effect decreases as D becomes larger.

Because the effects of social interaction and learning depend critically on the covariance between behavior and marker (D), we also need to know how they affect the covariance. Social interaction and imitation increase covariance between marker and behavior when the covariance is small. The reason is simple: individuals with the most common combinations of behavior and marker are more likely to interact with others with the same behavior and thus achieve a higher payoff.

We then represent population mixing due to intermarriage, relocation, and other factors with a migration phase that removes a proportion m of each group and replaces it with migrants drawn from neighboring groups. Clearly, such mixing will reduce the differences in the frequencies of both behavior and marker between neighboring groups. However, migration also has a less obvious and very important effect: as long as there is any difference in the frequencies of marker and behavior between neighboring groups, migration increases the covariance between marker and behavior within groups:

$$\Delta D = m\{\overline{D} - D + (p_1 - \bar{p}_1)(q_1 - \bar{q}_1)\} \tag{3}$$

where $\bar{p}_1$, $\bar{q}_1$, and $\overline{D}$ are the average frequencies of behavior and marker and the covariance between behavior and marker in neighboring groups that provide immigrants. To understand why mixing increases the covariance within groups, consider the case in which the frequency of marker and behavior is 0.9 in one group and 0.1 in a second group. Further suppose that the covariance between marker and behavior within both groups is zero, and therefore the marker is useless as a predictor of behavior. Now suppose that we mix the two groups completely. Most of the individuals coming into the first group will carry both marker and behavior 0, while those coming into the second will carry both marker and behavior 1. The frequency of both markers and both behaviors will be 0.5, but most (82%) of the individuals in the population will be either 1,1 or 0,0, with the result that markers are now good predictors of behavior within groups.

Finally, suppose that individuals sometimes acquire marker and behavior traits from different individuals, which leads to the randomization of behavior and marker—a process we term *recombination*. Recombination has no effect on the frequencies of behavior and marker, but it reduces the covariance between marker and behavior at a rate proportional to r.

Simulation Results

We have derived recursions that give the net effect of imitation, migration, and recombination on the frequencies of behavior and marker and the covariance between them. However, these recursions are too complex to solve analytically, and we have, therefore, relied on numerical simulation. We begin by describing simulations of the model when there are only two interacting populations. This system provides an intuition for the processes that sometimes give rise to marked groups. We then explore the parameter space of the model, varying e (the chance of interacting at random), m (migration), δ (the effects of social behavior on individual welfare), and r (the rate of recombination) to map the range of conditions under which marked groups arise. Finally, we generalize the model, allowing larger numbers of populations and a general coordination game structure. These analyses suggest that the simple model is relatively robust.

1. *Stable behavioral differences between groups usually become ethnically marked.* Social interaction alone can lead to the evolution of stable differences in behavior between two groups. People with more common behaviors achieve higher payoffs in the coordination game and are more likely to be imitated. Thus, if one behavior is initially common in one group and the alternative behavior is initially common in the other group, payoffs from social behavior coupled with imitation of the successful will cause the groups to become more different. If the diversifying effect of payoff-biased imitation is sufficiently strong compared with the homogenizing effect of migration, the two populations will reach an equilibrium at which behavior 1 is common in group 1 and behavior 0 in group 2. In contrast, if the rate of mixing is too high or if initially the same behavior is common in both populations, only one behavior will be present in both populations at equilibrium.

If stable behavioral differences between groups exist, each behavior can become associated with a different marker variant—behavior 1 will, for example, be associated with marker 0 and behavior 0 with marker 1. Figure 7.1 illustrates this dynamic. Initially behavior 1 is more common in population 1 and less common in population 2. Marker 0 is initially more common than marker 1 in *both* populations but relatively more common in population 2 than in population 1.

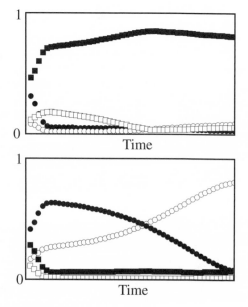

Figure 7.1. The frequencies of each of the four combinations of behavior and marker over time in each of two populations for $m = 0.025$, $e = 0.25$, and $r = 0.1$. The behaviors are denoted by the shape of the symbol, circle ($=0$) or square ($=1$), and the markers are denoted by color, black ($=0$) or white ($=1$). Initially behavior 1 (squares) has frequency 0.55 in population 1 and 0.45 in population 2. Marker 0 (black) is initially more common than marker 1 in both populations but relatively more common in population 1 ($q_{11} = 0.8$) than in population 2 ($q_{12} = 0.7$).

There is no initial covariance within populations. At first, rare-type disadvantage causes behavior 1 to become more common in population 1 and behavior 0 in population 2. At the same time, migration generates a negative covariance between marker and behavior so that behavior 1 tends to co-occur with marker 0 and marker 0 with behavior 1. This in turn strengthens the forces increasing the differences between the populations in frequencies of marker and behavior, which then generates greater covariance. This positive feedback process (figure 7.2) continues until a symmetrical equilibrium is reached at which a different behavior is common in each population and each behavior is associated with a different marker. The adaptive behaviors have become symbolically marked, even though the same marker was initially common in both groups.

However, migration and recombination oppose the positive feedback process described. Migration tends to make the two populations the same, equalizing the frequency of the markers in each population, and recombination destroys the covariance between marker and behavior. If recombination is strong, it dissipates the covariance between marker and behavior more rapidly than migration and imitation can create it. Even though the payoff advantage of being in the majority is sufficient to maintain behavioral differences between the two populations, these differences do not become ethnically marked. When individuals are unable to assort accurately on the basis of markers (e is large), the pattern is similar: stable group differences in behavior may emerge and persist, but selection on markers is too weak to generate covariance between marker and behavior.

The qualitative arguments are supported by systematic sensitivity analysis. We determined the range of parameters under which groups become marked by performing a large number of simulations. For each simulation we calculated the

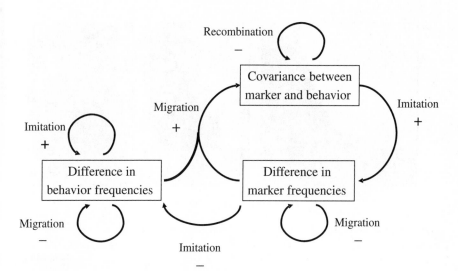

Figure 7.2. The feedback process that generates marked groups and the forces that oppose this process.

value of $\overline{D}$, the population average covariance between behavior and marker, averaged over the 100 simulations. We held parameter values constant at $m = 0.01$, $e = 0.3$, $r = 0.01$, $\delta = 0.5$ for parameters not varied in a run of simulations. Figure 7.3 summarizes these results. When biased imitation can maintain stable behavioral differences in the face of migration, stable marker differences evolve provided that (1) recombination (r) is not too strong and (2) individuals interact sufficiently often with individuals like themselves (e is not too high). There are no cases in which behavioral differences fail to evolve and marker differences manage to become stable.

2. *Spatial structure is needed to generate ethnic markers but not to maintain them.* Migration between groups generates the initial covariance essential for the evolution of ethnic markers. However, if individuals are able to use markers to assort accurately ($e \approx 1$), spatial structure is no longer necessary to maintain ethnic markers once such covariance arises (figure 7.4) and groups end up mixed together in space, but high covariance between markers and behaviors remains. This configuration can be a stable equilibrium only if r and e are very small. However, for somewhat larger values of r and e, there is a long transition period during which two ethnically marked types are present without spatial variation. A more complex model in which groups occupied different niches would likely be able to sustain spatially mixed ethnically marked groups in a wider range of circumstances. Also, we will demonstrate later that natural selection would reduce values of r and e if at all possible. This makes the possibility of the evolution of such spatially blended systems more likely. Such situations are an interesting and unexpected outcome of our model.

3. *Increasing the number of populations increases the range of initial conditions that give rise to ethnic markers.* Random starting conditions (random frequencies

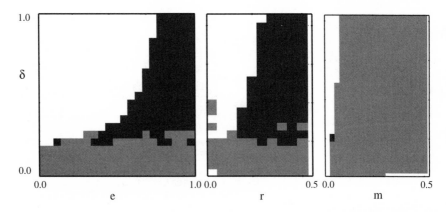

Figure 7.3. The evolution of stable marker differences. White regions are combinations of parameter values that produced both stable behavioral and marker differences (that is, these populations became ethnically marked). Black regions are cases in which behavioral differences were stable but marker differences were not (that is, these populations became culturally different but without ethnic markers). Gray regions are cases in which behavioral differences failed to evolve, typically because of strong migration.

of behavior and marker in each group) often lead to the evolution of behaviorally different and marked groups, and this result becomes more likely as more groups are added to the system (figure 7.5). The two-group system is most sensitive to starting conditions, as this case has the highest chance of randomly generating all groups with similar initial behavior frequencies.

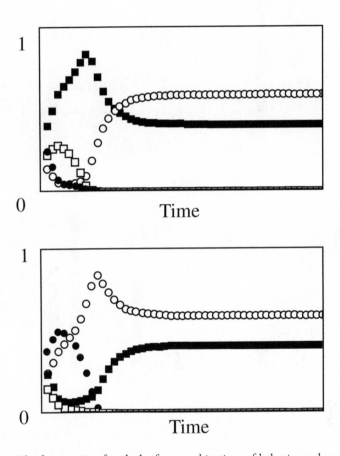

Figure 7.4. The frequencies of each the four combinations of behavior and marker over time in each of two populations. The behaviors are denoted by the shape of the symbol, circle ($=0$) or square ($=1$), and the markers are denoted by color, black ($=0$) or white ($=1$). The initial conditions and value of m are the same as in figure 7.1, but now assortment is perfect, $e=0.0$, and there is no recombination, $r=0.0$. As before, at first rare-type disadvantage causes the behavior 1 to become more common in population 1 and behavior 0 in population 2, and migration generates a negative correlatiion between marker 1 and behavior 0 (equation 4). However, because there is no recombination, this covariance builds up much more rapidly, especially in population 1, in which the initially relatively more common marker was also absolutely more common. The high correlation between marker and behavior combined with the accurate assortment eliminates rare-type disadvantage, and migration mixes the two groups until they are identical. Because the covariance increases more rapidly in population 1, the marker-behavior variant in population 2 experiences a transient advantage that is preserved at equilibrium.

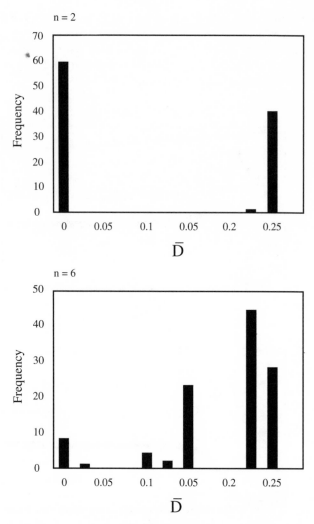

Figure 7.5. Equilibrium absolute values of $\overline{D}$ (covariance in the population as a whole) for simulations involving two groups (top, 100 simulations) and six groups (bottom, 100 simulations). Starting conditions were random with parameter values $m = 0.025$, $r = 0.10$, $e = 0.30$, $\delta = 0.50$. High $\overline{D}$ becomes more likely as the number of groups increases.

4. *Group differences are strongest at boundaries.* When more than two groups are arrayed in space, the correlation between marker and behavior $(R = D_k / \sqrt{U_k V_k})$ is greatest at the boundaries between culture areas. Figure 7.6 shows the steady state in ten populations arranged in a stepping-stone ring. This steady state results from an initial clinal distribution of behavior and marker frequencies with zero correlation between behavior and marker in each popu-lation. There is a region of three populations in the middle in which the

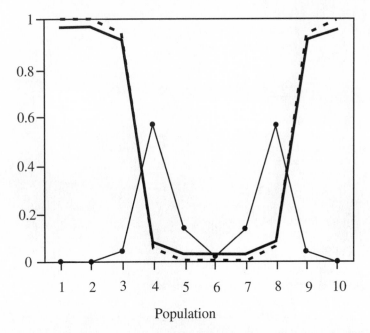

Figure 7.6. The steady state that arises from slightly clinal initial distributions of the frequencies of marker 1 and behavior 1 in ten populations arranged in a ring. Broken line, p_1; heavy solid line, q_1; light solid line, R.

frequency of marker 1 and behavior 1 is low and a region of three populations at the edges in which these frequencies are high (remember that the populations wrap around so that population 1 exchanges migrants with population 10). In both of these regions there is little or no correlation between marker and behavior. In between these regions are boundary areas in which frequencies are intermediate and there is substantial correlation between marker and behavior.

5. *A more general model of social interaction leads to similar results.* So far, we have assumed that social interaction can be modeled by a game of pure coordination with equal average payoffs for both equilibria. Symmetric, pure co-ordination games are very special because the basins of attraction of the two equilibria are the same size. To test whether our results were sensitive to this assumption, we ran a number of simulations in which we varied the parameters of the completely general two-person coordination game shown in table 7.2.

The results indicate that the system regularly evolves toward marked, be-haviorally distinct groups even when there are large deviations from the perfect coordination structure. Thus, our results do not depend in a sensitive way on the perfect nature of the game structure we have chosen. This suggests that any stable behavioral equilibria, regardless of their relative consequences for group or individual welfare, may become marked.

Table 7.2. Payoffs in a general two-person game with two stable equilibria

| | Player 2's behavior | |
Player 1's behavior	1	0
1	$1 + \delta + g$	$1 - h$
0	1	$1 + \delta$

Note: Payoffs shown for player 1; δ, g, and h are assumed to be positive.

Evolutionary Stability of the Parameters

This model depends on four parameters: m, δ, r, and e. The first two formalize assumptions about the ecology of the evolving populations. The second pair of parameters represents assumptions about human psychology. The simulation results indicate that social interactions in which common behaviors have high payoff will lead to the evolution of ethnic markers if both e and r are small, or, in other words, if people have a psychology that predisposes them to interact with individuals with the same marker as themselves and to acquire some markers and behaviors as a package. Natural selection will, all other things being equal, favor such a psychology (that is, selection will favor mutations that reduce the values of e and r). However, selection on other aspects of social learning and demands on interaction may restrict the extent to which selection can reduce these parameters.

Discussion

We have argued that ethnic markers do not function to allow individuals to direct altruism to others like themselves because such a system cannot resist invasion by cheaters who signal altruistic intent but then do not deliver. In contrast, ethnic markers can signal one's behavioral type when social interactions have a coordination structure because in such situations there is nothing to be gained from cheating. Both parties in the coordination setting gain the most when they honestly advertise their strategy, and as a result both the behavior and its advertisement spread when the successful are imitated. Axtell, Epstein, and Young (1999) have analyzed another model that is quite different structurally but works for similar reasons.

The intuition that ethnic markers and cooperation are related is not, however, without merit. Humans are peculiar in that we often cooperate with large numbers of unrelated individuals. As we have argued, the existence of ethnic markers alone cannot explain the scale of human cooperation. Yet we have shown that markers may evolve when individuals interact in a two-person coordination game, and we believe that any process that leads groups to occupy multiple stable equilibria may produce the same result. Two of us have argued at length elsewhere that human cooperation results from norms enforced by socially created rewards and punishments (Boyd and Richerson, 1990, 1992; Soltis, Boyd, and Richerson, 1995; Richerson and Boyd, 1998, 1999). If punishment is sufficiently costly, such systems can stabilize a very wide range of behavior. Then, competition between groups will lead to the spread of moral systems that enhance group survival, welfare, and expansion, including norms that lead to enhanced cooperation in economic and military activities.

As a result, we expect that systems of moral norms, some of which create group-beneficial cooperation, should come to be marked by ethnic markers by the process described. Punishment transforms the prisoner's dilemma structure of a cooperation problem into a coordination structure. The process we have described here can then lead to individuals selecting individuals with whom to cooperate on the basis of markers, but the markers themselves do not stabilize the cooperation.

Corollaries and Predictions

The goal of this kind of modeling study is to demonstrate the cogency of a deductive argument linking assumptions about microlevel social interactions to the empirically observable macrolevel social patterns that result. Accordingly, we conclude by describing several testable predictions of the model.

Our analysis of the evolutionary stability of e and r makes two predictions about the psychological tendencies of human beings:

1. *Individuals in marked communities should prefer interaction with similarly marked individuals.* Our analysis of the evolution of e, the rate at which individuals interact at random with respect to markers, suggests that natural selection or an analogous process operating on cultural rules for interaction should reduce e to zero, if possible. Thus, to the extent that e represents a psychological bias toward interacting with those who look like oneself rather than the ability or freedom to interact with ones like oneself, we expect members of marked communities to prefer individuals marked like themselves, at least when it comes to coordination interactions.

2. *Individuals in marked communities should acquire bundles of at least some norm and marker traits.* While the model does not suggest anything about the social learning of noncoordination behaviors and social markers, our analysis of the evolution of r, the rate of recombination of behavior and marker traits, predicts that, for our model to be relevant, individuals should acquire norm and marker traits as a bundle. They should also preserve these associations throughout substantial portions of their life spans. If this is not true, the process we describe here is unlikely to work.

The model makes three clear predictions about the nature of the distributions of marker traits and their relations to ethnic groups and their histories:

1. *Ethnic differences should be stronger at boundary regions than deep within ethnic territories.* Hodder (1977) suggests that this is true for some ethnoarcheological data from the Lake Baringo region of Kenya, but the data are

inadequate to test this prediction. The appropriate test would be examination of a large ethnic group, such as the Kikuyu of Kenya, which interacted at many border areas with a number of different ethnic groups. Another setting that holds promise for testing this prediction is fragmentary migration that brings smaller units of a larger ethnic population into contact with other ethnic groups. If these groups are on average more marked than their source populations, we may be able to conclude that interaction with the other ethnic groups has increased selection on markers and magnified initial differences in those settings.

2. *Norm and marker boundaries should coincide, while the distributions of other culture items may map onto one another differently.* Our model makes no predictions about the nature of all cultural traits and the distribution of ethnic markers. However, if this model is correct, a number of norm differences—on beliefs in inheritance, child rearing, household labor, and other categories of human life in which there are multiple coordinated solutions to the same problem—should correspond to the distributions of marker differences.

3. *Potential marker traits with the greatest initial differences should become marked first.* One test of this prediction would be to examine ethnographic settings in which two isolated source populations have contributed migrant groups that have since been in contact for some time. The source populations provide estimates of the initial differences in the migrant groups when they came into contact. The migrant groups provide estimates of the differences that might have grown from those initial differences. This prediction will earn support if the traits with greater differences between source populations appear to have led to marked traits in the contact groups.

NOTE

Supplementary material appears in the electronic edition of *Current Anthropology* 44 (2003) on the journal's web page (http://www.journals.uchicago.edu/CA/ home.html).

REFERENCES

Axelrod, R. M. 1984. *The evolution of cooperation.* New York: Basic Books.
Axtell, R. L., J. M. Epstein, & H. P. Young. 1999. The emergence of economic classes in an agent-based bargaining model. In: *Social dynamics.* S. Durlauf, & P. Young, eds. Santa Fe: Santa Fe Institute.
Barth, F. 1969. Introduction. In: *Ethnic groups and boundaries,* F. Barth, ed. Boston: Little, Brown.
Barth, F. 1981. *Features of person and society in Swat: Collected essays on Pathans.* London: Routledge and Kegan Paul.
Boyd, R., & P. J. Richerson. 1990. Group selection among alternative evolutionarily stable strategies. *Journal of Theoretical Biology* 145:331–342.
Boyd, R. & P. J. Richerson. 1992. Punishment allows the evolution of cooperation (or anything else) in sizable groups. *Ethology and Sociobiology* 13:171–195.
Gintis, H. 2000. *Game theory evolving.* Princeton, NJ: Princeton University Press.

Grafen, A. 1990. Do animals really recognize kin? *Animal Behaviour* 39:42–54.

Gumperz, J. J. 1982. *Discourse strategies*. Cambridge: Cambridge University Press.

Hamilton, W. D. 1964. The genetical evolution of social behaviour. *Journal of Theoretical Biology* 7:17–52.

Henrich, J., & F. Gil-White. 2001. The evolution of prestige: Freely conferred deference as a mechanism for enhancing the benefits of cultural transmission. *Evolution and Human Behavior* 22:165–196.

Hodder, I. R. 1977. The distribution of material culture items in Baringo District, western Kenya. *Man* 12:239–269.

Nave, A. 2000. Marriage and the maintenance of ethnic group boundaries: The case of Mauritius. *Ethnic and Racial Studies* 23:329–352.

Nettle, D., & R. J. M. Dunbar. 1997. Social markers and the evolution of reciprocal exchange. *Current Anthropology* 38:93–99.

Richerson, P. J., & R. Boyd. 1998. The evolution of human ultra-sociality. In: *Indoctrinability, ideology, and warfare*, I. Eibl-Eibesfeldt & F. Salter, eds. (pp. 71–95). New York: Berghahn Books.

Richerson, P. J., & R. Boyd. 1999. The evolutionary dynamics of a crude super organism. *Human Nature* 10:253–289.

Soltis, J., R. Boyd, & P. J. Richerson. 1995. Can group functional behaviors evolve by cultural group selection? An empirical test. *Current Anthropology* 36:473–494.

Trivers, R. L. 1971. The evolution of reciprocal altruism. *Quarterly Review of Biology* 46:35–57.

Van Den Berghe, P. L. 1981. *The ethnic phenomenon*. Westport, CT: Praeger.

PART 3

Human Cooperation, Reciprocity, and Group Selection

A number of years ago the Cambridge paleoanthropologist Rob Foley published a book on the evolutionary ecology of early hominins entitled *Another Unique Species*. The title was meant to capture the idea that while humans are unique in many ways, so too is every other species. We like the book very much, but perhaps the title is a bit misleading. Humans are, if you will allow us, "more unique" than any other primate. We are extreme outliers in our use of tools, in our ecological and geographical range, in the richness of our communication system, and so on and on. Perhaps the most singular feature of *Homo sapiens* is the scale on which humans cooperate. In most other species of mammals cooperation is limited to close relatives and (maybe) small groups of reciprocators. After weaning most individuals acquire virtually all of the food that they eat. There is little division of labor, no trade, and no large-scale conflict. Amend Hobbes to account for nepotism, and his picture of the state of nature is not so far off for other mammals. In contrast, people in even the simplest human societies regularly cooperate with many unrelated individuals. Sharing leads to substantial flows of food and other resources among different age and sex classes. Division of labor and trade are prominent features of every historically known human society, and archaeology indicates that such trade has a long history. Violent conflict among groups is also quite common. Since the development of agriculture 10,000 years ago, the scale of human cooperation has steadily increased so that most people on earth today are enmeshed in immense cooperative institutions like universities, business firms, religious groups, and nation states. Moreover, experimental work, both in psychology and economics, indicates that people have social preferences that incline them to such cooperation (see Fehr and

Fischbacher, 2003, for a review). In the laboratory, people behave altruistically in anonymous one-shot interactions, sometimes for very large stakes.

Thus, we have an evolutionary puzzle. At some time in the not so distant past, say 5 million years ago, our ancestors lived in small kin-based societies like other apes. Then, sometime between then and now, human psychology changed in such a way that large-scale cooperation became common. What were the evolutionary processes that gave rise to this change?

Ever since we started thinking about cultural evolution, we have thought that culture might provide the solution to this puzzle because it seems to generate lots of variation in social behavior among social groups. In other primate species there is little heritable variation among groups within a species. The behavior of groups depends on the habitat and ecology, the demographic structure, and the personalities of particular individuals. But these differences are small and ephemeral, and, as a consequence, group selection at the level of whole primate groups is not an important evolutionary force. In contrast, it is an empirical fact that there is much heritable cultural variation among human groups. Neighboring groups often have different languages, marriage systems, and property rights, and these differences persist for generations. This suggested to us that group selection might be a more important process shaping human behavior than the behavior of other animals. We have devoted quite a bit of our research effort to trying to gain a clearer understanding of this puzzle. This work is usefully divided into two parts.

Studies of cultural group selection. First, we have studied models of cultural group selection and attempted to collect empirical data necessary to determine whether the models are close to reality. We believe that the case for cultural group selection is strong.

Studies of the evolution of contingent cooperation. Many scholars in the evolutionary social science community believe that human cooperation is better explained by selection within groups that favored various forms of contingent cooperation. The idea is that during most of our evolutionary history, humans lived in small groups in which reciprocity and moralistic punishment supported cooperation. The psychological machinery that supported these behaviors "misfires" in the larger societies of the last 10,000 years. We have been skeptical about this argument because many other mammals live in small social groups, yet none of them shows very much evidence of contingent cooperation beyond pairwise reciprocity. It seemed to us that the advantages created by wider cooperation within groups like specialization, division of labor, risk spreading, and so on are huge, and lineages like ants and termites in which kin selection supports cooperation have been extremely successful. Thus, it seemed to us that if contingent cooperation could generate larger-scale cooperation, there ought to be lots of examples in nature. However, when we started thinking about this problem in the early 1980s, there was lots of work on the evolutionary theory of reciprocity among pairs of individuals, but very little about contingent cooperation in larger groups. So we undertook to develop theory in this area, and the results are reprinted here.

Studies of the Evolution of Contingent Cooperation

The modern theory of the evolution of reciprocity began in 1971 when Robert Trivers showed that contingent cooperation could be evolutionarily stable. His model goes roughly as follows: suppose that pairs of individuals interact repeatedly over time and that occasionally one member of a pair has the opportunity to provide a benefit, b, to the other at a cost, c, to itself. Now consider a population of *reciprocators* who help on the first interaction and keep helping as long as their partner helps. Trivers (apparently with help from W. D. Hamilton) showed that reciprocators can resist invasion by rare *defectors* who never help as long as the long-run benefit of mutual cooperation is greater than the short-run benefit that a defector gets by exploiting a cooperator. (Or, more formally, when $t(b-c) > b$, where t is the average number of helping opportunities for each pair of individuals.) This article has been widely cited and was the impetus for much empirical work on reciprocity.

However, there is a big problem with this analysis: when individuals interact repeatedly, reciprocity is evolutionarily stable, but so is everything else. Unbeknownst to Trivers and most other biologists working on reciprocity, game theorists in economics, political science, and mathematics had been working on the closely related problem of rational behavior in repeated games. As Trivers noted in his article, his model of reciprocity can be formalized as a repeated version of the famous prisoner's dilemma game. What Trivers apparently did not know is that by the late 1950s game theorists had proved that in a repeated prisoner's dilemma (or, in fact, in any repeated game in which players can strongly affect each others' payoffs) *any* pattern of behavior can be sustained by mutual self-interest, all cooperation for sure, but also all defection, or anything in between as long as interactions go on long enough. This important result was known as the "folk theorem" because nobody in the game theory community was exactly sure who first proved it, and though the theorem was widely known in that community, it wasn't actually published until 1986 (Fundenburg and Maskin, 1986). The basic logic of the folk theorem is simple. Suppose a strategy takes the form: do x, where x is some behavior, say alternating cooperate and defect, as long as the other guy does x. If the other guy does something else, defect forever. Once a strategy like this becomes common in a population, the only smart thing to do is x; otherwise, one will be punished by defection for the rest of the interaction. If interactions go on long enough, the costs of such punishment will exceed the short-run benefits of doing something other than x. Repeated interactions create the possibility of sanctions and any behavior that enough sanctioners are willing to sanction is an equilibrium. For the most part, the logic of the folk theorem applies to evolutionary theory, although a subtle and important difference affects the stability of punishment. We will return to this issue. The bottom line is that when everything is an equilibrium showing that reciprocity is an equilibrium too doesn't really tell you much. We need to know which equilibria are likely evolutionary outcomes and which are not.

In 1981 Robert Axelrod and W. D. Hamilton published an article in *Science* that showed that reciprocating strategies were, in fact, the most likely evolutionary outcome. Standard game theory assumes that people seek to maximize their average payoff. In evolutionary terms, this is equivalent to assuming that groups of interacting individuals are formed at random with respect to genotype. (When individuals interact at random, their actions do not change the relative fitness of other types in the population. Thus, all that matters is the effect of behavior on an individual's own fitness.) Reciprocators, or, more precisely, individuals with genes that cause them to reciprocate, are as likely to initially interact with defectors (i.e., individuals with defector genes) as are other defectors. This is not a bad assumption for a large, mobile mammal like humans, because there is ample gene flow among social groups and, to a rough approximation, individuals do interact at random. However, a better approximation is to assume a small tendency to interact with genetically similar individuals. Reciprocators are slightly more likely to interact with other reciprocators than defectors are. Axelrod and Hamilton showed that even small amounts of assortative interaction allowed reciprocal strategies to invade when rare and stabilized them when common. The reason is easy to see. When strategies interact at random, and defection is common, there is no chance that individuals carrying rare reciprocating genes will meet. So the long-run benefits associated with sustained cooperation are irrelevant. Reciprocators get exploited, and that is that. However, when there is some assortative interaction, rare reciprocators do occasionally meet, and if the long-run benefits of cooperation are big enough, even a small amount of assortment can cause the average fitness of reciprocators to exceed the average fitness of defectors. To see the strength of this effect, suppose that $b/c = 2$, helping behavior that would be favored only among full siblings. The following table calculates the amount of assortment necessary to cause reciprocating strategies to increase when rare. At even a modest number of interactions, the threshold value is very small. In dyads, a little kinship and a little repeat business can generate a lot of cooperation.

Expected number of interactions	1	3	7	15	49
Threshold value of r	.5	.25	.125	.0625	.02

Axelrod and Hamilton were also concerned that reciprocating strategies could do well in more complex social environments in which many different strategies were common. They famously championed a particular reciprocating strategy, tit-for-tat, showing that it did well in computer tournaments against a wide range of strategies. Subsequent research has shown that tit-for-tat is really not such a good strategy if individuals make mistakes. Other reciprocating strategies such as "contrite tit-for-tat" (Sugden, 1986; Boyd 1989) and "Pavlov" (Boerlijst, Nowak, and Sigmund, 1998) are really more robust. Nonetheless, their basic conclusion holds true. Given quite plausible assumptions, reciprocating strategies can increase when rare, can continue to increase under a range of assumptions, and can persist when common.

Axelrod and Hamilton's (1981) article, and most of the work that followed it, deals with reciprocity among *pairs* of individuals. Many authors interested in human behavior have assumed that the conclusions of this work can be extended to cooperation in larger groups (e.g., Trivers, 1971). We know from everyday experience that groups of people can organize contingent cooperation. Committees, sports teams and many similar groups work that way. So even though the theory applies to pairs, the general result seems to apply to larger groups. Several chapters included here resulted from checking to see if the theory of evolution of contingent cooperation applies to larger groups.

In our first effort (chapter 8), we extended the Axelrod-Hamilton analysis to groups of people repeatedly interacting in an n person prisoner's dilemma. During each interaction, individuals can cooperate producing a benefit, b/n, for all players including themselves at a cost, c, to themselves. Thus, if everyone cooperates, they achieve a long-run payoff, $t(b-c)$. As in the two-person case, however, defectors achieve a short-term payoff, now $b(n-1)/n$, by free-riding on the cooperative payoffs of others. We consider a family of reciprocating strategies that generalize tit-for-tat to larger groups. Namely, the strategy T_j cooperates on the first interaction and on subsequent interactions if j of the $n-1$ other individuals cooperated during the previous interaction. Thus, T_0 individuals always cooperate; T_{n-1} cooperate only if everyone else cooperated on the previous turn.

The equilibrium behavior of this model is qualitatively similar to the two-person case. As always, defection is evolutionarily stable. Contingent cooperation can be evolutionarily stable, but only if reciprocating strategies do not tolerate defection. A population in which the strategy T_{n-1} is common will resist invasion by rare mutant *defectors* if the long-run benefit of cooperation exceeds the short-term advantage of free-riding. However, none of the other more tolerant reciprocating strategies can resist invasion by defectors. For example, when T_{n-2}, the strategy that tolerates one defector in its group, is common, rare defectors will get the long-run benefits of cooperation without paying the cost and thus will increase in frequency. It turns out that strategies like T_{n-2} that tolerate a few defectors can persist in mixed stable equilibria with defectors, but interactions must go on for a very long time. Thus, like the two-person case, virtually any kind of behavior can be evolutionarily stable.

Our analysis of this model indicates that as groups get bigger, reciprocity becomes a much less likely evolutionary outcome. Once again, suppose that interacting groups are formed assortatively of relatives with degree of relatedness r. Then rare reciprocators using the potentially evolutionarily stable strategy T_{n-1} can invade if

$$\underbrace{(r(n-1)+1)(b/n)-c}_{\text{inclusive fitness}} + \underbrace{r^{n-1}(t-1)(b-c)}_{\text{reciprocity}} > 0$$

The first term on the right-hand side gives the inclusive fitness of rare reciprocators during the first interaction. If it is positive, cooperation pays even without reciprocity. The second term gives the increase in the fitness of reciprocators due to ongoing interactions in those groups in which reciprocation is sustained. As in the two-person case, this term increases linearly with the average number of interactions (t)—repeat business makes reciprocation pay. However, also notice that the second term decreases geometrically with group size because cooperation is sustained only in groups of all reciprocators.

Strategies supporting contingent cooperation in large groups have to achieve two competing desiderata. To be stable when common, they must be intolerant of defection; to increase when rare, there must be a substantial chance that groups will have enough reciprocators; otherwise, they can't be evolutionarily stable, as defectors will prosper. As groups get larger, this become geometrically more difficult.

A number of people have suggested (e.g., Bendor and Mookerjee, 1987) that this analysis *underestimates* the problems facing reciprocity in larger groups because contingent cooperation in large groups will be much more sensitive to errors than it is in pairs. This claim is true of the particular reciprocal strategies we analyzed, because a single error would lead to a collapse of cooperation in the group. However, we do not think that it is a robust effect because the reciprocating strategies in large groups can be modified to deal with errors in much the same way that two-person strategies can. For example, the n-person version of Pavlov would use the rule cooperate if everyone or no one cooperated on the last turn. Then an error would create universal defection, which, on the subsequent interaction, would then generate universal cooperation. Strategies analogous to *generous tit-for-tat* likely could also be designed to deal with errors in an n-person setting.

Colleagues have suggested to us that the n-person prisoner's dilemma is an extreme case because it assumes that noncooperators cannot be selectively excluded from enjoying the benefits of the cooperative act. For example, everybody gets the benefits of group defense whether they fight or not. Indeed, economists say that such goods are not "excludable." Perhaps in many instances of cooperation in groups, noncooperators can be excluded. Take the classic example of food sharing among hunter gatherers. In most foraging groups, successful hunters share their catch with the rest of their group, a behavior sometimes explained as a reciprocal arrangement that reduces risk of starvation. Couldn't earnest hunters easily exclude guys who don't hunt? Just don't give them a share of meat. Don't we need to consider models in which the fruits of cooperation are at least partly excludable? Maybe, but the problem is a little trickier than it first appears.

Excluding defectors is an example of a much more general phenomena. To prevent a defector from eating, somebody has to intervene when he reaches into the pot. That someone has to undertake a (perhaps) costly action that reduces the payoff of the defector and thus produces a benefit to the group as a whole. This is an example of what Trivers called "moralistic punishment" and applies to a much wider range of problems than excluding defectors from the fruits of cooperation. Even if the defectors cannot be excluded, punishment can create incentives for them to cooperate. Cowards may get the benefits of group defense, but they may also be shunned, beaten, or banished. The real question is under what conditions can selection favor moralistic punishment?

In chapter 9 we attempt to answer this question. The model assumes that individuals interact repeatedly in an n-person prisoner's dilemma. After each interaction, members have the opportunity to punish any other member of the group at a cost to themselves. We analyzed a variety of strategies, but here we begin by focusing on just two of them: moralistic punishers cooperate and punish defectors, and reluctant cooperators defect until they are punished, and then they cooperate. So that punishment could induce cooperation, we assume that the cost of being punished is greater than the cost of cooperating. Both types occasionally make mistakes and defect when they mean to cooperate. In this simple world, there are three types of stable equilibria. First, suppose reluctant cooperators are common in the population. They neither cooperate nor punish, so they achieve a payoff of zero. Rare mutant punishers will punish the $n - 1$ reluctant cooperators in their group and thereby induce them to cooperate over the long run. If the long-run benefit of being in a cooperative group is less than the one-time cost of punishing, reluctant cooperators are an ESS. However, if the long-run benefit is greater than the cost of punishing, moralistic punishment can invade even when groups are formed at random. The fact that the reluctant cooperators do better than the moralistic punishers in their group is unimportant when moralistic punishers are rare because the vast majority of reluctant cooperators are in groups without a punisher. As moralistic punishers increase in frequency, however, more and more reluctant cooperators find themselves in groups with a punisher, and as a consequence their relative fitness increases. Eventually the fitness of the two types equalizes at a stable polymorphic equilibrium at which the population is a mix of cooperative and noncooperative groups. At this equilibrium, cooperation arises as a consequence of private individual benefit. We jokingly referred to this as the "big man" equilibrium after the famous political/economic system common in New Guinea that it resembles. This model also has a second, quite different kind of equilibrium. Suppose that moralistic punishers are common. Now rare reluctant cooperators are always punished by every other member of their group during the first interaction, and as long as the cost of this punishment is less than the cost to moralistic punishers of punishing the occasional error, then punishment can sustain cooperation. However, it can also stabilize almost any other behavior. The long-run benefits of cooperation are irrelevant to the stability of this equilibrium. This is the folk theorem again. If almost everybody is going to punish individuals for some transgression then individuals must do what they want, no matter how foolish it is in any other terms.

We think these two very simple models capture a robust difference in contingent cooperation and moralistic punishment. Contingent cooperation strategies can be stable only if they insist that everyone in the group cooperate—otherwise, they can be exploited. However, since such strategies increase when rare with the greatest difficulty, they are not very likely evolutionary outcomes. Defecting equilibria are much more likely evolutionary outcomes. The directed punishment of moralistic strategies means that a small number of punishers can induce others to cooperate and thus achieve the long-run benefits of cooperation. If punishment is cheap enough that a single individual can induce all other group members to cooperate, then moralistic strategies can increase when rare. However, they can never spread to fixation precisely because only a few punishers are necessary, and as punishers become common, selection favors free riders who accept the benefits but don't do the police work necessary to generate them. We are quite doubtful that this kind of equilibrium is common in human groups. As Hobbes pointed out long ago, individual men have a similar capacity for inflicting harm. When I push you away from the food, you are likely to push back (weapons probably reduce differences in fighting ability—God created men, but Sam Colt made them equal, frontiersmen quipped). This problem does not afflict moralistic equilibria because defectors are rare and punishers are common. However, while moralistic punishment is stable, within-group evolutionary processes do not make it a likely evolutionary outcome. The fact that directed punishment requires only a few punishers is also responsible for the peculiar nature of moralistic equilibria. When moralistic punishers are common, mutant non-punishers have no effect on whether the group cooperates—all groups will be cooperative because there are plenty of punishers everywhere. Thus, while such equilibria are stable, individual natural selection has no reason to attach such punishment to group-beneficial cooperative behaviors.

The fact that there are always more than enough punishers at a punisher-cooperator equilibrium means that such equilibria can be invaded by "second-order free riders," individuals who cooperate from the first interaction but never punish. While much of the debate about moralistic punishment has focused on the problem of second-order free riders, we don't think it is a serious obstacle to evolution of cooperation in large groups. First of all, "metapunishment" can evolve, the punishment of nonpunishers. As we show in chapter 9, this can stabilize punishment. Many people believe that meta-punishment doesn't actually occur in real human societies. However, even if this is the case, other solutions to the second-order free rider problem are possible. If moralistic punishment is common, and punishments sufficiently severe, then cooperating will pay. As a result, most people may go through life without having to punish very much. On average, having a predisposition to punish may be cheap compared to a disposition to cooperate (in the absence of punishment). Thus, relatively weak evolutionary forces can maintain a moralistic predisposition. This argument is elaborated in chapter 10 in which it is shown that very small amounts of conformist social learning can stabilize moralistic punishment against second-order free riders, and in chapter 13 in which we show that group selection can also stabilize punish-ment. Finally, as Eric Smith and his colleagues have pointed out (Smith and Bliege Bird, 2000), punishing could be used to signal hard-to-observe personal qualities, giving punishers a private reward in the mating game, for example.

Cultural Group Selection

When we were graduate students during the late 1960s and early 1970s, it was quite common for biology texts to explain observed traits in terms of their benefit to the population or even the species. Reduced reproductive rates prevented overpopulation, and sexual reproduction maintained genetic variation necessary for the species to adapt. A key advance in biology over the last 40 years was to show that such explanations are mostly wrong. Natural selection does not normally lead to the evolution of traits that are for the good of the species, or population. With some interesting exceptions, selection favors traits that increase the reproductive success of individuals, or sometimes individual genes, and when there is a conflict between what is good for the individual and what is good for the species, or population, selection usually leads to the evolution of the trait that benefits the individual.

Many people mistakenly believe that this means that group selection is never important. In the early 1970s, an eccentric engineer named George Price published two articles (1970, 1972) that presented a genuinely new way to think about evolution. Price showed that selection can be thought of as a series of nested levels: among genes within an individual, among individuals within groups, and among groups. He discovered a very powerful mathematical formalism, now called the "Price covariance equation," for describing these processes. To keep things simple, let's suppose that there are two levels. Then the change in frequency of a gene undergoing selection is given by

$$\Delta q = V_G \beta_G + \overline{V_W \beta_W}$$

The first term gives the change due to selection between groups and is the product of the variance in frequency between groups (V_G) and the effect of a change in the frequency of the gene on the reproductive success of the group (β_G). This makes sense: β_G gives the effect of a change in gene frequency on group success, and V_G measures how different groups are. The second term, which gives the change in frequency due to changes within groups, has a similar form. It is the average over all groups of the product of the variance in frequency among individuals within the group (V_W) and the effect of a change in the frequency of the gene on the relative fitness of individuals within groups (β_W).

This equation makes it easy to see why selection does not lead to the evolution of traits that are beneficial to whole populations if there is any harm to individuals. A gene is beneficial to the group when increasing the frequency of the gene increases group fitness, or $\beta_G > 0$. If it is costly to the individual, then $\beta_W < 0$. The magnitude of these two terms depends on the details of the particular situation—you can't say anything in general. However, theory tells us that when groups are large, with even a small amount of migration among them, the variance between individuals (V_W) will be about n times bigger than the variance between groups (V_G; Rogers, 1990). Unless the group benefit is on the order of n times the cost, selection will eliminate the group-beneficial gene. But when this is the case, the trait is *individually* beneficial averaged over all groups.

However, this doesn't mean that group selection is unimportant. We have just seen that when groups of individuals interact over long periods of time, *any* behavior can be evolutionarily stable within groups. Moreover, multiple stable equilibria can also arise from the conformist tendency in social learning discussed in chapters, 1, 5, and 11. When lots of alternative equilibria exist, we need a theory that tells us which equilibrium will be the long-run evolutionary outcome—what game theorists call the equilibrium selection problem. We argue in several articles that selection among groups favors the most group-beneficial equilibrium. To see why this is plausible, consider the Price equation, and suppose that there are two inherited traits; both are stable within groups when common, but one leads to higher rates of group reproduction. This means that, as before, $\beta_G > 0$. Because both traits are favored by selection when they are common, each trait will be favored in some groups, so that the average value of β_W can be either positive or negative. However, as long as there is not too much migration, most of the groups will be near one equilibrium or the other. So the variance among groups will be much larger than the variance within groups, independent of group size. The reason for this discrepancy is simple: when traits are individually advantageous, selection and migration are working together to make all groups the same; the only process making groups different is genetic drift, which depends strongly on population size. When there are multiple equilibria, selection is driving groups toward different alternative stable equilibria, creating lots of stable between-group variation. Thus, selection between groups generates group-beneficial outcomes.

While the Price equation makes it easy to understand the logic of selection at the group level, it also conceals crucial details about population structure and the mode of intergroup competition. Evolutionary geneticists have studied a range of population structures ranging from "stepping stone" models in which groups exchange migrants with a small number of neighbors to "Wright Island" models in which all groups are connected by migration. Such models have incorporated two modes of intergroup competition: the group-beneficial trait can increase the productivity of the group so that it produces more emigrants, called "differential proliferation," or it can reduce the extinction rate of the group, called "differential extinction." The basic conclusion of theoretical work on the evolution of altruism is that these details don't matter much (e.g., Aoki, 1982; Rogers, 1990). However, when there are multiple equilibria, the population structure and modes of group competition matter a lot. In Boyd and Richerson (1990), we show that when there are multiple equilibria, and within-group adaptive processes (selection or selection-like biased cultural transmission) are strong, the equilibrium with the lowest extinction rate spreads under a wide range of conditions. Groups can be large and migration rates substantial. The main requirement is that habitats emptied by extinction are colonized by individuals drawn mostly from a single group. Interestingly, make this a differential proliferation model and group selection has no effect. The same process that preserves variation between groups prevents a steady trickle of immigrants from groups at the group-beneficial equilibrium from having much effect on groups at the other equilibrium.

Extinction, coupled with recolonization by a single other group, means that groups become crude "individuals" that reproduce their own group characteristics.

We also wanted to know whether intergroup competition will lead to change on the right time scales to explain observed rates of cultural evolution. Obviously, this depends on how often groups go extinct. So, working with Joseph Soltis, we estimated an upper bound on the rate of cultural evolution by this kind of intergroup competition using ethnographic data from New Guinea societies. This analysis (chapter 11) indicates that intergroup competition leads to the evolution of group-beneficial cultural traits on 500- to 1,000-year time scales, too slow to account for much cultural change. On the other hand, major change in social institutions is a slow process; witness the relatively slow growth in sophistication of complex societies over the past 5,000 years. The model may apply to conservative aspects of cultural change. Much historic and prehistoric cultural change has a time scale of a millennium or more.

Intergroup competition is not the only mechanism that can lead to the spread of group-beneficial cultural variants—a propensity to imitate successful neighbors can also lead to the spread of group-beneficial variants. Plausibly, people often know something about what goes on in neighboring groups. Now, suppose that neighboring groups are at different equilibria and that one of the equilibria is better, meaning that it makes people in that group better off. Then, behaviors could spread from groups at high payoff equilibria to neighboring groups at lower payoff equilibria because people imitate their more successful neighbors. To see whether this mechanism could actually work, we analyzed the model presented in chapter 12, and our results suggest that it can lead to the spread of group-beneficial beliefs as long as groups are connected to only a small number of neighboring groups (in a stepping stone population structure) so that the success of one group can affect neighbors enough to cause them to tip from one equilibrium to the other. The model also suggests that such spread can be rapid. Roughly speaking, it takes about twice as long for a group-beneficial trait to spread from one group to another as it does for an individually beneficial trait to spread within a group. This process is faster than intergroup competition because it depends on the rate at which individuals imitate new strategies, rather than the rate at which groups become extinct.

These models suggest that the evolution of cooperative norms is a side effect of rapid, cumulative cultural adaptation. Adaptation by cultural evolution brings significant benefits, especially in the climatic chaos of the later Pleistocene epoch. However, it also generates lots of variation between groups; thus, group selection is a much more important force in human cultural evolution than it is in genetic evolution. We think the best evidence from archaeology suggests that humans first began to rely on cumulative cultural adaptations roughly a half million years ago. If this inference is correct, humans have been living in social environments shaped by group selection for a long time. In chapter 14 (with Joe Henrich), we argue that in such social environments, ordinary natural selection will favor psychological mechanisms like empathy, guilt, and shame that make it more likely that individuals behave prosocially. The coevolutionary response of our innate social instincts to the selection pressures of living in rule-bound, prosocial tribal-scale communities substantially reshaped our social psychology.

In chapter 14 we argue that cultural group selection and moralistic punishment are both important to explaining cooperation. Cultural group selection will favor groups with high frequencies of moralistic punishment, and it helps ensure that moralistic punishment enforces functional norms. Moralistic punishment, as we have said, plays a considerable role in maintaining between-group variation on which cultural group selection acts. We believe that the tilt of the modeling results and of the empirical data distinctly favors what we call in this chapter the tribal social instincts hypothesis. At minimum, we believe that the case is sufficiently strong to lift the burden of proof that group selection hypotheses have labored under.

REFERENCES

Aoki, K. A. 1982. Condition for group selection to prevail over counteracting individual selection. *Evolution* 36:832–842.
Axelrod, R., & W. D. Hamilton. 1981. The evolution of cooperation. *Science* 211:1390–1396.
Bendor, J., & D. Mookherjee. 1987. Institutional structure and the logic of ongoing collective action. *American Political Science Review* 81:129–154.
Boerlijst, M. C., M. A. Nowak, & K. Sigmund. 1997. The logic of contrition. *Journal of Theoretical Biology* 185:281–293.
Boyd, R. 1989. Mistakes allow evolutionary stability in the repeated prisoner's dilemma game. *Journal of Theoretical Biology* 136:47–56.
Boyd, R. & P. J. Richerson. 1990. Group selection among alternative evolutionarily stable strategies. *Journal of Theoretical Biology* 145:331–342.
Fehr, E., & U. Fischbacher. 2003. The nature of human altruism. *Nature* 425:785–791.
Fudenberg, D., & E. Maskin. 1986. The folk theorem in repeated games with discounting or with incomplete information. *Econometrica* 54:533–556.
Price, G. R. 1970. Selection and covariance. *Nature* 227:520–521.
Price, G. R. 1972. Extension of selection covariance mathematics. *Annals of Human Genetics* 39:455–458.
Rogers, A. R. 1990. Group selection by selective emigration: The effects of migration and kin structure. *American Naturalist* 135:398–413.
Smith, E. A., & R. L. Bliege Bird. 2000. Turtle hunting and tombstone opening: Public generosity as costly signaling. *Evolution and Human Behavior* 21:245–261.
Sugden, R. 1986. *The Economics of Rights, Cooperation and Welfare*. Oxford: Basil Blackwell.
Trivers, R. L. 1971. The evolution of reciprocal altruism. *Quarterly Review of Biology* 46:35–37.

8 The Evolution of Reciprocity in Sizable Groups

Several lines of evidence suggest that sizable groups of people sometimes behave cooperatively, even in the absence of external sanctions against noncooperative behavior. For example, in many food foraging groups, game is shared among all members of the group regardless of who makes the kill (e.g., Kaplan and Hill, 1984; Lee, 1979; Damas, 1971). In many other stateless societies, men risk their lives in warfare with other groups (e.g., Meggit, 1977). There is also evidence that a great deal of cooperation takes place in contemporary state-level societies without external sanctions. For example, people contribute to charity, give blood, and vote—even though the effect of their own contributions on the welfare of the group is negligible. The groups benefiting are often very large and composed of very distantly related individuals. Perhaps the most dramatic examples of cooperation in contemporary societies are underground movements such as Poland's Solidarity in which people cooperate to achieve a common goal in opposition to all of the power of the modern state (see Olson, 1971, 1982, and Hardin, 1982, for further examples.) Because of the anecdotal nature of these data, it is possible to doubt any particular example. However, psychologists and sociologists have also shown that people cooperate under carefully controlled laboratory conditions, albeit for smaller stakes. For example, Marwell and Ames (1978, 1980) presented individual students with two alternative investments: a low return private investment in which profits accrued to the individual, and a higher return investment in which returns accrued to all group members whether they invested or not. Students invested in the group-beneficial investment at a much higher rate than that consistent with rational self-interest. (See Dawes, 1980, for a review of such experiments.)

The fact that people cooperate in sizable groups is puzzling from an evolutionary viewpoint. According to contemporary evolutionary theory, cooperative behavior can evolve only through one of two mechanisms: inclusive fitness effects (Hamilton, 1975) or reciprocity (Trivers, 1971). Inclusive fitness effects occur when social groups form so that cooperators are more likely to interact with other cooperators than with noncooperators. There has been controversy over what processes of group formation suffice to allow cooperation. Some authors (e.g., Maynard Smith, 1976) have argued that groups must be comprised of genetic relatives for cooperation to be favored. Others (e.g., Wilson, 1980; Wade, 1978) have argued that other mechanisms suffice. We believe that most authors would agree that inclusive fitness effects can give rise to cooperation among mammals only in relatively small groups. With the exception of humans, this prediction is supported by observations of mammalian social behavior. The relatively few animal societies that have levels of cooperation similar to those of humans are typically composed of close relatives (Wilson, 1975; Jarvis, 1981), while cooperation in large groups among humans includes cases where cooperators are virtually unrelated.

Cooperation may also arise through reciprocity when individuals interact repeatedly. Several related analyses (Axelrod, 1984; Axelrod and Hamilton, 1981; Brown, Sanderson, and Michod, 1982; Aoki, 1983; Peck and Feldman, 1986) suggest that cooperation can arise via reciprocity when *pairs* of individuals interact repeatedly. These results suggest that the evolutionary equilibrium in this setting is likely to be a contingent strategy with the general form "cooperate the first time you interact with another individual, but continue to cooperate only if the other individual also cooperates." Some authors have conjectured that reciprocity can lead to cooperation in larger groups through a similar mechanism (Trivers, 1971; Flinn and Alexander, 1982; Alexander, 1985, 1987:93ff). However, since there has been no explicit theoretical treatment of the evolution of behavior when there are repeated interactions in groups larger than two individuals, it is unclear whether this conjecture is correct.

The goal of this chapter is to clarify this issue by extending existing theory to explicitly include repeated interactions in large groups. We begin by reviewing the evolutionary models of the evolution of reciprocity. We then present a model of the evolution of reciprocal cooperation in sizable groups. An analysis of this model suggests that the conditions necessary for the evolution of reciprocity become extremely restrictive as group size increases.

Models of the Evolution of Reciprocal Cooperation

For the most part, evolutionary models of cooperation have been developed by biologists interested in explaining cooperative behavior among nonhuman animals. (See Wade, 1978; Uyenoyama and Feldman, 1980; Michod, 1982; Wilson, 1980, for reviews). These assume that individual differences in social behavior, including the strategies that govern individual behavior in potentially reciprocal social interactions, are affected by heritable genetic differences. They further assume that the outcome of potentially cooperative social interactions affects an individual's reproductive success. Successful behavioral strategies will, thus, increase in the population through natural selection. The question then is: under what conditions will natural selection favor behavioral strategies that lead to cooperation? The answer to this question should illuminate contemporary human cooperation to the extent that evolved propensities shape human behavior.

If behavioral strategies are transmitted culturally instead of genetically, evolutionary models also provide insight into the conditions under which cooperative behavior will arise in contemporary societies. Some authors (Axelrod, 1984; Brown et al., 1982, Maynard Smith, 1982; Pulliam, 1982; Boyd and Richerson, 1982, 1985) have constructed models, formally quite similar to the genetic ones, which assume that behavioral strategies are transmitted from one individual to another culturally, by teaching, imitation, or some other form of social learning. These models assume that the probability that a strategy is transmitted culturally is proportional to the average payoff associated with that strategy. There are many plausible ways in which this can occur. For example, it may be that people tend to imitate wealthy or otherwise successful individuals. (For discussions of the relationship between genetic and cultural evolution, see Cavalli-Sforza and Feldman, 1981; Lumsden and Wilson, 1981; and Boyd and Richerson, 1985).

The recent work of several authors (Boorman and Levitt, 1980; Axelrod, 1980, 1984; Axelrod and Hamilton, 1981; Brown et al., 1982; Aoki, 1983; Peck and Feldman, 1986; Boyd and Lorberbaum, 1987) suggests that natural selection may favor reciprocity when pairs of individuals interact a sufficiently large number of times. These models share many common features. Each assumes a population of individuals. Pairs of individuals sampled from this population interact a number of times. During each interaction, individuals may either cooperate (C) or defect (D). Table 8.1 gives the incremental effect of each interaction on the fitness of the members of a pair. This pattern of fitness payoffs defines a single period prisoner's dilemma; it means that cooperative behavior is altruistic in the sense that it reduces the fitness of the individual performing the cooperative behavior, but increases fitness of the other individual in the pair (Axelrod and Hamilton, 1981; Boyd, 1988). By assumption, each individual is characterized by an

Table 8.1. The incremental effect of interactions on the fitness of the members of a pair

| | | Player 2 | |
		C	D
	C	R, R	S, T
Player 1			
	D	T, S	P, P

Each player has the choice of two strategies, C for cooperate and D for defect. The pairs of entries in the table are the payoffs for players 1 and 2, respectively, associated with each combination of strategies. In the case of the prisoner's dilemma it is assumed that $T > R > P > S$, and $2R > S + T$.

inherited strategy that determines how it will behave. Strategies may be fixed rules like unconditional defection ("always defect"), or contingent ones like tit-for-tat ("cooperate during the first interaction; subsequently do whatever the other individual did last time"). The pair's two strategies determine the effect of the entire sequence of interactions on each pair member's fitness.

This literature produces three main conclusions about the evolution of reciprocity:

1. Reciprocating strategies, like tit-for-tat, that lead to mutual cooperation are successful if pairs of individuals are likely to interact many times. There is some dispute about what kinds of reciprocating strategies are most likely to be successful, and whether any pure strategy can be evolutionarily stable (Boyd and Lorberbaum, 1987; Hirshleifer and Martinez Coll, 1988). But it seems plausible there will be a stable equilibrium at which reciprocators are common whenever interactions last long enough.

2. A population in which unconditional defection is common can resist invasion by cooperative strategies under a wide range of conditions. When a population is mostly made up of individuals who never cooperate, and individuals are paired randomly, rare reciprocators are overwhelmingly likely to be paired with unconditional defectors. Reciprocators suffer because of their willingness to cooperate initially. In many situations, it is plausible that cooperative behavior is the derived condition. Thus, to explain the existence of reciprocal behavior, we must solve the puzzle of how reciprocating strategies increase when rare.

3. There seems to be a variety of plausible mechanisms that allow reciprocating strategies to increase when rare. Axelrod and Hamilton (1981; Axelrod, 1984) have shown that a very small degree of assortative group formation, when coupled with the possibility of prolonged reciprocity, allows strategies like tit-for-tat to invade noncooperative populations. Peck and Feldman (1986) have shown that the costs of cooperative behavior can be frequency dependent in such a way that cooperation increases when rare. Finally, Boyd and Lorberbaum (1987) show that if mutation or phenotypic variation is present, unconditional defection can be invaded even when groups are formed at random.

This theory suggests a robust conclusion: lengthy paired interactions favor reciprocity. We have suspected that this conclusion is sensitive to group size, for in larger groups, enforcing individuals bear the full cost of punishing defectors while the benefit of enforcement flows to the whole group. (See Boyd and Richerson, 1985, 228–230, for a simple game-theoretic presentation of this intuition.) Authors like R. D. Alexander (1985, 1987:93ff), however, have argued that reciprocity *can* lead to cooperation in sizable groups. Thus, we offer an explicit investigation of repeated interactions in groups larger than two.

Model Assumptions

Our model closely resembles evolutionary models of reciprocity in pairs. Suppose there is a population of individuals—each characterized by an inherited strategy. Groups are formed by sampling *n* individuals from the population who interact in a repeated n-person prisoner's dilemma. Each individual's payoff depends on his strategy and the strategies used by the $n-1$ other individuals in the group. The representation of any strategy in the next generation is a monotonically increasing function of the average payoff received by individuals playing that strategy during the previous period. (As argued by Brown et al., 1982, this assumption is consistent with haploid genetic inheritance of strategies and some simple forms of cultural transmission.) We then ask which strategies or combinations of strategies can persist.

We use an n-person prisoner's dilemma to model cooperation among a group of individuals (e.g., Schelling, 1978; Taylor, 1976; for alternative formulations, see Taylor and Ward, 1982; Hirshleifer, 1983). In any time period, each individual can choose either to cooperate (C) or to defect (D). An individual's payoff in a single time period depends on her own behavior and on the number of cooperators in the group. Let $V(C|i)$ and $V(D|i)$ be the payoffs to individuals choosing cooperation and defection, given that i of the n individuals in the group choose cooperation. The n-person prisoner's dilemma demands that these payoffs have the following properties:

1. In any interaction, each individual is better off choosing D, no matter what the other $n-1$ individuals in the group choose. Thus:

$$V(D|i) > V(C|i+1), \quad i=0,\ldots,n-1 \tag{1}$$

This assumption formalizes the notion that altruistic behavior is costly to the individual. If groups are formed at random, and interact only once, this assumption guarantees that cooperative behavior cannot evolve (Nunney, 1985).

2. If an individual switches from defection to cooperation, every other member of the group is better off. This requires that:

$$\begin{aligned} V(D|i+1) &> V(D|i) \\ V(C|i+1) &> V(C|i) \end{aligned} \quad i=0,\ldots,n-1 \tag{2}$$

This assumption formalizes the idea that cooperation benefits other members of the group.

3. The average fitness of individuals in the group increases if one switches from defection to cooperation. This requires:

$$\begin{aligned} &(i+1)V(C|i+1) + (n-i-1)V(D|i+1) \\ &> iV(C|i) + (n-i)V(D|i) \end{aligned} \tag{3}$$

where $i=0,\ldots,n-1$. This assumption formalizes the idea that the fitness benefits to the whole group from cooperative behavior exceed the fitness costs of cooperating.

We are free to choose the units in which payoffs are accounted. We can thus specify that $V(D|0)=0$ and $V(C|n)=B$, where B is a positive constant. When groups consist of only two individuals, these three conditions generate a slightly stronger form of the prisoner's dilemma than usual. That is, all three require that $T>R$, $P>S$, and $R>(T+S)/2>P$ rather than the two inequalities listed in table 8.1.

We derive many of our results here assuming that the payoff to each individual in a group during each interaction is a linear function of the number of individuals who cooperated during that interaction. Let the number of individuals choosing C during a particular turn be i. Then, the payoffs to individuals choosing C and D are:

$$V(C \mid i) = (B/n)i - c$$

and (4)

$$V(D \mid i) = (B/n)i$$

From the definition of the n-person prisoner's dilemma, it must be that $B > c$ and $c > B/n$. This model is identical to the linear model of social interactions used in most kin selection models. Economists and political scientists have used various versions of this model to represent the investment in public goods (Hardin, 1982), although Hirshleifer (1983) shows that nonlinear payoffs can strongly affect the advantages of cooperation. Two polar cases of the linear payoff model are of particular interest: the case in which B is constant with respect to n, and the case in which B is proportional to n. The first represents situations in which the benefits produced by a cooperative act are divided up among group members, so that increasing group size decreases the benefit per individual group member. The second case represents situations in which the benefits reaped by one individual do not reduce the benefits received by another.

Groups of n individuals are sampled from the population and interact repeatedly in the n-person prisoner's dilemma just described. The probability that a given group interacts more than t times is w^t, where w is a constant between zero and one. This assumption means that the expected number of interactions among the n individuals is $1/(1 - w)$. Thus, as w increases, so does the number of interactions between a group of n individuals. If $w \approx 0$, individuals usually interact only once. If $w \approx 1$, then individuals interact many times.

Each individual is characterized by an inherited "strategy" that specifies whether the individual will choose cooperation or defection during any time period based on the history of the group up to that point. In this analysis, we consider only the following strategies:

U: always defect.

T_a: cooperate on the first move and then cooperate on each subsequent move if a or more of the other $n - 1$ individuals in the group chose cooperation during the previous time period.

The set of strategies T_a is a generalization of tit-for-tat. In the n person case, there are $n - 1$ such contingent strategies (T_a with $a = 1, \ldots, n - 1$), one for each of the possible rules of the form "cooperate if a or more individuals cooperated on the last move." Taylor (1976) introduced this family of strategies. We begin by assuming that populations consist of only two strategies, U and T_a, in which a takes on some particular value. Later we will consider populations in which three or more strategies are present.

In populations in which only U and T_a are present, an individual's expected fitness depends only on his own strategy and on the number of T_a individuals among the other $n - 1$ individuals in its social group. To see this, consider the expected fitness of a T_a individual in a group in which j other individuals use the strategy T_a. The U individuals in such a group always play D. The T_a individuals always cooperate on the first interaction. They continue to cooperate as long as a or more of the other $n-1$ individuals cooperated last time. If $j \geq a$, the T_a individuals play C during every interaction. This means that during each time period the payoff to T_a individuals is $V(C|j+1)$. The effects of social interaction on the fitness of any particular individual depends on the number of time periods that individual's group interacts. If $j \geq a$, the average payoff to T_a individuals, over all groups with j other cooperators $F(T_a|j)$, is:

$$F(T_a|j) = V(C|j+1)(1 + w + w^2 + w^3 + \cdots)$$
$$= \frac{V(C|j+1)}{1-w} \tag{5}$$

If $j < a$, the T_a individuals cooperate during the first interaction and defect thereafter. This means that the payoff to T_a individuals is $V(C|j+1)$ during the first period and $V(C|0)$ during any subsequent periods. Thus,

$$F(T_a|j) = V(C|j+1) + V(D|n)(w + w^2 + w^3 + \cdots)$$
$$= V(C|j+1) + \frac{wV(D|0)}{1-w} \tag{6}$$

A similar argument shows that the expected payoff to U individuals in groups in which j of the other $n-1$ individuals are characterized by the strategy T_a is as follows:

$$F(U|j) = \begin{cases} \dfrac{V(D|j)}{1-w} & j > a \\[2mm] V(D|j) + \dfrac{wV(D|0)}{1-w} & j \leq a \end{cases} \tag{7}$$

After the episode of social behavior that generates these payoffs, individuals in the population reproduce. We assume that individual fitness is the sum of a baseline fitness W_0 and the payoff resulting from social interaction. We further assume that $W_0 \gg F(T_a|j)$, $F(U|j)$ for all values of j, meaning that selection acting on social behavior is weak. The expected fitness of T_a averaged over all the different kinds of groups, $W(T_a)$, is given by:

$$W(T_a) = \sum_{j=0}^{n-1} m(j|T_a)\{W_0 + F(T_a|j)\} \tag{8}$$

The term in braces is the expected fitness of a T_a individual in a group with j other T_a individuals. This term is multiplied by the probability that a T_a individual finds herself in such a group, $m(j|T_a)$, and is summed over all possible groups. Similarly, the expected fitness of an unconditional defector, $W(U)$, is the following:

$$W(U) = \sum_{j=0}^{n-1} m(j|U)\{W_0 + F(U|j)\} \tag{9}$$

where $m(j|U)$ is the probability that a U individual finds herself in a group in which there are j T_a individuals.

If the frequency of T_a in the population before social interaction is p, then the frequency before social interaction in the next generation, p', is:

$$p' = p + p(1-p)\frac{[W(T_a) - W(U)]}{\overline{W}} \tag{10}$$

where

$$\overline{W} = pW(T_a) + (1-p)W(U)$$

To determine the long-run evolutionary outcome, we determine what frequencies of T_a and U represent stable equilibria of the recursion (10).

Evolution of Reciprocity When Groups Are Formed Randomly

We begin by assuming that groups form randomly. This assumption means that individuals do not interact with genetic relatives, nor are they able to assort themselves based on observable phenotypic characteristics. In the special case of pairs, theory (reviewed earlier) suggests that strategies leading to reciprocal cooperation can evolve as long as individuals interact a large enough number of times. We want to know how increasing group size will affect this conclusion. We formalize this assumption by specifying that both $m(j|T_a)$ and $m(j|U)$ are binomial probability distributions with parameter p, labeled $m(j)$.

According to equation (10), the frequency of T_a will increase whenever the expected fitness of T_a, $W(T_a)$, is greater than the expected fitness of U, $W(U)$ (unless the population is at an equilibrium point, in which case there is no change). When groups are formed at random, the condition for T_a to increase has the following particularly simple and instructive form:

$$\sum_{j=0}^{a-1}[V(D|j) - V(C|j+1)]m(j) + \sum_{j=a+1}^{n-1}\frac{[V(D|j) - V(C|j+1)]m(j)}{1-w}$$
$$< \left[\frac{V(C|a+1)}{1-w} - V(D|a)\right]m(a) \tag{11}$$

where if the upper bound of the sum is less than the lower bound, the sum is zero by convention. This expression says that T_a individuals have a fitness advantage relative to U individuals only in groups in which a single additional defector will cause cooperation to collapse. For T_a to be favored by selection, the advantage it gains in such groups must be larger than the disadvantage T_a suffers in all other groups. To see this, consider each of the three terms in this expression. The first term represents the sum of the fitness advantages of U individuals in groups in which fewer than a of the other $n-1$ individuals are reciprocators, weighted by the probability that such groups form. In such groups, T_a individuals cooperate only once, and U individuals do not cooperate at all. The definition of the n-person prisoner's dilemma guarantees that $V(C|j+1) < V(D|j)$. This term is therefore always positive. The second term represents the average fitness advantage of unconditional defection in groups in which more than a of the other $n-1$ individuals are reciprocators. This term is multiplied by $1/(1-w)$, the expected number of interactions, because in such groups T_a individuals cooperate and U individuals defect for as long as the group persists. Again, this term is always positive. The right-hand side of the inequality gives the difference between the fitness of the two strategies in groups in which exactly a of the other $n - 1$ individuals are reciprocators, multiplied by the probability that such groups form. A T_a individual in such a group both cooperates and receives the benefits of cooperation of a other cooperators, $V(C|a+1)$, for as long as the group persists. Replacing that T_a individual with a U individual causes other reciprocators to cease cooperating after the first interaction. This term cannot be positive unless the fitness of a cooperator in such a group is greater than the fitness of a defector in a group of n defectors, that is, $V(C|a+1) > 0$. Suppose that this condition is satisfied. Then, if the expected number of interactions is large enough (i.e., w is close enough to one), T_a individuals will have an advantage relative to U individuals in groups in which a of the other $n - 1$ individuals are reciprocators. For T_a to be favored by selection, the advantage that T_a individuals gain in such groups must exceed the advantage to U individuals in all other groups.

With this result in mind, consider the equilibrium behavior of this model. The frequency of the two strategies in the population will not change when $p' = p$. Values of p that satisfy this condition are equilibrium values, denoted $\hat{p}$. Since there is no migration or mutation, $\hat{p} = 1$ (all T_a individuals) and $\hat{p} = 0$ (all U individuals) are always equilibrium values of equation (10). There may be other equilibria at which both U and T_a are present in the population. At these polymorphic (or "interior") equilibria, the average fitness of the two strategies must be equal. An equilibrium is stable if the population returns to the same equilibrium frequency after small perturbations. Stable equilibria are interesting because they tell us something about what kinds of strategies, or mixes of strategies, can persist in the long run. Unstable equilibria are also interesting because they give information about the range of initial conditions that can result in various long-run outcomes. Such an analysis yields the following results.

A population in which U is common can resist invasion by any reciprocating strategy, T_a. This is true for all values of w. As in the two-person case, a population that is all unconditional defectors can resist invasion by any reciprocal strategy we consider. When unconditional defection is very common and groups are formed randomly, most groups contain n unconditional defectors. The few T_a individuals in the population will be in groups in which all other individuals are unconditional defectors. These solitary reciprocators cooperate once and thereafter defect. The average fitness of unconditional defectors will always be higher than the average fitness of any reciprocal strategy, because $V(D|0) > V(C|1)$.

A population in which T_{n-1} is common can resist invasion by unconditional defection if, and only if, w is sufficiently large. It is the only reciprocal strategy that has this property. T_{n-1} is the reciprocating strategy that is completely intolerant of defection. Individuals using T_{n-1} will cooperate only if every other individual cooperated during the previous time period. Strategies that continue to cooperate despite one or more defections (T_a, $0 < a < n - 1$) cannot be evolutionarily stable when groups form randomly. When T_a is common, the great majority of unconditional defectors will be isolated in groups in which the other $n - 1$

individuals are all reciprocators. Unless $a = n - 1$, the T_a individuals in such groups will continue to cooperate despite the defector. Since $V(D|n-1) > V(C|n)$, unconditional defectors will have higher average fitness than reciprocators.

The parameter w is a measure of the number of times that individuals interact in groups. T_{n-1} is evolutionarily stable only if:

$$w > w_c = 1 - V(C|n)/V(D|n-1) \tag{12}$$

This relationship has a simple interpretation. Consider an individual in a group in which all other individuals use the strategy T_{n-1}. If this individual defects on every turn, his payoff will be $V(D|n-1)$ in the first time period and $V(D|0) = 0$ thereafter. If he instead cooperates, his payoff is $V(C|n)$ every period. Because the average number of interactions is $1/(1-w)$, condition (12) requires that the average payoff from choosing cooperation be greater than the average payoff from choosing defection—if cooperation is to resist invasion by individuals using U. More iterations mean more chance of satisfying this condition, all else being equal.

Assuming linear payoffs, the domain of attraction of T_{n-1} diminishes rapidly as group size increases. If pairs of individuals interact long enough, either unconditional defection or T_{n-1} can persist. How likely is it that a population will end up at the cooperative equilibrium? One approach to answering this question is to determine the domain of attraction of the two equilibria. An equilibrium's domain of attraction is the set of initial frequencies that begin trajectories ending at that equilibrium. The bigger the domain of attraction of an equilibrium, the more likely it is, in some sense, that a population will end up there. (Later we will consider a second approach to answering this question.)

We have not been able to determine the domains of attraction for the two fixed equilibria in general. We have found them, however, in the special case in which the payoffs are linear functions of the number of defectors. Only two stable equilibria exist in this special case, $\hat{p} = 0$ and $\hat{p} = 1$. There is also a single unstable polymorphic equilibrium. The frequency of reciprocators at the internal equilibrium, p_c, is (Appendix, part 1):

$$p_c = \left(\frac{c - B/n}{w(B-c)/(1-w)} \right)^{1/n} \tag{13}$$

If the initial frequency is higher than p_c, then the population eventually will consist of all reciprocating (T_{n-1}) individuals. If the initial frequency of cooperators is less than p_c, the population eventually will be comprised of all U individuals.

To interpret equation (13), remember that the expected fitness of the two strategies must be equal at any polymorphic equilibrium. The term $c - B/n$ is the difference in fitness between U and T_{n-1} individuals during the first interaction. The term $w(B-c)/(1-w)$ is the fitness advantage of T_{n-1} relative to U when the other $n-1$ members of the group are reciprocators. The critical frequency of T_{n-1} individuals necessary for selection to favor T_{n-1} thus is simply the ratio of the incremental benefit to the incremental cost of defecting during the first interaction raised to the $1/n$ power. Because the incremental benefit increases as the expected number of interactions becomes large (i.e., as $w \to 1$), the threshold frequency of cooperators necessary for cooperation to increase approaches zero (i.e., $p_c \to 0$). The domain of attraction for the unconditional defection equilibrium thus shrinks toward zero. Raising the ratio to the power $1/n$, however, means that the threshold frequency of cooperators necessary for cooperators to be favored, p_c, increases as group size increases. This effect occurs because the probability of forming cooperative groups diminishes geometrically as group size increases when groups are formed at random.

Figure 8.1 illustrates the magnitude of this effect by showing the values of p_c for various parameter combinations. For small groups, cooperators need increase to only a small fraction of the population for selection to favor cooperation. For even modest-sized groups, however, the cooperative strategy T_{n-1} must reach substantial frequency before this strategy increases. For large groups, virtually the entire population must consist of cooperators before the cooperative strategy can increase.

In populations composed of T_a $(n-1 > a > 0)$ and U, there is a single stable, internal equilibrium as long as w is large enough, $c < B(a+1)/n$, and payoffs are linear. Of the set of reciprocating strategies we have considered, we have found that only T_{n-1} can resist invasion by rare unconditional defectors (U). We also

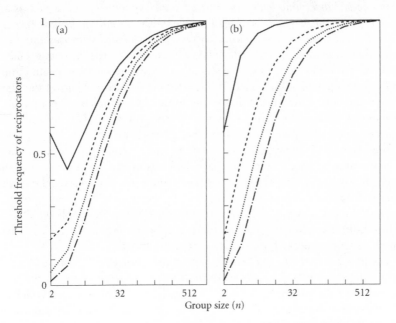

Figure 8.1. This figure presents the threshold frequency of T_{n-1} that must be exceeded for this strategy to increase (i.e., p_c) as a function of group size (n) for four values of w: 0.9 (——), 0.99 (- - -), 0.999 (····), and 0.9999 (-----). These values of w correspond to 10, 100, 1000, and 10000 interactions, on average, between pairs of individuals. (a) The incremental benefit to individual due to one cooperator is proportional to group size ($B = 1.141n$), (b) the incremental benefit is constant ($B = 2$).

found, however, that T_{n-1} is unlikely to increase when rare. It would be interesting to know whether there are any stable internal equilibria at which more tolerant cooperative strategies (T_a, $a < n - 1$) and unconditional defection coexist. It seems plausible that the threshold frequency necessary to get such strategies started in a population might be lower.

It turns out that there are two internal equilibria, one stable and the other unstable, as long as (see the Appendix, part 2):

$$p_d = \frac{B(a + 1)/n - c}{B - c} > 0 \tag{14}$$

and

$$w > \frac{c - B/n}{(c - B/n) \text{ Prob } (j < a \mid p = p_d) + a(B/n) \text{ Prob } (j = a \mid p = p_d)}$$

The frequency of T_a at the stable internal equilibrium, p_s, is always greater than p_d, and the frequency of T_a at the unstable equilibrium, p_u, is less than p_d. If the initial frequency of T_a is less than p_u, the population will eventually consist of all unconditional defectors. When the initial frequency of T_a is greater than p_u, the frequency of T_a eventually will stabilize at p_s. When w is less than this critical value, the only equilibria are monomorphic for T_a or U.

Numerical determination of the internal stable equilibria suggests that as a decreases (1) the frequency of the strategy T_a at the stable internal equilibrium decreases, (2) the threshold frequency of T_a necessary for T_a to be favored decreases, and (3) the threshold value of w necessary for the internal equilibria to exist increases. One can determine the frequency of the two strategies at these polymorphic equilibria by finding the values of p for which $W(U) = W(T_a)$. Figure 8.2 shows the results of numerical determinations of these equilibrium values for several combinations of parameter values. When a is almost $n - 1$, reciprocators will allow only a few defectors before defecting themselves. In this case, the frequency of the reciprocating strategy, T_a, is high, but so is the threshold frequency of T_a necessary to get cooperation started. Note also that when a is near $n - 1$, the internal equilibrium may be stable even when w is fairly small. As a decreases, the reciprocating strategy tolerates a larger number of defectors. This greater tolerance decreases the frequency of the cooperative individuals at the stable equilibrium, the threshold frequency of T_a necessary to get cooperation started. As a decreases, w must be large in order for a stable equilibrium to exist at all.

Populations at stable equilibria involving two strategies, T_a and U, ($n - 1 > a > 0$), can resist invasion by rare individuals using any other reciprocating strategy, T_b where a ≠ b. So far we have limited our analysis to populations in which only two strategies are present. This omission might be important. Assuming w is sufficiently large, it is relatively easier for cooperation to get started when cooperating individuals are quite tolerant. But tolerant strategies can achieve only a low frequency at equilibrium. Suppose that such an equilibrium is reached. If less tolerant individuals could then invade, the population might reach a new equilibrium at which cooperators existed in higher frequency. If this could happen repeatedly, then the cooperators might eventually achieve a high frequency through a sort of ratchet mechanism.

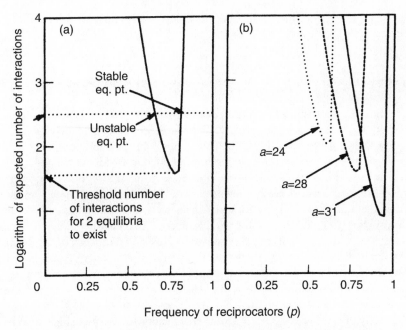

Figure 8.2. Plots of the two internal equilibria in populations characterized by two strategies, T_a and U, for various parameter values for $n = 32$ and $B = 2$. Part (a) shows how to determine the values of the two internal equilibria for a given value of $1/(1 - w)$. Part (b) shows how these values are affected by changes in the parameter a, the cooperation threshold of reciprocators.

It turns out, however, that a population at a stable polymorphic equilibrium involving U and T_a can resist invasion by any other rare reciprocating strategy, T_b. For a third strategy to invade, its expected fitness must be greater than the fitness of either of the two common strategies that are themselves equal. When the invading strategy is sufficiently rare, expected fitness of T_b individuals can be calculated assuming that the other $n - 1$ individuals in their groups are drawn from the equilibrium population. It turns out that (see the Appendix, part 3) any invading type has lower fitness than the common reciprocating strategy, T_a. To see this, suppose that $b > a$, so that the invading strategy is less tolerant of defection than the reciprocating strategy common at the equilibrium. First, recall that T_a individuals have higher fitness than unconditional defectors only in groups in which there are a other T_a individuals. In all other groups, unconditional defectors have the advantage. Now consider the fitness of T_b individuals. If there are a T_a individuals in the group, a T_b individual does almost as poorly as an unconditional defector, because her defection causes cooperation to collapse. In groups with any other composition, T_b individuals either act and thus suffer like T_a individuals, or they defect after one interaction—thus beating the T_a individuals but losing to the unconditional defectors. The strategy T_b therefore can neither capture the benefits of long-term cooperation in groups in which there are a threshold number of cooperators nor exploit the cooperation of the common reciprocators as effectively as unconditional defection. The Appendix shows that a similar logic holds for $a > b$.

The Evolution of Reciprocity When Groups Form Assortatively

Nonrandom interaction plays an important role in Axelrod's (1984) influential view of the evolution of reciprocity. Like most evolutionary analyses of reciprocity (but see Peck and Feldman, 1986; Boyd and Lorberbaum, 1987), Axelrod's study indicates reciprocating strategies such as tit-for-tat cannot increase when rare if individuals interact at random. Axelrod shows, however, that reciprocal strategies can increase when rare if individuals pair assortatively, meaning that individuals using reciprocating strategies are more likely paired with other reciprocators than chance alone would dictate. In genetic models, such assortative social interactions could arise if individuals tend to interact with genetic relatives. If w is large, even a very small amount of assortative interaction will allow reciprocating strategies to increase. Thus, in the two-person case, there is a synergistic relationship between kin selection and reciprocity in which small amounts of kin selection greatly facilitate the evolution of cooperation through reciprocity. We now consider whether this synergistic relationship changes as group size increases.

Once again suppose that payoffs are linear and that there are only two strategies: reciprocators who cooperate as long as a or more others also cooperate, T_a, and unconditional defection, U. Also suppose that groups are formed so that the probability that a T_a individual is in a group in with j other T_a individuals is:

$$m(j \mid T_a) = \binom{n-1}{j}[r + (1-r)p]^j[(1-r)(1-p)]^{n-j-1} \tag{15}$$

where p is the frequency of T_a in the population before group formation, and r is a measure of assortment (e.g., the relatedness coefficient of kin selection theory). The probability that an unconditional defector finds himself in a group in which j of the other $n-1$ individuals are T_a is:

$$m(j \mid U) = \binom{n-1}{j}[(1-r)p]^j[r + (1-r)(1-p)]^{n-j-1} \tag{16}$$

This model is meant to capture the general notion of assortative group formation in a mathematically tractable form. It is consistent with some genetic models—for example, a model in which strategies are inherited as haploid sexual traits and group members are half siblings. There are many other plausible modes of group formation that will not yield exactly this pattern of group formation—for example, groups of full siblings. Because the contingent strategies we consider cause payoffs to be highly nonlinear functions of the number of reciprocators, experience with kin selection models (Cavalli-Sforza and Feldman, 1978) suggests that different patterns of group formation may yield different results. Our model nonetheless has generality when used to determine the conditions under which a reciprocating strategy can invade a population in which all defection is common because many of these alternative models of assortative group formation become approximately equivalent to equations (15) and (16) when one strategy is rare.

With these assumptions, one can show that T_a can increase when rare only when

$$(B/n)[(n-1)r+1] - c + \frac{w}{1-w}\sum_{j=a}^{n-1}[B(j+1)/n - c]m(j\,|\,T_a) > 0 \qquad (17)$$

"inclusive fitness effect"   "reciprocity effect"

As the frequency of reciprocators, p, approaches zero, the probability that a reciprocator finds itself in a group with j other reciprocators, $m(j\,|\,T_a)$, becomes approximately

$$m(j\,|\,T_a) \approx \binom{n-1}{j}r^j(1-r)^{n-1-j} \qquad (18)$$

Selection can favor cooperative behavior when there is assortative social inter-action even with no possibility of reciprocity, because cooperators are more likely than defectors to benefit from the cooperation. The first term on the left-hand side of (17) represents this inclusive fitness effect (Hamilton, 1975). This term indicates that even if w is zero, T_a can increase as long as the inclusive fitness of T_a individuals is higher than that of unconditional defectors. In the present context, the most interesting cases are ones in which the first term is negative, meaning that cooperation could not be favored without reciprocity. The second term on the left-hand side of (17) gives the effect of reciprocity when reciprocators are rare. The added benefit received by reciprocators in groups in which there are more than a reciprocators is the increase in fitness per interaction $(B(j+1)/n - c)$ times the number of additional interactions during which reciprocators receive the benefit $(w/(1-w))$. Reciprocity will aid the spread of strategies like T_a as long as benefits produced by cooperation in a group of $a+1$ cooperators exceed the costs $(B(a+1)/n - c > 0)$.

There is a striking synergistic relationship between kin selection and reci-procity when pairs of individuals interact (Axelrod and Hamilton, 1981). A small degree of assortative social interaction, coupled with the possibility of long-term reciprocal relationships, can lead to extensive cooperation in situations in which neither factor alone would cause cooperation. This synergy diminishes very rapidly as group size increases according to (17). When r is small and a is a substantial fraction of $n-1$, the reciprocity effect in (17) becomes approximately proportional to the probability that a of the other $n-1$ individuals in the group are reciprocators. When r is small and $a/(n-1) \gg r$, this probability diminishes very rapidly as n increases. The clearest case is when $a = n - 1$. For a given B, c, and r, the expected number of interactions after the first must increase as $(1/r)^{n-1}$ for the magnitude of the reciprocity effect to remain constant.

Figure 8.3 illustrates the dramatic nature of this effect. It plots the threshold values of $1/(1-w)$ necessary for T_a to increase when rare as given by expression (17). We see that assortative group formation may play a significant role in getting reciprocal cooperation started when groups are small. For example, for $n = 3$ and

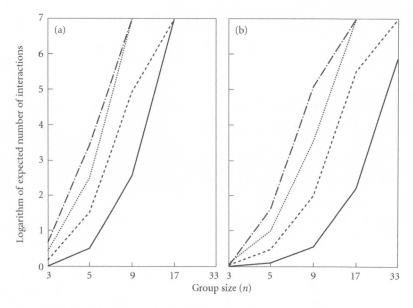

Figure 8.3. This figure presents the threshold values of $1/(1 - w)$ that must be exceeded if the strategy T_a is to increase when rare as a function of group size (n) for four values of r: 1/4 (———), 1/8 (- - -), 1/16 (····), and 1/32 (--·--). (a) $a = n - 1$: (b) $a = (3/4)(n - 1)$.

$a = 2$, even very small amounts of assortment (e.g., $r = 1/32$) will cause selection to favor reciprocity even when w is quite small (e.g., individuals interact roughly 10 times). When groups are larger, however, no amount of assortment will cause selection to favor reciprocity unless w reaches extremely high values. Consider $n = 16$ and $a = 15$. When $r = 1/2$, cooperation is favored without reciprocity. When $r = 1/4$, individuals must interact roughly 10 million times if reciprocity is to be favored. When $a < n - 1$, the qualitative picture is similar. T_a can increase when rare under a somewhat wider range of group sizes, but it remains true that the reciprocity effect diminishes rapidly as group size increases.

Conclusion

Reciprocity is likely to evolve only when reciprocating groups are quite small. Previous research based on the repeated two-person prisoner's dilemma game indicates that pairwise reciprocity will often evolve. Here we have modeled social interaction within groups of n individuals as a repeated n-person prisoner's dilemma game and asked under what conditions will selection favor strategies leading to reciprocal cooperation. In general, increasing the size of interacting social groups reduces the likelihood that selection will favor reciprocating strategies. For quite small groups, the results parallel the two-person case. For larger groups, however, the conditions under which reciprocity can evolve become extremely restrictive. This result satisfies the natural historian's conventional wisdom: large, cooperative groups composed of distantly related individuals are unusual in nature. But it leaves human cooperation unexplained.

Reciprocal strategies must satisfy two competing desiderata to succeed. First, to persist when common, they must prevent too many defectors in the population from receiving the benefits of long-term cooperation. The threshold number of cooperators thus must be a substantial fraction of group size. Second, to increase when rare, there must be a substantial probability that the groups with the threshold number of cooperators will form. This problem is not great when pairs of individuals interact; a relatively small degree of assortative group formation will allow reciprocating strategies to increase. As groups becomes larger, however, this desideratum can be satisfied only if the threshold number of cooperators is fairly small, or the degree of assortment in the formation of groups is large.

Our model omits many features that may be important in potentially cooperative social interactions. We suspect that three of the most important missing features are as follows:

1. *No internal sanctions.* We precluded the possibility that individuals could directly punish defectors. A cooperator in the n-person prisoner's dilemma can punish a defector only by withholding future cooperation—which also punishes other cooperators. Cooperation might flourish under a wider range of conditions if cooperators could focus punishment on defectors alone.

2. *No internal structure.* Our groups have no internal structure. Cooperation might arise in larger groups if individuals interact in some kind of network or hierarchy.

3. *Oversimplified game structure.* Much of our analysis presumed linear payoffs. Several authors have argued that other games may be equally important for our understanding of cooperation. Hirshleifer (1983) has shown that the nature of the payoff schedule as a function of number of cooperators has important effects on motivation to cooperate. Kelley and Thibaut (1978) discuss a large array of mixed-motive games that characterize various social interactions, and Taylor and Ward (1982) argue that the n-person version of the game "chicken" is essential to understanding cooperation. It may be that the prisoner's dilemma with linear payoffs is particularly demanding for the evolution of cooperation and that other models would allow the evolution of cooperation in sizable groups under a wider range of conditions.

Omitting these features certainly argues for caution in interpreting our results. But including these features would not necessarily allow reciprocity to evolve in large groups. It is especially unclear what peculiarities of the human case allow us to violate the generalization to which both theory and the natural history of nonhuman animals point: the evolution of large cooperative societies normally depends more on kin selection than reciprocity. Elsewhere we argue that cultural analogs of kin and group selection are indeed promising mechanisms to explain human cooperation (Boyd and Richerson, 1982, 1985, chs. 7 and 8). Campbell (1983) hypothesizes that effects like those we have modeled here suffice to explain the scale of cooperation observed in simpler human societies, but not in the state-level societies of the last 5,000 years. The range of plausible arguments is still quite broad. But the sharp decline in the tendency of reciprocity to evolve as a function of group size, and the apparent rarity of cooperation in large groups of nonkin in nature, commands attention. At the very least, our results suggest we should view with substantial skepticism and subject to more searching analysis explanations of human cooperation based on reciprocity.

APPENDIX

1. With linear payoffs, and w large enough, there is a single, unstable internal equilibrium at which the frequency of T_{n-1} is given by equation (13). At any interior equilibrium, $W(T_a) = W(U)$. With linear payoffs, this requires that

$$(B/n - c)\left[1 - w \sum_{j=a+1}^{n-1} m(j)\right] + w(B(a + 1)/n - c)m(a) = 0 \qquad (A1)$$

If w is large enough that (12) is satisfied, and $a = n - 1$, this equation can be satisfied only for one value of p, that given in (13). Since both of the boundary equilibria are stable when (12) is satisfied, the interior equilibrium is unstable.

2. If $n - 1 > a > 0$, payoffs are linear, and both conditions in (14) are satisfied, then there are two interior equilibria $\hat{p} = p_u$ and $\hat{p} = p_s$ such that $p_u < p_d < p_s$. $\hat{p} = p_u$ is unstable, and $\hat{p} = p_s$ is stable. Equation (A1) can be rewritten as follows:

$$h(p) = w(c - B/n)(1 - I_p(a, n - a)) + wB(a/n)m(a) = c - B/n \qquad (A2)$$

where $I_p(x, y)$ is the incomplete beta function. First, notice that $c - B/n > h(1) = w(c - B/n) > h(0)$. Next, differentiating $h(p)$ with respect to p (A2) yields this:

$$\frac{d}{dp}h(p) = w(n - 1 - a)m(a - 1)[B(a + 1)/n - c - p(B - c)] \qquad (A3)$$

If $B(a+1)/n - c < 0$, $h(p)$ is monotonically decreasing, and therefore there are no values of p in the interval $(0,1)$ that satisfy (A2), thus no interior equilibria exist. If $B(a+1)/n - c > 0$, $h(p)$ is unimodal with a maximum at $p = p_d$, where p_d has the value given in (14). Thus, if $h(p_d) > c - B/n$, there are two values of p that satisfy (A1), and if $h(p_d) < c - B/n$, there are none. Clearly for small enough $w, h(p_d) < c - B/n$, and thus there are no interior equilibria. Similarly, since $h(p_d) > w(c - B/n)$, for w close enough to one, $h(p_d) > c - B/n$, and there are two interior equilibria. Further, since $h(p_d)$ is a linear function of w, there is some value of w, w_d, such that there are no interior equilibria for $0 < p < p_d$, and there are two interior equilibria for $p_d < p < 1$. By solving (A1) for w and setting $p = p_d$, one obtains the expression for w_d given in the text.

From (10), the derivative of p' with respect to p evaluated at an interior equilibrium point, L, is the following:

$$L = 1 + \frac{a(w/(1 - w))m(a \,|\, p = \hat{p})[B(a + 1)/n - c - \hat{p}(B - c)]}{W_0 + F(U|p = \hat{p})} \qquad (A4)$$

Thus, if $\hat{p} < p_d$, $L > 1$, and the equilibrium is unstable. If $\hat{p} > p_d$, $L < 1$. As long as W_0 is large enough, $L > - 1$, and thus the equilibrium is stable.

3. Populations at stable equilibria involving two strategies, T_a and U $(n - 1 > a > 0)$, can resist invasion by rare individuals using any other reciprocating strategy, T_b where $a \neq b$.

When T_b is sufficiently rare, we can ignore the probability that groups with more than one T_b individual will occur. This means that the fitness of T_b individuals will depend only on j, the number of T_a individuals in their group. First, suppose that $a > b$. Then for $j \geq a$, or $b > j$, $F(T_a|j) = F(T_b|j)$. For $a > j \geq b$, $F(T_a|j) = B(j+1)/n - c$, while $F(T_b|j) = B(j+1)/n - c - w(c - B/n) < F(T_a|j)$ by definition. Thus, in this case, the expected fitness of the invading type is lower than that of the common reciprocator. Next, suppose that $a < b$. Then for $j \geq b$, or $a > j$, $F(T_a|j) = F(T_b|j)$. For $b > j > a$, $F(T_a|j) = [B(j+1)/n - c]/(1-w)$, while $F(T_b|j) = [B(j+1)/n - c] + w[Bj/n]/(1-w) > F(T_a|j)$ for values of w consistent with the existence of an interior equilibrium. For $j = a$, $F(T_a|j) = [B(j+1)/n - c]/(1-w)$, while $F(T_b|j) = [B(j+1)/n - c] + wBj/n < F(T_a|j)$ for values of w consistent with the existence of an interior equilibrium. Thus,

$$W(T_b) - W(T_a) = wm(a)\left[B(a/n) - \frac{B(a+1)/n - c}{1-w}\right] + \sum_{j=a+1}^{b-1} wm(j)\frac{B/n - c}{1-w}$$

(A5)

By using (11) to eliminate terms containing $m(a)$, (A5) becomes:

$$W(T_b) - W(T_a) = w\sum_{j=b}^{n-1} m(j)(B/n - c)/(1-w) + w\sum_{j=0}^{a-1} m(j)(B/n - c),$$

which is always less than zero.

NOTE

We thank Joan Silk and John Wiley for extremely useful comments on previous drafts of this chapter.

REFERENCES

Alexander, R. D. 1985. A biological interpretation of moral systems. *Zygon* 20:3–20.
Alexander, R. D. 1987. *The biology of moral systems.* New York: Aldine De Gruyter.
Aoki, K. 1983. A quantitative genetic model of reciprocal altruism: A condition for kin or group selection to prevail. *Proceedings of the National Academy of Sciences (USA)* 80:4065–4068.
Axelrod, R. 1980. Effective choice in the prisoner's dilemma. *Journal of Conflict Resolution* 24:3–25.
Axelrod, R. 1984. *The evolution of cooperation.* New York: Basic Books.
Axelrod, R., & W. D. Hamilton. 1981. The evolution of cooperation. *Science* 211:1390–1396.
Boorman, S., & P. Levitt. 1980. *The genetics of altruism.* New York: Academic Press.
Boyd, R. 1988. Is the repeated prisoner's dilemma game a good model of reciprocal altruism? *Ethology and Sociobiology* 9:211–221.
Boyd, R., & J. Lorberbaum. 1987. No pure strategy is evolutionarily stable in the repeated prisoner's dilemma game. *Nature* 327:58–59.
Boyd, R., & P. J. Richerson, 1982. Cultural inheritance and the evolution of cooperative behavior. *Human Ecology* 10:325–352.

Boyd, R., & P. J. Richerson. 1985. *Culture and the evolutionary process*. Chicago: University of Chicago Press.

Brown, J. S., M. J. Sanderson, & R. E. Michod. 1982. Evolution of social behavior by reciprocation. *Journal of Theoretical Biology* 99:319–339.

Campbell, D. T. 1983. Two routes beyond Kin selection to ultra-sociality: Implications for the humanities and the social sciences. In: *The nature of prosocial development: Theories and strategies*. D. Bridgeman, ed. (pp. 11–39). New York: Academic Press.

Cavalli-Sforza, L. L., & M. W. Feldman. 1978. Darwinian selection and altruism. *Theoretical Population Biology* 14:268–280.

Cavalli-Sforza, L. L., & M. W. Feldman. 1981. *Cultural transmission and evolution*. Princeton, NJ: Princeton University Press.

Damas, D. 1971. The Copper Eskimos. In: *Hunter gatherers Today*, M. G. Bicchieri, ed. (3–50). New York: Holt, Reinhardt, and Winston.

Dawes, R. 1980. Social dilemmas. *Annual Review of Psychology* 31:169–193.

Flinn, M. V., & R. D. Alexander. 1982. Culture theory: The developing synthesis from biology. *Human Ecology* 10:383–400.

Hamilton, W. D. 1975. Innate social aptitudes of man: an approach from evolutionary genetics. In *Biosocial Anthropology*, R. Fox, ed. (pp. 133–53) London: Malaby Press.

Hardin, R. 1982. *Collective action*. Baltimore: Johns Hopkins University Press.

Hirshleifer, J. 1983. The voluntary provision of public goods: From the weakest link to the best shot. *Public Choice* 41:371–386.

Hirshleifer, J., & J. Martinez-Coll. 1988. What strategies can support the evolutionary emergence of cooperation? *Journal of Conflict Resolution* 32:367–398.

Jarvis, J. U. M. 1981. Eusociality in a mammal: Cooperative breeding in naked mole rat colonies. *Science* 212:571–573.

Kaplan, H., & K. Hill. 1985. Food sharing among ache foragers: Tests of explanatory hypotheses. *Current Anthropology* 26:223–245.

Kelley, H. H., & J. W. Thibaut. 1978. *Interpersonal relations: A theory of interdependence*. New York: Wiley.

Lee, R. 1979. *The Dobe! Kung*. New York: Holt, Reinhardt, and Winston.

Lumsden, C. J., & E. O. Wilson. 1981. *Genes, mind, and culture*. Cambridge, MA: Harvard University Press.

Marwell, G., & R. E. Ames. 1979. Experiments on the provision of public goods, I. Resources, interest, group size, and the free-rider problem. *American Journal of Sociology* 84:1335–1361.

Marwell, G., & R. E. Ames. 1980. Experiments on the provision of public goods, II Provision points, stakes, experience, and the free-rider problem. *American Journal of Sociology* 85:926–937.

Maynard Smith, J. 1976. Group selection. *Quarterly Review of Biology* 51:277–283.

Maynard Smith, J. 1982. *Evolution and the theory of games*. London: Cambridge University Press.

Meggit, M. 1977. *Blood is their argument*. Palo Alto, CA: Mayfield.

Michod, R. E. 1982. The theory of kin selection. *Annual Review of Ecology and Systematics* 13:23–55.

Nunney, L. 1985. Group selection, altruism, and structured-deme models. *American Naturalist* 126:212–230.

Olson, M. 1971. *The logic of collective action*. Cambridge, MA: Harvard University Press.

Olson, M. 1982. *The rise and decline of nations*. New Haven, CT: Yale University Press.

Peck, J. R., & M. W. Feldman. 1986. The evolution of helping behavior in large, randomly mixed populations. *American Naturalist* 127:209–221.

Pulliam, H. R. 1982. A social-learning model of conflict and cooperation in human societies. *Human Ecology* 10:353–363.

Schelling, T. C. 1978. *Micromotives and macrobehavior*. New York: Norton.

Taylor, M. 1976. *Anarchy and cooperation*. New York: Wiley.

Taylor, M., & H. Ward. 1982. Chickens, whales and lumpy goods: Alternative models of public goods provision. *Political Studies* 30:350–370.

Trivers, R. L. 1971. The evolution of reciprocal altruism. *Quarterly Review of Biology* 46:35–57.

Uyenoyama, M., & M. W. Feldman. 1980. Theories of kin and group selection: A population genetics perspective. *Theoretical Population Biology* 17:380–414.

Wade, M. J. 1978. A critical review of the models of group selection. *Quarterly Review of Biology* 53:101–114.

Wilson, D. S. 1980. *The natural selection of population and communities*. Menlo Park, CA: Benjamin/Cummings.

Wilson, E. O. 1975. *Sociobiology: The new synthesis*. Cambridge, MA: Belknap/Harvard University Press.

9 Punishment Allows the Evolution of Cooperation (or Anything Else) in Sizable Groups

Human behavior is unique in that cooperation and division of labor occur in societies composed of large numbers of unrelated individuals. In other eusocial species, such as social insects, societies are made up of close genetic relatives. According to contemporary evolutionary theory, cooperative behavior can be favored by selection only when social groups are formed so that cooperators are more likely to interact with other cooperators than with noncooperators (Hamilton, 1975; Brown, Sanderson, and Michod, 1982; Nunney, 1985). It is widely agreed that kinship is the most likely source of such nonrandom social interaction. Human society is thus an unusual and interesting special case of the evolution of cooperation.

A number of authors have suggested that human eusociality is based on reciprocity (Trivers, 1971; Wilson, 1975; Alexander, 1987), supported by our more sophisticated mental skills to keep track of a large social system. It seems unlikely, however, that natural selection will favor reciprocal cooperation in sizable groups. An extensive literature (reviewed by Axelrod and Dion, 1989; also see Hirshleifer and Martinez-Coll, 1988; Boyd, 1988; Boyd and Richerson, 1989) suggests that cooperation can arise via reciprocity when *pairs* of individuals interact repeatedly. These results indicate that the evolutionary equilibrium in this setting is likely to be a contingent strategy with the general form: "cooperate the first time you interact with another individual, but continue to cooperate only if the other individual also cooperates." Several recent articles (Joshi, 1987; Bendor and Mookherjee, 1987; Boyd and Richerson, 1988, 1989) present models in which larger groups of individuals interact repeatedly in potentially cooperative situations. These analyses suggest that the conditions under which reciprocity can evolve become extremely restrictive as group size increases above a handful of individuals.

In most existing models, reciprocators retaliate against noncooperators by withholding future cooperation. In many situations other forms of retaliation are possible. Noncooperators could be physically attacked, be made the targets of gossip, or denied access to territories or mates. We will refer to such alternative forms of punishment as *retribution*. It seems possible that selection may favor cooperation enforced by retribution even in sizable groups of unrelated individuals because, unlike withholding reciprocity, retribution can be made only against noncooperators, and because the magnitude of the penalty imposed on noncooperators is not limited by an individual's effect on the outcome of cooperative behavior.

Here, we extend the theory of the evolution of cooperation to include the possibility of retribution. We review the evolutionary models of the evolution of reciprocity in sizable groups and present a model of the evolution of cooperation enforced by retribution. An analysis of this model suggests that retribution can lead to the evolution of cooperation in two qualitatively different ways.

1. If the long-run benefits of cooperation to a punishing individual are greater than the costs to that single individual of coercing all other individuals in a group to cooperate, then strategies that cooperate and punish noncooperators, strategies that cooperate only if punished, and, sometimes, strategies that cooperate but do not punish coexist at a stable equilibrium or stable oscillations.
2. If the costs of being punished are large enough, "moralistic" strategies that cooperate, punish noncooperators, and punish those who do not punish noncooperators can be evolutionarily stable.

We also show, however, that moralistic strategies can cause any individually costly behavior to be evolutionarily stable, whether or not it creates a group benefit. Once enough individuals are prepared to punish any behavior, even the most absurd, and to punish those who do not punish, then everyone is best off conforming to the norm. Moralistic strategies are a potential mechanism for stabilizing a wide range of behaviors.

Models of the Evolution of Reciprocity

Models of the evolution of reciprocity among pairs of individuals share many common features. Each assumes that there is a population of individuals. Pairs of individuals are sampled from this population and interact a number of times. During each interaction individuals may either cooperate (C) or defect (D). The incremental fitness effects of each behavior define a single period prisoner's dilemma, and, therefore, cooperative behavior is altruistic in the sense that it reduces the fitness of the individual performing the cooperative behavior but increases fitness of the other individual in the pair (Axelrod and Hamilton, 1981; Boyd, 1988). Each individual is characterized by an inherited *strategy* that determines how he will behave. Strategies may be fixed rules like unconditional defection ("always defect") or contingent ones like tit-for-tat ("cooperate during the first interaction; subsequently do whatever the other individual did last time"). The pair's two strategies determine the effect of the entire sequence of interactions on each pair member's fitness. An individual's contribution to the next generation is proportional to his fitness.

Analysis of such models suggests that lengthy interactions between pairs of individuals are likely to lead to the evolution of reciprocity. Reciprocating strategies, like tit-for-tat, leading to mutual cooperation, are successful if pairs of individuals are likely to interact many times. A population in which unconditional defection is common can resist invasion by cooperative strategies under a wide range of conditions. However, there seem to be a variety of plausible mechanisms that allow reciprocating strategies to increase when rare. Axelrod and Hamilton (1981) and Axelrod (1984) have shown that a very small degree of assortative group formation, when coupled with the possibility of prolonged reciprocity, allows strategies like tit-for-tat to invade noncooperative populations. Other mechanisms have been suggested by Peck and Feldman (1985), Boyd and Lorberbaum (1987), and Feldman and Thomas (1987).

Recent work suggests that these conclusions do not apply to larger groups. Joshi (1987) and Boyd and Richerson (1988) have independently analyzed a model in which n individuals are sampled from a larger population and then interact repeatedly in an n-person prisoner's dilemma. In this model, cooperation is costly to the individual, but beneficial to the group as a whole. This work suggests that increasing the size of interacting social groups reduces the likelihood that selection will favor reciprocating strategies. As in the two individual cases, if groups persist long enough, both reciprocal and noncooperative behavior are favored by selection when they are common. For large groups, however, the conditions under which reciprocity can increase when rare become extremely restrictive. Bendor and Mookherjee (1987) show that when errors occur, reciprocal cooperation may not be favored in large groups even if they persist forever. Boyd and Richerson (1989) derived qualitatively similar results in which groups were structured into simple networks of cooperation.

Intuitively, increasing group size places reciprocating strategies on the horns of a dilemma. To persist when common, they must prevent too many defectors in the population from receiving the benefits of long-term cooperation. Thus, reciprocators must be provoked to defect by the presence of even a few defectors. To increase when rare, there must be a substantial probability that the groups with less than this number of defectors will form. This problem is not great when pairs of individuals interact; a relatively small degree of assortative group formation will allow reciprocating strategies to increase. As groups become larger, however, both of these requirements can be satisfied only if the degree of assortment in the formation of groups is extreme.

This result should be interpreted with caution. Modeling social interaction as an n-person prisoner's dilemma means that the only way a reciprocator can punish a defector is by withholding future cooperation. There are two reasons to suppose that cooperation might be more likely to evolve if cooperators could retaliate in some other way. First, in the n-person prisoner's dilemma, a reciprocator who defects in order to punish defectors induces other reciprocators to defect. These defections induce still more defections. More discriminating retribution would allow defectors to be penalized without generating a cascade of defection. Second, in the n-person prisoner's dilemma the severity of the sanction is limited by an individual's effect on the whole group, which becomes diluted as group size increases. Other sorts of sanctions might be much more costly to defectors and therefore allow rare cooperators to induce others to cooperate in large groups.

There is also a problem with retribution. Why should individuals punish? If being punished is sufficiently costly, it will pay to cooperate. However, by assumption, the benefits of cooperation flow to the group as a whole. Thus, as long as administering punishment is costly, retribution is an altruistic act. Punishment is beneficial to the group but costly to the individual, and selection should favor individuals who cooperate but do not punish. This problem is sometimes referred to as the problem of "second-order" cooperation (Oliver, 1980; Yamagishi, 1986).

A recent article by Axelrod (1986) illustrates the problem of second-order cooperation. Axelrod analyzes a model in which groups of individuals interact for two periods. During the first period individuals may cooperate or defect in an n-person prisoner's dilemma, and in the second, individuals who cooperated on the first move have the opportunity to punish those individuals who did not cooperate at some cost to themselves. Axelrod shows that punishment may expand the range of conditions under which cooperation could evolve. However, the strategy of cooperating but not punishing was precluded. As Axelrod notes, such second-order defecting strategies would always do better because second-order punishment of nonpunishers is not possible.

The problem of second-order cooperation has been partly solved by Hirshleifer and Rasmusen (1989). They consider a game theoretic model in which a two-stage game consisting of a cooperation stage followed by a punishment stage is repeated a number of times. They show that if punishment is costless, then the strategy of cooperating, punishing noncooperators, and punishing nonpunishers is what game theorists call a "perfect equilibrium." (The perfect equilibrium is a generalization of the Nash equilibrium that is useful in repeated games. See Rasmusen, 1989, for an excellent introductory discussion of game theoretic equilibrium concepts.) Because it is a game theoretic model, it does not provide information about the evolutionary dynamics. It also seems possible that if the model were extended to an infinite number of periods, a similar strategy would be evolutionarily stable even if punishment is costly.

Here we consider evolutionary properties of an infinite period model of cooperation with the possibility of punishment that is similar to Hirshleifer and Rasmusen's. We will perform the analysis in three stages. First, we describe the basic structure of the model. Then, we consider populations in which there are cooperators who punish defection and a variety of strategies that initially defect and then respond to punishment in different ways. The goal is to investigate the evolutionary dynamics introduced by retribution without the complications introduced by second-order defection and second-order punishment. Finally, we consider the effects of these complications.

Description of the Model

Suppose that groups of size n are sampled from a large population and interact repeatedly. The probability that the group persists from one interaction to the next is w, and thus the probability that it persists for t or more interactions is w^{t-1}. Each interaction consists of two stages, a cooperation stage followed by a punishment stage. During the cooperation stage an individual can either cooperate (C) or defect (D). The incremental effect of a single cooperation stage on the fitness of an individual depends on that individual's behavior and the behavior of other members of the group as follows: let the number of *other* individuals choosing C during a particular turn be i. Then the payoffs to individuals choosing C and D are:

$$V(C|i) = (b/n)(i + 1) - c \tag{1}$$

$$V(D|i) = (b/n)i \tag{2}$$

where $b > c$ and $c > b/n$. Increasing the number of cooperators increases the payoff for every individual in the group, but each cooperator would be better off switching to defection. (This special case of the n-person prisoner's dilemma has been used in economics and political science to represent provision of public goods [Hardin, 1982]. It is also identical to the linear model of social interactions used in most kin selection models.) During the punishment stage any individual can punish any other individual. Punishing another individual lowers the fitness of the punisher an amount k and the fitness of the individual being punished an amount p.

Each individual is characterized by an inherited "strategy" that specifies how she will behave during any time period based on the history of her own behavior and the behavior of other members of the group up to that point. The strategy specifies whether the individual will choose cooperation or defection during the cooperation stage and which other individuals, if any, she will punish during the punishment stage. Strategies can be unconditional rules like the asocial rule "never cooperate/never punish." They can also be contingent rules like "always cooperate/punish all individuals who didn't cooperate during the cooperation stage."

We assume that individuals sometimes make errors. In particular, we suppose that any time an individual's strategy calls for cooperation, there is a probability $e > 0$ that the individual will instead defect "by mistake." This is the only form of error we investigate. Individuals who mean to defect always defect, and individuals always either punish or do not punish according to the dictates of their strategy.

Groups are formed according to the following rule: the conditional probability that any other randomly chosen individual in a group has a given strategy S_i, given that the focal individual also has S_i, is given by:

$$\Pr(S_i|S_i) = r + (1 - r)q_i \tag{3}$$

where q_i is the frequency of the strategy S_i in the population before social interaction, and $0 \le r \le 1$. The conditional probability that any other randomly chosen individual in a group has some other strategy S_j, given that the focal individual has S_i, is given by:

$$\Pr(S_j|S_i) = (1 - r)q_j \tag{4}$$

When $r = 0$, social interaction occurs at random. When $r > 0$, social interaction is assortative. There is a chance r of drawing an individual with the same strategy as the focal individual and a chance $1 - r$ of picking an individual at random from the population (who will also be identical to the focal individual with probability equal to the frequency of the focal individual's strategy in the population). If strategies are inherited as haploid sexual traits, r is just the coefficient of relatedness. For other genetic models, r is not exactly equal to the coefficient of relatedness. However, it is a good approximation for rare strategies and thus is useful for determining the conditions under which a rare reciprocating strategy can invade a population in which all defection is common.

After all social interactions are completed, individuals in the population reproduce. The probability of reproduction is determined by the results of social behavior. Thus, the frequency of a particular strategy, S_i, in the next generation, q_i', is given by:

$$q_i' = \frac{q_i W(S_i)}{\sum_j q_j W(S_j)} \tag{5}$$

where $W(S_i)$ is the average payoff of individuals using strategy S_i in all groups weighted by the probability that different types of groups occur. (As argued by Brown et al., 1982, this assumption is consistent with haploid genetic inheritance of strategies and some simple forms of cultural transmission.) We then ask, which strategies or combinations of strategies can persist?

Results

No Second-Order Defection

First, we analyze the evolutionary dynamics of retribution with second-order defection excluded. To do this, we consider a world in which only the following two strategies are possible.

Cooperator-punishers (P).   During each interaction (1) cooperate, and (2) punish all individuals who did not cooperate during the cooperation stage.

Reluctant cooperators (R_1).   Defect until punished once, then cooperate forever. Never punish.

We temporarily exclude strategies that cooperate but do not punish to eliminate the possibility of second-order defection. We also exclude strategies that continue to defect after one act of punishment. This latter assumption is *not* harmless. We show in the Appendix that if R_1 is replaced by unconditional defection, then (1) cooperation is much less likely to evolve, and (2) R_1 may not be able to invade a population in which unconditional defection is common. This analysis is justified for two reasons: first, it provides a best case for the evolution of cooperation, and, second, there is abundant empirical evidence that organisms do respond to punishment.

When groups are formed at random ($r = 0$), such a population can persist at one of three stable equilibria (or ESSs):

- All individuals are R_1—no one cooperates.
- All individuals are P—everyone cooperates.
- Most individuals are R_1, but a minority are P—most are induced to cooperate by the punishing few.

In what follows we describe and interpret the conditions under which each of these ESSs can exist. Proofs are given in the Appendix.

Reluctant cooperators resist invasion by the cooperating, punishing strategy whenever the cost to a cooperator-punisher of cooperating and punishing $n - 1$ reluctant cooperators exceeds the benefit to that punisher that results from the cooperation that is induced by his punishment. It can be shown that the responsive defecting strategy R_1 can be invaded by the cooperating, punishing strategy P as long as:

$$k(n - 1) + (c - b/n) < \frac{1}{1 - w}\left(w(b - c) - \frac{ek(n - 1)}{1 - e}\right) \tag{6}$$

initial cost of cooper-   long-run benefit induced
ating and punishing   by punishing

When cooperator-punishers are rare, and groups are formed at random, virtually all cooperator-punishers will find themselves in a group in which the other $n - 1$ individuals are defectors. The left-hand side of (6) gives the fitness loss associated with cooperating, and then punishing $n - 1$ defectors during the first interaction. The right-hand side of (6) gives the long-term net fitness benefit of the cooperation that results from punishment. The term $w(b - c)/(1 - w)$ is the long-term fitness benefit from the induced cooperation by R_1 individuals, and the term proportional to e is the long-run cost that results from having to punish erroneous defections. Thus, if this term is positive, P can invade if w is large enough.

If the cooperator-punisher strategy, P, can increase when rare, punishing is not altruistic. Retribution induces cooperation that creates benefits sufficient to compensate for its cost. The longer groups persist, the larger the benefit associated with cooperation. Thus, as long as error rates are low or the benefits of cooperation are large, longer interactions will permit cooperative strategies to invade, even if groups are formed at random. Also notice that the condition for R_1 to be invaded does not depend on p, the cost of being punished. As one would expect, increasing the group size or the error rate makes it harder for the cooperative strategy to invade.

The cooperating-punishing strategy, P, is evolutionarily stable as long as

$$p(n - 1) > c - b/n + \frac{ek(n - 1)}{(1 - w)(1 - e)} \tag{7}$$

cost of being punished   cost of cooperating
        and punishing

The first term on the right-hand side of (7) gives the cost of cooperating during one interaction; the term on the left-hand side is the cost of being punished by $n - 1$ other individuals, and the second term on the right-hand side is the cost of punishing mistakes over the long run. The rare R_1 individual suffers the cost of punishment but avoids the cost of cooperating on the first turn and the cost of punishing erroneous defection over the long run. Notice that this condition is independent of the long-run expected benefit associated with cooperation (because it does not contain terms of the form $b/(1 - w)$). It depends only on the cost of the cooperation to the individual and the costs of punishing and being punished. Thus, retribution can stabilize cooperation, but this stability does not result from the mutual benefits of cooperation.

There is a stable internal equilibrium at which both P and R_1 are present whenever (1) neither R_1 nor P are ESSs, or (2) R_1 is not an ESS but P is, and the condition (A14) given in the Appendix is satisfied. We have not been able to derive an expression for the frequency of P at the internal equilibrium. Figure 9.1 shows the frequency of P at this equilibrium determined numerically as a function of the expected number of interactions $(\log(1/(1 - w)))$ for various group sizes. When groups persist for only a few interactions, both P and R_1 are ESSs. Increasing the number of interactions eventually destabilizes R_1 and allows a stable internal equilibrium to exist. Further increases in the expected number of interactions destabilize P, leaving the internal equilibrium as the only stable equilibrium.

Without second-order defection, cooperation can persist at two qualitatively different equilibria: either cooperative strategies coexist with noncooperative strategies at a polymorphic equilibrium, or all individuals in the population are cooperative. When the cooperator-punisher strategy is very rare, it will increase whenever the benefit from long-run cooperation to an individual punisher exceeds the cost of the punishment necessary to induce reluctant cooperators to cooperate. As cooperator-punishers become more common, more reluctant cooperators find themselves in groups with at least one cooperator-punisher, and thus they enjoy the benefits of long-run cooperation without bearing the costs associated with punishing. Thus the relative fitness of cooperator-punishers declines. As cooperator-punishers become still more common, reluctant cooperators are punished more harshly during the initial interaction and their relative fitness declines.

Assortative group formation has both positive and negative effects on the conditions under which cooperator-punishers evolve. When there is assortative group formation, individuals are more likely to find themselves in groups with others like themselves than chance alone would dictate. Such assortment decreases the cost of cooperating and punishing because cooperators are more likely to receive the benefits that result from the cooperative acts of others than are noncooperators and because cooperator-punishers need to punish fewer noncooperators on the first interaction. However, assortment decreases the long-run benefit associated with punishment because cooperator-punishers are more likely to be punished for erroneous defection. (Assortment increases the amount of punishment that an inadvertently defecting cooperator-punisher receives.) The second effect becomes more pronounced the longer groups last because

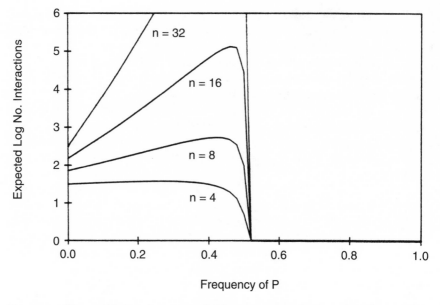

Figure 9.1. The equilibrium frequency of P for a given expected number of interactions for different group sizes ($n = 8$, 16, 32) assuming that $e = 0.001$. For these parameter values populations consisting of all P are always at a stable equilibrium. Populations without P individuals are also always an equilibrium, but it may be either stable or unstable. To find the polymorphic equilibria, pick a number of expected interactions and group size, and then determine the frequencies of P at which the horizontal line at that the value of $\log(1/(1 - w))$ intersects the curve at that value of n. If the horizontal line lies below the curve for some q_P, then the frequency of P increases; if it lies above the curve, the frequency of P decreases. Thus, if there is only one polymorphic equilibrium (e.g., $n = 4$, $\log(1/(1 - w)) = 1$), it is unstable and $q_P = 0$ is stable. If there are two polymorphic equilibria (e.g., $n = 16$, $\log(1/1 - w)) = 3$), the polymorphic equilibrium with the lower frequency of P is stable, and the other polymorphic equilibrium and $q_P = 0$ are both unstable. Finally, if there is no polymorphic equilibrium (e.g., $n = 8$, $\log(1/(1 - w)) = 4$), the only stable equilibrium is $q_P = 1$.

cooperator-punishers will make more errors. The negative effect will predominate whenever the following condition is satisfied:

$$(1 - e)(b/n + k) < \frac{ep}{1 - w} \qquad (8)$$

When expression (8) is satisfied, assortment increases the range of conditions under which R_1 is an ESS, decreases the range of conditions under which P is an ESS, and, if a stable internal equilibrium exists, decreases the frequency of P at that equilibrium. Note that the negative effects increase as the expected number of interactions increases. When (8) is not satisfied, increasing r decreases the range of parameters under which R_1 is an ESS, increases the range under which P is an ESS, and may either increase or decrease the frequency of P at internal equilibria.

Second-Order Defection

When punishers are common, cooperation is favored because cooperative individuals avoid punishment. Thus, if punishment is costly, punishment may be an altruistic act. It is costly to the individual performing the punishment but benefits the group as a whole. This argument suggests that individuals who cooperate, but do not punish, should be successful. In the previous model (and that of Axelrod, 1986) cooperators always punish noncooperators, and thus this conjecture could not be addressed. To allow for second-order defection, consider a model in which P and R_1 compete with the following strategy.

Easy-going cooperator (E): Always cooperate, never punish

When second-order defection is possible, neither E nor P is ever an ESS. A population in which P is common can always be invaded by E, because easygoing cooperators get the benefits of cooperation without incurring the cost of enforcement. A population in which E is sufficiently common can always be invaded by R_1, because reluctant cooperators can enjoy the benefits of cooperation without fear of punishment.

R_1 is an ESS whenever punishment does not pay (i.e., [6] is not satisfied). At this ESS, there is no cooperation because reluctant cooperators behave as unconditional defectors. If the long-run benefits of cooperation to an individual are not sufficient to offset the cost of coercing all the other members of the group to cooperate, the noncooperators can resist invasion by punishing or cooperating strategies. Persistent noncooperation is not the only possible outcome, however, under this condition. If P can resist invasion by R_1 (i.e., [7] is satisfied), then simulation studies indicate that there may be persistent oscillations involving all three strategies. Such oscillations seem to require that the cost of being punished is much greater than the cost of punishing $(p \gg k)$ and the benefits of cooperation barely exceed the cost $(b \approx c)$.

If punishment does pay, the long-run outcome is a mix of reluctant cooperators who coexist with cooperator-punishers and, sometimes, easygoing cooperators. This can happen in three different ways:

- There can be a stable mix of reluctant cooperators and cooperator-punishers. Such a stable equilibrium exists anytime there is a stable polymorphic equilibrium on the $R_1 - P$ boundary in the absence of E. If, in addition, P is not an ESS in the absence of E, this mixture of reluctant cooperators and cooperator-punishers is the only stable equilibrium, and numerical simulations suggest that the polymorphic equilibrium is globally stable. Thus, at equilibrium, populations will consist of a majority of reluctant cooperators with a minority of cooperator-punishers. E cannot invade because rare E individuals often find themselves in groups without a cooperator-punisher and thus pay the cost of cooperation without receiving the long-run benefits of cooperation. Punishers in all groups received the benefits of long-term cooperation.

- If there is no polymorphic equilibrium on the $R_1 - P$ boundary (i.e., in the absence of E), then there is a single interior equilibrium point at which all three strategies are present. We have not been able to derive an expression for the frequencies of the three traits at these interior equilibria or determine when they are stable. Numerical simulation indicates that when an interior equilibrium exists, it is almost always stable.
- The mixture of all three strategies can oscillate. When P is stable in the absence of E, the frequencies of the three strategies may oscillate indefinitely. Simulation studies suggest that this outcome only occurs under relatively rare parameter combinations.

In each case, as group size increases, the average frequency of cooperative strategies typically declines to a quite low level. However, the average frequency of groups with at least one P individual, and therefore groups in which cooperation occurs over the long run, can remain at substantial levels even when groups are large. One must keep in mind, however, that this conclusion presupposes that individual punishers can afford to punish every noncooperator in the group. A model in which the capacity to punish is limited would presumably stabilize at some higher frequency of punishers as group size increased.

Moralistic Strategies

The results of the previous section suggest that strategies that attempt to induce cooperation through retribution can always be invaded when they are common by strategies that cooperate but do not punish. However, such is not the case. Consider the following strategy.

> **Moralists (M)**: Always cooperate, and punish individuals who are not in "good standing." Individuals are in good standing if they have behaved according to M since the last time they were punished or the beginning of the interaction.

Thus, moralists punish individuals who do not cooperate. But they also punish those who do not punish noncooperators and those who do not punish nonpunishers. Each M individual punishes others at most once per turn. Once an individual is punished, he can avoid further punishment by cooperating, punishing noncooperators, and punishing nonpunishers (thus returning to good standing).

Moralists can resist invasion by reluctant cooperators (R_1) whenever the following is true

$$(n-1)p\left(1 - \frac{w}{1-w}(1 - (1-e)^{n-1})\right) > c - b/n + \frac{e(n-1)k}{(1-w)(1-e)} \qquad (9)$$

$$\underbrace{\qquad\qquad\qquad\qquad\qquad\qquad}_{\text{cost of being punished}} \qquad \underbrace{\qquad\qquad\qquad\qquad}_{\substack{\text{cost of cooperating and} \\ \text{punishing}}}$$

The left-hand side of inequality (9) gives the cost to an R_1 individual of being punished. It is proportional to the number of interactions because such reluctant cooperators are punished every time there is an error. The right-hand side is the cost of cooperating and punishing. Thus, as long as the error rate is not exactly zero, moralists can resist invasion by R_1 under a wider range of conditions than can P.

Moralists can resist invasion by easygoing cooperators (E) whenever the following condition is satisfied:

$$(1 - (1 - e)^n)wp > ek \tag{10}$$

If errors occur only infrequently ($ne \ll 1$), then this condition simplifies to become $nwp > k$. Thus, unless punishing is much more costly than being punished, moralists can resist invasion by easygoing cooperators.

In fact, as Hirshleifer and Rasmusen (1989) have pointed out, moralistic aggression of this kind is a recipe for stabilizing *any* behavior. Notice that neither condition (9) or (10) involves terms representing the long-run benefits of cooperation (i.e., terms of the form $b/(1 - w)$). When M is common, rare individuals deviating from M are punished; otherwise, they have no effect on the behavior of the group. Thus, as long as being punished by all the other members of the group is sufficiently costly compared to the individual benefits of not behaving according to M, M will be evolutionarily stable. It does not matter whether or not the behavior produces group benefits. The moralistic strategy could require any arbitrary behavior—wearing a tie, being kind to animals, or eating the brains of dead relatives. Then M could resist invasion by individuals who refuse to engage in the arbitrary behavior unless punished, as long as condition (9) was satisfied (where $c - b/n$ is the cost of the behavior), and resist invasion by individuals who perform the behavior but do not punish others, as long as (10) is satisfied.

Discussion

Our results suggest that problems of second-order cooperation can be overcome in two quite different ways: first, even though retribution creates a group benefit, it need not be altruistic. If defectors respond to punishment by a single individual by cooperating, and if the long-run benefits to the individual punisher are greater than the costs associated with coercing other group members to cooperate, then the strategy that cooperates and punishes defectors can increase when rare and will continue to increase until an interior equilibrium is reached. At this equilibrium, the punishing strategy coexists with strategies that initially defect but respond to punishment by cooperating and, sometimes, strategies that cooperate but do not punish. For plausible parameter values, the punishing strategy is rarer than the other two strategies at such an equilibrium. However, since a single punisher is sufficient to induce cooperation, cooperating groups are nonetheless quite common.

Increasing group size reduces the likelihood that this mechanism will lead to the evolution of cooperation because it increases the cost of coercion. This effect, however, is not nearly so strong as previous models in which defection was punished by withdrawal of cooperation. In those models (Joshi, 1987; Boyd and Richerson, 1988, 1989), a linear increase in group size requires an exponential increase in the expected number of interactions necessary for cooperation to increase when rare. In the present model, the same increase in group size requires only a linear increase in the expected number of interactions.

Moralistic strategies that punish defectors, individuals who do not punish noncooperators, and individuals who do not punish nonpunishers can also overcome the problem of second-order cooperation. When such strategies are common, rare noncooperators are selected against because they are punished. Individuals who cooperate but do not punish are selected against because they are also punished. In this way, selection may favor punishment, even though the cooperation that results is not sufficient to compensate individual punishers for its costs.

It is not clear whether moralistic strategies can ever increase when rare. We have not presented a complete analysis of the dynamics of moralistic strategies because to do so in a sensible way would require the introduction of additional strategies, a consideration of imperfect monitoring of punishment, and a consideration of more general temporal patterns of interaction. We conjecture, however, that the dynamics will be roughly similar to the dynamics of P and R_1 in the case in which there is no stable internal equilibrium: both defecting and moralistic strategies will be evolutionarily stable. Increasing the degree of assortment will mean that moralists will have fewer defectors to punish but will be punished more when they err. Assortative social interaction will not interact with group benefits in a way that will allow moralistic strategies to increase.

It is also interesting that moralistic strategies stabilize any behavior. The conditions that determine whether M can persist when rare are independent of the magnitude of the group benefit created by cooperation. The moralistic strategy could stabilize any behavior equally well, whether it is beneficial or not. If our conjecture about the dynamics of M is correct, then the dynamics will not be strongly effected by whether or not the sanctioned behavior is group-beneficial.

This result is reminiscent of the "folk theorem" from mathematical game theory. This theorem holds that in the repeated prisoner's dilemma with a constant probability of termination (the case analyzed by Axelrod and most other evolutionary theorists), strategies leading to *any* pattern of behavior can be a game theoretic perfect equilibrium (Rasmusen 1989). The proof of this theorem relies on the fact that if there is enough time available (on average) for punishment, then individuals can be induced to adopt any pattern of behavior. Thus, in games without a known endpoint, game theory may predict that anything can happen. This result, combined with the fact that nobody lives forever, has led many economists to restrict their analyses to games with known endpoints. The diversity of equilibria here and in the nonevolutionary analysis can be regarded as a flaw or embarrassment for the analysis.

We prefer to take these results as telling us something about the evolution of social behavior. Games without a known endpoint seem to us to be a good model for many social situations. Although nobody lives forever, social groups often persist much longer than individuals. When they do, individuals can expect to be punished until their own last act. Even dying men are tried for murder, and in many societies one's family is also subject to retribution. If one accepts this argument, then it follows that moralistic punishment is inherently diversifying in the sense that many different behaviors may be stabilized in exactly the same environment. It may also provide the basis for stable among-group variation. Such stable among-group variation can allow group selection to be an important process (Boyd and Richerson 1985, 1990 a,b), leading to the evolution of behaviors that increase group growth and persistence.

Conclusion

Cooperation enforced by retribution is strikingly different from reciprocity in which noncooperation is punished by withdrawal of cooperation. We think two features of this system are interesting and warrant further study:

1. Cooperation may be possible in larger groups than is the case with reciprocity. This effect invites further study of the limitations on the ability of single individuals to punish and how coalitions of punishers might or might not be able to induce reciprocity in very large groups.

2. In the model studied here, punishers collect private benefit by inducing cooperation in their group that compensates them for punishing, while providing a public good for reluctant cooperators. There are often polymorphic equilibria in which punishers are relatively rare, generating a simple political division of labor reminiscent of the "big man" systems of New Guinea and elsewhere. This finding invites study of further punishment strategies. Consider, for example, strategies that punish but do not cooperate. Such individuals might be able to coerce more reluctant cooperators than cooperator-punishers and therefore support cooperation in still larger groups. If so, such models might help explain the evolution of groups organized by full-time specialized, "parasitical" coercive agents like tribal chieftains.

The importance of the study of retribution can hardly be underestimated. The evolution of political complexity in human societies over the last few thousand years depended fundamentally on the development of a variety of coercive strategies similar to those we have investigated here.

APPENDIX

SENSITIVITY OF THE MODEL TO THE RESPONSE TO PUNISHMENT

The effects of punishment on the evolution of cooperation are strongly affected by the extent to which a defector responds to punishment by cooperating. To see this, consider a game in which cooperator-punishers (P) compete with the following nonresponsive strategy.

Unconditional defectors (U): Never cooperate. Never punish.

Many of the evolutionary properties of the two-person repeated prisoner's dilemma can be derived considering a model in which only tit-for-tat (TFT, cooperate on the first move, and punish each defection by defecting) and ALLD (always defect) are present. Our strategies P and U seem like the natural generalizations of TFT and ALLD to the n-person game with punishment, and one might (as we did) expect that their evolutionary dynamics would be similar. This expectation is largely incorrect. Understanding why provides useful insight into the evolutionary effects of punishment. For simplicity, we assume that there are no errors $(e=0)$ throughout this section.

Let j be the number of the other $n-1$ individuals in the group who are P. The expected fitness of U individuals given j is:

$$V(U|j) = \frac{(b/n - p)j}{1 - w} \tag{A1}$$

Similarly, the expected fitness of P individuals given j is:

$$V(P|j) = \frac{b/n(j + 1) - c - (n - 1 - j)k}{1 - w} \tag{A2}$$

The expected fitness of U individuals averaged over all groups is:

$$W(U) = \sum_{j=0}^{n-1} m(j|U)V(U|j)$$

$$= E(j|U)\frac{b/n - p}{1 - w} \tag{A3}$$

where $m(j|U)$ is the probability that there are j other cooperator-punishers, given that the focal is an unconditional defector, and $E(j|U)$ is the expected value of j conditioned on the focal individual being U. An analogous calculation shows that

$$W(P) = \frac{(b/n)(E(j|P) + 1) - c - (n - 1 - E(j|P))k}{1 - w} \tag{A4}$$

When groups are formed at random $E(j|P) = E(j|U) = (n-1)q$ where q is the frequency of P in the population just before groups are formed. To determine when U is an ESS, let $q \to 0$ and determine when $W(U) > W(P)$. To determine when P is an ESS, let $q \to 1$ and determine when $W(U) < W(P)$. When groups are formed assortatively and P is rare, $E(j|P) = (n-1)r$ and $E(j|U) = 0$. Combining these expressions yields the condition for P to increase when rare (A6).

It follows from these expressions for the fitness of U and P that (1) unconditional defection is always an ESS, and (2) P is an ESS only if:

$$c - b/n < (n - 1)p \tag{A5}$$

The left-hand side of (A5) is the per period cost to an individual of cooperating, and the right-hand side is the per period cost of being punished by $n-1$ individuals.

Superficially these properties seem analogous to the competition between always-defect and tit-for-tat in the two-person model. Always-defect is always an ESS; tit-for-tat is an ESS only under certain conditions. However, notice that (A5) does not depend on the parameter w, which measures the average number of interactions. Thus, if (A5) is satisfied, P is stable even if individuals interact only once! In contrast, tit-for-tat is stable against always-defect only if w is large enough that the long-run benefit of reciprocal interaction is greater than the short-term benefit of cheating. Tit-for-tat is never stable if individuals interact only once.

The qualitative difference between the two models is made clearer if we consider the effect of assortative group formation. In the two-person case, assortative group formation makes it easier for tit-for-tat to increase when rare, and if w is near one, even a small amount of assortment is sufficient. In the present model, the punishing strategy, P, can increase when rare if the following is true:

$$(b/n)[r(n-1)+1]-c > (n-1)(1-r)k \qquad \text{(A6)}$$

$$\underbrace{}_{\text{inclusive fitness}} \qquad \underbrace{}_{\text{punishment}}$$

The left-hand side gives the inclusive fitness advantage of cooperators relative to defectors. If P individuals are sufficiently likely to interact with other P individuals $(r \to 1)$, then P can increase in frequency even when it is rare in the population because P individuals benefit from the cooperation of other P individuals in their groups. The right-hand side gives the effect of punishment on the fitness of P individuals. Notice that this term is always positive. This means that cooperation supported by punishment is *harder* to get started in a population than unconditional cooperation.

Why are these two models so different? In models without retribution, reciprocal strategies such as tit-for-tat are favored because they lead to assortative interaction of cooperators (Michod and Sanderson, 1985). Even if individuals are paired at random, the fact that tit-for-tat individuals convert to defection if they experience acts of defection from others, causes a nonrandom distribution of cooperative behavior: tit-for-tat individuals are more likely to receive the benefits of cooperation than are always-defect individuals. In contrast, in the present model, punishment has no effect on who receives the benefits of cooperative behavior. P individuals continue to cooperate while they punish, and U individuals do not respond to punishment by cooperating—they keep defecting. Models of reciprocity without punishment suggest that the strategy of punishing defectors by withdrawing cooperation is unlikely to work in large groups (Joshi, 1987; Boyd and Richerson, 1988). However, it is not unreasonable to imagine that a kind of conditional defector might respond to punishment by cooperating much as tit-for-tat responds to cooperation with more cooperation.

SHOULD DEFECTORS RESPOND TO PUNISHMENT?

Should defecting individuals respond to punishment by cooperating? To address this question, we consider the conditions in which R_1 can invade a population in which the strategy U is common. We further assume that groups are formed at random.

Unfortunately, the answer to this question does not depend on the fitness consequences of alternative behaviors alone. It also depends on what kinds of punishing strategies are maintained in the population by nonadaptive processes like mutation and nonheritable environmental variation. In a population in which only U and R_1 are present (and every individual accurately follows its strategy), U and R_1 will have the same expected fitness. Both will defect forever and never be punished because no punishing strategies are present. The strategies U and R_1 will have different expected fitnesses only if there are punishing strategies present in the population. If U is common, however, the expected fitness of any rare punishing strategy must be less than the expected fitness of U. This means that any punishing strategies present in the population must be maintained by nonadaptive processes like errors or mutation. R_1 may or may not be able to invade, depending on the mix of punishing strategies maintained by such forces.

We conjecture that the most plausible source of nonadaptive variation is mistakes about the behavioral context. Modelers typically assume that there is a single behavioral context, with given costs and benefits, and an unambiguous set of behavioral strategies. However, in the real world, there are many behavioral contexts, each with its own appropriate strategy. Before deciding how to behave, individuals must categorize a particular situation as belonging to one context or another. It seems plausible that individuals sometimes miscategorize situations in which punishment is not favored and thus mistakenly punish others. Suppose, for example, selection favors individual retaliation if others damage personal property. Then individuals might sometimes punish others who damage commonly held property because they mistakenly miscategorize the behavior.

To prove that R_1 may or may not be able to invade U, consider the second punishing strategy.

Timid punishers (T_1): Always cooperate. Punish each defector the first time it defects, but only the first time.

Suppose that both U and R occasionally mistakenly play one of the punishing strategies. This could occur because individuals mistake the behavioral context for one in which they would normally punish. The relative fitness of U and R_1 depends on which of these two punishing strategies is present. Suppose that individuals occasionally play T_1 by mistake. R_1 can invade if a focal R_1 individual has higher fitness than a focal U individual in groups with one T_1 individual among the other $n - 1$. In such groups,

$$W(U) = \frac{b/n}{1 - w} - p \tag{A7}$$

$$W(R_1) = \frac{b/n}{1 - w} - p + \frac{w}{1 - w}(b/n - c) \tag{A8}$$

Thus, U is always favored if cooperation is costly. In contrast, when P is present as a result of errors, the fitnesses of the two types are as follows:

$$W(U) = b/n - p + \frac{w}{1 - w}(b/n - p) \tag{A9}$$

$$W(R_1) = b/n - p + \frac{w}{1 - w}(2b/n - c) \tag{A10}$$

Thus, R_1 is favored whenever the costs of punishment exceed the cost of cooperating.

We think that this result is likely to be quite general. Consider a strategy that begins cooperating only after being punished some number of times. Such a strategy will have higher fitness than an unresponsive strategy only if the punishing strategies present in the population continue to punish on subsequent turns. If they do not, the unresponsive strategy gets the benefit without paying the cost. When should punishing strategies give up? The answer to this question depends on whether the defecting strategies will respond. If defecting strategies are unresponsive, costly punishment provides no benefits.

EQUILIBRIA WHEN R_1 AND P COMPETE

Let j be the number of P individuals among the other $n - 1$ individuals in a group. Then the expected fitnesses of the two types are:

$$W(P) = (1 - e)[(b/n)(E(j|P) + 1) - c] - k[n - 1 - (1 - e)E(j|P)]$$
$$- epE(j|p) + \frac{w}{1 - w}[(1 - e)(b - c) - ek(n - 1) - epE(j|P)] \quad \text{(A11)}$$

$$W(R_1) = [(b/n)(1 - e) - p]E(j|R_1)$$
$$+ \frac{w}{1 - w}[(1 - e)(b - c) - epE(j|R_1)]$$
$$- \frac{w\Pr(j = 0|R_1)}{1 - w}(1 - e)(b - c) \quad \text{(A12)}$$

where $\Pr(j = 0|R_1)$ is the probability that an R_1 individual finds himself in a group with exactly zero P individuals.

When groups are formed at random, $E(j|P) = E(j|R_1) = (n - 1)q$ and $\Pr(j = 0|R_1) = (1 - q)^{n-1}$, where q is the frequency of P. Making these substitutions leads to the following condition for R_1 to increase:

$$(k + p)(n - 1)(1 - q) - \frac{w}{1 - w}(1 - q)^{n-1}(b - c)$$
$$- b/n + c - p(n - 1) + \frac{ek(n - 1)}{(1 - w)(1 - e)} > 0 \quad \text{(A13)}$$

The condition for R_1 to be an ESS (7) is derived by setting $q = 0$ in (A13). The condition for P to be an ESS (6) is derived by setting $q = 1$ in (A13).

To derive the necessary conditions for a stable internal equilibrium, first notice that the left-hand side of (A13) is a concave function with, at most, a single internal maximum. Thus, if neither R_1 or P is an ESS, then there is a single internal equilibrium point. If R_1 is not an ESS but P is, then there are two internal equilibria, one stable and the other unstable, if, and only if, the value of the left-hand side at that maximum is greater than zero. The value of q that maximizes the left-hand side of (A13) can be found by differentiation. Substituting this back into (A13) yields the following necessary condition for the existence of two internal equilibria:

$$\left(\frac{(1 - w)(k + p)}{w(b - c)}\right)^{1/(n-2)}(n - 2)(k + p) >$$
$$p(n - 1) - c + b/n - \frac{ek(n - 1)}{(1 - w)(1 - e)} \quad \text{(A14)}$$

If this condition is not satisfied, then P is the only ESS.

To derive the condition for R_1 to increase when groups are formed assortatively, let $E(j|P) = (n - 1)(r + (1 - r)q)$ and $E(j|R_1) = (n - 1)(1 - r)q$ and proceed in the same way.

EQUILIBRIA WHEN R_1, E, AND P COMPETE

Let i and j be the numbers of E and P individuals among the other $n - 1$ individuals. Here is the equation:

$$W(P) = (1 - e)[(b/n)(E(i|P) + E(j|P) + 1) - c]$$
$$- k[n - 1 - (1 - e)(E(i|P) + E(j|P))] - epE(j|P)$$
$$+ \frac{1}{1 - w}[(1 - e)(b - c) - ek(n - 1) - epE(j|P)] \quad \text{(A15)}$$

$$W(R_1) = (b/n)(1 - e)(E(i|R_1) + E(j|R_1)) - pE(j|R_1)$$
$$+ \frac{w}{1 - w}\Pr(j > 0|R_1)[(1 - e)(b - c) - epE(j|R_1 \& j > 0)]$$
$$+ \frac{w}{1 - w}(1 - e)(b/n)\Pr(j = 0|R_1)E(i|R_1 \& j = 0) \quad \text{(A16)}$$

$$W(E) = (1 - e)[(b/n)(E(i|E) + E(j|E) + 1) - c] - epE(j|E)$$
$$+ \frac{w}{1 - w} \Pr(j > 0|E)[(1 - e)(b - c) - epE(j|E \& j > 0)]$$
$$+ \frac{w}{1 - w}(1 - e)\Pr(j = 0|E)[(b/n)(E(i|E \& j = 0) + 1) - c] \tag{A17}$$

Assume that groups are formed at random so that $E(j|E) = E(j|P) = E(j|R_1) = (n - 1)q_P$, $E(i|E) = E(i|P) = E(i|R_1) = (n - 1)q_E$, $\Pr(j = 0|E) = \Pr(j = 0|R_1) = (1 - q_P)^{(n - 1)}$, and $E(i|R_1 \& j = 0) = E(i|E \& j = 0) = (n - 1)(q_E/(1 - q_P))$ where q_E and q_P are the frequencies of E and P. When $q_E = 1$, $W(E) < W(R_1)$ and when $q_P = 1$, $W(P) < W(E)$.

First, we derive conditions for the existence of an internal equilibrium and show that if such an equilibrium exists, it is unique.

It is useful to define the following functions, which give the difference in fitness between each pair of strategies as a function of q_P and q_E:

$$d_{PE}(q_P, q_E) = W(P) - W(E) \tag{A18}$$

$$d_{RE}(q_P, q_E) = W(R_1) - W(E) \tag{A19}$$

$$d_{PR}(q_P, q_E) = W(P) - W(R_1) \tag{A20}$$

Using equations (A15), (A16), and (A17) and the assumption of random group formation yields the following expression for d_{RE}:

$$d_{RE}(q_P, q_E) = -p(1 - e)(n - 1)q_P$$
$$+ (1 - e)(c - b/n)\left(1 + \frac{w}{1 - w}(1 - q_P)^{n-1}\right) \tag{A21}$$

Notice that the relative fitness of R_1 and E depends only on q_P. Further, note that (1) $d_{RE}(0, q_E) > 0$; (2) $d_{RE}(1, q_E) < 0$ as long as $c - b/n < (n - 1)p$, which is true by assumption; and (3) d_{RE} is a monotonically decreasing function of q_P. Thus, the value of q_P at equilibrium is unique and can be found by finding the root of $d_{RE} = 0$ as shown in figure A1. Let this value of q_P be $\tilde{q}_P$.

Once again, using equations (A15), (A16), and (A17) and the assumption of random group formation yields the following expression for d_{PE}:

$$d_{PE}(q_P, q_E) = \frac{-ke(n - 1)}{1 - w} + k(1 - e)(n - 1)(1 - q_E - q_P)$$
$$+ \frac{wb(n - 1)(1 - e)(1 - q_P)^{n-2}(1 - q_P - q_E)}{n(1 - w)}$$

Assume that q_E is fixed at some value. Then

$$d_{PE}(1 - q_E, q_E) = \frac{-ke(n - 1)}{1 - w} < 0$$

and

$$d_{PE}(0, q_E) = \frac{-(n - 1)ke}{1 - w} + (1 - q_E)(1 - e)(n - 1)\left(\frac{wb}{n(1 - w)} - k\right)$$

Thus, $d_{PE}(0, q_E) > 0$ if

$$q_E < 1 - \frac{ke}{(1 - e)(w(b/n) - k(1 - w))} \tag{A22}$$

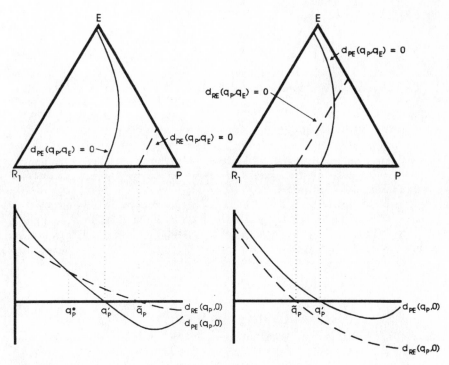

Figure A1. This figure illustrates the logic of the proofs given in this section. The left-hand pair of figures represents a situation in which there is a single polymorphic equilibrium on the $R_1 - P$ boundary. The lower figure shows $d_{RE}(q_P,0)$ and $d_{PE}(q_P,0)$. These curves intersect only once since there is a single polymorphic equilibrium. Thus, we know that $q'_P < \tilde{q}_P$. The upper figure shows how the forms of $d_{RE}(q_P,q_E)$ and $d_{PE}(q_P,q_E)$ guarantee that there is no internal equilibrium in this case. The right-hand pair of figures represents the situation in which there is no polymorphic equilibrium on the $R_1 - P$ boundary because P increases for all values of $q_P < 1$.

and

$$w(b/n) - k(1-w) > 0 \qquad (A23)$$

Otherwise, $d_{PE} \le 0$. Differentiating shows that d_{PE} is a convex function of q_P. Thus, if (A22) and (A23) are satisfied, $d_{PE}(q_P,q_E)=0$ has a unique root for each q_P as illustrated in figure A1. Let this root be $q'_P(q_E)$. Increasing q_E leads to a decrease in $q'_P(q_E)$. Thus, there is a internal equilibrium value if, and only if, $q'_P(0) > \tilde{q}_P$, and if it exists, such an equilibrium is unique. This result is shown graphically in figure A1.

We know that $d_{RE}(q_P)$ is monotonically decreasing and has one root in the interval $(0, 1)$ whenever R_1 is potentially present, and that $d_{PE}(q_P)$ has at most one root and is monotonically decreasing in the interval that contains the root.

Next, we show that if there is no stable polymorphic equilibrium on the $P - R_1$ boundary in the absence of E, then there is an internal equilibrium. If there is no stable equilibrium on the boundary in the absence of E, it follows from the results of the previous section that

$$d_{PR}(q_P,0) > 0$$

for all q_P Next, note that

$$d_{PR}(q_P, q_E) = d_{PE}(q_P, q_E) - d_{RE}(q_P, q_E) \qquad (A24)$$

Thus, there is an internal equilibrium since $d_{PE}(q_P, 0) > d_{RE}(q_P, 0)$ for all values of q_P. This situation is shown in the right-hand pair of figures in figure A1.

Next, we show that if there is a stable, polymorphic boundary equilibrium such that $q_E = 0$ and $W(R_1) > W(P)$ for $q_P = 1$, then there is no internal equilibrium. Let q_P^* be the frequency of P at a polymorphic equilibrium on the $P - R_1$ boundary. Then $d_{PR}(q_P^*, 0) = 0$, which implies that $d_{PE}(q_P^*, 0) = d_{RE}(q_P^*, 0)$. The fact that the equilibrium is stable in the absence of E implies that $\partial d_{PR}(q_P, 0)/\partial q_P < 0$ at q_P^*. Since $d_{PR}(1, 0) < 0$, it follows that $d_{PE}(q_P, 0) < d_{RE}(q_P, 0)$ for $q_P^* < q_P < 1$. But this means that $q_P(0) < \tilde{q}_P$, and, therefore, there is no internal equilibrium as shown in the left-hand pair of figures in figure A1.

It is important to note that there may be no internal equilibrium even if $W(R_1) < W(P)$. When this is the case, there is a second, unstable internal equilibrium on the $R_1 - P$ boundary. Anytime that $d_{PE} = d_{RE} < 0$ at this equilibrium, there will be no internal equilibrium, and numerical studies suggest that this is what actually occurs at the vast majority of parameter combinations.

M IS AN ESS AGAINST P AND E

Assume that M is common. When groups are formed at random, M can resist invasion by rare R_1 individuals if the average payoff of M in groups with $n - 1$ other M individuals, $V(M|n - 1)$ is greater than the average payoff of R_1 in groups in which the other $n - 1$ individuals are M, $V(R_1|n - 1)$:

$$V(M|n - 1) = \frac{1}{1 - w}((b - c)(1 - e) - e(n - 1)(k + p)) \qquad (A25)$$

$$V(R_1|n - 1) = (n - 1)(b/n)(1 - e) - (n - 1)p$$
$$+ \frac{w}{1 - w}[(1 - e)(b - c) - p(n - 1)(1 - (1 - e)^n)] \qquad (A26)$$

Substituting these expressions and simplifying yields condition (9). Similarly, the expected fitness of an E individual in a group of $n - 1$ M individuals, $V(E|n - 1)$, is:

$$V(E|n - 1) = (1 - e)(b - c) - e(n - 1)p$$
$$+ \frac{w}{1 - w}[(1 - e)(b - c) - e(n - 1)p - p(n - 1)(1 - (1 - e)^n)] \qquad (A27)$$

This expression is used to determine when $V(M|n - 1) > V(E|n - 1)$ yields equation (10).

NOTE

We thank Alan Rogers for useful comments and for carefully checking every result in this chapter. Joel Peck also provided helpful comments.

REFERENCES

Alexander, R. D. 1987. *The biology of moral systems.* New York: Aldine De Gruyter.
Axelrod, R. 1984. *The evolution of cooperation.* New York: Basic Books.

Axelrod, R. 1986. The evolution of norms. *American Political Science Review* 80: 1095–1111.

Axelrod, R., & D. Dion. 1989. The further evolution of cooperation. *Science* 232: 1385–1390.

Axelrod, R., & W. D. Hamilton. 1981. The evolution of cooperation. *Science* 211: 1390–1396.

Bendor, J., & D. Mookherjee. 1987. Institutional structure and the logic of ongoing collective action. *American Political Science Review* 81:129–154.

Boyd, R. 1988. Is the repeated prisoner's dilemma game a good model of reciprocal altruism? *Ethology and Sociobiology* 9:211–221.

Boyd, R. 1987. Mistakes allow evolutionary stability in the repeated prisoner's dilemma game. *Journal of Theoretical Biology* 136:47–56.

Boyd, R., & J. Lorberbaum. 1987. No pure strategy is evolutionarily stable in the repeated prisoner's dilemma game. *Nature* 327:58–59.

Boyd, R., & P. J. Richerson. 1985. *Culture and the evolutionary process*. Chicago: University of Chicago Press.

Boyd, R., & P. J. Richerson. 1988. The evolution of reciprocity in sizable groups. *Journal of Theoretical Biology* 132:337–356.

Boyd, R., & P. J. Richerson. 1989. The evolution of indirect reciprocity. *Social Networks* 11:213–236.

Boyd, R., & P. J. Richerson. 1990a. Culture and cooperation. In: *Beyond self interest*, Jane Mansbridge, ed. (pp. 111–132). Chicago: University of Chicago Press.

Boyd, R., & P. J. Richerson. 1990b. Group selection among alternative evolutionarily stable strategies. *Journal of Theoretical Biology* 145:331–324.

Brown, J. S., M. J. Sanderson, & R. E. Michod. 1982. Evolution of social behavior by reciprocation. *Journal of Theoretical Biology* 99:319–339.

Feldman, M. W., & E. R. C. Thomas. 1987. Behavior dependent contexts for repeated plays of the repeated prisoner's dilemma II: Dynamical aspects of the evolution of cooperation. *Journal of Theoretical Biology* 129:297–314.

Hamilton, W. D. 1975. Innate social aptitudes of man: An approach from evolutionary genetics. In: *Biosocial anthropology*, R. Fox, ed. (pp. 135–232). London: Malaby.

Hardin, R. 1982. *Collective action*, Baltimore: Johns Hopkins University Press.

Hirshleifer, D., & E. Rasmusen. 1989. Cooperation in a repeated prisoner's dilemma with ostracism. *Journal of Economic Behavior and Organization*, 12: 87–106.

Hirshleifer, D., & J. Martinez-Coll. 1988. What strategies can support the evolutionary emergence of cooperation? *Journal of Conflict Resolution* 32:367–398.

Joshi, N. V. 1987. Evolution of cooperation by reciprocation within structured demes. *Journal of Genetics* 1:69–84.

Michod, R. E., & M. J. Sanderson. 1985. Behavioural structure and the evolution of cooperation. In: *Evolution: Essays in honor of John Maynard Smith*, P. J. Greenwood and M. Slatkin, eds. Cambridge: Cambridge University Press (pp 95–104).

Nunney, L. 1985. Group selection, altruism, and structured deme models. *American Naturalist* 126:212–230.

Oliver, P. 1980. Rewards and punishments as selective incentives for collective action: Theoretical investigations. *American Journal of Sociology* 85:1356–1375.

Peck, J., & M. W. Feldman. 1985. The evolution of helping behavior in large, randomly mixed populations. *American Naturalist* 127:209–221.

Rasmusen, E. 1989. *Games and information: An introduction to game theory*. Oxford: Basil Blackwell.

188 HUMAN COOPERATION, RECIPROCITY, GROUP SELECTION

Trivers, R. 1971. The evolution of reciprocal altruism. *Quarterly Review of Biology* 46:35–57.

Wilson, E. O. 1975. *Sociobiology: The new synthesis*, Cambridge, MA: Belknap/Harvard University Press.

Yamagishi, T. 1986. The provisioning of sanctioning as a public good. *Journal of Personality and Social Psychology* 51:100–116.

10   Why People Punish Defectors

Weak Conformist Transmission Can Stabilize Costly Enforcement of Norms in Cooperative Dilemmas

With Joseph Henrich

In many societies, humans cooperate in large groups of unrelated individuals. Most evolutionary explanations for cooperation combine kinship (Hamilton, 1964) and reciprocity ("reciprocal altruism," Trivers, 1971). These mechanisms seem to explain the evolution of cooperation in many species including ants, bees, naked mole rats, and vampire bats. However, because social interaction among humans often involves large groups of mostly unrelated individuals, explaining cooperation has proved a tricky problem for both evolutionary and rational choice theorists. Evolutionary models of cooperation using the repeated n-person prisoner's dilemma predict that cooperation is not likely to be favored by natural selection if groups are larger than around 10, unless relatedness is very high (Boyd and Richerson, 1988). As group size rises above 10, to 100 or 1000, cooperation is virtually impossible to evolve or maintain with only reciprocity and kinship.[1]

Many students of human behavior believe that large-scale human cooperation is maintained by the threat of punishment. From this view, cooperation persists because the penalties for failing to cooperate are sufficiently large that defection "doesn't pay." However, explaining cooperation in this way leads to a new problem: why do people punish noncooperators? If the private benefits derived from punishing are greater than the costs of administering it, punishment may initially increase but cannot exceed a modest frequency (Boyd and Richerson, 1992). Individuals who punish defectors provide a public good, and thus can be exploited by nonpunishing cooperators if punishment is costly. Second-order free riders cooperate in the main activity but cheat when it comes time to punish noncooperators. As a consequence, second-order free riders receive higher payoffs than punishers do, and thus punishment is not evolutionarily stable. Adding third- (third-order punishers punish second-order free riders) or higher-order punishers only pushes the problem back to higher orders. Solving this problem is important because there is widespread agreement that the threat of punishment plays an important role in the maintenance of cooperation in many human societies.

Social scientists have explained the maintenance of punishment in three ways: (1) many authors assume that a state or some other external institution does the punishing; (2) others assume punishing is costless (McAdams, 1997; Hirshleifer and Rasmussen, 1989); and (3) a few scholars incorporate a recursive punishing method in which punishers punish defectors, individuals who fail to punish defectors, individuals who fail to punish nonpunishers, and so on in an infinite regress (Boyd and Richerson, 1992; Fundenberg and Maskin, 1986). However, none of these solutions is satisfactory. While it is useful to assume institutional enforcement in modern contexts, it leaves the evolution and maintenance of punishment unexplained because at some point in the past there were no states or institutions. Furthermore, the state plays a very small role in many contemporary small-scale societies that nonetheless exhibit a great deal of cooperative behavior. This solution avoids the problem of punishment by relocating the costs of punishment outside the problem. The second solution, instead of relocating the costs, assumes that punishment is costless. This seems unrealistic because any attempt to inflict costs on another must be accompanied by at least some tiny cost—and any nonzero cost lands both genetic evolutionary and rational choice approaches back on the horns of the original punishment dilemma. The third solution, pushing the cost of punishment out to infinity, also seems unrealistic. Do people really punish people who fail to punish other nonpunishers, and do people punish people who fail to punish people, who fail to punish nonpunishers of defectors and so on, ad infinitum? Although the infinite recursion is cogent, it seems like a mathematical trick.

Conformist Transmission in Social Learning Can Stabilize Punishment

In this chapter, we argue that the evolution of cooperation and punishment are plausibly a side effect of a tendency to adopt common behaviors during enculturation. Humans are unique among primates in that they acquire *much* of their behavior from other humans via social learning. However, both theory and evidence suggest that humans do not simply copy their parents, nor do they copy other individuals at random (Henrich and Boyd, 1998; Takahasi, 1998; Harris, 1998). Instead, people seem to use social learning rules like "copy the successful" (termed pay-off biased or prestige-biased transmission; see Henrich and Gil-White, 2001) and "copy the majority" (termed conformist transmission; Boyd and Richerson, 1985; Henrich and Boyd, 1998), which allow them to shortcut the costs of individual learning and experimentation and leapfrog directly to adaptive behaviors. These specialized social learning mechanisms provide a generalized means of rapidly sifting through the wash of information available in the social world and inexpensively extracting adaptive behaviors. These social learning shortcuts do not always result in the best behaviors, nor do they prevent the acquisition of maladaptive behaviors. Nevertheless, when averaged over many environments and behavioral domains (e.g., foraging, hunting, social interaction, etc.), these cultural transmission mechanisms provide fast and frugal means to acquire complex, highly adaptive behavioral repertoires.

Both theoretical and empirical research indicates that conformist transmission plays an important role in human social learning. We have already shown that a heavy reliance on conformist transmission outcompetes both unbiased (i.e., vertical) transmission and individual learning under a wide range of conditions (Henrich and Boyd, 1998), and especially when problems are difficult. Second, empirical research by psychologists, economists, and sociologists shows that people are likely to adopt common behaviors across a wide range of decision domains. Although much of this work focuses on easy perceptual tasks (Asch, 1951) and confounds normative conformity (going with the popular choice to avoid appearing deviant) with conformist transmission (using the popularity of a choice as an indirect measure of its worth), more recent work shows that social learning and conformist transmission are important in difficult individual problems (Baron, Vandello, and Brunsman, 1996; Insko, Smith, Alicko, Wade, and Taylor, 1985; Campbell and Fairey, 1989), voting situations (Wit, 1999), *and* cooperative dilemmas (Smith and Bell, 1994).

Conformist transmission can stabilize costly cooperation without punishment but only if it is very strong. All other things being equal, payoff-biased transmission causes higher payoff variants to increase in frequency, and thus cooperation is not evolutionarily stable under plausible conditions—because not cooperating leads to higher payoffs than cooperating. Thus, payoff-biased transmission, alone, suffers the same problem as natural selection in genetic evolution. However, under conformist transmission individuals preferentially adopt common behaviors and thus increase the frequency of the most common behavior in the population. Thus, if cooperation is common, conformist transmission will oppose payoff-biased transmission and, as long as cooperation is not too costly, maintain cooperative strategies in the population. However, if the costs of cooperation are substantial, it is less likely that conformist transmission will be able to maintain cooperation.

A quite different logic applies to the maintenance of punishment. Suppose that both punishers and cooperators are common and that being punished is costly enough that cooperators have higher payoffs than defectors. Rare invading second-order free riders who cooperate but do not punish will achieve higher payoffs than punishers because they avoid the costs of punishing. However, because defection does not pay, the only defections will be due to rare mistakes, and thus the *difference* between the payoffs of punishers and second-order free riders will be relatively small. Hence, conformist transmission is more likely to stabilize the punishment of noncooperators than cooperation itself. As we ascend to higher-order punishing, the difference between the payoffs to punishing versus nonpunishing decreases geometrically toward zero because the occasions that require the administration of punishment become increasingly rare. Second-order punishing is required only if someone erroneously fails to cooperate, and then someone else erroneously fails to punish that mistake. For third-order punishment to be necessary, yet another failure to punish must occur. As the number of punishing stages (i) increases, conformist transmission, no matter how weak, will at some stage overpower payoff-biased imitation and stabilize common i-th order punishment. Once punishment is stable at the i-th stage, payoffs will favor strategies that punish at the $(i - 1)$-th order, because common punishers at the i-th order will punish nonpunishers at stage $i - 1$. Stable punishment at stage $(i - 1)$ means payoffs at stage $i - 2$ will favor punishing strategies, and so on down the cascade of punishment. Eventually, common first-order punishers will stabilize cooperation at stage 0.

It is important to see that the stabilization of punishment is, from the gene's point of view, a maladaptive side effect of conformist transmission. If there were genetic variability in the strength of conformist transmission (α) and cooperative dilemmas were the *only* problem humans faced, then conformist transmission might never evolve. However, human social learning mechanisms were selected for their capability to efficiently acquire adaptive behaviors over a wide range of behavioral domains and environmental circumstances—from figuring out what foods to eat, to deciding what kind of person to marry—precisely because it is costly for individuals to determine the best behavior. Hence, we should expect conformist transmission to be important in cooperation as long as distinguishing cooperative dilemmas from other kinds of problems is difficult, costly, or error-prone. Looking across human societies, we find that cooperative dilemmas come in an immense variety of forms, including harvest rituals among agriculturalists, barbasco fishing among Amazonian peoples, warfare, irrigation projects, taxes, voting, meat sharing, and anti-smoking pressure in public places. It is difficult to imagine a cognitive mechanism capable of distinguishing cooperative circumstances from the myriad of other problems and social interactions that people encounter.

In what is to come, we formalize this argument. Our goal is to demonstrate the soundness of our reasoning and show how very weak conformist transmission can stabilize cooperation and punishment. After demonstrating this, we will describe how cooperation, once it is stabilized in one group, can spread across many populations via *cultural group selection*. We will also briefly show how genes for prosocial behavior may eventually spread in the wake of cultural evolution.

A Cultural Evolutionary Model of Cooperation and Punishment

In this model, a large number of groups, each consisting of N individuals, are drawn at random from a very large population. Individuals within each group interact with one another in an $i + 1$ stage game. The first stage is a one-shot cooperative dilemma, which is followed by i stages in which individuals can punish others. We number the first, cooperative stage as 0 and the punishment stages as $1, \ldots, i$. The behavior of individuals during each stage is determined by a separate culturally acquired trait with two variants, P (prosocial variant) and NP (not prosocial variant).

During the initial cooperative dilemma, individuals can either "cooperate"—contribute to a public good—or "defect"—not contribute and free-ride on the contributions of others. Each cooperator pays a cost C to contribute a benefit B ($B > C$) to the group—this B is divided equally among *all* group members. Defectors do not pay the cost of cooperation (C) but do share equally in the total benefits. The variable p_0 represents the frequency of individuals in the population with the cooperative variant in stage 0. People with the cooperative variant "intend" to cooperate but mistakenly defect with probability e. Individuals who have the defecting variant always defect. This makes sense because, in the real world, people may intend to cooperate but fail to for some reason. For example, a friend who plans to help you move may forget to show up or have car trouble en route. Defectors, however, are unlikely to mistakenly show up on moving day and start carrying boxes. We will assume errors are rare, so the value of e is small.

During the first punishment stage, individuals can punish those who defected during the cooperation stage. Doing this reduces the payoff of the individuals who are punished by an amount ρ, at a cost of ϕ to the punisher ($\phi < \rho < C$). Individuals with the punishing (P) variant at this stage intend to punish but mistakenly fail to punish with probability e. Nonpunishers, those with the NP-variant at stage 1, do nothing. We use p_1 to stand for the frequency of first-stage punishers (i.e., individuals who have the P-variant at stage 1), and ($1 - p_1$) gives the frequency of first-stage free riders.

During the second punishment stage, individuals with the P-variant punish those who did not punish the noncooperators during the previous stage with probability ($1 - e$) and mistakenly fail to punish with probability e. And as before, punishment costs punishers ϕ to administer and costs those being punished an amount ρ. Those with the NP-variant at stage 2 do not punish. Let p_2 be the frequency of second-stage punishers. At stage 3, individuals with the P-variant will punish individuals from stage 2 who failed to punish nonpunishers from stage 1. The costs of punishment remain the same. Those with the NP-variant in stage 3 will not punish anyone from stage 2. The pattern repeats as one descends to stage i in table 10.1 (p_i gives the frequency of punishers at stage i). Because the interaction ends after stage i, individuals who fail to punish on stage i cannot be punished. Note that the trait that controls individual behavior at each stage has only two variants, and the values of variants at different stages are

Table 10.1. Dichotomous traits for cooperation and punishment

Stage	Frequency of P-variant	P-variant	NP-variant
0	p_0	Cooperate	Defect
1	p_1	Punish defectors	Do not punish defectors
2	p_2	Punish nonpunishers at stage 1	Do not punish nonpunishers at stage 1
3	p_3	Punish nonpunishers at stage 2	Do not punish nonpunishers at stage 2
i	p_i	Punish nonpunishers at stage $i - 1$	Do not punish nonpunishers at stage $i - 1$

independent—so an individual could cooperate at stage 0 (have the P-variant), not punish at stage 1 (NP-variant), and punish at stage 2 (P-variant).

After all the punishments are complete, cultural transmission takes place. As we explained earlier, two components of human cognition create forces that change the frequency of the different variants: payoff-biased and conformist-biased imitation. Equation (1) gives the change in the frequency of stage 1 cooperators as a consequence of payoff-biased and conformist transmission (see Henrich, 1999).

$$\Delta p_0 = p_0(1 - p_0) \underbrace{[(1 - \alpha)\beta(b_C - b_D)}_{\text{Payoff} - \text{biased}} + \underbrace{\alpha(2p_0 - 1)]}_{\text{Conformist}} \tag{1}$$

The parameter α varies from 0 to 1 and represents the strength of conformist transmission in human psychology *relative* to payoff-biased transmission. We will generally assume α is positive but small. Practically speaking, α must be less than 0.50, because otherwise beneficial variants would never spread—once a variant became common, it would remain common no matter how deleterious. The second term in equation (1), labeled "conformist," varies in magnitude from $-\alpha$ to $+\alpha$ and is the component of the overall bias contributed by conformist transmission. In the term labeled "payoff-biased," the symbols b_C and b_D are the payoffs to cooperators and defectors, respectively. The quantity $(b_C - b_D)$, which we label Δb_0, gives the difference in payoffs between cooperation (P-variant) and defection (NP-variant) in stage 0. More generally, Δb_i is the difference in payoffs between the P- and NP-variants during the i-th stage. The parameter β normalizes the quantity Δb_i so that it varies between -1 and $+1$, and therefore $\beta = 1/|\Delta b_i|_{max}$. Thus, the term labeled "payoff-biased" varies between $-(1 - \alpha)$ and $+(1 - \alpha)$ and represents the component of the overall bias contributed by payoff-biased transmission.

The expected payoffs, b, to the P- and NP-variant at each stage depend on the rate of errors, the costs of cooperation and punishment, and the frequency of cooperators and punishers in the population. At stage 0, cooperators receive an average payoff of b_C, while defectors receive an average payoff of b_D:

$$\begin{aligned}
b_C &= (1 - e)(p_0 B(1 - e) - C + e(p_0 B - N p_1 \rho)), \\
b_D &= (1 - e)(p_0 B - N p_1 \rho), \\
\Delta b_0 &= b_C - b_D = (1 - e)(N p_1 (1 - e)\rho - C)
\end{aligned} \tag{2}$$

Also as we mentioned, the term Δb_0 gives the difference in payoffs between the two variants that control stage 0 behavior.

A Heuristic Analysis

Let us first analyse equation (1) by asking under what conditions will transmission favor cooperation ($\Delta p_0 > 0$) in the absence of stage 1 punishers ($p_1 = 0$). In this case, $\Delta b_0 = -C(1 - e)$, which is always negative; hence, payoff-biased transmission never favors cooperation in the absence of punishment. So, to give cooperation its best chance, we assume that by some stochastic fluctuations, the frequency of cooperators ends up near one. How big does α have to be so that conformist transmission overpowers payoff-biased transmission and increases the frequency of cooperators? The frequency of cooperators increases when

$$\alpha_0 > \frac{1}{1 + \beta_0 C(1 - e)} \tag{3}$$

where α_i (here, $i = 0$) is the minimum value of α that favors the spread or maintenance of the P-variant at stage i ($\Delta p_i > 0$). With no punishment, $\beta_i = 1/|\Delta b_i|_{max}$ means $\beta_0 = 1/(C(1 - e))$. As a consequence, α_0 must be greater than 0.50, and as we mentioned earlier, $\alpha_i > 0.50$ seems extremely unlikely because such high values would prevent the diffusion of novel practices—cultures would be entirely static (see Henrich, 2001). Hence, conformist transmission, operating directly on cooperative strategies, is unlikely to maintain cooperation in the absence of punishment.

Now, let us examine the conditions under which first-stage punishment will increase in frequency. Again, the change in the frequency of first-stage punishers, Δp_1, is affected by both payoff-biased and conformist transmission:

$$\Delta p_1 = p_1(1 - p_1)[(1 - \alpha)\beta(b_{P1} - b_{NP1}) + \alpha(2p_1 - 1)] \tag{4}$$

The payoffs (bs) to punishment and nonpunishment depend on the cost of punishing (ϕ) and of being punished (ρ), as well as the chance of mistakenly not punishing (e). The subscript P1 indicates the P-variant at stage 1, while NP1 indicates the NP-variant at stage 1.

$$\begin{aligned}
b_{P1} &= -(1 - e)N\phi(1 - p_0 + p_0 e) - eNp_2\rho(1 - e), \\
b_{NP1} &= -Np_2(1 - e)\rho, \\
\Delta b_1 &= b_{P1} - b_{NP1} = -N(1 - e)(\phi(1 - (1 - e)p_0) - p_2(1 - e)\rho)
\end{aligned} \tag{5}$$

Assuming that there is only one punishment stage ($i = 1$), and that cooperators and stage 1 punishers are initially common ($p_0 = 1$ and $p_1 = 1$), then $\Delta b_1 = -N(1 - e)e\phi$. If errors are rare enough that terms involving e^2 are negligible, then $\Delta b_1 \approx -Ne\phi$. Thus, the difference in payoff between the P-variant and the NP-variants at stage 1 is just the cost of punishing cooperators who make errors. If $e < (1/N)$, which is plausible unless groups are very large, then Δb_1 is less than ϕ—and smaller than Δb_0 because $\phi < \rho < C$. Note that, when $i > 0$, $\beta = 1/(N(1 - e)(\rho(1 - e) + e\phi))$, so the threshold value of α necessary to stabilize cooperation in a two-stage game α_1, is:[2]

$$\alpha_1 = \frac{\phi e}{\rho(1 - e) + 2\phi e} \approx \frac{e\phi}{\rho} \tag{6}$$

Equation (6) tells us that α_1 depends only on the error rate and the ratio of the cost of punishing to the cost of being punished. It also says that unless punishing is much more costly than being punished ($2\phi e > \rho$), the threshold strength of conformism necessary to maintain first-stage punishment is small and less than the amount of conformism necessary to stabilize 0-th stage cooperation ($\alpha_0 > \alpha_1 \approx e$).

If we do the same analysis for stage 2, we get the following expressions for Δp_2 and Δb_2:

$$\Delta p_2 = p_2(1 - p_2)[(1 - \alpha)\beta\Delta b_2 + \alpha(2p_2 - 1)] \tag{7}$$

where

$$\Delta b_2 = b_{P2} - b_{NP2} = -(1 - e)N[\phi(1 - p_1(1 - e))$$
$$\times (1 - p_0^N(1 - e)^N) - p_3(1 - e)\rho] \tag{8}$$

The first term inside the square brackets in equation (8) is proportional to the number of individuals who did not punish during stage 1 $(1 - p_1(1 - e))$ and to the probability that there was at least one defector during stage 0: $(1 - p_0^N(1 - e)^N)$. The quantity $p_0(1 - e)$ is the expected frequency of cooperators who did not make a mistake; thus, $(p_0(1 - e))^N$ gives the probability that a group contains all cooperators who did not make a mistake—so, to get the probability that a group contains at least one defector, we simply subtract this probability from one. The second term inside the brackets is the cost of being punished during stage 2 for failing to punish during stage 1. If no third-stage punishers exist $(p_3 = 0)$, and first-stage punishers and cooperators are initially very common, then $\Delta b_2 \approx -(eN)^2\phi$. Note, the difference in payoffs, Δb_2, is a factor of eN smaller than Δb_1, but the strength of conformist transmission remains constant. Calculating the required size of α_2 we get:

$$\alpha_2 = \frac{N\phi e^2}{\rho(1 - e) + e\phi} \approx \frac{e\phi}{\rho}Ne \tag{9}$$

Equation (9) demonstrates that $0 < \alpha_2 < \alpha_1 < \alpha_0 = \frac{1}{2}$. In this case $\alpha_2 \approx Ne\alpha_1$.

If we repeat this calculation for games with more punishment stages, we find that, although punishment during the last stage of the game is never favored by payoff-biased transmission *alone*, any positive amount of conformist transmission $(\alpha > 0)$ will, for some finite number of stages, overcome payoff-biased transmission and stabilize punishment. For any value i $(i > 0)$, the amount of conformist transmission required to stabilize punishment at the i-th stage is:

$$\alpha_i = \frac{\phi e(Ne)^{i-1}}{\rho(1 - e) + e\phi(1 + (Ne)^{i-1})} \approx \frac{e\phi}{\rho}(Ne)^{i-1}. \tag{10}$$

Equation (10) shows that the minimum amount of conformism necessary to stabilize punishment during the last stage, α_i, gets smaller and smaller for greater values of i (assuming $e < 1/N$).

Once conformist transmission overcomes payoff-biased transmission and stabilizes punishment at stage i, punishment at the stage $i - 1$ will be stabilized because nonpunishers at stage $i - 1$ will be punished by frequent punishers during stage i. Once punishing strategies are common and stable at stage $i - 1$, frequent punishers at $i - 1$ will cause payoff-biased transmission to favor the prosocial variant at stage $i - 2$. In most cases, a combination of punishment and conformist transmission will eventually stabilize cooperation at stage 0. However, if C is sufficiently greater than $N\rho(1 - e)$, then stable punishment at stage 1 will not be able to overcome the costs of cooperation at stage 0, and cooperation will not be maintained, despite stable, high-frequency first-stage punishers.

Formal Stability Analysis

A more rigorous local stability analysis of the complete set of recursions supports the heuristic argument just given. Consider the set of $i+1$ difference equations where $\Delta p_j (j = 0, 1, \ldots, i$; see the Appendix) provides the dynamics of the behavioral traits at each stage. The cooperative equilibrium point $(p_0 = 1, p_1 = 1, \ldots, p_i = 1)$ is locally stable under two distinct conditions:

Stability Condition 1

When $i > 0$ and $C < \rho(1 - e)N + (eN)^i \phi$, the cooperative equilibrium is locally stable when:

$$\lambda_0 = -\alpha + (1 - e)(1 - \alpha)\beta\phi(Ne)^i < 0 \tag{11}$$

where $\beta = 1/(N(1 - e)((1 - e) + e\phi))$. First, note that if $\alpha = 0$, the cooperative equilibrium is never stable because all the parameters involved are always positive. However, as long as α is positive and $e < 1/N$, then the system of equations will be stable for some finite value of i. Substituting in the value of β, and solving equation (11) for α, we find that the minimum value of α is:

$$\alpha_i > \frac{e\phi(Ne)^{i-1}}{\rho(1 - e) + e\phi(1 + (Ne)^{i-1})} \tag{12}$$

which is the same value, given in equation (10), derived using a less formal argument.

Stability Condition 2

However, if $C > \rho(1 - e)N + (eN)^i \phi$ and $i > 0$, then the cooperative equilibrium is stable when:

$$\lambda_0 = -\alpha + (1 - \alpha)(1 - e)\beta(C - (1 - e)N\rho) < 0 \tag{13}$$

If we then solve this for the values of α that create a stable cooperative equilibrium, we find:

$$\alpha_i > \frac{\beta(1 - e)(C - (1 - e)N\rho)}{1 + \beta(1 - e)(C - (1 - e)N\rho)} \tag{14}$$

Under stability condition 2, $\beta = 1/(C(1 - e))$, so:[3]

$$\alpha_i > \frac{1 - [N\rho(1 - e)/C]}{2 - [N\rho(1 - e)/C]} \tag{15}$$

The term $N\rho(1 - e)/C$ is always between zero and one, so the required α is always less than $\frac{1}{2}$. This means that, even when the expected costs of being punished by everyone does *not* exceed the cost of cooperation (or the cost saved by defecting), the cooperative equilibrium can still be favored. Intuitively, this is the case in which conformist transmission and punishment combine to overcome the cost of cooperation. As with the previous condition, however, it is con-formist transmission that stabilizes i-th stage punishment, which stabilizes first-stage punishment.

At first, stability condition 2 may seem strange, but the world is seemingly full of cases in which the costs of being punished seem insufficient to explain the observed degree of cooperation. Hence, this may illuminate such things as why Americans pay too much in taxes (i.e., more than they should assuming most people pay because they fear punishment; Skinner and Slemrod, 1985), why Americans wait in line, why the Aché share meat (Kaplan and Hill, 1985), and why people bother going to the voting booth (Mueller, 1989)—all of which seem overly cooperative, given the expected penalty. As we show, this may be important from a cultural group selection perspective because groups that minimize the costs of punishing and being punished (ρ and ϕ), while still main-taining cooperation, will do better than those that rely heavily on punishment to maintain cooperation.

Once Cooperation Is Stabilized, It Can Spread by Cultural Group Selection

By itself, the present model does not provide an explanation for human coop-eration. We have shown that, under plausible conditions, a relatively weak conformist tendency can stabilize punishment and therefore cooperation. How-ever, noncooperation and nonpunishment are also an equilibrium of the model, and we have given no reason, so far, why most populations should stabilize at the cooperative equilibrium rather than the noncooperative equilibrium. However, when there are multiple stable cultural equilibria with different average payoffs, *cultural group selection* can lead to the spread of the higher payoff equilibrium. As we have demonstrated, cultural evolutionary processes will cause groups to exist at different behavioral equilibria. This means that different groups have differ-ent expected payoffs (due to different degrees of economic production, for ex-ample). The expected payoff of individuals from cooperative groups is $b \approx (1 - e)(B - C - eN(\phi + \rho(1 + i)))$, while the expected payoff of individuals in noncooperative/nonpunishing groups is zero. Thus, cooperative groups will have a higher average payoff as long as the benefits of cooperation are bigger than the costs of cooperation and punishment. The combination of conformism and payoff-biased transmission must also be strong enough to maintain stable co-operation in the face of migration between groups. Such persistent differences between groups creates the raw materials required by cultural group selection.

Cultural group selection can operate in a number of ways to spread proso-cial behaviors. Cooperative groups will have higher total production and con-sequently, more resources that can support more rapid population growth relative to noncooperative groups. Or cooperative groups may be better able to marshal and supply larger armies than noncooperative groups and hence be more successful in warfare and conquest. However, although these factors may be important (see Bowles, 2000), another, slightly subtler, cultural group selec-tion process may also be significant. Payoff-biased imitation means people will preferentially copy individuals who get higher payoffs. The higher an individual's payoff, the more likely that individual is to be imitated. If individuals have occasion to imitate people in neighboring groups, people from cooperative populations will be preferentially imitated by individuals in noncooperative populations because the average payoff to individuals from cooperative populations is much higher than the average payoff of individuals in noncooperative populations. Boyd and Richerson (2000) have shown that, under a wide range of conditions (and fairly quickly), this form of cultural group selection will deterministically spread group-beneficial behaviors from a single group (at a group-beneficial equilibrium) through a meta-population of other groups, which were previously stuck at a more individualistic equilibrium.

Culturally Evolved Cooperation May Cause Genes for Prosocial Behavior to Proliferate

Once the cooperative equilibrium becomes common, it is plausible that natural selection acting on genetic variation will favor genes that cause people to cooperate and punish—because such genes decrease an individual's chance of suffering costly punishment. This could arise in many ways. Individuals might develop a preference for cooperative or punishing behaviors that increases their likelihood of acquiring such behaviors. Or, alternatively, natural selection might increase the reliance on conformist transmission, making people more likely to acquire the most frequent behavior.

Here, we analyze the case in which the probability of mistakenly defecting or not punishing, e, varies genetically. We assume that cultural evolution is much faster than genetic evolution, which implies that the population exists at a culturally evolved cooperative equilibrium. Further assume that while most individuals still make errors at the rate e, rare mutant individuals have a slightly different error probability of $e'(=e-\varepsilon)$, where ε is small ($|\varepsilon| \ll e$). If we assume that an individual's average payoff, b, is proportional to her average genetic fitness, then we can ask whether prosocial mutants will spread. The expected fitnesses for the two types, F and F_m ("m" for mutant), and the difference between them, ΔF, are as follows (assuming $i > 0$):[4]

$$F \approx (1-e)(B - C - eN(\phi + \rho(1-e)(i+1)),$$
$$F_m \approx B(1-e) - C(1-e') - N(e\phi + e'\rho(1-e)(i+1)),$$
$$\Delta F = F_m - F = \varepsilon(N\rho(i+1) - C) \qquad (16)$$

When ΔF is positive, prosocial genes can invade. If $C < (1-e)N\rho + (eN)^i\phi$ (*stability condition* 1), then C is always less than $N\rho(1-e)(i+1)$, and prosocial genes are always favored. Once at fixation, these prosocial genes cannot be invaded by more error-prone, anti-social, individuals.

In stability condition 2, where $C > (1-e)N\rho + (eN)^i\phi$, prosocial genes are favored (for $i > 0$) when:

$$1 + \frac{(Ne)^i\phi}{N\rho(1-e)} < \frac{C}{N\rho(1-e)} < i+1 \qquad (17)$$

which is a wide range, since the smallest possible value of i is 1. However, there exists a range of conditions in which culturally evolved cooperation is stable, but prosocial genes cannot invade—in fact, anti-social genes (genes favoring more mistakes) may invade. This occurs when (for $i > 0$):

$$\underbrace{(i+1)}_{\substack{\text{No} \\ \text{prosocial}}} < \frac{C}{N\rho(1-e)} < \underbrace{\frac{(1-\alpha)}{1-2\alpha}}_{\text{Stability}} \tag{18}$$

When condition (18) holds, cultural transmission will stabilize cooperation, but prosocial genes will not be able to invade—instead, anti-social genes will be favored (i.e., ε is negative). Note, however, that the *minimum* value of α for this condition to exist requires $\alpha > 0.333$, which occurs when $i = 1$. Generally, we believe α is much smaller than this, but we will await the verdict of future empirical work. Interestingly, this anti-social invasion is likely to occur in the groups most favored by cultural group selection—those who maximize group payoff by minimizing punishment costs (and i), without destabilizing cooperation. Unfortunately, anti-social invasion will decrease average payoffs and may eventually destabilize cooperation. Further work on this gene-culture interaction will require coevolutionary models that combine both cultural and genetic evolutionary processes (perhaps using quantitative traits) and particularly the cultural group selection process we have described.

As we have begun to model it here, prosocial genes are not strongly selected against in noncooperative populations because error making, in terms of mistaken cooperation and punishment, occurs only when individuals adopt prosocial traits—defectors do not mistakenly cooperate. So, if the world is a mix of cooperative and noncooperative populations, prosocial genes will be favored in a wide range of circumstances in cooperative populations and will be comparatively neutral in noncooperative populations. It is possible that incorporating defector errors, in the form of mistaken cooperation or punishment, may affect this prediction. Furthermore, cooperation may not be a dispositional trait of individuals, but rather a specific behavior or value tied only to certain cultural domains. Some cultural groups, for example, may cooperate in fishing and house-building but not warfare. Other groups may cooperate in warfare and fishing but not house-building. Such culturally transmitted traits would have the form "cooperate in fishing," "cooperate in house-building," and "do not cooperate in warfare," rather than the more dispositional approach of simply "cooperate" versus "do not cooperate." If this is the case, then the migration and spread of prosocial genes becomes more difficult. As prosocial genes spread among groups with different stable cooperative domains, individuals with such genes would be more likely to mistakenly cooperate in noncooperative cultural domains. For example, in cultures where people cooperate in fishing but not warfare, individuals with prosocial genes may be more likely to mistakenly cooperate in warfare (and pay the cost), as well as less likely to mistakenly defect in cooperative fishing. We intend to pursue those avenues in subsequent work.

Conclusion

We have done three things in this chapter. First, we have shown that, if humans possess a psychological bias toward copying the majority, as well as a bias toward imitating the successful, then cultural evolutionary processes will stabilize co-operation and punishment for some finite number of punishment stages. Second, we discussed how, once cooperation is stable, a particular form of cultural group selection is likely to spread these group-beneficial cultural traits through human populations. And finally, we have demonstrated that prosocial genes, which cannot otherwise spread, can invade in the wake of these cultural evolutionary processes, under a wide range of conditions.

APPENDIX

For all i:

$$\Delta p_i = p_i(1 - p_i)[[(1 - \alpha)\beta(\Delta b_i) + \alpha)2p_i - 1)]]$$

Difference in payoff for $i = 0$:

$$\Delta b_0 = b_C - b_D = (1 - e)(Np_1(1 - e)\rho - C)$$

Difference in payoffs for $i > 0$:

$$\Delta b_i = b_{Pi} - b_{NPi} = -(1 - e)N(\phi(1 - p_{i-1}(1 - e))$$

$$\times \prod_{j=0}^{i-2} (1 - p_{j-2}^N(1 - e)^N) - p_{i+1}(1 - e)\rho)$$

where

$$(1 - e)^N = 1 + \sum_{j=1}^{N} \frac{(-1)^j N! e^j}{j!(N - j)!} \approx 1 - Ne$$

Thus,

$$\Delta b_i = b_{Pi} - b_{NPi} \approx -(1 - e)N(\phi(1 - p_{i-1}(1 - e))$$

$$\times \prod_{j=0}^{i-2} (1 - p_{j-2}^N(1 - Ne) - p_{i+1}(1 - e)\rho)$$

Eigenvalues for the system of $i + 1$ equations with punishment up to the i-th stage:

$$\lambda_0 = -\alpha + (1 - \alpha)(1 - e)\beta(C - (1 - e)N\rho),$$

$$\lambda_j = -\alpha + (1 - \alpha)(1 - e)\beta((eN)^j\phi - \rho N(1 - e)), 0 < j < i,$$

$$\lambda_i = -\alpha + (1 - \alpha)(1 - e)\beta(eN)^i\phi$$

When the dominant eigenvalue (that with the largest value) is less than zero, the system is locally stable at point $(p_0, p_i, \ldots, p_{i+1}) = (1, 1, \ldots, 0)$.

NOTES

We would like to thank Natalie Smith, Herbert Gintis, and the anonymous reviewers for their assistance and suggestions in preparing this chapter.

1. Two other explanations for cooperation go by the handles *by-product mutualism* (Brown, 1983) and *group selection* (Sober and Wilson, 1998). In by-product mutualism, individuals who "cooperate" get a higher payoff (have a higher expected fitness) than noncooperators. The cooperative contribution to the fitness of others is simply a by-product of narrow self-interest. That is, in the process of helping myself, I also help you "by accident." Hence, although this situation may abound in nature, it is not the situation we are interested in (and not cooperation by many definitions). And, while genetic group selection may explain some cooperation in nature (e.g., honeybees; see Seeley, 1995), we believe that gene flow rates between human populations, relative to selection, are too high to maintain the required variation between groups (Richerson and Boyd, 1998).

2. Note, under a small range of conditions, when $C > N(\rho(1-e) + e\phi)$, the system can still remain stable. Under these conditions, however, β becomes $1/C(1-e)$. For simplicity, we leave this nuance until later in the chapter.

3. Actually, there is a tiny range of $(N\rho(1-e) + \phi(eN)^i) < C < (N\rho(1-e) + N\phi e)$ under which β still equals $1/(N(1-e)(\phi(1-e) + e\phi))$. Nothing particularly interesting happens in this range, so we will not discuss it. Note, if $i = 1$, the range is nonexistent.

4. If conformist transmission alone can stabilize cooperation without any punishment $(i = 0)$, then $\Delta F < 0$, and prosocial genes will never spread.

REFERENCES

Asch, S. E. 1951. Effects of group pressure upon the modification and distortion of judgments. In: *Groups, leadership and men*, H. Guetzkow, ed. (pp. 39–76). Pittsburgh: Carnegie.

Baron, R., J. Vandello, & B. Brunsman. 1996. The forgotten variable in conformity research: Impact of task importance on social influence. *Journal of Personality and Social Psychology* 71:915–927.

Bowles, S. 2000. Individual interactions, group conflicts and the evolution of preferences. In: *Social Dynamics*, S. Durlauf, & P. Young, eds. Washington, DC: Brookings Institution.

Boyd, R., & P. J. Richerson. 1985. *Culture and the evolutionary process*. Chicago: University of Chicago Press.

Boyd, R., & P. J. Richerson. 1988. The evolution of reciprocity in sizable groups. *Journal of Theoretical Biology* 132:337–356.

Boyd, R., & P. J. Richerson. 1992. Punishment allows the evolution of cooperation (or anything else) in sizable groups. *Ethology and Sociobiology* 13:171–195.

Boyd, R., & P. J. Richerson. 2000. Norms and bounded rationality. In: *The adaptive tool box*, G. Gigerenzer, & R. Selten, eds. Cambridge, MA: MIT Press.

Brown, J. L. 1983. Cooperation—a biologist's dilemma. *Adv. Stud. Behav.* 13:1–37.

Campbell, J. D., & P. J. Fairey. 1989. Informational and normative routes to conformity: The effect of faction size as a function of norm extremity and attention to the stimulus. *Journal of Personality and Social Psychology* 57:457–468.

Fudenberg, D., & E. Maskin. 1986. The folk theorem in repeated games with discounting or with incomplete information. *Econometrica* 54:533–554.

Hamilton, W. D. 1964. The genetical evolution of social behavior. *Journal of Theoretical Biology* 7:1–52.

Harris, J. R. 1998. *The nurture assumption: Why children turn out the way they do*. New York: Touchstone.

Henrich, J. 2001. Cultural transmission and the diffusion of innovations: Adoption dynamics indicate that biased cultural transmission is the predominate force in behavioral change and much of sociocultural evolution. *American Anthropologist* 103:992–1013.

Henrich, J., & R. Boyd. 1998. The evolution of conformist transmission and the emergence of between-group differences. *Evolution and Human Behavior* 19:215–242.

Henrich, J., & F. Gil-White. 2001. The evolution of prestige. *Evolution and Human Behavior* 2211–32.

Hirshleifer, D., & E. Rasmusen. 1989. Cooperation in the repeated prisoner's dilemma with ostracism. *Journal of Economic, Behavior and Organization* 12:87–106.

Insko, C. A., R. H. Smith, M. D. Alicke, J. Wade, & S. Taylor. 1985. Conformity and group size: The concern with being right and the concern with being liked. *Personality and Social Psychology Bulletin* 11:41–50.

Kaplan, H., & K. Hill. 1985. *Current Anthropology* 26:223–245.

McAdams, R. H. 1997. The origin, development, and regulation of norms. *Michigan Law Review* 96:338.

Mueller, D. 1989. *Public choice II*. Cambridge: Cambridge University Press.

Richerson, P. J., & R. Boyd. 1998. The evolution of ultrasociality. In: *Indoctrinability, ideology and warfare*, I. Eibl-Eibesfeldt, & F. K. Salter, eds. (pp. 71–96). New York: Berghahn Books.

Seeley, T. D. 1995. *The wisdom of the hive*. Cambridge, MA: Harvard University Press.

Skinner, J., & J. Slemrod. 1985. An economic perspective on tax evasion. *National Tax Journal* 38:345–353.

Smith, J. M., & P. A. Bell. 1994. Conformity as a determinant of behavior in a resource dilemma. *Journal of Social Psychology* 134:191–200.

Sober, E., & D. S. Wilson. 1998. *Unto others: The evolution and psychology of unselfish behavior*. Cambridge, MA: Harvard University Press.

Takahasi, K. 1998. Theoretical aspects of the mode of transmission in cultural inheritance. *Theoretical Population Biology* 55:208–225.

Trivers, R. L. 1971. The evolution of reciprocal altruism. *Quarterly Review of Biology* 46:35–57.

Wit, J. 1999. Social learning in a common interest voting game. *Games and Economic Behavior* 26:131–156.

11   Can Group-Functional
Behaviors Evolve by Cultural
Group Selection?

An Empirical Test

With Joseph Soltis

Many anthropologists explain human behavior and social institutions in terms of group-level functions (Rappaport, 1984; Lenski and Lenski, 1982; Harris, 1979; Radcliffe-Brown, 1952; Aberle, Cohen, Davis, Levy, and Sutton, 1950; Malinowski, 1984 [1922]; Spencer, 1891). According to this view, beliefs, behaviors, and institutions exist because they promote the healthy functioning of social groups. Such functionalists believe that the existence of an observed behavior or institution is explained if it can be shown how the behavior or institution contributes to the health or welfare of the social group. Most functionalists in anthropology have not explained how group-beneficial beliefs and institutions arise or by what processes they are maintained (Turner and Maryanski, 1979). When functionalists do provide a mechanism for the generation or maintenance of group-level adaptations, it is usually in terms of selection among social groups.[1] Functionalists believe that societies have many functional prerequisites. Social groups whose culturally transmitted values, beliefs, and institutions do not provide for these prerequisites become extinct, leaving only those societies with functional cultural attributes as survivors. We refer to this process as "cultural group selection" because it involves the differential survival and proliferation of culturally variable groups.

Cultural group selection is analogous to genetic group selection but acts on cultural rather than genetic differences between groups. This distinction is important. We will argue that cultural variation is more prone to group selection than genetic variation and that this may explain why human societies, in contrast to those of other animals, are frequently cooperative on scales far larger than kin groups. More generally, recent theoretical work on the processes of cultural evolution shows that there are many parallels between cultural and genetic evolution but also some fundamental differences (Durham, 1991; Boyd and Richerson, 1985; Cavalli-Sforza and Feldman, 1981; Pulliam and Dunford, 1980). To date, empirical investigations focused on these processes are few (but see, e.g., Cavalli-Sforza, Feldman, Chen, and Dornbusch, 1982). In addition to conducting empirical studies specifically designed to investigate these processes, it is possible to use many of the data collected by social scientists for other purposes. Here we use a small part of the very rich ethnographic record produced by anthropologists to test the empirical plausibility of the process of cultural group selection.

As emphasized by Campbell (1965, 1975, 1983), cultural group selection requires that (1) there be cultural differences among groups, (2) these differences affect persistence or proliferation of groups, and (3) these differences be transmitted through time. If these three conditions hold, then, other things being equal, cultural attributes that enhance the persistence or proliferation of social groups will tend to spread. There is no guarantee, however, that this process will be sufficiently powerful to overcome other social processes that act to produce other outcomes. There are two problems with cultural group selection as an explanation for the existence of group-beneficial traits: maintenance of variation among groups and rate of adaptation.

Group-functional explanations may be in conflict with the fact that human choices are at least partly self-interested. To the extent that they can evaluate alternative beliefs and attitudes, self-interested organisms should adopt only beneficial attitudes and beliefs and reject those that are individually harmful. Thus, beliefs that are costly to the individual should diminish, while beliefs that are beneficial to individuals should spread. Extensive theoretical analysis suggests that group selection can counteract this process only if groups are very small and migration among groups is very limited (Eshel, 1972; Levin and Kilmer, 1974; Wade, 1978; Slatkin and Wade, 1978; Boorman and Levitt, 1980; Wilson, 1983; Aoki, 1982; Rogers, 1990). As a result, most evolutionary biologists and social scientists influenced by them (e.g., Chagnon and Irons, 1979) reject functionalist explanations.

Furthermore, Hallpike (1986) has argued that group extinction does not occur often enough to justify functionalist explanations. Group selection works by eliminating those societies that have deleterious practices or institutions. If it takes a particular number of extinctions to eliminate a deleterious ritual form, then it will take a greater number to eliminate the deleterious ritual form and a deleterious marriage practice. Still further extinctions will be required to cause other aspects of the society to become adaptive. Hallpike argues that human societies do not have high enough extinction rates for group selection to cause many different attributes to be adaptive at the group level simultaneously.

In the face of these objections, is there any justification for taking group-functional hypotheses seriously? Here we describe a theoretical model and present supporting data which show that a role for cultural group selection should not be ruled out. Boyd and Richerson (1985, chs. 7 and 8, 1990a, b) have analyzed mathematical models of group selection acting on culturally transmitted variation and have shown that cultural group selection will work if certain key assumptions are met. Ethnographic data from Papua New Guinea and Irian Jaya give credence to some of the assumptions that underpin the group-selection model. These data also allow us to estimate an upper bound on the rate of adaptation that could result from group selection. We argue that these data suggest that group selection is too slow to be used to justify the common practice of interpreting as group-beneficial the detailed aspects of particular cultures. However, the data do not exclude the possibility that group selection may account for the gradual evolution of some group-level adaptations, such as complex social institutions, over many millennia.

How Group Selection Can Work

We begin with the premise that individuals acquire various skills, beliefs, attitudes, and values from other individuals by social learning and that these "cultural variants," together with their genotypes and environments, determine their behavior. To understand why people behave as they do in a particular environment, we must know the skills, beliefs, attitudes, and values that they have acquired from others by cultural inheritance. To do this, we must account for the processes that affect cultural variation as individuals acquire cultural traits, use the acquired information to guide behavior, and act as models for others. What processes increase or decrease the proportion of persons in a society who hold particular ideas about how to behave? Here we will consider two kinds of processes: biased cultural transmission and selection among social groups.

Biased cultural transmission occurs when individuals preferentially adopt some variants relative to others. Individuals may be exposed to a variety of beliefs or behaviors, evaluate these alternatives according to their own goals, and preferentially imitate those variants that seem best to satisfy their goals. If many of the individuals in a population have similar goals, this process will cause the cultural variants that best satisfy these goals to spread. For example, if the two variants are more and less restrictive forms of food taboos and individuals prefer the broader diet that results from the less restrictive variant, then that variant will spread. This process, which is important in the spread of innovations (Rogers, 1983), often tends to cause groups living in similar environments to have similar behaviors.

However, biased cultural transmission can also maintain differences between groups of people living in similar environments. This can occur in two ways: first, a belief or behavior may be more attractive if it is more widely used than the alternatives. Many social behaviors have this character. For example, if food taboos are used as ethnic markers, then in a group in which the more restrictive taboo predominates, individuals may choose that taboo over the less restrictive one because the social benefits compensate for the nutritional costs. Game theory suggests that many kinds of social interactions, including bargaining, contests, and punishment-enforced norms, will generate an astronomical number of alternative equilibria. Second, when individuals are unable to evaluate the merits of alternative variants, they may instead use a simple rule of thumb such as adopting the most common variant. This conformist form of biased transmission causes the more common variant to increase. For example, if the majority of a group observes the more restrictive taboo, it will tend to increase.

When either common-type-advantage or conformity maintains differences among groups, group selection can be an important force. Consider a large population sub-divided into many smaller, partially isolated groups. Suppose that biased cultural transmission maintains cultural differences among these groups despite frequent contact and occasional intermarriage and that these cultural differences affect the welfare of the group. For example, groups in which restrictive food taboos are common may tend to harvest game at approximately the maximum sustainable yield, while groups in which less restrictive taboos are common overexploit their game resources and suffer significantly poorer nutritional status as a result. Further suppose that social groups are occasionally disrupted and their members dispersed to other local groups and that the rate at which this occurs depends on the overall welfare of the group. Such disruption and dispersal may be the result of population decline, social discord, or the actions of aggressive neighbors. Poor nutritional status will contribute to these risks. Thus, according to our hypothetical example, groups with less restrictive food taboos will, on average, be more likely to be broken up and dispersed. Finally, suppose that as some groups decline and disappear, other groups grow and eventually divide, forming new groups, and that the rate at which this occurs increases with the overall welfare of the group. Thus, the growing, dividing groups will tend to have more restrictive food taboos than declining ones, and restrictive food taboos will tend to spread as a result of selection among groups. Others have proposed at least implicitly similar models (e.g., Peoples, 1982; Divale and Harris, 1976; Irons, 1975).

This model of group selection differs from those analyzed in population biology in that biased transmission maintains variation among groups. Biologists have been concerned with whether group selection could allow the evolution of altruistic behavior. In these models, natural selection acts against altruistic behavior in every group, and this selection process tends to reduce variation among groups. The only process creating variation among groups is genetic drift, a very weak force. Thus, group selection can have little effect because groups are genetically very similar. In the model outlined here, it is assumed that various forms of biased transmission, potentially very strong individual-level forces, act to *maintain* differences among groups and group selection can predominate.

The form of group selection just outlined can be a potent force even if groups are usually very large. For a favorable cultural variant to spread, it must become common in an initial subpopulation. The rate at which this will occur through random driftlike processes (Cavalli-Sforza and Feldman, 1981) will be slow for sizable groups (Lande, 1986). However, this need occur only once. Thus, even if groups are usually large, occasional population bottlenecks may allow group selection to get started. Similarly, environmental variation in even a few subpopulations may provide the initial impetus for group selection. Some environments may lead groups to adopt group-beneficial traits because they are also individually advantageous. These practices may then spread by group selection into environments where they have only a group advantage. For example, restrictive food taboos may arise in a very heterogeneous environment in which it is important for individuals to specialize in narrow-range food-procurement strategies and only later spread by group selection to less heterogeneous environments where they mainly function to protect resources against the tragedy of the commons.

Unlike many genetic models, this form of group selection does not require that the people who make up groups die during group extinction. All that is required is the disruption of the group as a social unit and the dispersal of its members throughout the metapopulation. Such dispersal has the effect of cultural extinction, because dispersing individuals have little effect on the frequency of alternative behaviors in the future; in any one host subpopulation, they will be too few to tip it from one equilibrium maintained by convention or conformity to another.

Cultural group selection is very sensitive to the way in which new groups are formed. If new groups are mainly formed by individuals from a single preexisting group, then the behavior with the lower rate of extinction or higher level of contribution to the pool of colonists can spread even when it is rare in the metapopulation. If, instead, new groups result from the association of individuals from many other groups, group selection cannot act to increase the frequency of rare strategies.

Empirical Evidence

To justify using this model of cultural group selection we need data that allow us to answer three questions:

1. Do groups suffer disruption and dispersal at a rate high enough to account for the evolution of any important attributes of human societies?
2. Are new groups formed mainly by fission in groups that avoid extinction?
3. Are there transmissible cultural differences among groups that affect their growth and survival, and do these differences persist long enough for group selection to operate?

To address these questions we present data on group extinction rates, group formation, and cultural variability drawn from the ethnographic literature of Irian Jaya and Papua New Guinea. We have chosen this area because it offers high-quality ethnographic descriptions of peoples that had not been pacified by a colonial administration. Colonialism is suspected by some to increase rates of intergroup conflict in stateless societies, casting doubt on data from areas like the American Plains, where contact predated good ethnography. New Guinea is unique in the amount of good ethnography obtained within a few years of first contact with complex societies. We have focused on pre-state societies because they are characteristic of more of human history than more complex societies, and the basic institutions of human societies evolved under stateless conditions.

We have made an effort to sample as many ethnographies as possible, focusing on those dealing with pre-contact warfare among indigenous peoples. We have chosen to focus on warfare only because it is a conspicuous way in which groups become extinct and is likely to be recorded. Even where defeat in war is

the proximate cause of an extinction, a variety of other factors may have pre-
cipitated the event by causing the defeated group to decline in numbers. Extinction through war may be the common fate of groups that have declined for
some other reason.

We define a group as a territorial population that can conduct warfare as a
unit. An extinction is said to occur when (1) all members of a group are killed or
(2) members of a group are assimilated into another group either wholly or in
part. When a group is routed from its territory but remains intact as a social unit
(or its fate is unknown), then a forced migration, not an extinction, is said to
have occurred.

Group Extinction

To estimate the rate of group extinction for a region, three types of information
are needed: (1) the number of extinctions, (2) the number of years over which
the extinctions took place, and (3) the number of groups among which the
extinctions took place. We were able to assemble this information for five regions in Irian Jaya and Papua New Guinea.

The Mae Enga

The Mae Enga live in the Central Western Highlands, where population density
averages 40 to 43 persons/km^2 but reaches densities of over 100 persons/km^2
(Meggitt, 1962:158, 1977:1). The immediate causes of war (Meggitt, 1977:13)
are land disputes (58 percent), other property disputes (24 percent), homicide
(15 percent), and problems related to sexual jealousy (3 percent). Meggitt recorded a 50-year warfare history for 14 Mae Enga clans. In the 29 conflicts for
which the outcome was known, there were five extinctions. Extinctions did not
result from the killing of all group members; routed clan members were forced
to disperse and find refuge among other clans, often with kin (1977:15, 25–27).
There is evidence that these immigrants became culturally assimilated into their
host group, usually within a generation (Meggitt, 1965:31–35). Rapid assimilation occurred because true clan members received unqualified land rights, as well
as economic, ritual, and military aid. As Meggitt (1977:190) notes, "Members of
defeated and dispersed groups who have gone to live elsewhere have good political and economic reasons not to draw attention to their immigrant status but
instead try for relatively rapid absorption into the host clan.... In consequence,
the identities of extinguished clans or subclans are soon lost to public knowledge
and in time such groups drop out of the genealogies of their former phratries."

The Maring

The Maring live in the Central Highlands, an area of relatively low population
densities, averaging less than 20 persons/km^2 (Vayda, 1971:22). Wars are usually
triggered by a murder or attempted murder (56 percent of cases). The remaining
44 percent are fought over land, women, or theft (1971:4). Vayda's warfare
history concerns 32 clan-clusters and autonomous clans and has a depth of about

50 years (Andrew Vayda, personal communication). He mentions 14 wars in which victims were routed from their territories. Only in one case was there a clear extinction; the other groups eventually returned. However, in two of these cases routed clans reclaimed their territory only with the help of the Australian police and probably would have become extinct otherwise. Rappaport (1967:26) explains that members of vanquished groups who find refuge in another group do not maintain their autonomy: "the *de facto* membership of the living in groups with which they have taken refuge is converted eventually into *de jure* membership. Sooner or later the groups with which they have taken up residence will have occasion to plant *rumbin*, thus ritually validating their connection to the new territory and their new group."

The Mendi

The Mendi live in the Southern Highlands, where population density is 18 persons/km^2 (Meggitt, 1965:272). Ryan (1959) describes, for a 50-year period, the history of clan degeneration, extinction, and new group formation for a group of nine clans known as the Mobera-Kunjop. In this period there were three clan extinctions. In two cases, the clans were routed by warfare and absorbed by other groups; in the third a degenerating clan was eventually absorbed by another clan.

In two cases, vanquished groups did not suffer disruption but managed to remain functioning as an intact subclan in their host group. Ryan (1959:271) suggests that such accretionary subclans eventually become assimilated into their host clan: "The refugee group, consisting of sub-clan brothers and their families, may be large enough to assume the immediate status of a subclan.... Once the people have been accepted, granted land, and have settled down, there is almost no further differentiation made between them and the original subclans." However, individual nonagnates suffer discrimination from members of their host clans (Ryan, 1959). They are less likely to receive bridewealth support (which normally comes from fellow subclan members) than are true group members, and therefore refugees have reason to want to assimilate into their host group: "Although it is asserted that acceptance is complete ... marriage figures indicate that non-agnatic men tend to marry later than agnatic clan members, more of them marry only once, and more of them have only one wife at a time" (p. 269).

The Fore and Usufura

Berndt (1962) recorded detailed descriptions of war involving groups in four adjacent linguistic regions of the Eastern Highlands—the Fore, the Usufura, the Jate, and the Kamano. Fore population density is approximately 15 persons/km^2 and that of the Usufura 27 persons/km^2 (Berndt, 1962:20). No values are given for the other linguistic groups. Berndt recorded one extinction during the 10-year period preceding his research. The group was routed in warfare and dispersed into several different districts in the area. The number of groups involved is slightly ambiguous; Berndt indicates that his warfare data are most complete for only 8 districts in the area but mentions some 24 districts in his accounts of warfare.

The Tor

The Tor region is located on the northern coast of Irian Jaya (Oosterwal, 1961). No density figures are provided. Oosterwal recorded a 40-year history for the 26 tribal territories in the Tor region. Four tribes suffered extinction either through peaceful absorption, military defeat and dispersal, or outright extermination (Oosterwal 1961:21–26). In one of the extinctions, Oosterwal is clear about the cultural assimilation of the extinct group: "Formerly the Mander language was only spoken by the Mander, but since the Foja have lived together with the Mander, they have adopted the Mander language entirely. Save for a small number of words, these Foja do not recollect any more of their own language. Their kinship terminology is also identical with that of the Mander" (p. 23).

Table 11.1 summarizes extinction rates for the five regions for which there were enough data to compute such estimates. We assume that the number of groups remains constant, which means that each extinction is followed by an immediate recolonization. To the extent that this assumption is wrong, extinction rates will be higher. We found no ethnographies that yielded an extinction rate of zero. In our sample, the percentage of groups suffering extinction each generation ranges from 1.6 percent to 31.3 percent.

It seems likely that other areas in New Guinea had similar group extinction rates. There is mention of group extinction in 54 percent (15/28) of the societies sampled. This is no doubt an underestimate, because the failure to mention an extinction in an ethnographic account of warfare does not necessarily mean that extinctions never occurred. In 89 percent (25/28) of the societies sampled, there is mention of either group extinction or forced migration (see table 11.2). The near ubiquity of extinction and forced migration in the ethnographic record suggests that high rates of extinction were common throughout Papua New Guinea and Irian Jaya before pacification.

New Group Formation

Group selection is most effective when new groups are made up of members of a single existing group rather than of members of many different groups. If new groups are formed when a single group generates a daughter group from among

Table 11.1. Summary of group extinction rates for five regions of Papua New Guinea and Irian Jaya

Region	Groups	Extinctions	Years	Percentage of groups extinct every 25 years	Source
Mae Enga	14	5	50	17.9	Meggitt (1977)
Maring	32	1–3	50	1.6–4.7	Vayda (1971)
Mendi	9	3	50	16.7	Ryan (1959)
Fore/Usufura	8–24	1	10	31.3–10.4	Berndt (1962)
Tor	26	4	40	9.6	Oosterwal (1961)

Table 11.2. Mentions of group extinction and forced migration in Papua New Guinea and Irian Jaya

People	Extinction	Migration	Source
Mae Enga	+	−	Meggitt (1977:14)
Huli	−	−	Glasse (1959)
Melpa	+	+	Strathern (1971:55–56, 67)
Raiapu Enga	+	+	Waddel (1972:37, 186, 263–65)
Wola	+	+	Sillitoe (1977:79)
Maring	+	+	Vayda (1971:11–13)
Ok	+	+	Morren (1986:266–67, 272–73, 278–79)
Kuma	+	+	Reay (1959:7, 27, 32)
Chimbu	−	+	Brown and Brookfield (1959:41, 61, 263–65)
Usufura	−	+	Berndt (1962:242)
Jate	+	+	Berndt (1962:253, 260–61)
Fore	−	+	Berndt (1962:236, 251, 257)
Auyana	+	+	Robbins (1982:213–14)
Kukukuku	−	+	Blackwood (1978:102)
Gahuku	−	+	Read (1955:253–54)
Arapesh	+	+	Tuzin (1976:63)
Abelam	−	+	Lea (1965:196, 205)
Mailu	−	+	Saville (1926)
Kiwai	+	+	Landtman (1970[1927]:148–49, 204)
Dugum Dani	+	+	Heider (1970:119–22)
Ilaga Dani	−	+	Sillitoe (1977:77)
Bokondini-Dani	−	+	Sillitoe (1977:76)
Jale	−	+	Koch (1974:79)
Kapauku	−	−	Pospisil (1963)
Tor	+	+	Oosterwal (1961:21–26, 48)
Jaqai	−	−	Boelaars (1981)
Marind-Anim	+	−	Ernst (1979:36)
Bena Bena	+	−	Langness (1964:174)

its own members, then the daughter group will preserve the cultural variants common in the mother group. Cultural variants that facilitate daughter-group formation will become more common in the region as a whole.

Societies in Irian Jaya and Papua New Guinea are characterized by a segmentary social system (Langness, 1964). When members of a social group become too numerous, the group may split into two similar groups. Conversely, when members of a social group become too few, they may be absorbed by another group at a lower segmentary level (Brown, 1978:184–185, 187–188). There are numerous anecdotal accounts of new group formation (e.g., Brown and Brookfield, 1959:57; Sillitoe, 1977:79; Vayda, 1971:17; Morren, 1986:269–270), but Meggitt (1962, 1965) and Ryan (1959) provide the most detailed descriptions of new group formation in two highland societies.

The Enga have a nested hierarchy of patrilineal descent groups. The phratry is the most inclusive, followed by the clan, the subclan, the patrilineage, and the family. Groups everywhere in the hierarchy may grow or decline over time, generate daughter groups, or become absorbed by other groups: "Groups may emerge, increase in size and take over different functions, and in doing so achieve higher status by becoming co-ordinate with groups that previously included them. In absorption, groups that are decreasing in numbers have to relinquish particular functions and descend to a lower level in the hierarchy. ...If the decline continues, the groups eventually vanish" (Meggitt, 1965:79). For a group to achieve or retain a particular position in the hierarchy, it must contain enough members to perform the functions appropriate to that position. For example, from 1900 onward, the population of one Enga clan began increasing noticeably until one of its two subclans could no longer support itself on its share of land and began encroaching on a neighboring clan's territory (Meggitt, 1965:62–63). In skirmishes with the neighboring clan, the subclan functioned as if it were a sovereign clan, fighting and negotiating homicide payments independently of the second subclan, which was itself trying to expand in another direction. Eventually members of the two subclans settled at opposite ends of the clan territory and behaved as members of separate clans by intermarrying.

Meggitt (1965:78–79) gives an account of two Laiapu Enga phratries demonstrating extinction and new group formation. Each phratry was initially made up of four territorial clans. One expanding clan of phratry A attacked and killed many members of two clans of phratry B. The survivors of the two clans fled to other clans, and the victorious clan occupied the abandoned territory. This successful clan was becoming so large as to achieve subphratry status (Meggitt, 1965:79). Ryan (1959) gives similar accounts of group extinction and new group formation in the Mendi Valley. When clans become too populous, they expand into new territory and an off-shoot subclan occupies it. The breakaway subclan attains clan status as it takes on more and more functions appropriate to a clan.

Cultural Variation among Groups

Group extinction and group fission will lead to cultural change only if there are transmissible cultural differences that affect the extinction rate or the proliferation rate. Unfortunately, there is little evidence about the amount of cultural variation among local groups because so few ethnographers study more than one local group. Furthermore, there is even less evidence about how differences between local groups are related to individual and group fitness in New Guinea ethnography, although there is quite good evidence from other areas that such variation exists (e.g., Kelly's [1985] study of the causes of Nuer expansion at the expense of the Dinka). Nor is there evidence about how long such differences can persist in New Guinea groups. Archaeological and linguistic data from small-scale societies elsewhere document many examples of group expansion by cultures with more effective social organization in which the differences persisted for many generations during the expansionary phase (e.g., Bettinger and Baumhoff's [1982] study of the Numic expansion from southeastern California across the Great Basin).

Here we review three detailed studies of cultural variation among local groups in New Guinea. Two of these studies focus on the Mountain Ok of Papua

New Guinea, while the third covers the lowland Tor region of northern Irian Jaya. Each of these studies suggests that there is substantial cultural variation among local groups.

The Mountain Ok

The Mountain Ok occupy the center of New Guinea and are made up of nine "tribes" based on ethnolinguistic affinities (Morren, 1986:180–181). Within these tribes are endogamous "communities," sometimes composed of several exogamous clans. Only 15 percent of marriages take place between members of different communities (Barth, 1971:176).

Ritual practice and belief vary considerably from community to community. Ritual knowledge, surrounded by secrecy, is fully shared by only a few elders in each community. It is transmitted at male initiations, where it is rationed out to initiates in steps. Barth argues that the ritual knowledge of different communities diverges because of error and innovation on the part of the few persons who control it. This produces intergroup variation in such things as the interpretation of important ritual symbols, the use of myths in ritual contexts, theories of conception, and the emphasis on symbolic constructions of human sexuality in ritual (Barth, 1987).

Sacred objects used in the initiation ritual take on different symbolic meaning in different communities (Barth, 1987:4–5). For example, fat from a wild male boar is emphatically "male" among both the Bimin-Kuskusmin and Baktaman of the Faiwolmin tribe. The pig's fat is mixed with various substances to form a red paint that is applied to the bodies of novices, except for their "female" parts. In communities of the Telefolmin tribe, however, the red paint signifies female menstrual blood. In fact, menstrual blood is sometimes added to the concoction, a practice which would be "completely destructive" to the integrity of the Faiwolmin rituals.

Modes in which cosmological ideas are communicated also differ among Ok communities. The Baktaman know almost no myths at all. A peripheral Ok community, the Mianmin, has a larger corpus of myths, but these are not central to their ritual events. The Bimin-Kuskusmin, in contrast, have an abundance of myths that are integrated into ritual (Barth, 1987:5–6).

Theories of conception differ among communities (Barth, 1987:13–15). Members of the Baktaman and neighboring communities believe that children spring from male semen that is nourished in the mother. Telefolmin males believe that children are created from a fusion of male and female substances; females believe that a fusion of male and female substances creates only the flesh and blood of a child, while the female's menstrual blood alone forms the bones. Other communities are characterized by still different theories of conception.

The Faiwolmin

Variation among communities within the Faiwolmin tribal area of the Ok region may provide an example of cultural variation that is linked to group survival. Barth (1971, cf. Morren 1986) argues that more elaborate, communal rituals and specialized cult houses lead to more centralized community organization, which increases the survivability of the communities embracing them, and that communities with less elaborate cultural forms and more dispersed settlement patterns are more likely to become extinct. Within the Faiwolmin tribal area, ritual organization and specialization find their most elaborate expression in the centralized communities (Barth, 1971:179–181). Male initiation is organized in seven grades through which males pass as age-sets. In western communities there are four such grades, and in the southeastern communities they range from four to one (p. 185). Different rituals take place in specialized cult houses. Most Faiwolmin communities contain three permanent cult houses as well as a communal men's house. As one moves east and southward from central Faiwolmin, the number of cult houses declines. Most of the southeastern communities contain only one permanent cult house, and some perform initiations in temporary structures.

There is also variation in social organization among Faiwolmin communities, following a similar west-to-east pattern of decreasing centralization (Barth, 1971:184–186). The centralized communities of the Faiwolmin form compact villages around several types of semipermanent cult houses, and several exogamous clans make up an isolated, largely endogamous political unit. In the east the population is dispersed within the community territory, shifting household locations at intervals because of soil depletion or fear of sorcery.

According to Barth, "The dispersed pattern without the cult houses ... clearly organizes a smaller population for defense, and their history of displacement would seem to demonstrate this disadvantage" (p. 189); "the greater centralization clearly also offers military advantages and has resulted in conquest and territorial expansion of the more highly centralized groups in a general south-eastward direction" (p. 186). He argues that the elaborate rituals and the concomitant communal centralization were first introduced to the Faiwolmin communities from the northwest, and the diffusion of these cultural forms created cultural variation among them. Finally, selection among groups increased the frequency of those cultural forms conferring the highest fitness on groups (p. 188):

> The distribution of [cultural] forms is thus generated by a number of simultaneously partly independent processes. A process of diffusion from an innovation centre ... seems to be taking place. Simultaneously, the organization of local cultural transmission is such that both loss and improvisation occur and new local variants emerge. Different ritual forms imply different community types; these again confront each other in warfare and compete and replace each other on the basis of their unequal defensive and offensive capacities.

If Barth is correct, this is an example of group selection increasing the cultural variants that enhance group survival. He considers the alternative hypothesis that ecological processes explain the smaller scale of social organization. Although he cannot completely rule out an ecological explanation, he clearly suggests that a ritual system that organizes more people and thus leads to a greater frequency of victory in violent conflicts is leading to the spread of more complex ritual (pp. 188–189).

The Tor

Significant cultural variation also existed between tribal territories of the Tor region (Oosterwal, 1961). The Tor region is divided into 26 tribal territories, but it has 8 separate languages (Oosterwal, 1961: appendix). Thus, many adjacent tribes speak different languages, although the most common language, that of the Berrick, is known by members of all tribes (Oosterwal, 1961:18). Oosterwal also notes these differences: "the three culture areas in the Tor district are very distinct. . . . [There are] differences in . . . kinship terminology, the kinship structure, the socio-religious aspect of culture, the way of counting, language-(dialect)-differentiations, and some aspects of material culture" (p. 46). These three "cultural areas," with associated kinship terminologies, are the Berrick, the Ittik and Mander, and the Segar and Naidjbeedj. Tribes in "transitional zones" have elements of all three cultural areas, and there is variation within each area (pp. 149–174). The terminology of the Berrick tribe emphasizes the age criterion (e.g., MoElSi is terminologically distinguished from MoYoSi) but often ignores the generational criterion (e.g., MoBr and SiSo call each other by the same term). The terminology of the second cultural area ignores the generational criterion to a far greater extent. In contrast to those of the previous two areas, cultures in the third region have a strong generational aspect in their terminology. There is also variation within each of these three broad areas. For example, the cousin terminology of the Berrick is of the Hawaiian type (all cross and parallel cousins called by the same terms as those for sisters), while the Waf and Goeammer (of the same culture area) use the Iroquois type (FaSiDa and MoBrDa called by the same terms but terminologically differentiated from parallel cousins and from sisters, parallel cousins commonly but not always classified with sisters).

Although it is difficult to show that the particular group extinctions that we have counted for the five regions are due to persistent cultural differences, there is abundant evidence in New Guinea and elsewhere that cultural differences do lead to the success of some groups and the decline of others. For example, among the Fore the practice of mortuary cannibalism caused the spread of the deadly disease kuru. According to Durham's (1991:411–413) account of this episode, ritual cannibalism was originally adopted by Fore women as a response to a shortage of game. Nevertheless, the spread of the disease as a by-product of this ritual innovation threatened Fore groups with extinction until modern medical teams intervened. This case points up the ambiguous role of rational choice in the group-selection process. Individual calculation of advantage may often run counter to group advantage, especially when acts of cooperation are involved. Rappaport (1979:100) called attention to the role of the sacred in concealing group-advantageous traits from ready attack by selfish reason. As the Fore experience with kuru illustrates, traits disadvantageous to groups (and to individuals in this case) may sometimes be concealed in the same way.

Knauft (1985) gives an example of an apparent group extinction in progress. The completely acephalous Gebusi were a small and declining group at the time of his study. The better-organized Bedamini, making use of the big-man style of political organization, were able to raid Gebusi villages, but the Gebusi were unable to organize an effective defense or a retaliatory response. The boundary

Gebusi villages most exposed to Bedamini raids were in the process of assimilating to Bedamini customs.

Knauft (1993) also provides examples of cultural differences among seven culture areas along New Guinea's south coast. He describes how the Marind-Anim system of mythico-religious affiliation supports intragroup peace and the organization of large-scale head-hunting raids against distant enemies. By contrast, the Purari head-hunt among themselves and are declining relative to their neighbors. The existence of considerable variation at the scale of language groups suggests a considerable time depth for these differences. Although this variation occurs among larger groups than we are concerned with here, it does show that variation in sociopolitical organization encoded in myth and religion has a strong effect on group success.

It is also important that cultural differences between groups persist on time scales sufficient for the operation of group selection. Although there is variation among local groups in New Guinea, there are no data bearing on the question of how long that variation persists. However, there is ample evidence for the long-term persistence of cultural differences among larger groups in other culture areas. For example, concepts such as *mana* and *tabu* typify political culture throughout Polynesia despite the fact that these societies have been isolated from each other for more than 1,000 years (Kirch, 1984). Egerton (1971) documents the existence of important differences among four tribal groups living in two different types of environment, inlcuding two tribes belonging to the Bantu and two to the Kalenjin language groups, which have been separated for thousands of years. He notes that tribal history is more important than contemporary environmental circumstances in explaining most of the variation in attitudes and values measured in his data. The roots of the 38 languages of Western American Indians go back 6,500 years, and cultural differences among close neighbors with different cultural history have persisted for long periods (Jorgensen, 1980:109). Belgium is divided by a stable linguistic boundary, with a Flemish North and a Walloon South (van den Berghe, 1981); despite the fact that there is no topographical separation, the linguistic frontier has persisted for 2,000 years. Such examples from archaeology and history can be multiplied at will. While they do not prove that cultural differences can persist at smaller scales as required by the model, they indicate that this assumption is plausible.

Discussion

Cultural group selection can explain the evolution of group-functional behaviors and institutions in human societies if two conditions are met: first, there must be some mechanism that preserves between-group variation so that group selection can operate. The model described provides one such mechanism, and we have here tested several of the model's basic assumptions against the ethnographic record to determine if those assumptions are empirically realistic. Second, group selection must be sufficiently rapid to explain observed patterns of cultural change. The data from Papua New Guinea and Irian Jaya allow us to estimate the maximum rate of adaptation through group selection. Thus, we can estimate a minimum time period in which the group-selection process can give rise to group-level adaptations. Cultural changes that have occurred on a longer time scale are possibly the result of group selection, cultural changes that have occurred on a shorter time scale are unlikely to have resulted directly from group selection, but they may be its indirect result. For example, cultural group selection may lead to the evolution of property rights, which lead to efficient allocations of resources, or of political institutions that lead to group-beneficial decisions.

Model Assumptions

The data from New Guinea provide some qualified support for the model of group selection described.

1. Group disruption and dispersal are common. Extinction rates per generation range from 2 percent to 31 percent, with a median of 10.4 percent in the five areas for which quantitative data are available, and the frequent mention of extinction elsewhere suggests that these rates are representative.
2. New groups are usually formed by fission of existing groups. The detailed picture from the Mae Enga and the Mendi is supported by anecdotal evidence from other ethnographies. We are not aware of any ethnographic report from New Guinea in which colonists of new land are drawn from multiple groups.
3. There is variation among local groups, but it is unknown whether this variation persists long enough to be subject to group selection and whether this variation is responsible for the differential extinction or proliferation of groups.

Rates of Change

The New Guinea data on extinction rates allow us to estimate the maximum rate of cultural change that can result from cultural group selection. For a given group extinction rate, the rate of cultural change depends on the fraction of group extinctions that are the result of heritable cultural differences among groups. If most extinctions are due to nonheritable environmental differences (e.g., some groups have poor land) or bad luck (e.g., some groups are decimated by natural disasters), then group selection will lead to relatively slow change. If most extinctions are due to heritable differences (e.g., some groups have a more effective system of resolving internal disputes), then group selection can cause rapid cultural change. The rate of cultural change will also depend on the number of different, independent cultural characteristics affecting group extinction rates. The more different attributes, the more slowly will any single attribute respond to selection among groups. By assuming that all extinctions result from a single heritable cultural difference (or tightly linked complex of differences) between groups, we can calculate the maximum rate of cultural change.

Such an estimate suggests that group selection is unlikely to lead to significant cultural change in less than 500 to 1,000 years. The length of time it takes

Table 11.3. Minimum number of generations necessary to change the fraction of groups in which a favorable trait is common assuming a particular extinction rate

Initial fraction favorable trait	Final fraction favorable trait	Extinction rate			
		1.6%	10.4%	17.9%	31%
0.1	0.9	192	40.0	22.3	11.8
0.01	0.99	570	83.7	46.6	24.8

Note: Extinction rates were chosen as follows: 1.6 percent (for the Maring) is the lowest estimate, 10.4 percent is the median extinction rate, 17.9 percent (for the Mae Enga) is the estimate based on the best data, and 31 percent (for the Fore/Usufura) is the highest estimate.

a rare cultural attribute to replace a common cultural attribute is one useful measure of the rate of cultural change. Suppose that initially a favorable trait is common in a fraction q_0 of the groups in a region. Then the number of generations (t) necessary for it to become common in a fraction q_t of the groups can be estimated (see Appendix). The time necessary for different parameters is given in table 11.3. If we take the median extinction rate as representative, these results suggest that group selection could cause the replacement of one cultural variant by a second, more favorable variant in about 40 generations, or roughly 1,000 years. If we take the extinction rate calculated using the best data, those from the Mae Enga, this time is cut roughly in half. These calculations assume that colonizing groups are selected at random from the population. If group proliferation is as selective as group extinction, then the time is again cut in half, reducing the substitution time (based on the median extinction rate), once again, from 1,000 to 500 years. Not all extinctions and new group formations result from heritable cultural differences. Since the New Guinea ethnographic data are not sufficient to estimate the extent to which cultural variation influences group extinctions, it is not possible to make an estimate of the actual strength of group selection in New Guinea. If such estimates were possible, we expect that they would show that actual rates are considerably below the maximum. The maximum rate is nevertheless useful as an upper bound on the kinds of evolutionary events that cultural group selection might explain.

Our estimate of the maximum rate of adaptation suggests that group selection is too slow to account for the many cases of cultural change that occur in less than 500 to 1,000 years. For example, according to Feil (1987) the arrival of the sweet potato in the highlands of New Guinea sometime in the eighteenth century led to many important cultural changes. The introduction of the horse to the Great Plains of North America in the 1500s led to the evolution of the culture complex of the Plains Indians in less than 300 years. If the rates of group extinction estimated for New Guinea are representative of small-scale societies, cultural changes such as these cannot be explained in group-functional terms. There has not been enough time for group selection to have driven a single cultural attribute to fixation, even if that attribute had a strong effect on group survival. Processes based on individual decisions are likely to account for such episodes of rapid evolution (see Smith and Winterhalder, 1992; Boyd and Richerson, 1985). Such processes will not lead to group-functional outcomes except in certain special circumstances (see n. 1). It is possible that situations in which a trait or trait complex that increases the scale of cooperation is spreading such as the one Barth posits for the Faiwolmin do show rapid cultural group selection in progress. If the arrival of the sweet potato a few centuries ago did provide the subsistence basis for larger and more complex societies, we might expect to observe group selection in the early to middle stages of the spread of newly advantageous forms of social organization (Golson and Gardner, 1990; Feil, 1987).

These results also suggest that group selection cannot justify the practice of interpreting many different aspects of a culture as group-beneficial. A given extinction rate will lead to slower change if many different, unrelated aspects of the culture affect group survival. Suppose that both beliefs about food consumption and beliefs about spatial organization affect group survival. Then, unless each extinction occurs in a group in which both deleterious beliefs about food consumption *and* deleterious beliefs about spatial organization are common, some extinctions have no effect on the fraction of groups with deleterious beliefs about food, and some extinctions have no effect on the fraction of groups with deleterious beliefs about spatial organization. Thus, a given number of extinctions must lead to slower evolution of each character than would be the case if only one of the characters affected group survival. If group selection can cause the substitution of a single trait in 500 to 1,000 years, the rate for many traits will be substantially longer. We know from linguistic and archaeological evidence that related cultural groups that differ in many cultural attributes have often diverged from a single ancestral group in the past few thousand years. Thus, there has not been enough time for group selection to have produced the many attributes that distinguish one culture from another.

It is important to understand that slow does not necessarily mean weak. When individual decision making is in opposition to group function in every group, then the relatively slow group-selection process will be too weak to favor group-functional behaviors. But when social interaction results in many alternative stable social arrangements, then individual decision making *maintains* differences among groups. If the resulting variation is linked to group fitness, then group selection will proceed. For example, consider the response to an environmental change such as the opening of New Guinea to trade with Europeans. Initially, changes in the costs and benefits of alternative beliefs and values will cause rapid cultural change, soon leading to a new sociopolitical equilibrium in each culture. But if there are many alternative equilibria, the nature of each new equilibrium may depend on existing norms and values. As long as the resulting differences affect group survival, selection among groups will continue. Over a millennium or so, New Guinea societies with a better political adaptation to world contact will replace those with a poorer adaptation.

Thus, it follows that these results do not preclude interpreting *some* aspects of contemporary cultures in terms of their benefit to the group. The model demonstrates that under the right conditions group selection can be an important process, and the data from New Guinea suggest that some of these conditions are empirically realistic. The data also suggest that the rates of group extinction are high enough to cause a small number of traits with substantial effects on group welfare to evolve on time scales that characterize some aspects of cultural change. Group selection cannot explain why the many details of Enga culture differ from the many details of Maring culture. It *might* explain the existence of geographically widespread practices that allow large-scale social organization in the New Guinea highlands, practices that evolved along with, and perhaps allowed, the transition from band-scale societies to the larger-scale societies that exist today.

Cultural group selection provides a potentially acceptable explanation for the increase in scale of sociopolitical organization in human prehistory and history precisely because it is so slow. Scholars convinced of the overwhelming power of individual-level processes have real difficulty in explaining slow, long-term historical change. Anatomically modern humans appear in the fossil record about 90,000 years ago, yet there is no evidence for symbolically marked boundaries (perhaps indicative of a significant sociopolitical unit encompassing an "ethnic" group of some hundreds to a few thousand individuals) before about 35,000 years ago (Mellars and Stringer, 1989). The evolution of simple states from food-producing tribal societies took about 5,000 years, and that of the modern industrial state took another 5,000. Evolutionary processes that lead to change on 10- or 100-year time scales cannot explain such slow change unless they are driven by some environmental factor that changes on longer time scales. In contrast, the more or less steadily progressive trajectory of increasing scale of sociopolitical complexity over the past few tens of thousands of years indeed is consistent with adaptation by a relatively slow process of group selection.

These results should be interpreted with caution. It is important to remember that we have estimated a maximum rate of change for group selection on the basis of the assumptions that observed differences among local groups are heritable and that they are persistent. Unless both assumptions are satisfied, group selection will be less important than our results indicate. It is also important to keep in mind that we have studied only one form of group selection—competition among small, culturally heterogeneous groups. Other plausible group-selection processes might lead to more rapid change. For example, one cultural region may encroach upon another along a frontier, constantly capturing additional land and gradually expanding its domain. The Nuer and Dinka formed such a system before they were both overtaken by European colonists (Kelly, 1985). In state-level societies, we have to allow for internal group selection via the extinction and proliferation of subgroups, such as ruling classes, interest groups, firms, and the like, as well as selection among states themselves (Hannan and Freeman, 1989). Some economists have considered business failure and proliferation rates sufficient to drive group selection of these units (Alchian, 1950; Nelson and Winter, 1982). The development of collective decision-making institutions like bureaucracies and legislatures may permit group-functional behaviors to be deliberately adopted by state-level societies. These processes might act at a much faster rate than we have estimated on the basis of tribal institutions.

In conclusion, these data suggest that group selection cannot explain rapid cultural change or the many differences between related cultures. However, they also show that group selection, perhaps in concert with other processes, is a plausible mechanism for the evolution of widespread attributes of human societies over the long run.

APPENDIX: Time for Trait Substitution

Assume that there are two cultural variants—deleterious and advantageous. Each is at a local equilibrium under the influence of within-group processes. Groups are connected by the mixing of individuals, and there are many such groups. Groups in which the advantageous variant is common never go extinct. A fraction e of the groups in which the deleterious variant is common suffer an extinction each generation. The dynamics of this system are quite complicated because the frequency of advantageous variants within subpopulations in which that variant is common depends, to a small degree, on the frequencies of both variants in the population as a whole. However, if both variants are in local equilibrium, even when there is only a single population in which they are common, then it is roughly correct to regard the subpopulations as individuals and use formulas from population genetics (see Boyd and Richerson, 1990b for a fuller treatment). Then, if the advantageous trait is common in a fraction q of the groups in the region, after one generation

$$q' = \frac{q}{(1-q)(1-e) + q}$$

and the frequency after t generation is

$$q_t = \frac{q_0}{(1-q_0)(1-e)^t + q_0}$$

Solving this for t yields

$$t = \frac{\ln\left[\frac{q_0(1-q_t)}{(1-q_0)q_t}\right]}{\ln(1-e)}$$

which was used to generate table 11.3.

NOTES

We thank Philip Newman, Paul Sillitoe, Andrew Vayda, Mark Allen, and Bob Rechtman for help in locating data used in this analysis. Joan Silk, Timothy Earle, Eric Smith, Paul Allison, Lore Ruttan, Mark Jenike, Alan Rogers, Monique Borgerhoff Mulder, and an anonymous referee provided very useful comments on earlier drafts of this chapter. Members of the University of Bielefeld's Center for Interdisciplinary Research project on the Biological Foundations of Human Culture provided a constructively critical audience for an early version (special thanks are due its director, Peter Weingart). Jonathan Turner convinced us that state-level institutions are different from tribal ones.

Some authors (e.g., Harris, 1979) have suggested that the self-interested choices of individuals will result in group-beneficial behavior. However, this claim is not cogent—group-beneficial behavior will not result from individual choice except as a side effect of other processes or in certain limited circumstances. For example, many authors have suggested that food taboos exist because they prevent overexploitation of ecological resources. To keep things simple, let us suppose that individuals must choose to observe a particular taboo or not and that individuals who observe this taboo forgo a satisfying and nutritious food item. Choosing to ignore the taboo has a positive effect on individuals' own welfare and, by assumption, a negative effect on the welfare of the group. However, unless the group is very small, the personal effect will be much larger than the effect on the group, and thus choosing to ignore the taboo will better serve individuals' goals, even if their goals include the welfare of the group. This effect is at the heart of both rational-strategy and evolutionary arguments against the easy development of group-beneficial behavior. The effect is not a matter of cognitive capacity, as writers such as Harris seem to imply. Rational strategists are assumed to have unlimited cognitive capacity, whereas evolved creatures are the products of blind selective sorting, but the essential problem is the same; both rational strategists and evolved creatures are expected to act in their own self-interest.

Group-beneficial behavior may result from self-interested individual choice under certain circumstances. First, since individual and group benefit are often correlated, individual choice may often produce group-beneficial outcomes as a side effect (see Sugden, 1986, for several examples). Second, markets will lead to an "efficient" allocation of economic resources if the state or some other external authority enforces contracts, external effects such as air pollution are not present, and a number of other conditions are satisfied. The allocation is efficient only in the sense that no one can be made better off without someone else's being made worse off— the distribution of wealth that results could be extremely deleterious to the survival of the society. Clearly, most aspects of culture are not regulated by markets or prices, even in contemporary societies. Third, rational planning by leaders or institutions may also lead to group-beneficial outcomes. While the extent to which political institutions can ever be modeled as acting in the common interest is debatable, it is clear that most aspects of culture are not the result of rational planning. Finally, individuals may choose group-beneficial activities if they value those activities for their own sake, not because they benefit the group (Margolis, 1982; Batson, 1991). For example, men may fight to defend the group if they value heroism in battle. However, one is left with explaining how men come to have such preferences— otherwise, the explanation is that people choose group-beneficial behaviors because they like to do so. Thus, we do not deny that people make group-beneficial choices. We are claiming that when such choices occur, they cannot be the result of mainly self-interested choice.

REFERENCES

Aberle, D. F., A. K. Cohen, A. K. Davis, M. J. Levy, & F. X. Sutton. 1950. The functional prerequisites of a society. *Ethics* 60:100–111.

Alchian, A. A. 1950. Uncertainty, evolution, and economic theory. *Journal of Political Economy* 58:211–221.

Aoki, K. 1982. A condition for group selection to prevail over counteracting individual selection. *Evolution* 36:832–842.

Barth, F. 1971. Tribes and intertribal relations in the Fly head-waters. *Oceania* 41:171–191.

Barth, F. 1987. *Cosmologies in the making: A generative approach to cultural variation in inner New Guinea.* Cambridge: Cambridge University Press.

Batson, C. D. 1991. *The altruism question: Toward a social psychological answer.* Hillsdale, NJ: Lawrence Erlbaum.

Berndt, R. 1962. *Excess and restraint.* Chicago: University of Chicago Press.

Bettinger, R. L., & M. A. Baumhoff. 1982. The Numic spread: Great Basin cultures in competition. *American Antiquity* 47:485–503.

Blackwood, B. 1978. *Kukukuku of the Upper Watut.* Pitt Rivers Museum Monograph Series 2.

Boelaars, J. H. M. C. 1981. *Head-hunters about themselves.* The Hague: Martinus Nijhoff.

Boorman, S., & P. Levitt. 1980. *The genetics of altruism.* New York: Academic Press.

Boyd, R., & P. J. Richerson. 1985. *Culture and the evolutionary process.* Chicago: University of Chicago Press.

Boyd, R., & P. J. Richerson. 1990a. Culture and cooperation. In: *Beyond self-interest,* J. Mansbridge, ed. Chicago: University of Chicago Press.

Boyd, R., & P. J. Richerson. 1990b. Group selection among alternative evolutionary stable strategies. *Journal of Theoretical Biology* 145:331–342.

Brown, P. 1978. *Highland peoples of New Guinea.* Cambridge: Cambridge University Press.

Brown, P., & H. C. Brookfield. 1959. Chimbu land and society. *Oceania* 30:1–75.

Campbell, D. T. 1965. Variation and selective retention in sociocultural evolution. In: *Social change in developing areas: A reinterpretation of evolutionary theory,* H. Barringer, G. I. Blanksten, & R. W. Mack, eds. (pp. 19–49). Cambridge: Schenkman.

Campbell, D. T. 1975. On the conflicts between biological and social evolution and between psychology and moral tradition. *American Psychologist* 30:1103–1126.

Campbell, D. T. 1983. Two routes beyond kin selection to ultra-sociality: Implications for the humanities and the social sciences. In: *The nature of prosocial development: Theories and strategies,* D. Bridgeman, ed. (pp. 11–39). New York: Academic Press.

Cavalli-sforza, L. L., & M. W. Feldman. 1981. *Cultural transmission and evolution.* Princeton, NJ: Princeton University Press.

Cavalli-sforza, L. L., M. W. Feldman, K. H. Chen, & S. M. Dornbusch. 1982. Theory and observation in cultural transmission. *Science* 218:19–27.

Chagnon, N., & W. Irons. 1979. *Evolutionary biology and human social behavior: An anthropological perspective.* North Scituate: Duxbury Press.

Divale, W., & W. Harris. 1976. Population, warfare, and the male supremacist complex. *American Anthropologist* 78:521–38.

Durham, W. H. 1991. *Coevolution: Genes, culture, and human diversity.* Stanford: Stanford University Press.

Egerton, R. B. 1971. *The individual in cultural adaptation: A study of four East African peoples.* Berkeley: University of California Press.

Ernst, T. M. 1979. Myth, ritual, and population among the Marind-Anim. *Social Analysis* 1:34–52.

Eshel, I. 1972. On the neighborhood effect and the evolution of altruistic traits. *Theoretical Population Biology* 3:258–277.

Feil, D. K. 1987. *The evolution of highland Papua New Guinea societies.* Cambridge: Cambridge University Press.

Glasse, R. M. 1959. Revenge and redress among the Huli. *Mankind* 5:273–289.

Golson, J., & D. S. Gardner. 1990. Agricultural and sociopolitical organization in New Guinea Highlands prehistory. *Annual Review of Anthropology* 19:395–417.

Hallpike, C. R. 1986. *The principles of social evolution.* Oxford: Clarendon Press.

Hannan, M. T., & J. Freeman, 1979. *Cultural materialism: The struggle for a science of culture.* New York: Random House.

Hannan, M. T., & J. Freeman. 1989. *Organizational ecology.* Cambridge, MA: Harvard University Press.

Harris, M. *Cultural Materialism.* New York: Random House.

Heider, K. G. 1970. *The Dugum Dani: A Papuan culture in the highlands of West New Guinea*. Viking Fund Publications in Anthropology 49.

Irons, W. 1975. *Yomut Turkmen: A study of social organization among a Central Asian Turkic-speaking population*. University of Michigan Museum of Anthropology Anthropological Paper 58.

Lorgensen, J. C. 1980. *Western Indians*. San Francisco: W. H. Freeman.

Kelly, R. C. 1985. *The Nuer conquest: The structure and development of an expansionist system*. Ann Arbor: University of Michigan Press.

Kirch, P. 1984. *The evolution of Polynesian chiefdoms*. Cambridge: Cambridge University Press.

Knauft, B. M. 1985. *Good company and violence: Sorcery and social action in a lowland New Guinea society*. Berkeley: University of California Press.

Knauft, B.M. 1993. *South Coast New Guinea cultures: History, comparison, dialectic*. Cambridge: Cambridge University Press.

Koch, K. F. 1974. *War and peace in Jalemo*. Cambridge, MA: Harvard University Press.

Lande, R. 1986. The dynamics of peak shifts and the pattern of morphological evolution. *Paleobiology* 12:343–354.

Landtman, G. 1970 (1927). *The Kiwai Papuans of British New Guinea: A nature-born instance of Rousseau's ideal community*. New York: Johnson Reprint Company.

Langness, L. L. 1964. Some problems in the conceptualization of highlands social structures. *American Anthropologist* 66:162–182.

Lea, D. A. M. 1965. The Abelam: A study in local differentiation. *Pacific Viewpoint* 6:191–214.

Lenski, G., & J. Lenski. 1982. *Human societies: An introduction to macrosociology*. New York: McGraw-Hill.

Levin, B. R., & W. L. Kilmer. 1974. Interdemic selection and the evolution of altruism: A computer simulation study. *Evolution* 28:527–545.

Malinowski, B. 1984 (1922). *Argonauts of the western Pacific*. Prospect Heights: Waveland Press.

Margolis, H. 1982. *Selfishness, altruism, and rationality: A theory of social choice*. Chicago: University of Chicago Press.

Meggitt, M. J. 1962. Growth and decline of agnatic descent groups among the Mae Enga of the New Guinea highlands. *Ethnology* 1:158–165.

Meggitt, M. J. 1965. *The lineage system of the Mae Enga of New Guinea*. New York: Barnes and Noble.

Meggitt, M. J. 1977. *Blood is their argument*. Mountain View, CA: Mayfield.

Mellars, P., & C. Stringer, eds. 1989. *The human revolution: Behavioral and biological perspectives on the origin of modern humans*. Edinburgh: Edinburgh University Press.

Morren, G. 1986. *The Miyanmin: Human ecology of a Papua New Guinea society*. Ann Arbor: University of Michigan Research Press.

Nelson, R., & S. Winter. 1982. *An evolutionary theory of economic change*. Cambridge, MA: Harvard University Press.

Oosterwal, G. 1961. *People of the Tor*. Assen: Royal van Gorcum.

Peoples, J. G. 1982. Individual or group advantage? A reinterpretation of the Maring ritual cycle. *Current Anthropology* 23:291–310.

Pospisil, L. 1963. *Kapauku Papuan economy*. Yale University Publications in Anthropology 67.

Pulliam, H. R., & C. Dunford. 1980. *Programmed to learn: An essay on the evolution of culture*. New York: Columbia University Press.

Radcliffe-Brown, A. R. 1952. *Structure and function in primitive society*. London: Cohen and West.

Rappaport, R. A. 1967. Ritual regulation of environmental relations among a New Guinea people. *Ethnology* 6:17–30.

Rappaport, R. A. 1979. *Ecology, meaning, and religion*. Richmond, CA: North Atlantic Books.

Rappaport, R. A. 1984. *Pigs for the ancestors*. New Haven, CT: Yale University Press.

Read, K. E. 1955. Morality and the concept of person among the Gahuku-Gama. *Oceania* 25:233–282.

Reay, M. 1959. *The Kuma: Freedom and conformity in the New Guinea Highlands*. Melbourne: Melbourne University Press.

Richerson, P. J., & R. Boyd. 1989. The role of evolved predispositions in cultural evolution, or human sociobiology meets Pascal's wager. *Ethology and Sociobiology* 10:195–219.

Robbins, S. 1982. *Auyana: Those who held on to home*. Seattle: University of Washington Press.

Rogers, A. R. 1990. Group selection by selective emigration: The effects of migration and kin structure. *American Naturalist* 135:398–413.

Rogers, E. 1983. 3rd edition. *Diffusion of innovations*. New York: Free Press.

Ryan, D. 1959. Clan formation in the Mendi Valley. *Oceania* 29:257–289.

Saville, W. J. V. 1926. *In unknown New Guinea*. London: Seely, Service.

Sillitoe, P. 1977. Land shortage and war in New Guinea. *Ethnology* 16:71–82.

Slatkin, M., & M. J. Wade. 1978. Group selection on a quantitative character. *Proceedings of National Academy of Sciences, U.S.A.* 75:3531–3534.

Smith, E., & B. Winterhalder. 1992. Natural selection and decision making: Some fundamental principles. In: *Evolutionary ecology and human behavior*. Smith and Winterhalder, eds. (pp. 25–60). Hawthorne: Aldine de Gruyter.

Spencer, H. 1891. *Essays: Scientific, political, and speculative*. 3 vols. London: Williams and Norgate.

Strathern, A. 1971. *The rope of Moka*. Cambridge: Cambridge University Press.

Sugden, R. 1986. *The economics of rights, cooperation, and welfare*. Oxford: Blackwell Scientific Publications.

Turner, J. H., & A. Maryanski. 1979. *Functionalism*. Menlo Park, CA: Benjamin/Cummings.

Tuzin, D. F. 1976. *The Ilahita Arapesh*. Berkeley: University of California Press.

Van Den Berghe, P. L. 1981. *The ethnic phenomenon*. New York: Elsevier.

Vayda, A. P. 1971. Phases of the process of war and peace among the Marings of New Guinea. *Oceania* 42:1–24.

Waddel, E. 1972. *The mound builders*. Seattle: University of Washington Press.

Wade, M. J. 1978. A critical review of group selection models. *Quarterly Review of Biology* 53:101–114.

Wilson, D. S. 1983. The group selection controversy: History and current status. *Annual Review of Ecology and Systematics* 14:159–187.

12 Group-Beneficial Norms Can Spread Rapidly in a Structured Population

Many culturally transmitted norms are group-beneficial (Sober and Wilson, 1998): property rights encourage productive effort, rules against murder and assault encourage civil order, norms governing the filling of political offices reduce the chances of civil war, and product standards, building codes, and rules of professional conduct allow more efficient commerce. For most of human history, states were weak or nonexistent, and norms were not enforced by external sanctions. Nonetheless, norms were important regulators of social order, and while in modern states black-letter laws also further many of the same ends as informal norms, the evidence is that informal custom still plays a very important role in regulating behavior (Ellickson, 1991).

The *persistence* of group-beneficial norms is easily explained. When people interact repeatedly, behavior can be rewarded or punished, and such incentives can stabilize almost any behavior once there is consensus about what is normative. People conform to normative behavior in order to gain rewards or avoid punishment. The provision of rewards and punishments can be explained in several ways: first, if interactions are repeated indefinitely, punishing or rewarding also can be normative behaviors, and violators of that norm can be punished or rewarded as well (Boyd and Richerson, 1992a). Second, even if interactions do not go on indefinitely (or equivalently, people cannot remember large number of interactions), the relative disadvantage suffered by those who enforce social norms compared with those who do not rapidly becomes small as the number of interactions increases and is easily balanced by even a weak tendency to imitate the common type (Henrich and Boyd, 2001). (Of course, strong conformism can also explain the maintenance of norms without punishment; Boyd and Richerson, 1985.) Finally, punishment may be individually beneficial if it is a costly signal of an individual's qualities as a mate or coalition partner (Bleige Bird, Smith, and Bird, 2001). Several authors suggest that the stability of such norms explains human cultural diversity—distinct groups represent alternative, stable equilibria in a complex, repeated "game of life" (Boyd and Richerson, 1992b; Binmore, 1994; Cohen, 2001).

The fact that group-beneficial norms can persist does not explain why such norms are widely observed. While punishment and reward can stabilize group-beneficial norms, they can also stabilize virtually any behavior (Fundenberg and Maskin, 1986; Boyd and Richerson, 1992a). We can be punished if we lie or steal, but we can also be punished if we fail to wear a tie or refuse to eat the brains of dead relatives. Thus, we need an explanation for why populations should be more likely to wind up at a group-beneficial equilibrium than one of the vastly greater number of stable but non-group-beneficial equilibria. Put another way, if social diversity results from many stable social equilibria, then social evolution must involve shifting among alternative stable equilibria. Group-beneficial equilibria will be common only if the process of equilibrium selection tends to pick out group-beneficial equilibria.

Currently, there are two different kinds of models of equilibrium selection, but neither provides a plausible explanation for the widespread existence of group-beneficial norms.

Within-group models of equilibrium selection (Kandori, Mailath, and Rob, 1993; Ellison, 1993; Young, 1998; Samuelson, 1997) consider the effects of random processes that act within groups to change the frequency of alternative behavioral strategies. In finite populations, sampling variation will affect patterns of interaction and replication, which in turn will lead to random fluctuations in the frequencies of types through time. As long as some mutation-like process acts to maintain variation, the probability that the population will be in any state will eventually converge to a stationary distribution. If mutation rates are low and populations are of reasonable size, most of the probability mass of the stationary distribution will pile up around the stable equilibrium of the deterministic dynamic model that has the largest basin of attraction. Since there is no necessary relationship between the size of a basin of attraction and whether it is group beneficial, within-group models do not predict that group-beneficial norms will be common. Within-group models also suffer from two other related problems. First, it takes a very long time for populations to shift from one equilibrium to another unless the number of interacting individuals is very small. Second, these models provide no mechanism for cumulative irreversible social change because populations are assumed to be in stochastic steady state, randomly wandering back and forth between alternative equilibria.

Between-group models posit that equilibrium selection results from the competition between groups near alternative stable equilibria. These models assume that groups at more efficient equilibria are less likely to go extinct, or more able to compete with other groups in military or economic contests. This kind of group selection process leads to the evolution of group-beneficial equilibria even when groups are large, and there is substantial migration between groups (Boyd and Richerson, 1982, 1990). However, given observed rates of group extinction, the spread of group-beneficial equilibria will occur too slowly to account for much observed social evolution. Calculations based on empirical data on the social extinction of small groups in highland New Guinea suggest that even though rates of extinction are appreciable, the time scale for the substitution of one norm by a better one is on the order of a millennium (Soltis, Boyd, and Richerson, 1995). Moreover, these models also lack any mechanism that allows for the efficient recombination of group-beneficial innovations occurring in different groups, and thus cannot easily account for the cumulative nature of social change over the last 10,000 years.

Here, we show that when the standard replicator dynamic model of evolutionary game theory is embedded in a spatially structured population, group-beneficial equilibria can spread rapidly and innovations can readily recombine to form beneficial new combinations. The basic logic of this result is simple: evolutionary game theory is applicable to human social evolution when behavioral strategies are transmitted by imitation, and people who have achieved high payoffs are most likely to be imitated. Strategies that have high average payoffs will increase in frequency, in most cases eventually leading to a stable evolutionary equilibrium state. If the payoff structure of social interactions leads to multiple stable equilibria and a population is structured, partially isolated groups can be stabilized at different equilibria with different average payoffs. Consequently, behaviors can spread from groups at high payoff equilibria to neighboring groups at lower payoff equilibria because people imitate their more successful neighbors. Such spread can be rapid because it depends on the rate at which individuals imitate new strategies, rather than the rate at which groups become extinct.

In what follows, we first derive the dynamic equations that govern replicator dynamics in a spatially structured population. We then show that these equations can lead to the rapid spread of group-beneficial traits under plausible conditions. Finally, we show that this process readily leads to the recombination of different group-beneficial traits that arise in different populations.

Replicator Dynamics in a Structured Population

In many situations, people have important social interactions shaped by social norms with one group of people but know about the behavior, and the norms that regulate it, of a larger group of people. People interact every day with the members of their local group—they exchange food, labor, and land; aid others in need; marry and care for children—transactions that are regulated by social norms that define property rights and moral obligations. However, people also often know about the behavior of others in neighboring groups. They know that we can marry our cousins here, but over there they cannot; or anyone is free to pick fruit here, while there fruit trees are owned by individuals. With this kind of population structure, payoffs are determined by the composition of the local group, but cultural traits can diffuse among groups.

To generalize evolutionary game theory to allow for this kind of population structure, consider a population that is subdivided into n large groups in which frequent social interaction occurs. Individuals are characterized by one of k strategies. The proportion of people in group d who have strategy i is p_{id}, and the vector of frequencies in group d is $\mathbf{p}_d$. Social interaction generates a payoff, $W_i(\mathbf{p}_d)$ for individuals with behavior i in group d that depends on individuals' own strategy and the strategies of other members of their group because frequent social interaction occurs with other group members.

To allow for the possibility of cultural diffusion between groups, we adopt the following model of cultural transmission: during each time period, each individual from group f encounters an individual, their "model," from group d with probability m_{df} and observes that individual's strategy and payoff from social interaction during that period. We will assume that $m_{ff} > \Sigma_{d \neq f} m_{df}$ so that most encounters occur within social groups. After the encounter, individuals may imitate the strategy of their model.

We assume that individuals are more likely to imitate if their model has a higher payoff than they do. More formally, if an individual with behavior i from group f encounters an individual with behavior j from group d, individual i switches to j with this probability:

$$Pr(j|i,j) = \tfrac{1}{2}(1 + \beta(W_j(\mathbf{p}_d) - W_i(\mathbf{p}_f))) \tag{1}$$

where β is a positive parameter that scales payoffs so that $0 \leq Pr(j|i,j) \leq 1$ for all $\mathbf{p}_d$ and $\mathbf{p}_f$. Equation (1) implies that individuals sometimes switch to a lower payoff strategy, unlike some recent derivations of replicator dynamics (Borgers and Sarin, 1997; Schlag, 1998; Gale, Binmore, and Samuelson, 1995). We think this model is preferable because it captures the effect of uncertainty about the payoffs of others, and because it allows diffusion between groups even when there are no payoff differences, a conservative feature that reduces the effect of population structure.

Then the frequency of behavior i in group f, p'_{if}, after one time period is given by equation (2):

$$p'_{if} = \sum_d m_{df}\left[p_{if} \sum_j p_{jd}\frac{1}{2}(1 + \beta(W_i(\mathbf{p}_f) - W_j(\mathbf{p}_d))) \right.$$
$$\left. + p_{id} \sum_j p_{jf}\frac{1}{2}(1 + \beta(W_i(\mathbf{p}_d) - W_j(\mathbf{p}_f))) \right] \tag{2}$$

The first sum inside the square brackets gives the probability that an individual with trait i in group f remains the same, and the second sum gives the probability that someone who is not i initially converts to i. Some algebraic manipulation yields the following expression for the change in the frequency of behavior i in population f:

$$p'_{if} - p_{if} = \delta p_{if}\left(1 - \sum_{\substack{d \neq f}}^{n}\frac{1}{2}m_{df}\right)$$
$$+ \sum_{\substack{d \neq f}}^{n}\frac{1}{2}m_{df}[\delta p_{id} + (p_{id} - p_{if})(1 + \beta(W(\mathbf{p}_d) - W(\mathbf{p}_f)))] \tag{3}$$

where $\delta p_{if} = \beta p_{if}(W_i(\mathbf{p}_f) - W(\mathbf{p}_f))$ is the replicator dynamic equation for strategy i in group f and is the canonical description of strategy dynamics in evolutionary game theory. Thus, when individuals imitate only members of their own group

$(m_{df} = 0, d \neq f)$, equation (3) says that imitation within each group causes behaviors with the highest payoff *relative* to others in the group to increase in frequency—effects on average payoff within a group are irrelevant. When there is contact between different groups, however, the effect of a behavior on average group payoff can become important. The second term in equation (3) includes the effect of diffusion between groups that differ in trait frequency. When payoffs do not effect imitation ($\beta = 0$), this term includes only passive diffusion. However, when individuals with higher payoffs are more likely to be imitated, there is a net flow of strategies from groups with high average payoff to groups with lower average payoff.

How Group-Beneficial Equilibria Spread

Next, we show how this effect can lead to the spread of group-beneficial equilibria. Consider a simple model in which there are two strategies, 1 and 2. For example, strategy 1 might be a norm forbidding cousin marriage, while strategy 2 is the norm allowing free choice of a spouse. Within each group, individuals who deviate from the common norm suffer because they are punished by other group members. The norm forbidding cousin marriage might lead to higher average payoff due to the formation of wider political alliances. We formalize these ideas by assuming that the payoff to an individual with behavior 1 in group d is $W_1(p_{1d}) = 1 + s(p_{1d} - p) + gp_{1d}$ and the payoff to an individual using behavior 2 is $W_2(p_{1d}) = 1 + gp_{1d}$. Thus, each strategy has a higher relative payoff when common. The unstable equilibrium that divides the two basins of attraction is p. The parameter s measures the magnitude of the difference in payoffs of the two strategies, and g measures the effect of behavior 1 on average payoff. We assume that $g > 0$, so that groups in which behavior 1 is common have higher average payoff. For example, a norm against cousin marriage might lead to more alliance formation among clans within the group. Finally, for simplicity, we assume that social groups are arranged in a ring so individuals imitate only members of their own group and the two neighboring groups. (So that $m_{df} = m$ for the two neighbors of group f and zero otherwise.)

For a novel group-beneficial trait to evolve, two things must occur. First, it must become common in one population, and second it must spread from that population to others. Various random processes may cause the initial shift of one population to the group-beneficial equilibrium. In finite populations, sampling variation in who is imitated (Gale et al., 1995) or in patterns of interaction (Kandori et al., 1993; Ellison, 1993; Young, 1998) can lead to random fluctuations in trait frequencies that can tip populations into the basin of attraction of the group-beneficial equilibrium. Randomly varying environments can lead to similar shifts (Price, Turelli, and Slatkin, 1993) in populations. Finally, individual learning can be conceptualized as a process in which individuals use data from the environment to infer the best behavior. Learning experiences of individuals within a population may often be correlated, because they are utilizing the same data. Thus, random variation in such correlated learning experiences could also cause equilibrium shifts in large populations. We do not model these processes here.

To see how imitation of the successful can lead to the spread of group-beneficial strategies, assume that one of these unmodeled processes causes the group-beneficial strategy to become common in one group, while the other strategy remains common in the rest of the groups. Then, if enough individuals in the two neighboring groups imitate behavior 1, these groups will be tipped into its basin of attraction, and the group-beneficial trait will increase in those two groups. This process is illustrated in figure 12.1. Trait 1 is initially common in population $i-1$. In the neighboring population i, trait 2 is common, and thus within-group imitation tends to decrease the frequency of trait 1. However, individuals in population i are more likely to imitate individuals in population $i-1$ than in population $i+1$, so extra-group imitation tends to increase the frequency of trait 1 in group i. If this latter process is sufficiently strong, it can tip population i into trait 1's basin of attraction. If this occurs, the process will be repeated in group $i+1$, then group $i+2$, and so on, with behavior 1 spreading throughout the population in a wave-like fashion. This process is formally similar to one recent model of the third phase of Wright's shifting balance theory (Gavrilets, 1995), but is unlike that model in two ways. First, the underlying dynamic processes arise from differential imitation, not changes in demography. Second, because the multiple equilibria arise from frequency-dependent social interaction, not underdominance, the process modeled here leads to the spread of the group-beneficial trait for a wide range of parameters (figure 12.2).

It is important to see that the spread of the group-beneficial trait depends crucially on the assumption that people imitate strategies that lead to success in neighboring groups, but will lower their payoff in their own group where different norms are enforced. In this simple model, a type that restricted imitation to its own group would replace the type of imitation assumed here. We think our assumption is plausible nonetheless. Empirically, the tendency to imitate the successful has been observed in a wide variety of contexts (see Henrich and Gil-White, 2001). This tendency makes sense adaptively. The world is complex and hard to understand. It is very difficult in many situations to connect behavior to outcomes with much confidence. An individual observes that in the neighboring group they never marry cousins and that they are much better off. His neighbors say that the gods punish those who marry cousins, and they have had much greater success in warfare lately. Of course, the individual knows that it will cause trouble to forbid a marriage that both his daughter and his brother want, but maybe it will be worth it. The same kinds of uncertainties beset us in the modern world despite vastly greater information-gathering capacity. In the early 1990s it was commonplace to attribute Japan's economic success to encouragement of long-term investment, their "just in time" inventory practices, or to their quality circles, and all of these practices were imitated by American firms and policy makers. We have argued at length (Boyd and Richerson, 1985) that cultural transmission rules like *imitate the successful* and *imitate the common type* should be seen as adaptations for dealing with this kind of uncertainty. We have a propensity to imitate the successful *because* it is often very difficult to decide what is the best behavior. These learning rules are shortcuts that on average allow us to acquire lots of useful information but may, as in the model in this chapter, sometimes lead us astray.

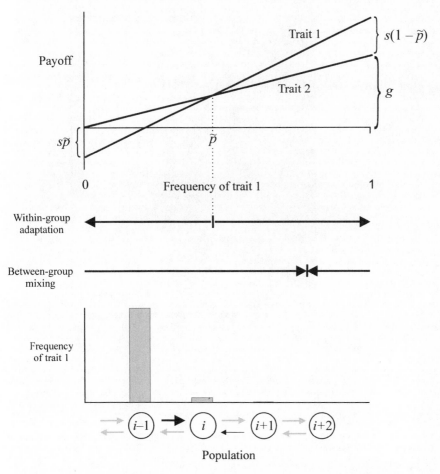

Figure 12.1. This graph illustrates the assumed payoff structure and why it can lead to the spread of group-beneficial traits. The top panel plots the payoffs to traits 1 and 2 as a function of the frequency of trait 1 in their local group. Each trait has a higher relative payoff when it is common, but increasing the frequency of trait 1 raises the payoff of all group members. As a result, within-group imitation increases the frequency of trait 1 above the threshold frequency p and increases the frequency of trait 2 below that threshold. The lower panel shows the state of a part of a population in which trait 1 is initially common in group $i-1$ and trait 2 is common in all other groups. In group i, individuals are more likely to imitate people in population $i-1$ than in population $i+1$ because the former have higher payoffs than the latter. Thus, between-group imitation tends to increase the frequency of trait 1 in population i. If this effect is strong enough, it can tip group i into the basin of attraction of trait 1 and cause the spread of this group-beneficial trait.

Figure 12.2 plots combinations of the parameters m, s, p, and g that lead to the spread of the group-beneficial strategy. It indicates that the group-beneficial strategy fails to spread under three circumstances. If there is too much mixing between neighboring groups, the beneficial strategy cannot persist in the initial population; it is swamped by the flow of behavior 2 from the neighboring groups.

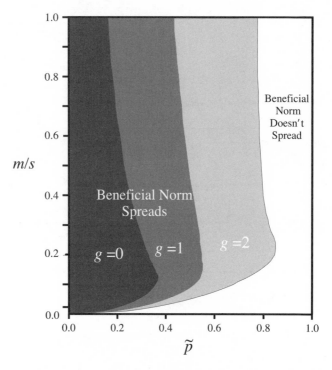

Figure 12.2. This graph shows the range of parameters over which the beneficial norm spreads to all groups, eliminating the alternative norm, given that the beneficial norm is initially common in a single group. The vertical axis gives the ratio of m, the probability that individuals interact with others from one of the neighboring groups, to s, the rate of change due to imitation within groups. The horizontal axis plots p, the unstable equilibrium that separates the basins of attraction of group-beneficial and nongroup-beneficial equilibria in isolated groups. The shaded areas give the combinations of m/s and p that lead to the spread of the group-beneficial strategy for three values of g. When $g = 0$, neither norm is group-beneficial. Larger values of g mean that the group-beneficial norm leads to a greater increase in average payoff. When m is small, the group-beneficial norm cannot spread because there is not enough interaction between neighbors for the beneficial effects of the norm to cause it to spread. Very large values of m prevent the spread of the group-beneficial norm because it cannot persist in the initial population. If the domain of attraction of the group-beneficial strategy is too small, the flow of strategies from successful groups to less successful groups does not tip neighboring groups into its basin of attraction. Increasing the degree to which strategy 1 is group-beneficial (i.e., the magnitude of g) enlarges the range of parameters that lead to the increase in that strategy. Here, the number of groups, n, was 32, but results are insensitive to n as long as it is sufficiently large. Very small values of n increase the range of parameters under which the group-beneficial trait spreads. These results are from simulation—if the group-beneficial trait had not spread to all groups after 10,000 time periods, we assumed it would not spread. To construct the graph, we chose values of m/s and then used an interval-halving algorithm to find the threshold value of p at which trait 1 did not spread.

If there is too little mixing, the group-beneficial behavior remains common in the initial population but cannot spread because there is not enough interaction between neighbors for the beneficial effects of the norm to cause it to spread. If the domain of attraction of the group-beneficial strategy is too small, the flow of ideas from successful groups to less successful groups may not be sufficient to tip neighboring groups into its basin of attraction. Increasing the degree to which strategy 1 is group-beneficial (i.e., the magnitude of g) enlarges the range of parameters that lead to the increase in that strategy.

The results plotted in figure 12.3 show that the group-beneficial trait spreads at a rate that is roughly comparable with the rate at which individually

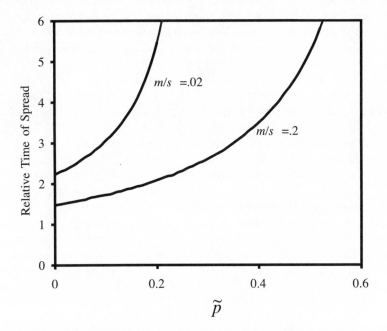

Figure 12.3. This figure plots a measure of the length of time necessary for the spread of the group-beneficial trait relative to the length of time necessary for the spread of an individually advantageous trait. In the simulations reported, the group-beneficial trait spreads from one group to the next at a constant rate after an initial transient period. Here, we plot the ratio of the time necessary to increase from a frequency of 0.1 to 0.9 in a single group at the boundary of the wave spreading at the constant rate divided by the length of time necessary for a purely advantageous trait with dynamics $\Delta p = sp(1-p)$ to spread from 0.1 to 0.9 in a single isolated population for two different values of the ratio m/s. As in figure 12.1, m is the probability of interacting with, and potentially imitating, an individual in each of the two neighboring groups. In both graphs, $g = 1.0$, and the parameter p is the unstable equilibrium that divides the basins of attraction of the group-beneficial trait and the other trait. These results indicate that spatial structure causes an initially individually disadvantageous but group-beneficial trait to spread on roughly the same time scale as a simple individually advantageous trait whose within-group dynamics are governed by the same rate parameter s.

beneficial traits spread within a single group under the influence of the same learning process. Thus, if an individually beneficial trait can spread within a population in 10 years, a group-beneficial trait will spread from one population to the next in 15–30 years, depending on the amount of mixing and the effect of the trait on average fitness. Game theorists have considered a number of mechanisms of equilibrium selection that arise because of random fluctuations in outcomes due to sampling variation and finite number of players (Kandori et al., 1993; Ellison, 1993; Young, 1998; Samuelson, 1997). These processes tend to pick out the equilibrium with the largest domain of attraction. However, unless spatial structure limits interactions to a small number of individuals, the rate at which this occurs in a large population is very slow. Similarly, group selection models appear to require unrealistically high group extinction rates to explain many examples of the spread of group-beneficial cultural traits (Boyd and Richerson, 1990; Soltis et al., 1995). In contrast, the process we describe here leads to the deterministic spread of the group-beneficial trait on roughly the same time scale as the same social learning processes cause individually beneficial traits to spread within groups.

Of course, we have not accounted for the processes that influence the rate at which the beneficial behavior initially becomes common in a particular group. However, if the conditions for spread are satisfied, the group-beneficial trait needs to become common only in a single group. If we imagine that group-beneficial traits mainly arise as a result of random processes in small populations, only the initial group, not the whole population, needs to be small, and the group must remain small only for long enough for random processes to give rise to an initial "group mutation," which can then spread relatively rapidly to the population as a whole. If we imagine that rare events, such as the emergence of uniquely charismatic reformers or alignment of the particular constellations of political forces, are required to affect a group-favoring innovation, the same considerations apply. Only one group need make the original innovation; any others with substantial cultural contact can rapidly acquire the trait by the mechanism we model here.

Recombination at the Group Level

The process described here readily leads to the recombination of group-beneficial strategies that initially arise in different groups. The exact combination of strategies necessary to support complex, adaptive social institutions would seem unlikely to arise through a single chance event. It is much more plausible that complex institutions are assembled in numerous small steps. Previous group selection models of equilibrium selection are analogous to the evolution of an asexual population in that they lack any mechanism that allows the recombination of beneficial strategies that arise in different populations and thus require innovations to occur sequentially in the same lineage. Within-group models in which equilibrium selection occurs through random sampling processes assume that the population has reached a stationary distribution, and thus while recombination is possible, there is no cumulative, irreversible change. By contrast, the present model allows recombination of different strategies and irreversible, cumulative change. To see this, consider a model in which strategies consist of two components (x, y), each with two values

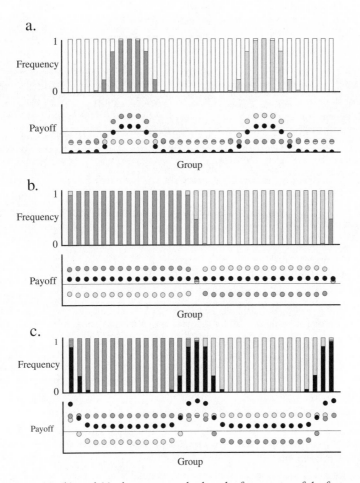

Figure 12.4. In (a), (b), and (c), the upper graph plots the frequencies of the four possible strategies as stacked bar graphs for each of 32 groups: $(0, 0)$ white, $(1, 0)$ light gray, $(0, 1)$ dark gray, and $(1, 1)$ black. The lower graph plots the payoff to each strategy net of the group effects in each group. The (—) line gives the payoff of $(0, 0)$ and the ($\bullet \ \bullet \ \bullet$) circles give the payoffs of the other three strategies. The parameters are $m = 0.02$, $s = 0.1$, $p = 0.4$, and $g = 2$. (a) Initially $(0, 1)$ is common in group 8 and $(1, 0)$ is common in group 24, and the two group-beneficial traits begin to spread. (b) When the two spreading fronts meet, the frequencies of $x = 1$ and $y = 1$ are one half, which means that the strategy $(1, 1)$ has the highest payoff. (c) Recombination at the individual level introduces strategy $(1, 1)$ into the boundary group, and strategy $(1, 1)$ then spreads deterministically, first in that group and then to adjacent groups.

(0, 1). Let p_d and q_d be the frequencies of $x=1$ and $y=1$ in group d, respectively. Let the payoff of an individual in group d be as follows:

$$W_d(x,y) = 1 + sx(p_d - p)$$
$$+ sy(q_d - p)$$
$$+ g(q_d + p_d) \qquad (4)$$

Thus, both $x=1$ and $y=1$ have an independent group-beneficial effect, and all four combinations of x and y can be stable equilibria in isolated groups. Finally, suppose that individuals occasionally learn the x component of their strategy from one individual and the y component from another, leading to recombination of behavioral strategies at the individual level. Once again suppose that the population is initially all strategy (0, 0), and that random shocks cause (1, 0) to become common in one population and (0, 1) common in a second population. Then, if conditions are right, both strategies will begin to spread (figure 12.4[a]). When the two waves meet, the frequency of $x=1$ is equal to one half and the frequency of $y=1$ is equal to one half at the boundary between the two expanding fronts. The outcome depends on the value of p. If $p < \frac{1}{2}$, the strategy (1, 1) has the highest payoff in the group on the boundary, increases deterministically in that group, and eventually spreads throughout the population as a whole (figure 12.4[b]). If $p > \frac{1}{2}$, the strategy (1, 1) has a lower payoff than (1, 0) or (0, 1), and the two waves form a stable boundary. However, in the boundary group, the most beneficial combination, (1, 1), has a relatively small payoff disadvantage compared with (0, 1), and (0, 1) is present at substantial frequency. In this situation, a shift to the most beneficial combination due to random shocks is much more likely than the shifts that were necessary to cause (0, 1) and (1, 0) to become common in the first place. Thus, existing group-beneficial traits will recombine more rapidly than new ones arise.

Conclusion

Many anthropologists and sociologists have long believed that human behavior is regulated by culturally transmitted norms in ways that promote the survival and growth of human societies. Economists and other rational choice theorists have been skeptical about such functionalist claims because there was no plausible mechanism to explain why such norms should be common. Social scientists influenced by evolutionary biology tend to share this skepticism based upon theoretical models and empirical findings suggesting that group selection is generally a weak force in nature. We believe that humans are an exception to this rule because cultural variation is much more susceptible to group selection than genetic variation. The cultural group selection hypothesis explains both why humans cooperate on such a large scale and why the pattern of this cooperation is so different from that of other ultrasocial animals (Richerson and Boyd, 1999). Human societies are based upon cooperation between nonrelatives, while kinship underlies cooperation and complex sociality in other taxa like the social insects.

Despite a general fit between the existing models of cultural group selection and the facts of human sociality, much uncertainty remains. Earlier work suggests that the differential survival of culturally distinctive groups can lead to the evolution of group-beneficial behavior under plausible circumstances, but that this process is quite slow and likely to produce historically contingent group-level adaptations (Boyd and Richerson, 1982, 1990; Soltis et al., 1995). Since the evolution of human social institutions does have a time scale of millennia and the resulting institutions are highly variable, such group selection processes may have had a role in shaping these institutions. On the other hand, some social institutions do diffuse from one society to another and on time scales shorter than a millennium. The spread of the joint stock company on time scales of a century is a recent example. Such events accord better with a mechanism like the one we model here.

We suspect that both differential survival and differential diffusion may affect the evolution of human social institutions. The operation of many social institutions is opaque even to the people who enact them (Nelson and Winter, 1982, ch. 5), and such institutions are even harder for outsiders to understand. In such cases, diffusion may be ineffective because actors cannot connect the attributes of particular institutions to their success, and this fact may explain why the path from the origins of agriculture to our complex modern industrial nations took some 10 millennia to traverse. Other institutions spread much more readily because their costs and benefits are more readily understood. Proselytizing religions, for example, take pains to be transparent to potential converts and thus may readily spread. The rate of diffusion of institutions may also be affected by how much people know about other societies. It is plausible that the spread of literacy and the development of ever better means of transportation have gradually increased the importance of the rapid processes based on borrowing relative to the slower ones based on group extinction. In the twentieth century, social institutions like central banks, soccer, and government bureaucracies have become all but universal in about a century. Nevertheless, globalization is incomplete; dramatic differences exist even between modern societies (Nisbett, Peng, Choi, and Norenzayan, 2001). Some elements of culture likely still have time scales of change measured in millennia.

NOTE

We thank Sam Bowles, Ernst Fehr, Daniel Friedman, Francisco Gil-White, Herb Gintis, Joe Henrich, Richard McElreath, Rajiv Sethi, David Sloan Wilson, and several anonymous reviewers for useful comments on the research reported here.

REFERENCES

Binmore, K. 1994. *Game theory and the social contract.* Cambridge, MA: MIT Press.
Bleige Bird, R., E. A. Smith, & D. Bird. 2001. The hunting handicap: Costly signaling in human foraging strategies. *Behavioral Ecology and Sociobiology*, 50:9–19.

Borgers, T., & R. Sarin. 1997. Learning through reinforcement and replicator dynamics. *Journal of Economic Theory* 77:1–14.

Boyd, R., & P. J. Richerson. 1982. Cultural transmission and the evolution of cooperative behavior. *Human Ecology* 10:325–351.

Boyd, R., & P. J. Richerson. 1985. *Culture and the evolutionary process*. Chicago: University of Chicago Press.

Boyd, R., & P. J. Richerson. 1990. Group selection among alternative evolutionarily stable strategies. *Journal of Theoretical Biology* 145:331–342.

Boyd, R., & P. J. Richerson. 1992a. How microevolutionary processes give rise to history. In: *Evolution and history*, M. H. Niteki, & D. V. Nitecki, eds. (pp. 179–209). Albany: State University of New York Press.

Boyd, R., & P. J. Richerson. 1992b. Punishment allows the evolution of cooperation (or anything else) in sizable groups. *Ethology and Sociobiology* 13:171–195.

Cohen, D. 2001. Cultural variation: Considerations and implications. *Psychological Bulletin* 127:451–471.

Ellickson, R. C. 1991. *Order without law: How neighbors settle disputes*. Cambridge, MA: Harvard University Press.

Ellison, D. 1993. Learning, social interaction, and coordination. *Econometrica* 61:1047–1071.

Fundenburg, D. S., & E. Maskin. 1998. The folk theorem in repeated games with discounting or with incomplete information. *Econometrica* 51:533–554.

Gale, J., K. G. Binmore, & L. Samuelson. 1995. Learning to be imperfect: The ultimatum game. *Games and Economic Behavior* 8:56–90.

Gavrilets, S. 1995. On phase three of the shifting balance theory. *Evolution* 50:1034–1041.

Henrich, J., & R. Boyd. 2001. Why people punish defectors: Weak conformist transmission can stabilize costly enforcement of norms in cooperative dilemmas. *Journal of Theoretical Biology* 208:79–89.

Henrich, J., & F. Gil-White. 2001. The evolution of prestige: Freely conferred deference as a mechanism for enhancing the benefits of cultural transmission. *Evolution and Human Behavior* 22:165–196.

Kandori, M., G. Mailath, & R. Rob. 1993. Learning, mutation, and long run equilibria in games. *Econometrica* 61:29–56.

Nelson, R. R., & S. G. Winter. 1982. *An evolutionary theory of economic change*. Cambridge, MA: Harvard University Press.

Nisbett, R. E., K. Peng, I. Choi, & A. Norenzayan. 2001. Culture and systems of thought: Holistic vs. analytic cognition. *Psychological Review* 108:291–310.

Price, T., M. Turelli, & M. Slatkin. 1993. Peak shifts produced by correlated response to selection. *Evolution* 4:280–290.

Richerson, P. J., & R. Boyd. 1999. The evolutionary dynamics of a crude super organism. *Human Nature* 10:253–289.

Samuelson, L. 1997. *Evolutionary games and equilibrium selection*. Cambridge, MA: MIT Press.

Schlag, K. H. 1998. Why imitate, and if so, how? A boundedly rational approach to multi-armed bandits. *Journal of Economic Theory* 78:130–156.

Sober, E., & D. S. Wilson. 1998. *Unto others*. Cambridge, MA: Harvard University Press.

Soltis, J., R. Boyd, & P. J. Richerson. 1995. Can group functional behaviors evolve by cultural group selection? An empirical test. *Current Anthropology* 36:473–494.

Young, P. 1998. *Individual strategy and social structure*. Princeton, NJ: Princeton University Press.

13 The Evolution of Altruistic Punishment

With Herbert Gintis and Samuel Bowles

Unlike any other species, humans cooperate with nonkin in large groups. This behavior is puzzling from an evolutionary perspective because cooperating individuals incur individual costs to confer benefits on unrelated group members. None of the mechanisms commonly used to explain such behavior allows the evolution of altruistic cooperation in large groups. Repeated interactions may support cooperation in dyadic relations (Axelrod and Hamilton, 1981; Trivers, 1971; Clutton-Brock and Parker, 1995), but this mechanism is unsustainable if the number of individuals interacting strategically is larger than a handful (Boyd and Richerson, 1998). Interdemic group selection can lead to the evolution of altruism only when groups are small and migration is infrequent (Sober and Wilson, 1998; Eshel, 1972; Aoki, 1982; Rogers, 1990). A third recently proposed mechanism (Hauert, De Monte, Hofbauer, and Sigmund, 2002) requires that asocial, solitary types outcompete individuals living in uncooperative social groups, an implausible assumption for humans.

Altruistic punishment provides one solution to this puzzle. In laboratory experiments, people punish noncooperators at a cost to themselves even in one-shot interactions (Fehr and Gächter, 2002; Ostrom, Gardner, and Walker, 1994), and ethnographic data suggest that such altruistic punishment helps to sustain cooperation in human societies (Boehm, 1993). It might seem that invoking altruistic punishment simply creates a new evolutionary puzzle: why do people incur costs to punish others and provide benefits to nonrelatives? However, here we show that group selection can lead to the evolution of altruistic punishment in larger groups because the problem of deterring free riders in the case of altruistic cooperation is fundamentally different from the problem of deterring free riders in the case of altruistic punishment. This asymmetry arises because the payoff disadvantage of altruistic cooperators relative to defectors is independent of the frequency of defectors in the population, whereas the cost disadvantage for those engaged in altruistic punishment declines as defectors become rare because acts of punishment become very infrequent (Sethi and Somanathan, 1996). Thus, when altruistic punishers are common, individual-level selection operating against them is weak.

To see why, consider a model in which a large population is divided into groups of size n. There are two behavioral types, contributors and defectors. Contributors incur a cost (c) to produce a total benefit (b) that is shared equally among group members. Defectors incur no costs and produce no benefits. If the fraction of contributors in the group is x, the expected payoff for contributors is $bx - c$ and the expected payoff for defectors is bx, so the payoff disadvantage of the contributors is a constant c independent of the distribution of types in the population. Now add a third type, "punishers," who cooperate and then punish each defector in their group, reducing each defector's payoff by p/n at a cost k/n to the punisher. If the frequency of punishers is y, the expected payoffs become $b(x+y) - c$ to contributors, $b(x+y) - py$ to defectors, and $b(x+y) - c - k(1 - x - y)$ to punishers. Contributors have higher fitness than defectors if punishers are sufficiently common that the cost of being punished exceeds the cost of co-operating $(py > c)$. Punishers suffer a fitness disadvantage of $k(1 - x - y)$ compared with nonpunishing contributors. Thus, punishment is altruistic and mere contributors are "second-order free riders." Note, however, that the payoff disadvantage of punishers relative to contributors approaches zero as defectors become rare because there is no need for punishment. In a more realistic model (like the one we show), the costs of monitoring or punishing occasional mistaken defections would mean that punishers have slightly lower fitness than contributors and that defection is the only one of these three strategies that is an evolutionarily stable strategy in a single isolated population. However, the fact that punishers experience only a small disadvantage when defectors are rare means that weak within-group evolutionary forces, such as mutation (Sethi and Somanathan, 1996) or a conformist tendency (Henrich and Boyd, 2001), can stabilize punishment and allow cooperation to persist. But neither produces a systematic tendency to evolve toward a cooperative outcome. Here we explore the possibility that selection among groups leads to the evolution of altruistic punishment when it could not maintain altruistic cooperation.

Suppose that more cooperative groups are less prone to extinction. Humans always live in social groups in which cooperative activities play a crucial role. In small-scale societies, such groups frequently become extinct (Soltis, Boyd, and Richerson, 1995). It is plausible that more cooperative groups are less subject to extinction because they are more effective in warfare, more successful in coin-suring, more adept at managing common resources, or for similar reasons. This means that, all other things being equal, group selection will tend to increase the frequency of cooperation in the population. Because groups with more punishers will tend to exhibit a greater frequency of cooperative behaviors (by both con-tributors and punishers), the frequency of punishers and cooperative behaviors will be positively correlated across groups. As a result, punishment will increase as a "correlated response" to group selection that favors more cooperative groups. Because selection within groups against punishment is weak when punishment is common, this process might support the evolution of substantial levels of punishment and maintain punishment once it is common.

To evaluate this intuitive argument we studied the following model using simulation methods. There are N groups. Local density-dependent competition maintains each group at a constant population size n. Individuals interact in a two-stage "game." During the first stage, contributors and punishers cooperate with probability $1 - e$ and defect with probability e. Cooperation reduces the payoff of cooperators by an amount c and increases the ability of the group to compete with other groups. For simplicity, we begin by assuming that cooperation has no effect on the individual payoffs of others, but does reduce the probability of group extinction. Defectors always defect. During the second stage, punishers punish each individual who defected during the first stage. After the second stage, individuals encounter another individual from their own group with probability $1 - m$ and an individual from another randomly chosen group with probability m. An individual i who encounters an individual j imitates j with probability $W_j/(W_j + W_i)$, where W_x is the payoff of individual x in the game, including the costs of any punishment received or delivered. Thus, imitation has two distinct effects: first, it creates a selection-like process that causes higher payoff behaviors to spread within groups. Second, it creates a migration-like process that causes behaviors to diffuse from one group to another at a rate proportional to m. Because cooperation has no individual-level benefits, defectors spread between groups more rapidly than do contributors or punishers. Group selection occurs through intergroup conflict (Bowles, 2001). In each time period, groups are paired at random, and with probability ε, intergroup conflict results in one group defeating and replacing the other group. The probability that group i defeats group j is $1/2(1 + (d_j - d_i))$, where d_q is the frequency of defectors in group q. This means that the group with more defectors is more likely to lose a conflict. Note that cooperation is the sole target of the resulting group selection process; punishment increases only to the extent that the frequency of punishers is correlated with that of cooperation across groups. Finally, with probability μ individuals of each type spontaneously switch into one of the two other types. Mutation and erroneous defection ensure that punishers will incur some punishment costs, even when they are common, thus placing them at a disadvantage with respect to the contributors.

Methods

Two simulation programs implementing the model were independently written, one by R. B. in Visual Basic, and a second by H. G. in Delphi. Code is available on request. Results from the two programs are highly similar. In all simulations there were 128 groups. Initially one group consisted of all altruistic punishers and the other 127 groups were all defectors. Various random processes could cause such an initial shift. Sampling variation in who is imitated (Gale, Binmore, and Samuelson, 1995) could increase the frequency of punishers. Randomly varying environments can lead to similar shifts (Price, Turelli, and Slatkin, 1993) in populations. Finally, individual learning can be conceptualized as a process in which individuals use data from the environment to infer the best behavior. Learning experiences of individuals within a population may often be correlated because they are using the same data. Thus, random variation in such correlated learning experiences could also cause equilibrium shifts in large populations. We do not model these processes here. Simulations were run for 2,000 time periods. The long-run average results plotted in figures 13.1–13.4 represent the average of frequencies over the last 1,000 time periods of 10 simulations.

Base case parameters were chosen to represent cultural evolution in small-scale societies. We set the time period to be 1 year. Because individually beneficial cultural traits, such as technical innovations, diffuse through populations in 10–100 years (Rogers, 1983), we set the cost of cooperation, c, and punishing, k, so that traits with this cost advantage would spread in 50 time periods

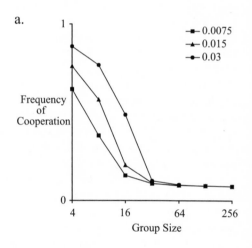

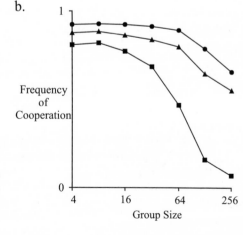

Figure 13.1. The evolution of cooperation is strongly affected by the presence of punishment. (a) The long-run average frequency of cooperation (i.e., the sum of the frequencies of contributors and punishers) as a function of group size when there is no punishment ($p = k = 0$) for three different conflict rates, 0.075, 0.015, and 0.003. Group selection is ineffective unless groups are quite small. (b) When there is punishment ($p = 0.8$, $k = 0.2$), group selection can maintain cooperation in substantially larger groups.

a.

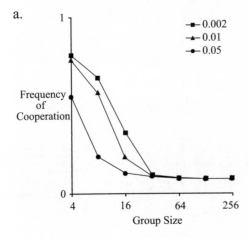

b.

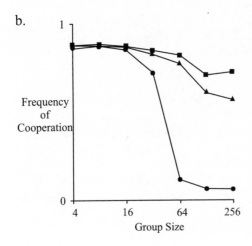

Figure 13.2. The evolution of cooperation is strongly affected by rate of mixing between groups. (a) The long-run average frequency of cooperation (i.e., the sum of the frequencies of contributors and punishers) as a function of group size when there is no punishment ($p=k=0$) for three mixing rates, 0.002, 0.01, and 0.05. Group selection is ineffective unless groups are quite small. (b) When there is punishment ($p=0.8$, $k=0.2$), group selection can maintain cooperation in larger groups for all rates of mixing. However, at higher rates of mixing, cooperation does not persist in the largest groups.

($c=k=0.2$). To capture the intuition that in human societies punishment is more costly to the punishee than to the punisher, we set the cost of being punished to four times the cost of punishing ($p=0.8$). We assume that erroneous defection is relatively rare ($e=0.02$). The migration rate, m, was set so that in the absence of any other evolutionary forces (i.e., $c=p=k=e=\varepsilon=0$), passive diffusion will cause two neighboring groups that are initially as different as possible to achieve the same trait frequencies in ≈ 50 time periods ($m=0.01$), a value that approximates the migration rates in a number of small-scale societies (Harpending and Rogers, 1986). We set the value of the mutation rate so that the long-run average frequency of an ordinary adaptive trait with payoff advantage c is ≈ 0.9 ($\mu=0.01$). This means that mutation maintains considerable variation, but not so much as to overwhelm adaptive forces. We assume that the average group extinction rate is consistent with a recent estimate of cultural

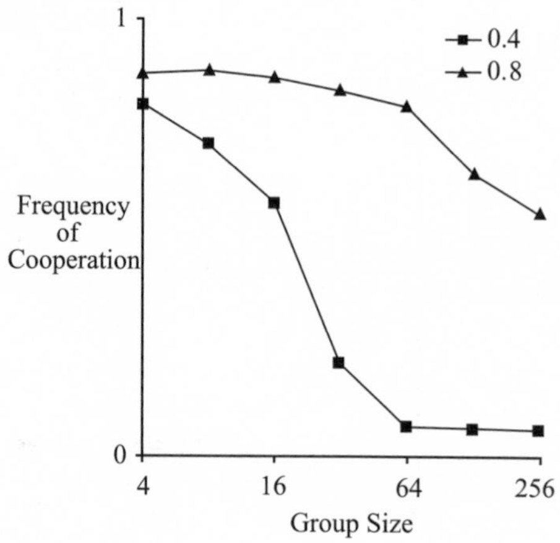

Figure 13.3. The evolution of cooperation is sensitive to the cost of being punished (p). Here we plot the long-run average frequency of cooperation with the base case cost of being punished ($p=0.8$) and with a lower value of p. Lower values of p result in much lower levels of cooperation.

extinction rates in small-scale societies, ≈ 0.0075 (Soltis et al., 1995). Because only one of the two groups entering into a conflict becomes extinct, this implies that $\varepsilon = 0.015$.

Results

Simulations using this model indicate that group selection can maintain altruistic punishment and altruistic cooperation over a wider range of parameter values than group selection will sustain altruistic cooperation alone. Figure 13.1 compares the long-run average levels of cooperation with and without punishment for a range of group sizes and extinction rates. If there is no punishment, our simulations replicate the standard result: group selection can support high frequencies

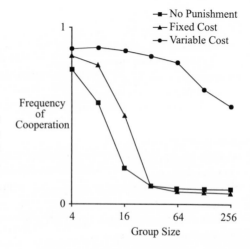

Figure 13.4. Punishment does not aid in the evolution of cooperation when the costs born by punishers are fixed, independent of the number of defectors in the group. Here we plot the long-run average frequency of cooperation when the costs of punishing are proportional to the frequency of defectors (variable cost), fixed at a constant cost equal to the cost of cooperating (c), and when there is no punishment.

of cooperative behavior only if groups are quite small. However, adding punishment sustains substantial amounts of cooperation in much larger groups. As one would expect, increasing the rate of extinction increases the long-run average amount of cooperation.

In this model, group selection leads to the evolution of cooperation only if migration is sufficiently limited to sustain substantial between-group differences in the frequency of defectors. Figure 13.2 shows that when the migration rate increases, levels of cooperation fall precipitously. When punishers are common, defectors do badly, but when punishers are rare, defectors do well. Thus, the imitation of high payoff individuals creates a selection-like adaptive force that acts to maintain variation between groups in the frequency of defectors. However, if there is too much migration, this process cannot maintain enough variation between groups for group selection to be effective.

The long-run average amount of cooperation is also sensitive to the cost of being punished (figure 13.3). When the cost of being punished is at base case value ($p=4k$), even a modest frequency of punishers will cause defectors to be selected against, and, as a result, there is a substantial correlation between the frequency of cooperation and punishment across groups. When the cost of being punished is twice the cost of cooperation ($p=2k$), punishment does not sufficiently reduce the relative payoff of defectors, and the correlation between the frequency of cooperators and punishers declines. Lower correlations mean that selection among groups cannot compensate for the decline of punishers within groups, and eventually both punishers and contributors decline.

It is important to see that punishment leads to increased cooperation only to the extent that the costs associated with being a punisher decline as defectors become rare. Monitoring costs, for example, must be paid whether or not there are any defectors. When such costs are substantial, or when the probability of mistaken defection is high enough that punishers bear significant costs even when defectors are rare, group selection does not lead to the evolution of altruistic punishment (figure 13.4). However, because people live in long-lasting social groups and language allows the spread of information about who did what, it is plausible that monitoring costs may often be small compared with enforcement costs. This result also leads to an empirical prediction: people should be less inclined to pay fixed than variable punishment costs if the mechanism outlined here is responsible for the psychology of altruistic punishment.

Further sensitivity analyses suggest that these results are robust. In addition to the results described, we have studied the sensitivity of the model to variations in the remaining parameter values. Decreasing the mutation rate substantially increases the long-run average levels of cooperation. Random drift-like processes have an important effect on trait frequencies in this model. Standard models of genetic drift suggest that lower mutation rates will cause groups to stay nearer the boundaries of the state space (Crow and Kimura, 1970), and our simulations confirm this prediction. Increasing mutation rate, on average, increases the amount of punishment that must be administered and therefore increases the payoff advantage of second-order free riders compared with altruistic punishers. Increasing e, the error rate, reduces the long-run average amount of cooperation. Reducing the number of groups, N, adds random noise to the results. We also tested the sensitivity of the model to three structural changes. We modified the payoffs so that each cooperative act produces a per capita benefit of b/n for each other group member and modified the extinction model so that the probability of group extinction is proportional to the difference between warring groups in average payoffs including the costs of punishment, rather than simply the difference in frequency of cooperators. The dynamics of this model are more complicated because now group selection acts against punishers because punishment reduces mean group payoffs. However, the correlated effect of group selection on cooperation still tends to increase punishment as in the original model. The relative magnitude of these two effects depends on the magnitude of the per capita benefit to group members of each cooperative act, b/n. For reasonable values of b ($2c$, $4c$, and $8c$), the results of this model are qualitatively similar to those shown. We also investigated a model in which cooperation and punishment are characters that vary continuously from zero to one. An individual with cooperation value x behaves like a cooperator with probability x and like a defector with probability $1 - x$. Similarly, an individual with a punishment value y behaves like a punisher with probability y and like a nonpunisher with probability $1 - y$. New mutants are uniformly distributed. The steady-state mean levels of cooperation in this model are similar to the base model. Finally, we studied a model without extinction analogous to a recent model of selection among stable equilibria because of biased imitation (Boyd and Richerson, 2002). Populations are arranged in a ring, and individuals imitate only individuals drawn from the neighboring two groups. Cooperative acts produce a per capita benefit b/n so that groups with more cooperators have higher average payoff, and thus cooperation will, all other things being equal, tend to spread because individuals are prone to imitate successful neighbors. We could find no reasonable parameter combination that led to significant long-run average levels of cooperation in this last model.

Discussion

We have shown that although the logic underlying altruistic cooperation and altruistic punishment is similar, their evolutionary dynamics are not. In the absence of punishment, within-group adaptation acts to decrease the frequency of altruistic cooperation, and as a consequence weak drift-like forces are insufficient to maintain substantial variation between groups. In groups in which altruistic punishers are common, defectors are excluded, and this maintains variation in the amount of cooperation between groups. Moreover, in such groups punishers bear few costs, and punishers decrease only very slowly in competition with contributors. As a result, group selection is more effective at maintaining altruistic punishment than altruistic cooperation.

These results suggest that group selection can play an important role in the cultural evolution of cooperative behavior and moralistic punishment in humans. The importance of group selection is always a quantitative issue. There is no doubt that selection among groups acts to favor individually costly, group-beneficial behaviors. The question is always, is group selection important under plausible conditions? With parameter values chosen to represent cultural evolution in small-scale societies, cooperation is sustained in groups on the order of 100 individuals. If the "individuals" in the model represent family groups (on grounds that they migrate together and adopt common practices), altruistic punishment could be sustained in groups of 600 people, a size much larger than typical foraging bands and about the size of many ethno-linguistic units in nonagricultural societies. Group selection is more effective in this model than in standard models for two reasons: first, in groups in which defectors are rare, punishers suffer only a small payoff disadvantage compared with contributors, and, as a result, variation in the frequency of punishers is eroded slowly. Second, payoff-biased imitation maintains variation among groups in the frequency of cooperation, because in groups in which punishers are common, defectors achieve a low payoff and are unlikely to be imitated.

It would be possible to construct an otherwise similar genetic model in which natural selection played the same role that payoff-biased imitation plays in the present model, and there is little doubt that for analogous parameter values the results for such a genetic model would be very similar to the results presented here. However, such a choice of parameters would not be reasonable for a genetic model because natural selection is typically much weaker than migration for small, neighboring social groups of humans. Our results (figure 13.2) suggest that for parameters appropriate for a genetic model, the group selection process modeled here will not be effective. It should be noted, however, that the genetic evolution of moral emotions might be favored by ordinary natural selection in social environments shaped by cultural group selection (Richerson and Boyd, 1998; Bowles, Choi, and Hopfensitz).

NOTE

We thank Ernst Fehr, Marc Feldman, Daniel Friedman, Gerd Gigerenzer, Francisco Gil-White, Peter Hammerstein, Joe Henrich, and Richard McElreath for useful comments on previous drafts. R.B. also thanks the members of Evolutionary Theory Seminar in the University of California, Los Angeles, Anthropology Department. We thank the MacArthur Foundation for its generous funding of the research network that funded some of this work.

REFERENCES

Aoki, K. 1982. A condition for group selection to prevail over counteracting individual selection. *Evolution* 36:832–842.

Axelrod, R., & W. D. Hamilton. 1981. The evolution of cooperation. *Science* 211:1390–1396.

Boehm, C. 1996. Emergency decisions, cultural selection mechanics, and group selection. *Current Anthropology* 37:763–793.

Bowles, S. 2001. In *Social dynamics*, S. Durlauf, & P. Young, eds. (pp. 155–190). Cambridge, MA: MIT Press.

Bowles, S., J-K. Choi, & A. Hopfensitz. 2003. The coevolution of individual behaviors and group level institutions. *Journal of Theoretical Biology* 223:135–147.

Boyd, R., & P. J. Richerson. 1988. The evolution reciprocity in sizable groups. *Journal of Theoretical Biology* 132:337–356.

Boyd, R., & P. J. Richerson. 2002. Group beneficial norms spread rapidly in a structured population. *Journal of Theoretical Biology* 215:287–296.

Clutton-Brock, T. H., & G. A. Parker. 1995. Punishment in animal societies. *Nature* 373:209–216.

Crow, J., & M. Kimura. 1970. *An introduction to population genetics theory*. New York: Harper and Row.

Eshel, I. 1972. On the neighborhood effect and the evolution of altruistic traits. *Theoretical Population Biology* 3:258–277.

Fehr, E., & S. Gachter. 2002. Altruistic punishment in humans. *Nature* 415:137–140.

Gale, J., K. Binmore, & L. Samuelson. 1995. Learning to be imperfect: The ultimatum game. *Games and Economic Behavior* 8:56–90.

Harpending, H., & A. Rogers. 1986. *Evolution* 40:1312–1327.

Hauert, C., S. De Monte, J. Hofbauer, & K. Sigmund. 2002. Volunteering as red queen mechanism for cooperation in public goods game. *Science* 296:1129–1132.

Henrich, J. 2001. Cultural transmission and the diffusion of innovations: Adoption dynamics indicate that biased cultural transmission is the predominate force in behavioral change and much of sociocultural evolution. *American Anthropologist* 103:992–1013.

Henrich, J., & R. Boyd. 2001. Why people punish defectors: Weak conformist transmission can stabilize costly enforcement of norms in cooperative dilemmas. *Journal of Theoretical Biology* 208:79–89.

Ostrom, E., J. Gardner, & R. Walker. 1994. *Rules, games, and common-pool resources*. Ann Arbor: University of Michigan Press.

Price T., M. Turelli, & M. Slatkin. 1993. Peak shifts produced by correlated response to selection. *Evolution* 47:280–290.

Richerson, P. J., & R. Boyd. 1998. The evolution of human ultra-sociality. In: *Indoctrinability, Ideology, and Warfare: Evolutionary Perspectives*, I. Eibl-Eibisfeldt, & F. Salter, eds. (pp. 71–96). New York: Berghahn Books.

Rogers, E. M. 1983. *Diffusion of innovations*. New York: Free Press.

Rogers, A. R., & H. C. Harpending. 1986. Migration and genetic drift in human populations. *Evolution* 40:1312–1327.

Rogers, A. R. 1990. Group selection by selective emigration: The effects of migration and kin structure. *American Naturalist* 135:398–413.

Sethi, R., & E. Somanathan. 1996. The evolution of social norms in common property resource use. *American Economic Review* 86:766–788.

Sober, E., & D. S. Wilson. 1998. *Unto others: The evolution and psychology of unselfish behavior*. Cambridge, MA: Harvard University Press.

Soltis, J., R. Boyd, & P. J. Richerson. 1995. Can group functional behaviors evolve by cultural group selection? An empirical test. *Current Anthropology* 36:473–494.

Trivers, R. L. 1971. The evolution of reciprocal altruism. *Quarterly Review of Biology* 46:35–57.

14 Cultural Evolution of Human Cooperation

With Joseph Henrich

Cooperation[1] is a problem that has long interested evolutionists. In both the *Origin* and *Descent of Man*, Darwin worried about how his theory might handle cases such as the social insects in which individuals sacrificed their chances to reproduce by aiding others. Darwin could see that such sacrifices would not ordinarily be favored by natural selection. He argued that honeybees and humans were similar. Among honeybees, a sterile worker who sacrificed her own reproduction for the good of the hive would enjoy a vicarious reproductive success through her siblings. Humans, Darwin (1874:178–179) thought, competed tribe against tribe as well as individually, and their "social and moral faculties" evolved under the influence of group competition:

> It must not be forgotten that although a high standard of morality gives but slight or no advantage to each individual man and his children over other men of the tribe, yet that an increase in the number of well-endowed men and an advancement in the standard of morality will certainly give an immense advantage to one tribe over another. A tribe including many members who, from possessing in a high degree the spirit of patriotism, fidelity, obedience, courage, and sympathy, were always ready to aid one another, and to sacrifice themselves for the common good, would be victorious over most other tribes; and this would be natural selection.

More than a century has passed since Darwin wrote, but the debate among evolutionary social scientists and biologists is still framed in similar terms—the conflict between individual and prosocial behavior guided by selection on individuals versus selection on groups. In the meantime social scientists have developed various theories of human social behavior and cooperation—rational choice theory takes an individualistic approach while functionalism analyzes the group-advantageous aspects of institutions and behavior. However, unlike more traditional approaches in the social sciences, evolutionary theories seek to explain both contemporary behavioral patterns and the origins of the impulses, institutions, and preferences that drive behavior.

In this chapter we refer to "culture" as the information stored in individual brains (or in books and analogous media) that was acquired by imitation of, or teaching by, others. Because culture can be transmitted forward through time from one person to another and because individuals vary in what they learn from others, culture has many of the same properties as the genetic system of inheritance but also, of course, many differences. The formal import of the analogies and disanalogies has been worked out in some analytical detail (e.g., Cavalli-Sforza and Feldman, 1981; Boyd and Richerson, 1985). We also subscribe to Price's approach to the concept of group selection. Heritable variation between entities can appear at any level of organization, and any level above the individual merits the term group selection (Henrich, 2004a; Hamilton, 1975; Price, 1972; Sober and Wilson, 1998). Here we focus on the more conventional notion that selection on variation between fairly large social units counts as group selection. In fact, we have in mind, like Darwin and Hamilton, selection among tribes of at least a few hundred people, so we are referring to the *cultural analog* of what is sometimes called interdemic group selection.

Theories of Cooperation

We draw evidence about cooperation from many sources. Ethnographic and historical sources include diverse religious doctrines, norms, and customs, as well as folk psychology. Anthropologists and historians document an immense diversity of human social organizations, and most of these are accompanied by moral justifications, if often contested ones. Johnson and Earle (2000) provide a good introduction to the vast body of data collected by sociocultural anthropologists. Some important empirical topics are the focus of sophisticated work. For example, the cross-cultural study of commons management is already a well-advanced field (Baland and Platteau, 1996), drawing upon the disciplines of anthropology, political science, and economics.

Human Cooperation Is Extensive and Diverse

Human patterns of cooperation are characterized by a number of features:

- *Humans are prone to cooperate, even with strangers.* Many people cooperate in anonymous one-shot prisoner's dilemma games (Marwell and Ames, 1981) and often vote altruistically (Sears and Funk, 1990). People begin contributing substantially to public goods sectors in economic experiments (Ostrom, 1998; Falk, Fehr, and Fischbacher, 2002). Experimental results accord with common experience. Most of us have traveled in foreign cities, even poor foreign cities filled with strange people for whom our possessions and spending money are worth a small fortune, and found risk of robbery and commercial chicanery to be small. These observations apply across a wide spectrum of societies, from small-scale foragers to modern cities in nation states (Henrich, 2004a).

- *Cooperation is contingent on many things.* Not everyone cooperates. Aid to distressed victims increases substantially if a potential altruist's empathy is engaged (Batson, 1991). Being able to discuss a game beforehand and to make promises to cooperate affects success (Dawes, van de Kragt, and Orbell, 1990). The size of the resource, technology for exclusion and exploitation of the resource, and similar gritty details affect whether cooperation in commons management arises (Ostrom, 1990:202–204). Scientific findings correspond well to personal experience. Sometimes people cooperate enthusiastically, sometimes reluctantly, and sometimes not at all. People vary considerably in their willingness to cooperate even under the same environmental conditions.

- *Institutions matter.* People from different societies behave differently because their beliefs, skills, mental models, values, preferences, and habits have been inculcated by long participation in societies with different institutions. In repeated play common property experiments, initial defections induce further defections until the contribution to the public good sector approaches zero. However, if players are allowed to exercise strategies they might use in the real world (e.g., to punish those who defect), participation in the commons stabilizes a substantial degree of cooperation (Fehr and Gächter, 2002), even in one-shot (nonrepeated) contexts. Strategies for successfully managing commons are generally institutionalized in sets of rules that have legitimacy in the eyes of the participants (Ostrom, 1990, ch. 2). Families, local communities, employers, nations, and governments all tap our loyalties with rewards and punishments and greatly influence our behavior.

- *Institutions are the product of cultural evolution.*[2] Richard Nisbett's group has shown how people's affective and cognitive styles become intimately entwined with their social institutions (Cohen and Vandello, 2001; Nisbett and Cohen, 1996; Nisbett, Peng, Choi, and Norenzayan, 2001). Because such complex traditions are so deeply ingrained, they are slow both to emerge and decay. Many commons management institutions have considerable time depths (Ostrom, 1990, ch. 3). Throughout most of human history, institutional change was so slow as to be almost imperceptible by individuals. Today, change is rapid enough to be perceptible. The slow rate of change of institutions means that different populations experiencing the same environment and using the same technology often have quite different institutions (Kelly, 1985; Salamon, 1992).

- *Variation in institutions is huge.* Already with its very short list of societies and games, the experimental ethnography approach has uncovered striking differences (Henrich et al., 2001; Nisbett et al., 2001). Plausibly, design complexity, coordination equilibria, and other phenomena generate multiple evolutionary equilibria and much historical contingency in the evolution of particular institutions (Boyd and Richerson, 1992a); consider how different communities, universities, and countries solve the same problems differently.

Evolutionary Models Can Explain the Nature of Preferences and Institutions

These facts constrain the theories we can entertain regarding the causes of human cooperation. For example, high levels of cooperation are difficult to reconcile with the rational choice theorist's usual assumption of self-regarding preferences, and the diversity of institutional solutions to the same environmental problems challenges any theory in which institutions arise directly from universal human nature. The "second-generation" bounded rational choice theory, championed by Ostrom (1998), has begun to address these challenges from within the rational choice framework. These approaches add a psychological basis and institutional constraints to the standard rational choice theory. Experimental studies verify that people do indeed behave quite differently from rational selfish expectations (Fehr and Gächter, 2002; Batson, 1991). Although psychological and social structures are invoked to explain individual behavior and its variation, an explanation for the origins and variation in psychology and social structure is not part of the theory of bounded rationality.

Evolutionary theory permits us to address the origin of preferences. A number of economists have noted the neat fit between evolutionary theory and economic theory (Hirshleifer, 1977; Becker, 1976). Evolution explains what organisms want, and economics explains how they should go about getting what they want. Without evolution, preferences are exogenous, to be estimated empirically but not explained. The trouble with orthodox evolutionary theory is that its predictions are similar to predictions from selfish rationality, as we will see. At the same time, unvarnished evolutionary theory does do a good job of explaining most other examples of animal cooperation. To do a satisfactory job of explaining why *humans* have the unusual forms of social behavior depicted in our list of stylized facts, we need to appeal to the special properties of *cultural* evolution and more broadly to theories of culture-gene coevolution (Henrich and Boyd, 2001; Richerson and Boyd, 1998, 1999; Henrich, 2004a).

Such evolutionary models have both intellectual and practical payoffs. The intellectual payoff is that evolutionary models link answers to contemporary puzzles to crucial long timescale processes. The most important economic phenomenon of the past 500 years is the rise of capitalist economies and their tremendous impact on every aspect of human life. Expanding the timescale a bit, the most important phenomena of the last 10 millennia are the evolution of ever-more complex social systems and ever-more sophisticated technology following the origins of agriculture (Richerson, Boyd, and Bettinger, 2001). A satisfactory explanation of both current behavior and its variation must be linked to such long-run processes, where the times to reach evolutionary equilibria are measured in millennia or even longer spans of time. More practically, dynamism of the contemporary world creates major stresses on institutions that manage cooperation. Evolutionary theory will often be useful because it will lead to an understanding of how to accelerate institutional evolution to better track rapid technological and economic change. Nesse and Williams (1995) provide an analogy in the context of medical practice.

Evolutionary Models Account for the Processes That Shape Heritable Genetic and Cultural Variation through Time

Evolutionary explanations are *recursive*. Individual behavior results from an interaction of inherited attributes and environmental contingencies. In most species, genes are the main inherited attributes; however, inherited cultural information is also important for humans. Individuals with different inherited attributes may develop different behaviors in the same environment. Every generation, evolutionary processes—natural selection is the prototype—impose environmental effects on individuals as they live their lives. Cumulated over the whole population, these effects change the pool of inherited information, so that the inherited attributes of individuals in the next generation differ, usually subtly, from the attributes in the previous generation. Over evolutionary time, a lineage cycles through the recursive pattern of causal processes once per generation, more or less gradually shaping the gene pool and thus the succession of individuals that draw samples of genes from it. Statistics that describe the pool of inherited attributes (e.g., gene frequencies) are basic state variables of evolutionary analysis. They are what change over time.

Note that in a recursive model, we explain individual behavior and population-level processes in the same model. Individual behavior depends, in any given generation, on the gene pool from which inherited attributes are sampled. The pool of inherited attributes depends in turn upon what happens to a population of individuals as they express those attributes. Evolutionary biologists have a long list of processes that change the gene frequencies, including natural selection, mutation, and genetic drift. However, no organism experiences natural selection. Organisms either live or die, reproduce or fail to reproduce, for concrete reasons particular to the local environment and the organism's own particular attributes. If, in a particular environment, some *types* of individuals do better than others, and if this variation has a heritable basis, then *we* label as "natural selection" the resulting changes in gene frequencies of populations. We use abstract categories like selection to describe such concrete events because we wish to build up some useful generalizations about evolutionary process. Few would argue that evolutionary biology is the poorer for investing effort in this generalizing project.

Although some of the processes that lead to cultural change are very different from those that lead to genetic change, the logic of the two evolutionary problems is very similar. For example, the cultural generation time is short in the case of ideas that spread rapidly, but modeling the evolution of such cultural phenomena (e.g., semiconductor technology) presents no special problems (Boyd and Richerson, 1985:68–69). Similarly, human choices include ones that modify inherited attributes directly, rather than indirectly, by natural selection. These

"Lamarckian" effects are easily added to models, and the models remain evolutionary so long as rationality remains bounded (Young, 1998). Such models easily handle continuous (nondiscrete) traits, low-fidelity transmission, and any number of "inferential transformations" that might occur during transmission (Henrich and Boyd, 2002; Cavalli-Sforza and Feldman, 1981; Boyd and Richerson, 1985). The degenerate case of omniscient rationality, of course, needs no recursion because everything happens in the first generation (instantly in a typical rational choice model). The study of how genetically and culturally inherited elements impose bounds on choice is a natural extension of the concept of bounded rationality (Boyd and Richerson, 1993).

Evolution Is Multilevel

Evolutionary theory is always *multilevel*; at a minimum, it keeps track of properties of individuals, like their genotypes, and of the population, such as the frequency of a particular gene. Other levels also may be important. Individual's phenotypes are derived from many genes interacting with each other and the environment. Populations may be structured (e.g., divided into social groups with limited exchanges of members). Thus, evolutionary theories are systemic, integrating every part of biology. In principle, everything that goes into causing change through time plays its proper part in the theory.

This in-principle completeness led Ernst Mayr (1982) to speak of "proximate" and "ultimate" causes in biology. Proximate causes are those that physiologists and biochemists generally treat by asking *how* an organism functions. These are the causes produced by individuals with attributes interacting with environments and producing effects upon them. Do humans use innate cooperative propensities to solve commons problems, or do they have only self-interested innate motives? Or are the causes more complex than either proposal? Ultimate causes are evolutionary. The ultimate cause of an organism's behavior is the history of evolution that shaped the gene pool from which our samples of innate attributes are drawn. Evolutionary analyses answer *why* questions. Why do human communities typically solve at least some of the commons dilemmas and other cooperation problems on a scale unknown in other apes and monkeys? Human-reared chimpanzees are capable of many human behaviors, but they nevertheless retain many chimpanzee behaviors and cannot act as full members of a human community (Savage-Rumbaugh and Lewin, 1994; Gardner, Gardner, and Van Cantfort, 1989). Thus, we know that humans have different innate influences on their behavior than chimpanzees do, and these must have arisen in the course of the two species' divergence from our common ancestor.

In Darwinian evolutionary theories, the ultimate sources of cooperative behavior are classically categorized into three evolutionary processes operating at different levels of organization (for a framework unifying these classical divisions, see Henrich, 2004a):

- *Individual-level selection*. Individuals and the variants they carry are obviously a locus of selection. Selection at this level favors selfish individuals who are evolved to maximize their own survival and reproductive success. Pairs of self-interested actors can cooperate when they interact repeatedly (Axelrod and Hamilton, 1981; Trivers, 1971). Alexander (1987) argued that such reciprocal cooperation can also explain complex human social systems, but most formal modeling studies make this proposal doubtful (Leimar and Hammerstein, 2001; Boyd and Richerson, 1989). Still, some version of Alexander's *indirect reciprocity* is perhaps the most plausible alternative to the cultural group selection hypothesis that we champion here. Most such proposals beg the question of how humans and not other animals can take massive advantage of indirect reciprocity (e.g., Nowak and Sigmund, 1998). Smith (2003) proposes to make language the key.[3]

- *Kin selection*. Hamilton's (1964) articles showing that kin should cooperate to the extent that they share genes identical by common descent are one of the theoretical foundations of sociobiology. Kin selection can lead to cooperative social systems of a remarkable scale, as illustrated by the colonies of termites, ants, and some bees and wasps. However, most animal societies are small because individuals have few close relatives. It is the fecundity of insects, and in one case rodents, that permits a single queen to produce huge numbers of sterile workers and hence large, complex societies composed of close relatives (Campbell, 1983).
- *Group selection*. Selection can act on any pattern of heritable variation that exists (Price, 1972). Darwin's model of the evolution of cooperation by intertribal competition is perfectly plausible, as far as it goes. The problem is that genetic variation between groups other than kin groups is hard to maintain unless the migration between groups is very small or unless some very powerful force generates between-group variation (e.g., Aoki, 1982; Slatkin and Wade, 1978; Wilson, 1983). In the case of altruistic traits, selection will tend to favor selfish individuals in all groups, tending to aid migration in reducing variation between groups. Success of kin selection in accounting for the most conspicuous and highly organized animal societies (except humans) has convinced many, but not all, evolutionary biologists that group selection is of modest importance in nature (for a group selectionist's view of the controversy, see Sober and Wilson, 1998). It is also important to note that the problem of maintenance of between-group variation applies *only* to altruistic/cooperative traits, not to social behavior in general. Nearly all evolutionary biologists would agree that group selection is likely to be important for any social interaction with multiple stable equilibria, such as those coordination situations mentioned by Smith (2003).

We could make this picture much more complex by adding higher and lower levels of structure. Many examples from human societies will occur to the reader, such as gender. Indeed, Rice (1996) has elegantly demonstrated that selection on genes expressed in the different sexes sets up a profound conflict of interest between these genes. If female *Drosophila* are prevented from evolving defenses, male genes will evolve that seriously degrade female fitness. The genome is full of such conflicts, usually muted by the fact that an individual's genes are forced by the evolved biology of complex organisms to all have an equal shot at being represented in one's offspring. Our own bodies are a group-selected community of genes organized by elaborate "institutions" to ensure fairness in genetic transmission, such as the lottery of meiosis that gives each chromosome of a pair a fair chance at entering the functional gamete (Maynard Smith and Szathmáry, 1995).

Culture Evolves

In theorizing about human evolution, we must include processes affecting *culture* in our list of evolutionary processes alongside those that affect genes. Culture is a system of inheritance. We acquire behavior by imitating other individuals much as we get our genes from our parents. A fancy capacity for high-fidelity imitation is one of the most important derived characters distinguishing us from our primate relatives (Tomasello, 1999). We are also an unusually docile animal (Simon, 1990) and unusually sensitive to expressions of approval and disapproval by parents and others (Baum, 1994). Thus, parents, teachers, and peers can rapidly, easily, and accurately shape our behavior compared to training other animals using more expensive material rewards and punishments. Finally, once children acquire language, parents and others can communicate new ideas quite economically. Our own contribution to the study of human behavior is a series of mathematical models of what we take to be the fundamental processes of cultural evolution (e.g., Boyd and Richerson, 1985). Application of Darwinian methods to the study of cultural evolution was forcefully advocated by Campbell (1965, 1975). Cavalli-Sforza and Feldman (1981) constructed the first mathematical models to analyze cultural recursions. The list of processes that shape cultural change includes:

- *Biases.* Humans do not passively imitate whatever they observe. Rather, cultural transmission is biased by decision rules that individuals apply to the variants they observe or try out. The rules behind such selective imitation may be innate or the result of earlier imitation or a mixture of both. Many types of rules might be used to bias imitation. Individuals may try out a behavior and let reinforcement guide acceptance or rejection, or they may use various rules of thumb to reduce the need for costly trials and punishing errors. Rules like "copy the successful," "copy the prestigious" (Henrich and Gil-White, 2001; Boyd and Richerson, 1985), or "copy the majority" (Boyd and Richerson, 1985; Henrich and Boyd, 1998) allow individuals to acquire rapidly and efficiently adaptive behavior across a wide range of circumstances and play an important role in our hypothesis about the origins of cooperative tendencies in human behavior (Henrich and Boyd, 2001).
- *Nonrandom variation.* Genetic innovations (mutations, recombinations) are random with respect to what is adaptive. Human individual

innovation is guided by many of the same rules that are applied to biasing ready-made cultural alternatives. Bias and learning rules have the effect of increasing the rate of evolution relative to what can be accomplished by random mutation, recombination, and natural selection. We believe that culture originated in the human lineage as an adaptation to the Plio-Pleistocene ice-age climate deterioration, which includes much rapid, high-amplitude variation of just the sort that would favor adaptation by nonrandom innovation and biased imitation (Richerson and Boyd, 2000a, b).

- *Natural selection.* Since selection operates on any form of heritable variation and imitation and teaching are forms of inheritance, natural selection will influence cultural as well as genetic evolution. However, selection on culture is liable to favor different behaviors than selection on genes. Because we often imitate peers, culture is liable to selection at the subindividual level, potentially favoring pathogenic cultural variants—selfish memes (Blackmore, 1999). On the other hand, rules like conformist imitation have the opposite effect. By tending to suppress cultural variation within groups, such rules protect variation between them, potentially exposing our cultural variation to much stronger group selection effects than our genetic variation (Soltis, Boyd, and Richerson, 1995; Henrich and Boyd, 1998). Human patterns of cooperation may owe much to cultural group selection.

Evolutionary Models Are Consistent with a Wide Variety of Theories

Evolutionary theory prescribes a method, not an answer, and a wide range of particular hypotheses can be cast in an evolutionary framework. If population-level processes are important, we can set up a system for keeping track of heritable variation and the processes that change it through time. Darwinism as a method is not at all committed to any particular picture of how evolution works or what it produces. Any sentence that starts with "evolutionary theory predicts" should be regarded with caution.

Evolutionary social science is a diverse field (Borgerhoff Mulder, Richerson, Thornhill, and Voland, 1997; Laland and Brown, 2002). Our own work, which emphasizes an ultimate role for culture and for group selection on cultural variation, is controversial. Many evolutionary social scientists assume that culture is a strictly proximate phenomenon, akin to individual learning (e.g., Alexander, 1979), or is so strongly constrained by evolved psychology as to be virtually proximate (Wilson, 1998). As Alexander (1979:80) puts it, "Cultural novelties do not replicate or spread themselves, even indirectly. They are replicated as a consequence of the behavior of vehicles of gene replication." We think both theory and evidence suggest that this perspective is dead wrong. Theoretical models show that the processes of cultural evolution can behave differently in critical respects from those only including genes, and much evidence is consistent with these models.

Most evolutionary biologists believe that individually costly group-beneficial behavior can arise only as a side effect of individual fitness maximization. We have noted the problems with maintaining variation between groups in theory and the seeming success of alternative explanations. Many, but by no means all, students of evolution and human behavior have followed the argument against group selection forcefully articulated by Williams (1966).[4]

However, *cultural* variation is more plausibly susceptible to group selection than is genetic variation. For example, if people use a somewhat conformist bias in acquiring important social behaviors, variation between groups needed for group selection to operate is protected from the variance-reducing force of migration between groups (Boyd and Richerson, 1985, 2002; Henrich and Boyd, 2001).

Evolution of Cooperative Institutions

Here we summarize our theory of institutional evolution, developed elsewhere in more detail (Richerson and Boyd, 1998, 1999), which is rooted in a mathematical analysis of the processes of cultural evolution and is consistent with much empirical data. We make limited claims for this particular hypothesis, although we think that the thrust of the empirical data as summarized by the stylized facts are much harder on current alternatives. We make a much stronger claim that a dual gene-culture theory of some kind will be necessary to account for the evolution of human cooperative institutions.

Understanding the evolution of contemporary human cooperation requires attention to two different timescales: first, a long period of evolution in the Pleistocene epoch shaped the innate "social instincts" that underpin modern human behavior. During this period, much genetic change occurred as a result of humans living in groups with social institutions *heavily influenced by culture*, including cultural group selection (Richerson and Boyd, 2001). On this timescale, genes and culture *coevolve*, and cultural evolution is plausibly a leading rather than lagging partner in this process. We sometimes refer to the process as "culture-gene coevolution." Then, only about 10,0000 years ago, the origins of agricultural subsistence systems laid the economic basis for revolutionary changes in the scale of social systems. Evidence suggests that genetic changes in the social instincts over the last 10,000 years are insignificant. Evolution of complex societies, however, has involved the relatively slow cultural accumulation of institutional "work-arounds" that take advantage of a psychology evolved to cooperate with distantly related and unrelated individuals belonging to the same symbolically marked "tribe" while coping more or less successfully with the fact that these social systems are larger, more anonymous, and more hierarchical than the late Pleistocene tribal-scale systems.[5]

Tribal Social Instincts Hypothesis

Our hypothesis is premised on the idea that selection between groups plays a much more important role in shaping culturally transmitted variation than it does in shaping genetic variation. As a result, humans have lived in social environments characterized by high levels of cooperation for as long as culture has played an important role in human development. To judge from the other living apes, our remote ancestors had only rudimentary culture (Tomasello, 1999) and lacked cooperation on a scale larger than groups of close kin (Boehm, 1999). The difficulty of constructing theoretical models of group selection on genes favoring cooperation matches neatly with the empirical evidence that cooperation in most social animals is limited to kin groups. In contrast, rapid cultural adaptation can lead to ample variation among groups whenever multiple stable social equilibria arise. At least two cultural processes can maintain multiple stable equilibria: (1) conformist social learning and (2) moralistic enforcement of norms. Such models of group selection are relatively powerful because they require only the social, not physical, extinction of groups. Formal theoretical models suggest that conformism is an adaptive heuristic for biasing imitation under a wide variety of conditions (Boyd and Richerson, 1985, ch. 7; Henrich and Boyd, 1998; Simon, 1990), and both field and laboratory work provide empirical support (Henrich, 2001). Models of moralistic punishment (Boyd and Richerson, 1992b; Boyd, Gintis, Bowles, and Richerson, 2003; Henrich and Boyd, 2001) lead to multiple stable social equilibria and to reductions in noncooperative strategies if punishment is prosocial. As a consequence, we believe, a growing reliance on cultural evolution led to larger, more cooperative societies among humans over the last 250,000 years or so.

Ethnographic evidence suggests that small-scale human societies are subject to group selection of the sort needed to favor cooperation at a tribal scale. Soltis et al. (1995) analyzed ethnographic data on the results of violent conflicts among Highland New Guinea clans. These conflicts fairly frequently resulted in the social extinction of clans. Many of the details of this process are consistent with cultural group selection. For example, social extinction does not mean physical elimination of the entire group. Quite the contrary, most people survive defeat but flee as refugees to other groups, into which they are incorporated. This sort of extinction cannot support genetic group selection because so many of the defeated survive and because they would tend to carry their unsuccessful genes into successful groups, rapidly running down variation between groups. However, the effects of conformist cultural transmission combined with moralistic punishment makes between-group cultural variation much less subject to erosion by migration and within-group success of uncooperative strategies than is true in the case of acultural organisms.

The New Guinea cases had little information regarding the cultural variants that might have been favored by cultural group selection. Other examples are more informative in this regard. Kelly (1985) has worked out in detail the way bridewealth customs in the Nuer and Dinka, cattle-keeping people of the Southern Sudan, led to the Nuer maintaining larger tribal systems. These larger tribes, in turn, allowed the Nuer to field larger forces than Dinka in disputes between the two groups. As a result, the Nuer expanded rapidly at the expense of the Dinka in the nineteenth and early twentieth centuries. Here, as in New Guinea, many Dinka lineages survived these fights and were often assimilated into Nuer tribes, a process, again, highly hostile to group selection on genes. The larger ethnographic corpus suggests that the sort of intergroup conflict described by Soltis and Kelly is very common, if not ubiquitous (Keeley, 1996; Otterbein, 1970). Darwin's picture of a group selection process operating at the level of competing, symbolically marked tribal units with the outcome determined by differences in "patriotism, fidelity, obedience, courage, sympathy" (Darwin, 1874, ch. 5) and the like can work, but only upon cultural—not genetic— variation for such traits.

Consistent with this argument, evidence suggests that people in late Pleistocene human societies cooperated on a tribal scale (Bettinger, 1991:203–205; Richerson and Boyd, 1998). "Tribe" is sometimes used in a technical sense to include only societies with fairly elaborate institutions for organizing cooperation among distantly related and unrelated people. We apply the term to any institution that organizes interfamilial cooperation, even if it is rather simple and the amount of cooperation organized modest. Definitional issues aside, our claim is controversial because the archaeological record permits only weak inferences about social organization and because the spectrum of social organization in ethnographically known hunter-gatherers is very broad (Kelly, 1995). At the simple end of the spectrum are "family-level" societies (Johnson and Earle, 2000; Steward, 1955), such as the Shoshone of the Great Basin and !Kung of the Kalahari. Because these two groups are so simply organized, some scholars used them as an archetypal model for Paleolithic societies (Kelly, 1995:2). However, such groups are likely poor examples of the "average" Paleolithic society because they inhabit and have adapted to marginal environments using subsistence strategies quite different from any known from the Paleolithic (R. Bettinger, personal communication). Also, we believe that the ethnographic societies used to exemplify the family level of organization actually have tribal institutions of some sophistication.

Much evidence suggests that typical Paleolithic societies were more complex than the Shoshone or the !Kung. Many late Pleistocene societies emphasized big game hunting, often in resource-rich environments, rather than the plant foods emphasized in the marginal environments inhabited by Kalahari foragers and the Shoshone. For example, the Kalahari foragers (along with the Aranda in the Australian desert) anchor the low end of the distribution with respect to plant biomass found in regions of 23 ethnographically known nomadic foraging groups (Kelly, 1995:122). As Steward (1955) reports, big game hunting in ethnographic cases typically involves cooperation on a larger scale than plant collecting and small game hunting; thus, we should expect societies in the late Pleistocene to be more, not less, socially complex than the !Kung and Shoshone. In any case we think it an error to try to identify an archetypal Pleistocene society; most likely last glacial societies spanned as large or larger a spectrum of social organization as ethnographically known cases. Art and settlement size (several hundred people) at Upper Paleolithic sites in France and Spain suggest that these societies were toward the complex end of the foraging spectrum (Price and Brown, 1985). In Central Europe, the palisades and large housing structures look much more like those of the Northwest Coast Indians or big-men social forms of New Guinea than those of the !Kung or Shoshone (Johnson and Earle, 2000).

Moreover, despite the marginality of their environment, the archetypal family-level societies do have tribal-scale institutions for dealing with environmental uncertainty (Wiessner, 1984). For example, the Shoshonean peoples of the North American Great Basin foraged for most of the year in nuclear family units. Resources in the basin were not only sparse but widely scattered, militating against aggregation into larger units during much of the year. Although such bands were generally politically autonomous, they were at least tenuously linked into larger units. In regard to the Shoshoneans, Steward (1955:109) remarks that the "nuclear families have always co-operated with other families in various ways. Since this is so, the Shoshoneans, like other fragmented family groups, represent the family level of sociocultural integration only in a relative sense." Winter encampments of 20 or 30 families were the largest aggregations among Shoshoneans; however, these were not formal organizations but rather aggregations of convenience. Aside from visiting, some cooperative ventures, such as dances (fandangos), rabbit drives, and occasional antelope drives, were organized during winter encampments. The number of families that a given family might camp with over a period of years was also not fixed, although people preferred to camp with people speaking the same dialect (R. Bettinger, personal communication). Steward's picture of the simplicity of the Shoshone has been challenged. Thomas, Pendleton, and Cappannari (1986:278) observe that, at best, Steward's characterization applied only to limiting cases, as, indeed, his frank use of them to imperfectly exemplify an ideal type suggests. Murphy and Murphy (1986), citing the case of the Northern Shoshone and Bannock, argue that the unstructured fluidity of Shoshonean society conceals a sophisticated adaptation to the sparse and uncertain resources of the Great Basin. The Shoshoneans maintained peace among themselves over a very large region, enabling families and small groups of families to move over vast distances in response to local feast and famine. When local resources permitted and necessity required, they were able to assemble considerable numbers of people for collective purposes. Murphy and Murphy cite the formation of war parties numbering in the hundreds to contest bison hunting areas with the Blackfeet. Indeed, the Shoshone and their relatives were relatively recent immigrants to the Great Basin who pushed out societies that were probably socially more complex but less well adapted to the sparse Great Basin environment (Bettinger and Baumhoff, 1982). Murphy and Murphy summarize by saying "the Shoshone are a 'people' in the truest sense of the word" (p. 92). Compared to our great ape relatives, and presumably our remoter ancestors, Shoshonean families maintained generally friendly relations with a rather large group of other families, could readily strike up cooperative relations with strangers of their ethnic group, and organized cooperative activities on a considerable scale.

We believe that the human capacity to live in larger-scale forms of tribal social organization evolved through a coevolutionary ratchet generated by the interaction of genes and culture. Rudimentary cooperative institutions favored genotypes that were better able to live in more cooperative groups. Those individuals best able to avoid punishment and acquire the locally relevant norms were more likely to survive. At first, such populations would have been only slightly more cooperative than typical nonhuman primates. However, genetic changes, leading to moral emotions, like shame and a capacity to learn and internalize local practices, would allow the cultural evolution of more sophisticated institutions that in turn enlarged the scale of cooperation. These successive rounds of coevolutionary change continued until eventually people were equipped with capacities for cooperation with distantly related people, emotional attachments to symbolically marked groups, and a willingness to punish others for transgression of group rules. Mechanisms by which cultural institutions might exert forces tugging in this direction are not far to seek. People are likely to discriminate against genotypes that are incapable of conforming to cultural norms (Richerson and Boyd, 1989; Laland, Kumm, and Feldman, 1995). People who cannot control their self-serving aggression ended up exiled or executed in small-scale societies and imprisoned in contemporary ones. People whose social skills embarrass their families will have a hard time attracting mates. Of course, selfish and nepotistic impulses were never entirely suppressed; our genetically transmitted evolved psychology shapes human cultures, and, as a result, cultural adaptations often still serve the ancient imperatives of inclusive genetic fitness. However, cultural evolution also creates new selective environments that *build cultural imperatives into our genes*.

Paleoanthropologists believe that human cultures were essentially modern by the Upper Paleolithic, 50,000 years ago (Klein, 1999, ch. 7), if not much earlier (McBrearty and Brooks, 2000). Thus, even if the cultural group selection process began as late as the Upper Paleolithic, such social selection could easily have had extensive effects on the evolution of human genes through this process. More likely, Upper Paleolithic societies were the culmination of a long period of coevolutionary increases in a tendency toward tribal social life.[6]

We suppose that the resulting "tribal instincts" are something like principles in the Chomskian linguists' "principles and parameters" view of language (Pinker, 1994). Innate principles furnish people with basic predispositions, emotional capacities, and social dispositions that are implemented in practice through highly variable cultural institutions, the parameters. People are innately prepared to act as members of tribes, but culture tells us how to recognize who belongs to our tribes; what schedules of aid, praise, and punishment are due to tribal fellows; and how the tribe is to deal with other tribes: allies, enemies, and clients. The division of labor between innate and culturally acquired elements is poorly understood, and theory gives little guidance about the nature of the synergies and trade-offs that must regulate the evolution of our psychology (Richerson and Boyd, 2000a). The fact that human-reared apes cannot be socialized to behave like humans guarantees that some elements are innate. Contrarily, the diversity and sometimes rapid change of social institutions guarantee that much of our social life is governed by culturally transmitted rules, skills, and even emotions. We beg the reader's indulgence for the necessarily brief and assertive nature of our argument here. The rationale and ethnographic support for the tribal instincts hypothesis are laid out in more detail in Richerson and Boyd (1998, 1999); for a review of the broad spectrum of empirical evidence supporting the hypothesis, see Richerson and Boyd (2001).

Work-around Hypothesis

Contemporary human societies differ drastically from the societies in which our social instincts evolved. Pleistocene hunter-gatherer societies were comparatively small, egalitarian, and lacking in powerful institutionalized leadership. By contrast, modern societies are large, inegalitarian, and have coercive leadership institutions (Boehm, 1993). If the social instincts hypothesis is correct, our innate social psychology furnishes the building blocks for the evolution of complex social systems, while simultaneously constraining the shape of these systems (Salter, 1995). To evolve large-scale, complex social systems, cultural evolutionary processes, driven by cultural group selection, take advantage of whatever support these instincts offer. For example, families willingly take on the essential roles of biological reproduction and primary socialization, reflecting the ancient and still powerful effects of selection at the individual and kin level. At the same time, cultural evolution must cope with a psychology evolved for life in quite different sorts of societies. Appropriate larger-scale institutions must regulate the constant pressure from smaller groups (coalitions, cabals, cliques) to subvert rules favoring large groups. To do this, cultural evolution often makes use of "work-arounds." It mobilizes the tribal instincts for new purposes. For example, large national and international (e.g., great religions) institutions develop ideologies of symbolically marked inclusion that often fairly successfully engage the tribal instincts on a much larger scale. Military and religious organizations (e.g., Catholic Church), for example, dress recruits in identical clothing (and haircuts) loaded with symbolic markings and then subdivide them into small groups with whom they eat and engage in long-term repeated interaction. Such work-arounds are often awkward compromises, as is illustrated by the existence of contemporary societies handicapped by narrow, destructive loyalties to small tribes (West, 1941) and even to families (Banfield, 1958). In military and religious organizations, excessive within-group loyalty often subverts higher-level goals. If this picture of the innate constraints on current institutional evolution is correct, it is evidence for the existence of tribal social instincts that buttress the uncertain inferences from ethnography and archaeology about late Pleistocene societies. Complex societies are, in effect, grand natural social-psychological experiments that stringently test the limits of our innate dispositions to cooperate. We expect the social institutions of complex societies to simulate life in tribal-scale societies in order to generate cooperative "lift." We also expect that complex institutions will accept design compromises to achieve such "lift," which would be unnecessary if innate constraints of a specifically tribal structure were absent.

Coercive Dominance

The cynics' favorite mechanism for creating complex societies is command backed up by force. The conflict model of state formation has this character (Carneiro, 1970), as does Hardin's (1968) recipe for commons management.

Elements of coercive dominance are no doubt necessary to make complex societies work. Tribally legitimated self-help violence is a limited and expensive means of altruistic coercion. Complex human societies have to supplement the moralistic solidarity of tribal societies with formal police institutions. Otherwise, the large-scale benefits of cooperation, coordination, and division of labor would cease to exist in the face of selfish temptations to expropriate them by individuals, nepotists, cabals of reciprocators, organized predatory bands, greedy capitalists, and classes or castes with special access to means of coercion. At the same time, the need for organized coercion as an ultimate sanction creates roles, classes, and subcultures with the power to turn coercion to narrow advantage. Social institutions of some sort must police the police so that they will act in the larger interest to a measurable degree. Indeed, Boehm (1993) notes that the egalitarian social structure of simple societies is itself an institutional achievement by which the tendency of some to try to dominate others on the typical primate pattern is frustrated by the ability of the individuals who would be dominated to collaborate to enforce rules against dominant behavior. Such policing is never perfect and, in the worst cases, can be very poor. The fact that leadership in complex systems always leads to at least some economic inequality suggests that narrow interests, rooted in individual selfishness, kinship, and, often, the tribal solidarity of the elite, always exert an influence. The use of coercion in complex societies offers excellent examples of the imperfections in social arrangements traceable to the ultimately irresolvable tension of more narrowly selfish and more inclusively altruistic instincts.

While coercive, exploitative elites are common enough, we suspect that no complex society can be based purely on coercion for two reasons: (1) coercion of any great mass of subordinates requires that the elite class or caste be itself a complex, cooperative venture; (2) defeated and exploited peoples seldom accept subjugation as a permanent state of affairs without costly protest. Deep feelings of injustice generated by manifestly inequitable social arrangements move people to desperate acts, driving the cost of dominance to levels that cripple societies in the short run and often cannot be sustained in the long run (Insko et al., 1983; Kennedy, 1987). Durable conquests, such as those leading to the modern European national states, Han China, or the Roman Empire, leaven raw coercion with other institutions. The Confucian system in China and the Roman legal system in the West were far more sophisticated institutions than the highly coercive systems sometimes set up by predatory conquerors and even domestic elites.

Segmentary Hierarchy

Late Pleistocene societies were undoubtedly segmentary in the sense that supraband ethnolinguistic units served social functions. The segmentary principle can serve the need for more command and control by hardening lines of authority without disrupting the face-to-face nature of proximal leadership in egalitarian societies. The Polynesian ranked lineage system illustrates how making political offices formally hereditary according to a kinship formula can help deepen and strengthen a command and control hierarchy (Kirch, 1984). A common method of deepening and strengthening the hierarchy of command and control in complex societies is to construct a nested hierarchy of offices, using various mixtures of ascription and achievement principles to staff the offices. Each level of the hierarchy replicates the structure of a hunting and gathering band. A leader at any level interacts mainly with a few near-equals at the next level down in the system. New leaders are usually recruited from the ranks of subleaders, often tapping informal leaders at that level. As Eibl-Eibesfeldt (1989) remarks, even high-ranking leaders in modern hierarchies adopt much of the humble head-man's deferential approach to leadership. Henrich and Gil-White's (2001) work on prestige provides a coevolutionary explanation for this phenomenon.

The hierarchical nesting of social units in complex societies gives rise to appreciable inefficiencies (Miller, 1992). In practice, brutal sheriffs, incompetent lords, venal priests, and their ilk degrade the effectiveness of social organizations in complex societies. Squires (1986) dissects the problems and potentials of modern hierarchical bureaucracies to perform consistently with leaders' intentions. Leaders in complex societies must convey orders downward, not just seek consensus among their comrades. Devolving substantial leadership responsibility to subleaders far down the chain of command is necessary to create small-scale leaders with face-to-face legitimacy. However, it potentially generates great friction if lower-level leaders either come to have different objectives than the upper leadership or are seen by followers as equally helpless pawns of remote leaders. Stratification often creates rigid boundaries so that natural leaders are denied promotion above a certain level, resulting in inefficient use of human resources and a fertile source of resentment to fuel social discontent.

On the other hand, failure to articulate properly tribal-scale units with more inclusive institutions is often highly pathological. Tribal societies often must live with chronic insecurity due to intertribal conflicts. One of us once attended the *Palio*, a horse race in Siena in which each ward, or *contrada*, in this small Tuscan city sponsors a horse. Voluntary contributions necessary to pay the rider, finance the necessary bribes, and host the victory party amount to a half a million dollars. The *contrada* clearly evoke the tribal social instincts: they each have a totem—the dragon, the giraffe—special colors, rituals, and so on. The race excites a tremendous, passionate rivalry. One can easily imagine medieval Siena in which swords clanged and wardmen died, just as they do or did in warfare between New Guinea tribes (Rumsey, 1999), Greek city-states (Runciman, 1998), inner-city street gangs (Jankowski, 1991), and ethnic militias.

Exploitation of Symbolic Systems

The high population density, division of labor, and improved communication made possible by the innovations of complex societies increased the scope for elaborating symbolic systems. The development of monumental architecture to serve mass ritual performances is one of the oldest archaeological markers of emerging complexity. Usually an established church or less formal ideological umbrella supports a complex society's institutions. At the same time, complex societies exploit the symbolic ingroup instinct to delimit a quite diverse array of culturally defined subgroups, within which a good deal of cooperation is routinely achieved. Ethnic group–like sentiments in military organizations are often most strongly reinforced at the level of 1,000–10,000 or so men (British and German regiments, U.S. divisions; Kellett, 1982). Typical civilian symbolically marked units include nations, regions (e.g., Swiss cantons), organized tribal elements (Garthwaite, 1993), ethnic diasporas (Curtin, 1984), castes (Srinivas, 1962; Gadgil and Guha, 1992), large economic enterprises (Fukuyama, 1995), and civic organizations (Putnam, Leonardi, and Nanetti, 1993).

How units as large as modern nations tap into the tribal social instincts is an interesting issue. Anderson (1991) argues that literate communities, and the social organizations revolving around them (e.g., Latin literates and the Catholic Church), create "imagined communities," which in turn elicit significant commitment from members of the community. Since tribal societies were often large enough that some members were not known personally to any given person, common membership would sometimes have to be established by the mutual discovery of shared cultural understandings, as simple as the discovery of a shared language in the case of the Shoshone. The advent of mass literacy and print media—Anderson stresses newspapers—made it possible for all speakers of a given vernacular to have confidence that all readers of the same or related newspapers share many cultural understandings, especially when organizational structures such as colonial government or business activities really did give speakers some institutions in common. Nationalist ideologists quickly discovered the utility of newspapers for building imagined communities, typically several contending variants of the community, making nations the dominant quasi-tribal institution in most of the modern world.

Many problems and conflicts revolve around symbolically marked groups in complex societies. Official dogmas often stultify desirable innovations and lead to bitter conflicts with heretics. Marked subgroups often have enough tribal cohesion to organize at the expense of the larger social system. The frequent seizure of power by the military in states with weak institutions of civil governance is probably a by-product of the fact that military training and segmentation, often based on some form of patriotic ideology, are conducive to the formation of *relatively* effective large-scale institutions. Wherever groups of people interact routinely, they are liable to develop a tribal ethos. In stratified societies, powerful groups readily evolve self-justifying ideologies that buttress treatment of subordinate groups, ranging from neglectful to atrocious. White American Southerners had elaborate theories to justify slavery, and pioneers everywhere found the brutal suppression of Indian societies legitimate and necessary. The parties and interest groups that vie to sway public policy in democracies have well-developed rationalizations for their selfish behavior. A major difficulty with loyalties induced by appeals to shared symbolic culture is the very language-like productivity possible with this system. Dialect markers of social subgroups emerge rapidly along social fault-lines (Labov, 2001). Charismatic innovators regularly launch new belief and prestige systems, which sometimes make radical claims on the allegiance of new members, sometimes make large claims at the expense of existing institutions, and sometimes grow explosively. Or larger loyalties can arise, as in the case of modern nationalisms overriding smaller-scale loyalties, sometimes for the better, sometimes for the worse. The ongoing evolution of social systems can develop in unpredictable, maladaptive directions by such processes (Putnam, 2000). The worldwide growth of fundamentalist sects that challenge the institutions of modern states is a contemporary example (Marty and Appleby, 1991). If T. Wolfe (1965) is right, mass media can be the basis of a rich diversity of imagined subcommunities using such vehicles as specialized magazines, newsletters, and websites. The potential of deviant subgroups, such as sectarian terrorist organizations, to use modern media to create small but highly motivated imagined communities is an interesting variant on Anderson's theory. Ongoing cultural evolution is impossible to control wholly in the larger interest, at least impossible to control completely, and forbidding free evolution tends to deprive societies of the "civic culture" that spontaneously produces so many collective benefits.

Legitimate Institutions

In small-scale egalitarian societies, individuals have substantial autonomy, considerable voice in community affairs, and can enforce fair, responsive—even self-effacing—behavior and treatment from leaders (Boehm, 1999). At their most functional, symbolic institutions, a regime of tolerably fair laws and customs, effective leadership, and smooth articulation of social segments can roughly simulate these conditions in complex societies. Rationally administered bureaucracies, lively markets, the protection of socially beneficial property rights, widespread participation in public affairs, and the like provide public and private goods efficiently, along with a considerable amount of individual autonomy. Many individuals in modern societies feel themselves part of culturally labeled tribal-scale groups, such as local political party organizations, that have influence on the remotest leaders. In older complex societies, village councils, local notables, tribal chieftains, or religious leaders often hold courts open to humble petitioners. These local leaders, in turn, represent their communities to higher authorities. To obtain low-cost compliance with management decisions, ruling elites have to convince citizens that these decisions are in the interest of the larger community. As long as most individuals trust that existing institutions are reasonably legitimate and that any felt needs for reform are achievable by means of ordinary political activities, there is considerable scope for large-scale collective social action.

Legitimate institutions, however, and trust of them, are the result of an evolutionary history and are neither easy to manage or engineer. Social distance between different classes, castes, occupational groups, and regions is objectively great. Narrowly interested tribal-scale institutions abound in such societies. Some of these groups have access to sources of power that they are tempted to use for parochial ends. Such groups include, but are not restricted to, elites. The police may abuse their power. Petty administrators may victimize ordinary citizens and cheat their bosses. Ethnic political machines may evict historic elites from office but use chicanery to avoid enlarging their coalition.

Without trust in institutions, conflict replaces cooperation along fault lines where trust breaks down. Empirically, the limits of the trusting community define the universe of easy cooperation (Fukuyama, 1995). At worst, trust does not extend outside family (Banfield, 1958), and potential for cooperation on a larger scale is almost entirely foregone. Such communities are unhappy as well as poor. Trust varies considerably in complex societies, and variation in trust seems to be the main cause of differences in happiness across societies (Inglehart and Rabier, 1986). Even the most efficient legitimate institutions are prey to manipulation by small-scale organizations and cabals, the so-called special interests of modern democracies. Putnam et al.'s (1993) contrast between civic institutions in

Northern and Southern Italy illustrates the difference that a tradition of functional institutions can make. The democratic form of the state, pioneered by Western Europeans in the last couple of centuries, is a powerful means of creating generally legitimate institutions. Success attracts imitation all around the world. The halting growth of the democratic state in countries ranging from Germany to sub-Saharan Africa is testimony that legitimate institutions cannot be drummed up out of the ground just by adopting a constitution. Where democracy has taken root outside of the European cultural orbit, it is distinctively fitted to the new cultural milieu, as in India and Japan.

Conclusion

The processes of cultural evolution quite plausibly led to group selection being a more powerful force on cultural rather than genetic variation. The cultural system of inheritance probably arose in the human lineage as an adaptation to the increasingly variable environments of the recent past (Richerson and Boyd, 2000a, b). Theoretical models show that the specific structural features of cultural systems, such as conformist transmission, have ordinary adaptive advantages. We imagine that these adaptive advantages favored the capacity for a system that could respond rapidly and flexibly to environmental variation in an ancestral creature that was not particularly cooperative. As a by-product, cultural evolution happened to favor large-scale cooperation. Over a long period of coevolution, cultural pressures reshaped "human nature," giving rise to innate adaptations to living in tribal-scale social systems. Humans became prepared to use systems of legitimate punishment to lower the fitness of deviants, for example. We believe that the cultural explanation for human cooperation is in accord with much evidence, as summarized by stylized facts about human cooperation with which we introduced our remarks. More detailed surveys of the concordance of our conjectures with various bodies of data may be found in Richerson and Boyd (1999, 2001) and Richerson, Boyd, and Paciotti (2002).

Regardless of the fate of any particular proposals, we think that explanations of human cooperation have to thread some rather tight constraints. They have to somehow finesse the awkward fact that humans, at least partly because of our ability to cooperate with distantly related people in large groups, are a huge success yet quite unique in our style of social life. If a mechanism like indirect reciprocity works, why have not many social species used it to extend their range of cooperation? If finding self-reinforcing solutions to coordination games is mostly what human societies are about, why do not other animals have massive coordination-based social systems? If reputations for pairwise cooperation are easy to observe or signal (but unexploitable by deceptive defectors), why have we found no other complex animal societies based on this principle? By contrast, we do find plenty of complex animal societies built on the principle of inclusive fitness.

The unique pattern of cooperation of our species suggests that human cooperation is likely to derive from some other unique feature or features of human life. Advanced capacities for social learning are also unique to humans; thus, culture is, prima facie, a plausible key element in the evolution of human cooperation. Our argument depends upon the existence of culture and group selection on cultural variation. Since sophisticated culture is unique to humans, we do not expect this mechanism to operate in other species. Ours is not the only hypothesis that passes this basic test. For example, E. Smith's (2003) signaling hypothesis depends upon language, another unique feature of the human species. E. Hagen made a similar proposal in his comment on our background paper. He argued that the inventiveness of humans combined with language as a cheap communication device adapts us to solve problems of cooperation. We think that hypotheses in this vein, like Alexander's proposed indirect reciprocity mechanism, cannot be decisively rejected, but they are far from completely specified. What is it that biases invention and cheap talk in favor of cooperative rather than selfish ends? The intuition that cheap talk, symbolic rewards, and clever institutions are in themselves sufficient to explain human cooperation probably comes from the common experience that people do find it rather easy to use such devices to cooperate (e.g., Ostrom, Gardner, and Walker, 1994). The difficult question is whether these are backed up by unselfish motives on the part of at least some people. A literal interpretation of experiments such as those of Fehr and Gächter (2002) and Batson (1991) suggests that unselfish motives play important roles. However, unselfish motives may be a proximal evolutionary result of an ultimate indirect reciprocity sort of evolutionary process rather than the result of a group selection mechanism. Those who attempt deception in a world of clever cooperators may simply expose their lack of cleverness, so that the best strategy is an unfeigned willingness to cooperate. The data that cultural group selection is an appreciable process (Soltis et al., 1995) are also not definitive, since they could be weak relative to some competing process of the indirect reciprocity sort.

Another complication is that hypotheses leaning on language, technology, and intelligence are appealing to phenomena with considerable cultural content. The evolution of technology and the diffusion of innovations are cultural processes that depend upon institutions and a sophisticated social psychology (Henrich, 2001). Both the cultural and genetic evolution of our cognitive capacities (some of which gave rise to language) likely emerged from a culture-gene coevolutionary process (Henrich and McElreath, 2002; Tomasello, 1999). Thus, these hypotheses are not, we submit, clean alternatives to the cultural group selection hypothesis, absent further specification. In the future, we expect that competing hypotheses will be developed in sufficient detail that more precise comparative empirical tests will be possible.

For example, even if innatist linguists are correct that much of what we need to know to speak is innate, we wonder why more is not innate? Why is it that mutually unintelligible languages arise so rapidly? Would not we be better off if everyone spoke the same common entirely innate language? Not necessarily. Very often people from distant places are likely to have evolved different ways of doing things that are adaptive at home but not abroad. Similarly, avoiding listening to people is a wise idea if they are proposing a behavior deviant from locally prevailing coordination equilibria. Cultural evolution can run up adaptive *barriers* to communication quite readily if listening to foreigners makes one liable to acquire erroneous ideas (McElreath, Boyd, and Richerson 2003). Dialect evolution seems to be a highly nuanced system for regulating communication within languages as well as between them, although the adaptive significance of dialect is hardly well worked out (Laboy, 2001). Interestingly, in McElreath et al.'s model, using a symbolic signal to express a willingness to cooperate cannot support the evolution of a symbolic marker of group membership because defectors as well as potential cooperators will be attracted by the signal. A symbolic system can be used to communicate intention to cooperate only if potential cooperative partners can exchange trustworthy signals. Once symbolic markers became sufficiently complex as to be unfakable by defectors *and* a sufficiently large pool of relatively anonymous but trustworthy signalers exist, *then* cheap signals will be useful. Dialect is difficult to fake although cheap to use, and once some level of cooperation on a proto-tribal scale was possible, proto-languages might have come under selection to create unfakable signals of group membership that imply an intention to cooperate. We suspect that language could have evolved only in concert with a measure of trust of other speakers rather than being an unaided generator of trust. To the extent that cooperation is the game, one has no interest in listening to speakers whose messages are self-serving. Think of how annoying we find telemarketer's speech acts. Sociolinguists make much of the concept that speech is a cooperative system and argue that the empirical structure of conversation is consistent with this assumption (Wardhaugh, 1992). Language seems to presuppose cooperation as much as it in turn facilitates cooperation.

That technology, like language, is one of the major components of the human adaptation is undeniable. It opens up opportunities to gain advantage to cooperation in hunting and defense and to exploit the possibilities of the division of labor. What is less well understood is the extent to which technology is likely a product of large-scale social systems. Henrich (2004b) has analyzed models of the "Tasmanian Effect." At the time of European contact, the Tasmanians had the simplest toolkit ever recorded in an extant human society; it was, for example, substantially simpler than the toolkits of ethnographically known foragers in the Kalahari and Tierra del Fuego, as well as those associated with human groups from the Upper Paleolithic. Archaeological evidence indicates that Tasmanian simplicity resulted from both the gradual loss of items from their own pre-Holocene toolkit and the failure to develop many of the technologies that subsequently arose only 150 km to the north in Australia. The loss likely began after the Bass Strait was flooded by rising post-glacial sea levels (Jones, 1995). Henrich's analysis indicates that imperfect inference during social learning, rather than stochastic loss due to drift-like effects, is the most likely reason for this loss. This suggests that to maintain an equilibrium toolkit as complex as those of late Pleistocene hunter-gatherers likely required a rather large population of people who interacted fairly freely so that rare, highly skilled performances, spread by selective imitation, could compensate for the routine loss of skills due to imperfect inference. Neanderthals and perhaps other archaic human populations had large brains but simple toolkits. The Tasmanian effect may explain why. Archaeology suggests that Neanderthal population densities were lower than those of the modern humans that replaced them in Europe and that they had less routine contact with their neighbors, as evidenced by shorter distance movement of high-quality raw materials from their sources compared to those for modern humans (Klein, 1999).

The proposal that human intelligence is at the root of human cooperation is difficult to evaluate because of the ambiguity in what we might mean by intelligence in a comparative context (Hinde, 1970:659–663). As the Tasmanian Effect illustrates, *individual* human intelligence is only a part, and perhaps only a small part, of being able to create complex adaptive behaviors. In fact, we think "intelligence" plays little role in the emergence of many human complex adaptations. Instead, humans seem to depend upon socially learned strategies to finesse the shortcomings of their cognitive capabilities (Nisbett and Ross, 1980). The details of human cognitive abilities apparently vary substantially across cultures because culturally transmitted cognitive styles differ (Nisbett et al., 2001). Although we share the common intuition that humans are individually more intelligent than even our very clever fellow apes, we are not aware of any experiments that sufficiently control for our cultural repertoires to be sure that it is correct. The concept of "intelligence" in individual humans perhaps makes little sense apart from their cultural repertoires: humans are smart in part because they can bring a variety of "cultural tools" (e.g., numbers, symbols, maps, various kinematic models) to bear on problems. A hunter-gatherer would seem an incredibly stupid college professor, but college professors would seem equally dense if forced to try to survive as hunter-gatherers (a few knowledgeable anthropologists aside). Even abilities as seemingly basic as those related directly to visual perception vary across cultures (Segall, Campbell, and Herskovits, 1966). Second, *intelligence* implies a means to an end, not an end in itself. Individual intelligence ought to serve the ends of both cooperation and defection. We suspect that actually defection, requiring trickery and deception, is better served by intelligence than cooperation. Game theorists assuming perfect, but selfish, rationality predict that humans should defect in the one-shot anonymous prisoner's dilemma, just as evolutionary biologists predict that dumb beasts using evolved predispositions will. Whiten and Bryne (1988) characterized our social intelligence as "Machiavellian," implying that it does indeed serve deception equally with honesty. However, just as humans punish altruistically, they seem also to exert their political intelligence altruistically (e.g., Sears and Funk, 1990), biasing the evolution of institutions accordingly. On the basis of our brain size compared to that of other apes, Dunbar (1992) predicts that human groups ought to number around 50. Hunter-gatherer co-residential bands do number around 50, but culturally transmitted *institutions* web together bands to create tribes typically numbering a few hundred to a few thousand people, as we have seen. Human political systems do seem to exceed in scale anything predicted on the basis of enhanced Machiavellian talents (supposing that such talents can on average increase social scale at all). The institutional basis of these systems is not far to seek. For example, Wiessner (1984) describes how institutions of ceremonial exchange of gifts knit the famous !Kung San bands into a much larger-risk pooling cooperative. Australian aboriginal groups show similar functional patterns, which are built out of quite different and substantially more elaborate sets of cultural practices (Peterson, 1979). Underpinning such individual-to-individual bond making is likely the kind of generalized trust that co-ethnics have for one another. If Murphy and Murphy (1986) are correct about the Northern Shoshone, a society of thousands constituted a functional "people" engaging in mutual aid in a hostile and uncertain environment on the basis of little more than a common language. In his classic ethnography of the Nuer, Evans-Pritchard (1940) describes how simple tribal institutions can knit herding people into tribes numbering tens of thousands, much larger than was possible among hunter-gathers. The size of hunter-gatherer societies was evidently limited by low population density, not by their relatively unsophisticated institutions. Third, Henrich and Gil-White (2001) propose that human prestige systems are an adaptation to facilitate cultural transmission. Social learning means that the returns to effort in individual learning potentially result in gains for many subsequent social learners who do not have to "reinvent the wheel." If extra individual effort in acquiring better ideas pays off in prestige and if prestige leads to fitness advantages, then the social returns to effortful individual learning will in part be reflected in private returns to individual learners. Group selection on prestige systems may further enlarge the returns to investment in individual learning and bring returns up to a level that reflects the group optimum amount of effort in individual learning. If this mechanism operates, human intelligence may have been enhanced by social selection emanating from institutions of prestige.[7]

We propose that group selection on cultural variation is at the heart of human cooperation, but we certainly recognize that our sociality is a complex system that includes many linked components. Surely, without punishment, language, technology, individual intelligence and inventiveness, ready establishment of reciprocal arrangements, prestige systems, and solutions to games of coordination, our societies would take on a distinctly different cast, to say the least. Human sociality no doubt has a number of components that were necessary to its evolution and are necessary to its current functions. If such is the case, prime mover explanations giving pride of place to a single mechanism are vain to seek. Thus, a major constraint on explanations of human sociality is its systemic structure. Explanations have to have a plausible historical sequence tracing how the currently interrelated parts evolved, perhaps piecemeal. And explanations have to account for the current functional and dysfunctional properties of human social systems. We are far from having completed this task.

NOTES

1. "Cooperation" has a broad and a narrow definition. The broad definition includes all forms of mutually beneficial joint action by two or more individuals. The narrow definition is restricted to situations in which joint action poses a dilemma for at least one individual such that, at least in the short run, that individual would be better off not cooperating. We employ the narrow definition in this chapter. The "cooperate" versus "defect" strategies in the prisoner's dilemma and commons games anchor our concept of cooperation, making it more or less equivalent to the term "altruism" in evolutionary biology. Thus, we distinguish "coordination" (joint interactions that are "self-policing" because payoffs are highest if everyone does the same thing) and division of labor (joint action in which payoffs are highest if individuals do different things) from cooperation.

2. We refer to cultural evolution as changes in the pool of cultural variants carried by a population of individuals as a function of time and the processes that cause the changes.

3. It is not obvious that language potentiates indirect reciprocity. Whereas superficially language may seem to promote the exchange of high-quality information required for indirect reciprocity to favor cooperation, this addition merely changes the question slightly to one of why individuals would cooperate in information sharing; language merely recreates the same public goods dilemma. Lies about hunting success, for example, are difficult to check and often ambiguous. Among the Gunwinggu (Australian foragers), members of one band often lied to members of other bands about their success to avoid having to share meat (Altman and Peterson, 1988).

4. Several prominent modern Darwinians, Hamilton (1975), Wilson (1975: 561–562), Alexander (1987:169), and Eibl-Eibesfeldt (1982), have given serious consideration to group selection as a force *in the special case* of human ultra-sociality. They are impressed, as we are, by the organization of human populations into units that engage in sustained, lethal combat with other groups, not to mention other forms of cooperation. The trouble with a straightforward group selection hypothesis is our mating system. We do not build up concentrations of intrademic relatedness like social insects, and few demic boundaries are without considerable intermarriage. Moreover, the details of human combat are more lethal to the hypothesis of genetic group selection than to the human participants. For some of the most violent groups among simple societies, wife capture is one of the main motives for raids on neighbors, a process that could hardly be better designed to erase genetic variation between groups and stifle genetic group selection.

5. We are aware that much controversy surrounds the use of microevolutionary models to explain macroevolutionary questions. Our thoughts on the issues are summarized in Boyd and Richerson (1992a).

6. It would be a mistake to assume that complex technology is a prerequisite for tribal-level forms of social organization. At the time of European discovery, the Tasmanians had a technology substantially simpler than that of many Upper Paleolithic peoples: they lacked bone tools, composite spears, bows, arrows, spear throwers, and fish hooks, etc. Yet they lived in multiband groups, which controlled territories. Intertribal trade, warfare, and raiding were all commonplace (Jones, 1995). The last 4,000 years of the Tasmania archaeological record do not look much different from many middle Paleolithic sites.

7. Similarly, as Smith (2003) notes, Hawkes hypothesizes that men contribute to hunting success to "show off" and that showing off earns men reproductive success in terms of sexual favors from women. Contrary to what Hawkes supposes, this system is a possible focus of cultural group selection. In many hunter-gatherer groups, meat is very widely shared and hunters often do not control its distribution. Personal favors granted to a successful hunter as recompense for effort will benefit all who share his kills. Showing that individuals who contribute heavily to the common good are rewarded is not evidence that group-selected effects are absent. In the end, group selection can succeed only if altruistic individuals on average do better than selfish ones. The fact that hunters are not allowed to bargain with consumers of their kills and yet are rewarded by consumers anyway is at least as consistent with the operation of group selection as with a competing individualist explanation.

REFERENCES

Alexander, R. D. 1979. Darwinism and human affairs. The Jessie and John Danz Lectures, Seattle: University of Washington Press.

Alexander, R. D. 1987. The biology of moral systems: Foundations of human behavior. Hawthorne, NY: Aldine de Gruyter.

Altman, J., & N. Peterson. 1988. Rights to game and rights to cash among contemporary Australian hunter-gatherers. In: *Hunters and gatherers, vol. 2: Property, power, and ideology*, T. R. Ingold, D. Riches, & J. Woodburn, ed. (pp. 75–94). Oxford: Berg.

Anderson, B. R. O'G. 1991. Imagined communities: Reflections on the origin and spread of nationalism. Rev. and extended ed. London: Verso.

Aoki, K. 1982. A condition for group selection to prevail over counteracting individual selection. *Evolution* 36:832–842.

Axelrod, R., & W. D. Hamilton. 1981. The evolution of cooperation. *Science* 211:1390–1396.

Baland, J. -M., & J. P. Platteau. 1996. Halting degradation of natural resources: Is there a role for rural communities? Oxford: Oxford University Press.

Banfield, E. C. 1958. The moral basis of a backward society. Glencoe, IL: Free Press.

Batson, C. D. 1991. The altruism question: Toward a social psychological answer. Hillsdale, NJ: Lawrence Erlbaum.

Baum, W. B. 1994. Understanding behaviorism: Science, behavior, and culture. New York: Harper Collins.

Becker, G. S. 1976. Altruism, egoism, and genetic fitness: Economics and sociobiology. *Journal of Economic Literature* 14:817–826.

Bettinger, R. L. 1991. *Hunter-gatherers: Archaeological and evolutionary theory. Interdisciplinary contributions to archaeology*. New York: Plenum.

Bettinger, R. L., & M. A. Baumhoff. 1982. The numic spread: Great Basin cultures in competition. *American Antiquity* 47:485–503.

Blackmore, S. 1999. The meme machine. Oxford: Oxford University Press.

Boehm, C. 1993. Egalitarian behavior and reverse dominance hierarchy. *Current Anthropology* 34:227–254.

Boehm, C. 1999. *Hierarchy in the forest: The evolution of egalitarian behavior*. Cambridge, MA: Harvard University Press.

Borgerhoff Mulder, M., P. J. Richerson, N. W. Thornhill, & E. Voland. 1997. The place of behavioral ecological anthropology in evolutionary science. In: *Human by nature: Between biology and the social sciences*, P. Weingart, S. D. Mitchell, P. J. Richerson, & S. Maasen, eds. (pp. 253–282). Mahwah, NJ: Lawrence Erlbaum.

Boyd, R., H. Gintis, S. Bowles, & P. J. Richerson. 2003. The evolution of altruistic punishment. *Proceedings of the National Academics of Sciences (USA)* 100:3531–3535.

Boyd, R., & P. J. Richerson. 1985. *Culture and the evolutionary process*. Chicago: University of Chicago Press.

Boyd, R., & P. J. Richerson. 1989. The evolution of indirect reciprocity. *Social Networks* 11:213–236.

Boyd, R., & P. J. Richerson. 1992a. How microevolutionary processes give rise to history. In: *History and evolution*, M. H. Nitecki, & D. V. Nitecki, eds. (pp. 178–209). Albany: SUNY Press.

Boyd, R., & P. J. Richerson. 1992b. Punishment allows the evolution of cooperation (or anything else) in sizable groups. *Ethology and Sociobiology* 13:171–195.

Boyd, R., & P. J. Richerson. 1993. Rationality, imitation, and tradition. In: *Nonlinear dynamics and evolutionary economics*, R. H. Day, & P. Chen, eds. (pp. 131–149). New York: Oxford University Press.

Boyd, R., & P. J. Richerson. 2002. Group beneficial norms spread rapidly in a structured population. *Journal of Theoretical Biology* 215:287–296.

Campbell, D. T. 1965. Variation and selective retention in socio-cultural evolution. In: *Social change in developing areas: A reinterpretation of evolutionary theory*, H. R.

Barringer, G. I. Blanksten, & R. W. Mack, eds. (pp. 58–79). Cambridge, MA: Schenkman.

Campbell, D. T. 1975. On the conflicts between biological and social evolution and between psychology and moral tradition. *American Psychology* 30:1103–1126.

Campbell, D. T. 1983. The two distinct routes beyond kin selection to ultrasociality: Implications for the humanities and social sciences. In: *Nature of prosocial development: Theories and strategies*, D. L. Bridgeman, ed. (pp. 71–81). New York: Academic.

Carneiro, R. L. 1970. A theory for the origin of the state. *Science* 169:733–738.

Cavalli-Sforza, L. L., & M. W. Feldman. 1981. Cultural transmission and evolution: A quantitative approach. *Monographs in Population Biology* 16. Princeton, NJ: Princeton University Press.

Cohen, D., & J. Vandello. 2001. Honor and "faking" honorability. In: *Evolution and the capacity for commitment*, R. M. Nesse, ed. (pp. 163–185). New York: Russell Sage.

Curtin, P. D. 1984. *Cross-cultural trade in world history: Studies in comparative world history*. Cambridge: Cambridge University Press.

Darwin, C. 1874. *The descent of man*. Project Gutenberg, www.gutenberg.net.

Dawes, R. M., A. J. C. van de Kragt, & J. M. Orbell. 1990. Cooperation for the benefit of us—not me or my conscience. In: *Beyond self-interest*, J. J. Mansbridge, ed. (pp. 97–110). Chicago: University of Chicago Press.

Dunbar, R. I. M. 1992. Neocortex size as a constraint on group size in primates. *Journal of Human Evolution* 22:469–493.

Eibl-Eibesfeldt, I. 1982. Warfare, man's indoctrinability, and group selection. *Zeitschrift fur Tierpsychology* 67:177–198.

Eibl-Eibesfeldt, I. 1989. *Human ethology: Foundations of human behavior*. New York: Aldine de Gruyter.

Evans-Pritchard, E. E. 1940. *The Nuer: A description of the modes of livelihood and political institutions of a nilotic people*. Oxford: Clarendon.

Falk, A., E. Fehr, & U. Fischbacher. 2002. Approaching the commons: A theoretical explanation. In: *The drama of the commons*, E. Ostrom, T. Dietz, N. Dolsak, et al., eds. (pp. 157–191). Washington, DC: National Academy Press.

Fehr, E., & S. Gächter. 2002. Altruistic punishment in humans. *Nature* 415:137–140.

Fukuyama, F. 1995. *Trust: Social virtues and the creation of prosperity*. New York: Free Press.

Gadgil, M., & R. Guha. 1992. *This fissured land: An ecological history of India*. Delhi: Oxford University Press.

Gardner, R. A., B. T. Gardner, & T. E. Van Cantfort. 1989. *Teaching sign language to chimpanzees*. Albany: SUNY Press.

Garthwaite, G. R. 1993. Reimagined internal frontiers: Tribes and nationalism—Bakhtiyari and Kurds. In: *Russia's Muslim frontiers: New directions in cross-cultural analysis*, D. F. Eickelman, ed. (pp. 130–148). Bloomington: Indiana University Press.

Hamilton, W. D. 1964. Genetic evolution of social behavior. I. II. *Journal of Theoretical Biology* 7:1–52.

Hamilton, W. D. 1975. Innate social aptitudes of man: An approach from evolutionary genetics. In: *Biosocial anthropology*, R. Fox, ed. (pp. 115–132). New York: Wiley.

Hardin, G. 1968. The tragedy of the commons. *Science* 162:1243–1248.

Hawkes, K. 1991. Showing off: Tests of an hypothesis about men's foraging goals. *Ethology and Sociobiology* 12:29–54.

Henrich, J. 2001. Cultural transmission and the diffusion of innovations: Adoption dynamics indicate that biased cultural transmission is the predominate force in behavioral change. *American Anthropologist* 103:992–1013.

Henrich, J. 2004a. Cultural group selection, coevolutionary processes and large-scale cooperation. *Journal of Economic Behavior and Organization* 53:127–143.

Henrich, J. 2004b. Demography and cultural evolution: Why adaptive cultural processes produced maladaptive losses in Tasmania, *American Antiquity* 69:197–214.

Henrich, J., & R. Boyd. 1998. The evolution of conformist transmission and the emergence of between-group differences. *Evolution and Human Behavior* 19:215–241.

Henrich, J., & R. Boyd. 2001. Why people punish defectors: Weak conformist transmission can stabilize costly enforcement of norms in cooperative dilemmas. *Journal of Theoretical Biology* 208:79–89.

Henrich, J., & R. Boyd. 2002. On modeling cognition and culture: Why replicators are not necessary for cultural evolution. *Journal of Culture and Cognition* 2:67–112.

Henrich, J., R. Boyd, S. Bowles, et al. 2001. In search of *Homo economicus*: Behavioral experiments in 15 small-scale societies. *American Economic Review* 91:73–78.

Henrich, J., & F. J. Gil-White. 2001. The evolution of prestige: Freely conferred deference as a mechanism for enhancing the benefits of cultural transmission. *Evolution and Human Behavior* 22:165–196.

Henrich, J., & R. McElreath. 2002. Are peasants risk-averse decision makers? *Current Anthropology* 43:172–181.

Hinde, R. A. 1970. Animal behaviour: *A synthesis of ethology and comparative psychology.* 2nd ed. New York: McGraw-Hill.

Hirshleifer, J. 1977. Economics from a biological viewpoint. *J. Law Econ.* 20:1–52.

Inglehart, R., & J. -R. Rabier. 1986. Aspirations adapt to situations—But why are the Belgians so much happier than the French? A cross-cultural analysis of the subjective quality of life. In: *Research on the quality of life,* F. M. Andrews, ed. (pp. 1–56). Ann Arbor: Institute for Social Research, University of Michigan.

Insko, C. A., R. Gilmore, S. Drenan, et al. 1983. Trade versus expropriation in open groups. A comparison of two types of social power. *Journal of Personality and Social Psychology* 44:977–999.

Jankowski, M. S. 1991. Islands in the street: Gangs and American urban society. Berkeley: University of California Press.

Johnson, A. W., & T. K. Earle. 2000. *The evolution of human societies: From foraging group to agrarian state.* 2nd ed. Stanford: Stanford University Press.

Jones, R. 1995. Tasmanian archaeology: Establishing the sequences. *Annual Review of Anthropology* 24:423–446.

Keeley, L. H. 1996. *War before civilization.* New York: Oxford University Press.

Kellett, A. 1982. Combat motivation: The behavior of soldiers in battle. *International Series in Management Science/Operations Research.* Boston: Kluwer.

Kelly, R. C. 1985. *The nuer conquest: The structure and development of an expansionist system.* Ann Arbor University of Michigan Press.

Kelly, R. L. 1995. *The foraging spectrum: Diversity in hunter-gatherer lifeways.* Washington, DC: Smithsonian Institution Press.

Kennedy, P. M. 1987. *The rise and fall of the great powers: Economic change and military conflict from 1500 to 2000.* New York: Random House.

Kirch, P. V. 1984. *The evolution of the Polynesian chiefdoms: New studies in archaeology.* Cambridge: Cambridge University Press.

Klein, R. G. 1999. *The human career: Human biological and cultural origins.* 2nd ed. Chicago: University of Chicago Press.

Labov, W. 2001. *Principles of linguistic change: Social factors,* P. Trudgill, ed., vol. 29, *Language in society.* Malden, MA: Blackwell.

Laland, K. N., & G. R. Brown. 2002. *Sense and nonsense: Evolutionary perspectives on human behaviour.* Oxford: Oxford University Press.

Laland, K. N., J. Kumm, & M. W. Feldman. 1995. Gene-culture coevolutionary theory: A test case. *Current Anthropology* 36:131–156.

Leimar, O., & P. Hammerstein. 2001. Evolution of cooperation through indirect reciprocity. *Proceedings of the Royal Society of London, Series B* 268:745–753.

Marty, M. E., & R. S. Appleby. 1991. *Fundamentalisms observed: The fundamentalism project*, vol. 1. Chicago: University of Chicago Press.

Marwell, G., & R. E. Ames. 1981. Economist free ride: Does anyone else? *Journal of Public Economics* 15:295–310.

Maynard Smith, J., & E. Szathmáry. 1995. The major evolutionary transitions. *Nature* 374:227–232.

Mayr, E. 1982. *The growth of biological thought: Diversity, evolution, and inheritance.* Cambridge MA: Harvard University Press.

McBrearty, S., & A. S. Brooks. 2000. The revolution that wasn't: A new interpretation of the origin of modern human behavior. *Journal of Human Evolution* 39:453–563.

McElreath, R., R. Boyd, & P. J. Richerson. 2003. Shared norms can lead to evolution of ethnic markers. *Current Anthropology* 44:122–129.

Miller, G. J. 1992. *Managerial dilemmas: The political economy of hierarchy.* Cambridge: Cambridge University Press.

Murphy, R. F., & Y. Murphy. 1986. Northern Shoshone and Bannock. In: *Handbook of North American Indians: Great Basin*, W. L. d'Azevedo, ed. (pp. 284–307). Washington, DC: Smithsonian Institution Press.

Nesse, R. M., & G. C. Williams. 1995. *Why we get sick: The new science of darwinian medicine.* New York: Times Books.

Nisbett, R. E., & D. Cohen. 1996. *Culture of honor: The psychology of violence in the south. New directions in social psychology.* Boulder, CO: Westview Press.

Nisbett, R. E., K. P. Peng, I. Choi, & A. Norenzayan. 2001. Culture and systems of thought: Holistic versus analytic cognition. *Psychological Review* 108:291–310.

Nisbett, R. E., & L. Ross. 1980. *Human inference: Strategies and shortcomings of social judgment.* Englewood Cliffs, NJ: Prentice Hall.

Nowak, M. A., & K. Sigmund. 1998. Evolution of indirect reciprocity by image scoring. *Nature* 393:573–577.

Ostrom, E. 1990. *Governing the commons: The evolution of institutions for collective action.* Cambridge: Cambridge University Press.

Ostrom, E. 1998. A behavioral approach to the rational choice theory of collective action. *American Political Science Review* 92:1–22.

Ostrom, E., R. Gardner, & J. Walker. 1994. *Rules, games, and common-pool resources.* Ann Arbor: University of Michigan Press.

Otterbein, K. F. 1970. *The evolution of war: A cross-cultural study.* New Haven, CT: Human Relations Area Files Press.

Peterson, N. 1979. Territorial adaptations among desert hunter-gatherers: The !Kung and Austalians compared. In: *Social and ecological systems*, P. Burnham, & R. Ellen, eds. (pp. 111–129). New York: Academic.

Pinker, S. 1994. *The language instinct.* New York: Morrow.

Price, G. 1972. Extensions of covariance selection mathematics. *Annals of Human Genetics* 35:485–490.

Price, T. D., & J. A. Brown. 1985. *Prehistoric hunter-gatherers: The emergence of cultural complexity. Studies in archaeology.* Orlando: Academic.

Putnam, R. D. 2000. *Bowling alone: The collapse and revival of American community.* New York: Simon and Schuster.

Putnam, R. D., R. Leonardi, & R. Nanetti. 1993. *Making democracy work: Civic traditions in modern Italy.* Princeton, NJ: Princeton University Press.

Rice, W. R. 1996. Sexually antagonistic male adaptation triggered by experimental arrest of female evolution. *Nature* 381:232–234.

Richerson, P. J., & R. Boyd. 1989. The role of evolved predispositions in cultural evolution: Or sociobiology meets Pascal's Wager. *Ethology and Sociobiology* 10: 195–219.

Richerson, P. J., & R. Boyd. 1998. The evolution of human ultrasociality. In: *Indoctrinability, ideology, and warfare: Evolutionary perspectives*, I. Eibl-Eibesfeldt, & F. K. Salter, eds. (pp. 71–95). New York: Berghahn.

Richerson, P. J., & R. Boyd. 1999. Complex societies: The evolutionary origins of a crude superorganism. *Human Nature* 10:253–289.

Richerson, P. J., & R. Boyd. 2000a. Climate, culture, and the evolution of cognition. In: *The evolution of cognition*, C. Heyes, & L. Huber, eds. (pp. 329–346). Cambridge, MA: MIT Press.

Richerson, P. J., & R. Boyd. 2000b. The Pleistocene climate variation and the origin of human culture: Built for speed. In: *Evolution, culture, and behavior*, F. Tonneau, & N. S. Thompson, eds. (pp. 1–45). *Perspectives in Ethology* 13. New York: Kluwer Academic/Plenum.

Richerson, P. J., & R. Boyd. 2001. The evolution of subjective commitment to groups: A tribal instincts hypothesis. In: *Evolution and the capacity for commitment*, R. M. Nesse, ed. (pp. 186–220). New York: Russell Sage.

Richerson, P. J., R. Boyd, & R. L. Bettinger. 2001. Was agriculture impossible during the Pleistocene but mandatory during the Holocene? A climate change hypothesis. *American Antiquity* 66:387–411.

Richerson, P. J., R. Boyd, & B. Paciotti. 2002. An evolutionary theory of commons management. In: *The drama of the commons*, E. Ostrom, T. Dietz, N. Dolsak, et al., eds. (pp. 403–442). Washington, DC: National Academy Press.

Rumsey, A. 1999. Social segmentation, voting, and violence in Papua New Guinea. *Contemporary Pacific* 11:305–333.

Runciman, W. G. 1998. Greek hoplites, warrior culture, and indirect bias. *Journal of the Royal Anthropological Institute* 4:731–751.

Salamon, S. 1992. *Prairie patrimony: Family, farming, and community in the midwest*. Chapel Hill: University of North Carolina Press.

Salter, F. K. 1995. *Emotions in command: A naturalistic study of institutional dominance*. Oxford: Oxford University Press.

Savage-Rumbaugh, E. S., & R. Lewin. 1994. *Kanzi: The ape at the brink of the human mind*. New York: Wiley.

Sears, D. O., & C. L. Funk. 1990. Self-interest in Americans' political opinions. In: *Beyond self-interest*, J. Mansbridge, ed. (pp. 147–170). Chicago: University of Chicago Press.

Segall, M., D. Campbell, & M. J. Herskovits. 1966. *The influence of culture on visual perception*. New York: Bobbs-Merrill.

Simon, H. A. 1990. A mechanism for social selection and successful altruism. *Science* 250:1665–1668.

Slatkin, M., & M. J. Wade. 1978. Group selection on a quantitative character. *Proceedings of the Natural Academy of Sciences (USA)* 75:3531–3534.

Smith, E. A. 2003. Human cooperation: Perspectives from behavioral ecology. In: *Genetic and Cultural Evolution of Cooperation*, P. Hammerstein, ed. Cambridge: MIT Press, pp. 401–427.

Sober, E., & D. S. Wilson. 1998. *Unto others: The evolution and psychology of unselfish behavior*. Cambridge, MA: Harvard University Press.

Soltis, J., R. Boyd, & P. J. Richerson. 1995. Can group-functional behaviors evolve by cultural group selection: An empirical test. *Current Anthropology* 36:473–494.

Squires, A. M. 1986. *The tender ship: Governmental management of technological change.* Boston: Birkhäuser.

Srinivas, M. N. 1962. *Caste in modern India and other essays.* Bombay: Asia Publ. House.

Steward, J. H. 1955. *Theory of culture change: The methodology of multilinear evolution.* Urbana: University of Illinois Press.

Thomas, D. H., L. S. A. Pendleton, & S. C. Cappannari. 1986. Western Shoshone. In: *Handbook of North American Indians: Great Basin,* W. L. d'Azevedo, ed. (pp. 262–283). Washington, DC: Smithsonian Institution Press.

Tomasello, M. 1999. *The cultural origins of human cognition.* Cambridge, MA: Harvard University Press.

Trivers, R. L. 1971. The evolution of reciprocal altruism. *Quarterly Review of Biology* 46:35–57.

Wardhaugh, R. 1992. *An introduction to sociolinguistics.* 2nd ed. Oxford: Blackwell.

West, R. 1941. *Black lamb and crey falcon.* New York: Penguin.

Whiten, A., & R. W. Byrne. 1988. *Machiavellian intelligence: Social expertise and the evolution of intellect in monkeys, apes, and humans.* Oxford: Oxford University Press.

Wiessner, P. 1984. Reconsidering the behavioral basis for style: A case study among the Kalahari San. *Journal of Anthropological Archaeology* 3:190–234.

Williams, G. C. 1966. *Adaptation and natural selection: A critique of some current evolutionary thought.* Princeton, NJ: Princeton University Press.

Wilson, D. S. 1983. The group selection controversy: History and current status. *Annual Review of Ecology and Systematics* 14:159–188.

Wilson, E. O. 1975. *Sociobiology: The new synthesis.* Cambridge, MA: Belknap.

Wilson, E. O. 1998. *Consilience: The unity of knowledge.* New York: Knopf.

Wolfe, T. 1965. *The kandy-kolored tangerine-flake streamline baby.* New York: Farrar, Straus and Giroux.

Young, H. P. 1998. *Individual strategy and social structure: An evolutionary theory of institutions.* Princeton, NJ: Princeton University Press.

PART 4

Archaeology and Culture History

Historians and scientists do not always get along very well. Many historians view science as a procrustean enterprise whose practitioners insist on shoehorning complex historical phenomena into overly simple general laws. For their part, scientists often think that historians exaggerate the complexity and contingent nature of historical events, willfully refusing to see the order that underlies chaos of one thing after another. This debate is echoed in evolutionary biology where Steven J. Gould famously upheld a historicist version of organic evolution, a habit that made many mainstream evolutionary biologists hopping mad.

In our view, these debates are rooted in a mistaken view of evolutionary theory. Surely historical contingency plays a role in every sort of evolution from the cosmic to the cultural. The Big Bang was a singular event. So was the evolution of our unique species (and every other unique species, for that matter). However, evolutionary scientists do not try to jam this complexity into the straitjacket of general laws like those in physics. Instead, they aim to develop a toolkit of models and a collection of related empirical generalizations. The phenomena of evolution are not only complex but also diverse. No model and no empirical generalization is guaranteed to hold from one case to the next. Yet the lesson of biology is that this piecemeal approach to theory can yield deep insights. In chapter 19, we review the case for using a toolkit of simple models to explain complex and diverse phenomena like cultural evolution. Here we consider the role of theory-as-tools in understanding phenomena in which historical contingency plays a large, if not dominant, role.

Chapter 15 discusses why evolutionary processes give rise to history—meaning patterns of change with time in which the same initial conditions result in divergent evolutionary trajectories or in which change is nonstationary. The very simplest evolutionary models of adaptation by natural selection give rise to trajectories that converge on unique equilibria from divergent initial conditions. Add simple noise or simple oscillatory mechanisms in key processes and the change will never cease. But it is stationary and thus will remain "lawful." However, real evolutionary trajectories *do* diverge from identical starting points and do result in patterns whose statistical properties are not stationary, and this fact limits the predictive power of evolutionary theory. The "laws" of nature are, in effect, ever-changing. In this chapter, we suggest a number of means by which rather straightforward adaptive processes can result in divergent, nonstationary change. If the argument is correct, the scientists' tools should prove quite useful to historians even if what we provide is not laws. Just demonstrating how divergence and nonstationarity themselves arise shows how the scientific approach can illuminate historical questions at the most fundamental level.

In chapter 16 we consider the problem of constructing cultural phylogenies. Phylogenies are useful, among other things, for controlling for the effects of common history in scientific studies of organic and cultural evolution. In recent years, evolutionary biologists have made great technical strides in the science of phylogeny reconstruction, and these advances have promise for application to cultures. The difficulty is that cultures do not have the simple branching histories that characterize most biological species—cross-cultural diffusion occurs in every domain of culture. Whether this fact causes important problems for phylogenetic reconstruction is an open question. Historical linguists have long struggled with this problem with some success. Language trees are a much used starting point for cultural phylogeny reconstruction, despite their obvious limitations. In places like aboriginal western North America, groups with unrelated languages often have very similar subsistence systems and even similar political and social organization. Even the most conservative features of language change rapidly so that most historical linguists believe that phylogenetic reconstruction is possible only for the last few thousand years. Another approach is to consider the phylogeny of single traits or small, tightly knit clusters of traits rather than of cultures as a whole. However, such items may contain too little historical information for accurate reconstruction. Future methodological innovations may solve many of these problems. In the meantime, the difficulty of cultural phylogeny reconstruction illustrates an important point. Humans are one species; our genes and our culture tend to diffuse very widely. Local populations are seldom if ever isolated for any substantial length of time. Ideologues often want to use the concept of culture like the concept of race, imagining that their culture has a "pure" history. In fact, all cultures have tangled, messy histories, even messier than our genes, if that is possible.

Chapter 17 deals with a specific historical problem, the origins of agriculture. This phenomenon is typical of a number of problems in human evolution in that it is a particular nonstationary pattern: it is "progressive." Human technology and probably human social complexity have increased more or less steadily, if at greatly different rates, seemingly since our lineage branched from that of the other apes. The progressive pattern is especially marked during the last 250,000 years or so (McBrearty and Brooks, 2000). Many scholars are not puzzled by such patterns. To them, the obvious explanation is that evolution is the process of replacing antique, less adaptive traits with modern, more adaptive ones. The problem is that selective processes usually reach equilibria too rapidly to generate long-run progress on geological timescales. Evolution can produce steady progress only if the processes internal to the evolutionary process slow it down or if the pace of evolution is set by external environmental factors. The origin and spread of agriculture provides an interesting test case because it is among the most important events in human history, serving, as it still does, as the subsistence basis for the evolution of even more complex societies in the last few thousand years. Recently, the most popular explanations have been based on population pressure, the idea that humans turned to agriculture when population densities rose to the point that less intensive hunting and gathering techniques began to favor investment in agricultural production. In this chapter, we argue that population pressure acts at far too short a timescale to explain agricultural origins. As Malthus noted, population pressure builds appreciably on the generational timescale; if it paced cultural evolution, events would transpire at a much faster pace than archaeologists normally observe. Climate change is a better candidate to explain why agriculture first began appearing about 11,600 years ago. Recent advances in paleoclimatology have shown that last-glacial climates were exceedingly variable compared to the period since 11,600 years ago. Climates in the last glacial age were also mainly drier than modern ones and lower CO_2 may also have handicapped plant production. Agricultural subsistence is difficult in modern climates and takes several thousand years to evolve. Perhaps agriculture was impossible in the Pleistocene epoch.

Our main objective in this section is not to push particular answers to particular historical, archaeological, and paleoanthropological problems (Richerson and Boyd, 2001). Rather, we want to advertise to those who study historical problems that cultural evolutionary theory has tools that students of these phenomena need in their repertoire. Even when we cannot say much about how evolution works, we can often use a combination of theory and empiricism to estimate the rates of change characteristic of different processes. Quite elementary considerations can sometimes rule some processes in and some out as candidate explanations for a given event. Just as astronomers need the theory of nuclear physics to understand how stars evolve, so historians, archaeologists, and paleoanthropologists need the theory of cultural evolution to understand human evolutionary history.

REFERENCES

McBrearty, S., & A. S. Brooks. 2000. The revolution that wasn't: A new interpretation of the origin of modern human behavior. *Journal of Human Evolution* 39(5):453–563.
Richerson, P. J., & R. Boyd. 2001. Institutional evolution in the holocene: The rise of complex societies. In: *The origin of human social institutions*, W. G. Runciman, ed. (pp. 197–234). Oxford: Oxford University Press.

15 How Microevolutionary
Processes Give Rise
to History

Over the last decade a number of authors, including ourselves, have attempted to understand human cultural variation using Darwinian methods. This work is unified by the idea that culture is a system of inheritance: individuals vary in their skills, habits, beliefs, values, and attitudes, and these variations are transmitted to others through time by teaching, imitation, and other forms of social learning. To understand cultural change, we must account for the microevolutionary processes that increase the numbers of some cultural variants and reduce the numbers of others.

Social scientists have made a number of objections to this approach to understanding cultural change. Among these is the idea that culture can only be explained historically. Because the history of any given human society is a sequence of unique and contingent events, explanations of human social life, it is argued, are necessarily interpretive and particularistic. Present phenomena are best explained mainly in terms of past contingencies, not ahistorical adaptive processes that would erase the trace of history. Like other scientific (rather than historical) explanations of human cultures, the argument goes, Darwinian models cannot account for the lack of correlation of environmental and cultural variation, nor the long-term trends in cultural change.

In this chapter, we defend the Darwinian theories of cultural change against this objection by suggesting that several cultural evolutionary processes can give rise to divergent evolutionary developments, secular trends, and other features that can generate unique historical sequences for particular societies. We also argue that Darwinian theory offers useful tools for those interested in understanding the evolution of particular societies. Essentially similar processes act in the case of organic evolution. Darwinian theory is both scientific and historical.

The history of any evolving lineage or culture is a sequence of unique, contingent events. Similar environments often give rise to different evolutionary trajectories, even among initially similar taxa or societies, and some show very long-run trends in features such as size. Nonetheless, these historical features of organic and cultural evolution can result from a few simple microevolutionary processes.

A proper understanding of the relationship between the historical and the scientific is important for progress in the social and biological sciences. There is (or ought to be) an intimate interplay between the study of the unique events of given historical sequences and the generalizations about process constructed by studying many cases in a comparative and synthetic framework. The study of unique cases furnishes the data from which generalizations are derived, while the generalizations allow us to understand better the processes that operated on particular historical trajectories. We cannot neglect the close, critical study of particular cases without putting the database for generalization in jeopardy. Besides, we often have legitimate reasons to be curious about exactly how particular historical sequences, such as the evolution of *Homo sapiens*, occurred. On the other hand, it is from the study of many cases that we form a body of theory about evolutionary processes. No one historical trajectory contains enough information to obtain a very good grasp of the processes that affected its own evolution. Data are missing because the record is imperfect. The lineage may be extinct, and so direct observation is impossible. Even if the lineage is extant, experimentation may be impossible for practical or ethical reasons. Potential causal variables may be correlated in particular cases, so understanding their behavior may be impossible. The comparative method can often clarify such cases. "Scientists" need "historians" and vice versa.

Darwinian Models of Cultural Evolution

Over the past two decades, a number of scholars have attempted to understand the processes of cultural evolution in Darwinian terms. Social scientists (Campbell, 1965, 1975; Cloak, 1975; Durham, 1976; Ruyle, 1973) have argued that the analogy between genetic and cultural transmission is the best basis for a general theory of culture. Several biologists have considered how culturally transmitted behavior fits into the framework of neo-Darwinism (Pulliam and Dunford, 1980; Lumsden and Wilson, 1981; Boyd and Richerson, 1985; Richerson and Boyd, 1989b; Cavalli-Sforza and Feldman, 1983; Rogers, 1989). Other biologists and psychologists have used the formal similarities between genetic and cultural transmission to develop theories describing the dynamics of cultural transmission (Cavalli-Sforza and Feldman, 1973, 1981; Cloninger, Rice, and Reich, 1979; Eaves, Last, Young, and Martin, 1978). All of these authors are interested in a synthetic theory of process applying to how culture works in all cultures, including in other species that might have systems with a useful similarity to human culture. Note that this last broadly comparative concern is likely to be useful in dissecting the reasons why the human lineage originally became more cultural than typical mammals.[1]

The idea that unifies the Darwinian approach is that culture constitutes a system of inheritance. People acquire skills, beliefs, attitudes, and values from others by imitation and enculturation (social learning), and these "cultural variants," together with their genotypes and environments, determine their behavior. Since determinants of behavior are communicated from one person to another, individuals sample from and contribute to a collective pool of ideas that changes over time. In other words, cultures have similar population-level properties as gene pools, as different as the two systems of inheritance are in the details of how they work. (In one respect, the Darwinian study of cultural evolution is more Darwinian than the modern theory of organic evolution. Darwin not only used a notion of "inherited habits" that is much like the modern concept of culture but also thought that organic evolution generally included the property of the inheritance of acquired variation, which culture does and genes do not.)

Because cultural change is a population process, it can be studied using Darwinian methods. To understand why people behave as they do in a particular environment, we must know the nature of the skills, beliefs, attitudes, and values that they have acquired from others by cultural inheritance. To do this, we must account for the processes that affect cultural variation as individuals acquire cultural traits, use the acquired information to guide behavior, and act as models for others. What processes increase or decrease the proportion of people in a society who hold particular ideas about how to behave? We thus seek to understand the cultural analogs of the forces of natural selection, mutation, and drift that drive genetic evolution. We divide these forces into three classes: random forces, natural selection, and the decision-making forces.

Random forces are the cultural analogs of mutation and drift in genetic transmission. Intuitively, it seems likely that random errors, individual idiosyncrasies, and chance transmission play a role in behavior and social learning. For example, linguists have documented a good deal of individual variation in speech, some of which is probably random individual variation (Labov, 1972). Similarly, small human populations might well lose rare skills or knowledge by chance, for example, due to the premature deaths of the only individuals who acquired them (Diamond, 1978).

Natural selection may operate directly on cultural variation. Selection is an extremely general evolutionary process (Campbell, 1965). Darwin formulated a clear statement of natural selection without a correct understanding of *genetic* inheritance because it is a force that will operate on *any* system of inheritance with a few key properties. There must be heritable variation, the variants must affect phenotype, and the phenotypic differences must affect individuals' chances of transmitting the variants they carry. That variants are transmitted by imitation rather than sexual or asexual reproduction does not affect the basic argument, nor does the possibility that the source of variation is not random. Darwin imagined that random variation, acquired variation, and natural selection all acted together as forces in organic evolution. In the case of cultural evolution, this seems to be the case. It may well be, however, that behavioral variants favored by natural selection depend on the mode of transmission. The behaviors that maximize numbers of offspring may not be the same as those that maximize cultural influence on future generations (Boyd and Richerson, 1985).

Decision-making forces result when individuals evaluate alternative behavioral variants and preferentially adopt some variants relative to others. If many of the individuals in a population make similar decisions about variants, especially if similar decisions are made for a number of generations, the pool of cultural variants can be transformed. Naive individuals may be exposed to a variety of models and preferentially imitate some rather than others. We call this force biased transmission. Alternatively, individuals may modify existing behaviors or invent new ones by individual learning. If the modified behavior is then transmitted, the resulting force is much like the guided, nonrandom variation of "Lamarckian" evolution. Put differently, humans are embedded in a complex social network through which they actively participate in the creation and perpetuation of their culture.

The decision-making forces are derived forces (Campbell, 1965). Decisions require rules for making them, and ultimately the rules must derive from the action of other forces. That is, if individual decisions are not to be random, there must be some sense of psychological reward or similar process that causes individual decisions to be predictable, in given environments, at least. These decision-making rules may be acquired during an earlier episode of cultural transmission, or they may be genetically transmitted traits that control the neurological machinery for acquisition and retention of cultural traits. The latter possibility is the basis of the sociobiological hypotheses about cultural evolution (Alexander, 1979; Lumsden and Wilson, 1981). Some authors argue that the course of cultural evolution is determined by natural selection operating indirectly on cultural variation through the decision-making forces.

Like natural selection, the decision-making forces may improve the fit of the population to the environment. The criteria of fit depend on the nature of the underlying decision rules. This is easiest to see when the goals of the decision rules are closely correlated with fitness. If human foraging practices are adopted or rejected according to their energy payoff per unit time (optimal foraging theory's operational proxy for fitness), then the foraging practices used in the population will adapt to changing environments much as if natural selection were responsible. If the adoption of foraging practices is strongly affected by consideration of prestige, say, that associated with male success hunting dangerous prey, then the resulting pattern of behavior may be different. However, there will still be a pattern of adaptation to different environments but now in the sense of increasing prestige rather than calories.

What Makes Change Historical?

It has often been argued that historical scientific explanations are different in kind. Ingold (1985) gives two important versions of this argument. Some authors (e.g., Collingwood, 1946) argue that history is uniquely human because it entails conscious perception of the past. The second view (e.g., Trigger, 1978) is quite different and holds that history involves unique, contingent pathways from the past to the future that are strongly influenced by unpredictable, chance events. We focus on the latter view here. For example, capitalism arose in Europe rather than China, perhaps because medieval and early modern statesmen failed to create a unified empire in the West (McNeill, 1980), and marsupials dominate the Australian fauna perhaps because of Australia's isolation from other continents in which placental mammals chanced to arise. In contrast, it is argued, scientific explanations involve universally applicable laws. In evolutionary biology and in anthropology, these often take the form of functional explanations, in which only knowledge of present circumstances and general physical laws (e.g., the principles of mechanics) are necessary to explain present behavior (Mitchell and Valone, 1990). For example, long fallow horticulture is associated with tropical forest environments, perhaps because it is the most efficient subsistence technology in such environments (Conklin, 1969).

It has been argued, perhaps nearly as often, that this dichotomy is false. Eldredge (1989:9) provides a particularly clear and forceful example of a common objection: all material entities have properties that can change through time. Even simple entities like molecules are characterized by position, momentum, charge, and so on. If we could follow a particular water molecule, we would see that these properties changed through time—even the water molecule has a history, according to Eldredge. Yet everyone agrees that we can achieve a satisfactory scientific theory of water. Historical explanations, Eldredge argues, are just scientific explanations applied to systems that change through time. We are misled because chemists tend to study the average properties of very large numbers of water molecules.

This argument explains too much. Not all change with time is history in the sense intended by historically oriented biologists and social scientists. To see this, consider an electrical circuit composed of a voltage source, a capacitor, and a fluorescent light. Under the right conditions, the voltage will oscillate through time, and these changes can be described by simple laws. Are these oscillations historical? In Eldredge's view they are; the circuit has a history, a quite boring one, but a history nonetheless. Yet such a system does not generate unique and contingent trajectories. After the system settles down, one oscillation is just like the previous one, and the period and amplitude of the oscillations are not contingent on initial conditions. They are not historical in the sense that "one damn thing after another" (Elton 1967:40) leads to cumulative, but unpredictable change.

What then makes change historical? We think that two requirements capture much of what is meant by "history." These two requirements pose a more interesting and serious challenge for reconciling history with a scientific approach to explanation. Consider a system like a society or a population that changes through time both under the influence of internal dynamics and exogenous shocks. Then we suggest that the pattern of change is historical if the following statements apply.

1. *Trajectories are not stationary on the time scales of interest.* History is more than just change—it is change that does not repeat itself. On long enough timescales, the oscillations in the circuit become stationary, meaning that the chance of finding the system in any particular state becomes constant. Similarly,

random day-to-day fluctuations in the weather do not constitute historical change if one is interested in organic evolution because, on long evolutionary timescales there will be so many days of rain, so many days of sun, and so on. By choosing a suitably long period of time, we can construct a scientific theory of stationary processes using a statistical rather than strictly deterministic approach. In the case of nonstationary historical trajectories, a society or biotic lineage tends to become gradually more and more different as time goes by. There is no possibility of basing explanation on, say, a long-run mean about which the historical entity fluctuates in some at least statistically predictable way, because the mean calculated over longer and longer runs of data continues to change significantly. One of the most characteristic statistical signatures of nonstationary processes is that the variance they produce grows with time rather than converging on a finite value. Note that a process that is historical in one spatio-temporal frame *may* not be in another. If we are not too interested in a specific species or societies in given time periods, we can often average over longer periods of time or many historical units to extract ahistorical generalizations. Any given water molecule has a history, but it is easy to average over many of them and ignore this fact.

2. *Similar initial conditions give rise to qualitatively different trajectories.* Historical change is strongly influenced by happenstance. This requires that the dynamics of the system must be path-dependent; isolated populations or societies must tend to diverge even when they start from the same initial condition and evolve in similar environments. Thus, for example, the spread of a favored allele in a series of large populations is not historical. Once the allele becomes sufficiently common, it will increase at first exponentially and then slowly, asymptotically approaching fixation. Small changes in the initial frequencies, population size, or even degree of dominance will not lead to qualitative changes in this pattern. In separate but similar environments, populations will converge on the favored allele. Examples of convergence in similar environments are common—witness the general similarity in tropical forest trees and many of the behaviors of the long fallow cultivators who live among them the world over. On the other hand, there are also striking failures of convergence—witness the many unique features of Australian plants, animals, and human cultures. The peculiar hanging leaves of eucalypts, the bipedal gait of kangaroos, and the gerontocratic structure of Australian aboriginal societies make them distinctively different from the inhabitants of similar temperate and subtropical dry environments on other continents.

It is important not to blur the distinction between simple trajectories and true historical change; it is easy to see how evolutionary processes like natural selection give rise to simple, regular change like the spread of a favored allele or subsistence practice. However, it is not so easy to see how such processes give rise to unique, contingent pathways. Scientists tout the approach to steady states and convergence in similar situations as evidence for the operation of natural "laws," so it seems natural to conclude that a lack of stationarity and convergence are evidence of processes that cannot be subsumed in the standard conceptions of science. Our argument is that things are not at all that simple. There is every reason to expect that perfectly ordinary scientific processes, ordinary in the sense that they result from natural causes and are easily understood by conventional methods, regularly generate history in the sense defined by these two criteria.

How Do Adaptive Processes Give Rise to History?

Let us begin with the two most straightforward answers to this question. First, it could be that most evolutionary change is random. Much change in organic evolution may be the result of drift and mutation, and much change in cultural evolution may result from analogous processes. The fact that drift is a very slow process would explain long-term evolutionary trends. Raup (1977) and others argue that random-walk models produce phylogenies that are remarkably similar to real ones. The fact that cultural and genetic evolutionary change is random would allow populations in similar environments to diverge from each other. It seems likely that some variation in both cases evolves mainly under the influence of nonadaptive forces—for example, much of the eukaryotic genome does not seem to be expressed and evolves under the influence of drift and mutation (Futuyma, 1986:447). Similarly, the arbitrary character of symbolic variation suggests that nonadaptive processes are likely to be important in linguistic change and similar aspects of culture. In both cases, isolated populations diverge at an approximately constant rate on the average. However, to understand why a particular species is characterized by a particular DNA sequence, or why a particular people use a particular word for mother, one must investigate the sequence of historical events that led to the current state.

It is also possible that historical change is generated by abiotic environmental factors (Valentine and Moores, 1972). Long-term trends in evolution could result from the accurate adaptive tracking of a slowly changing environment. For example, during the last hundred million years, there has been a long-term increase in the degree of armoring of many marine invertebrates living on rocky substrates and a parallel increase in the size and strength of feeding organs among their predators (Vermeij, 1987; Jackson, 1988). It is possible that these biotic trends have been caused by long-run environmental changes over the same period—for example, an increase in the oxygen content of the atmosphere (Holland, 1984). Similarly, beginning perhaps as much as 17,000 years ago, humans began a shift from migratory big game hunting to sedentary, broad-spectrum, more labor-intensive foraging, finally developing agriculture about 7,000 years ago (Henry, 1989). Many authors (e.g., Reed, 1977) have argued that the transition from glacial to interglacial climate that occurred during the same period is somehow responsible for this change. Similarly, differences among populations in similar environments may result from the environments actually being different in some subtle but important way. For example, Westoby (1989) has argued that some of the unusual features of the Australian biota result from the continent-wide predominance of highly weathered, impoverished soils on this relatively undisturbed continental platform. Perhaps the failure of agriculture to develop in or diffuse to aboriginal Australia, despite many favorable preconditions and the presence of cultivators just across the Torres Strait, also reflects poor soils.

It is more difficult to understand how adaptive processes like natural selection can give rise to historical trajectories. There are two hurdles: first, adaptive processes in both organic and cultural evolution appear to work on rather short timescales compared to the timescales of change known from the fossil record, archaeology, and history. Theory, observation, and experiment suggest that natural selection can lead to change that is much more rapid than any observed in the fossil record (Levinton, 1988:342–347). For example, the African Great Lakes have been the locus of spectacular adaptive radiations of fishes amounting to hundreds of highly divergent forms from a few ancestors in the larger lakes (Lowe-McConnell, 1975). The maximum timescales for these radiations, set by the ages of the lakes and not counting that they may have dried up during the Pleistocene epoch, are only a few million years. The radiation in Lake Victoria (about 200 endemic species) seems to have required only a few hundred thousand years. Adaptive cultural change driven by decision-making forces can be very fast indeed, as is evidenced by the spread of innovations (Rogers, 1983). It is not immediately clear how very short timescale processes such as these can give rise to longer-term change of the kind observed in both fossil and archaeological records, unless the pace of change is regulated by environmental change. In the absence of continuing, long-term, nonstationary environmental change, adaptive processes seem quite capable of reaching equilibria in relatively short order. In other words, both cultural and organic evolution seem, at first glance, to be classic scientific processes that produce functional adjustments too rapidly to account for the slow historical trajectories we actually observe.

Second, it is not obvious why adaptive processes should be sensitive to initial conditions. Within anthropology the view that adaptive processes are ahistorical in this sense underpins many critiques of functionalism. Many anthropologists claim that it is self-evident that cultural evolution is historical and that, therefore, adaptive explanations (being intrinsically equilibrist and ahistorical) must be wrong. For example, Hallpike (1986) presents a variety of data that show that peoples living in similar environments often have quite different social organization, and historically related cultures often retain similar social organizations despite occupying radically different environments. Because functionalist models predict a one-to-one relationship between environment and social organization, he argues, these data falsify the functionalist view. Indeed, functionalists like Cohen (1974:86) expect to see history manifest only in the case of functionally equivalent symbolic forms. Biologists have generally been more aware that a population's response to selection depends on phylogenetic and developmental constraints and, therefore, that evolutionary trajectories are, at least to a degree, path-dependent. Nonetheless, lack of convergence is sometimes used to argue the lack of importance of natural selection. Should selection not cause populations exposed to similar environments to converge on similar adaptations? Certainly, some striking convergences from unlikely ancestors do exist.

Here we argue that path dependence and long-term change are likely to be consequences of any adaptive process analogous to natural selection. Our claims are rather general and are thus independent of the nature of the transmission process (genetic or cultural) and of the details of development. Let us begin with an especially simple example of genetic evolution. Consider a large population of organisms in which individuals' phenotypes can be represented as a number of quantitative characters. Let us assume that there are no constraints on what can evolve due to properties of the genetic system itself. One model with this property assumes that the distribution of additive genetic values[2] for each character is Gaussian, that there are no genetic correlations among characters, that no genotype-environment interactions exist, and that mutation maintains a constant amount of heritable variation for each character. Further, assume that the fitness of each individual depends only on its own phenotype, not on the frequency of other phenotypes or the population density, and there is no environmental change. With these assumptions, it can be shown that the change in the vector of mean values for each character is along the gradient of the logarithm of average fitness (Lande, 1979). In other words, the mean phenotype in the population changes in the direction that maximizes the increase in the average fitness of the population. This is the sort of situation in which selection, and similar processes in the cultural system, ought to produce optimal adaptations in the straightforward manner depicted in elementary textbooks.

In this simple model the evolutionary trajectory of the population will be completely governed by the shape of average fitness as a function of mean phenotype. If the adaptive topography has a unique maximum, then every population will evolve to the same equilibrium mean phenotype, independent of its starting position, and, once there, be maintained by stabilizing selection. On the other hand, if there is more than one local maximum, different equilibrium outcomes are possible depending on initial condition. The larger the number of local maxima, the more path-dependent the resulting trajectories will be (see figure 15.1).

Unfortunately, we do not know what real adaptive topographies look like, and, as Lande (1986) has pointed out, there is little chance that we will be able to determine their shape empirically. In evolutionary texts, adaptive topographies are commonly depicted as a smooth three-dimensional surface with a small number of local maxima. However, if evolutionary "design problems" are similar to the engineering ones, this picture is misleading. Experience with engineering design problems suggests that real adaptive topographies are often extremely complex, with long ridges, multiple saddle points, and many local optima—more akin to the topographic map of a real mountain range than the smooth textbook surfaces.

A computer design problem discussed by Kirkpatrick, Gelatt, and Vecchi (1983) provides an excellent example. Computers are constructed from large numbers of interconnected circuits, each with some logical function. Because the size of chips is limited, circuits must be divided among different chips. Because signals between chips travel more slowly and require more power than signals within chips, designers want to apportion circuits among chips so as to minimize the number of connections between them. For even moderate numbers of circuits, there is an astronomical number of solutions to this problem. Kirkpatrick et al. present an example in which the 5,000 circuits that make up the IBM 370 microprocessor were to be divided between two chips. Here there are about 10^{1503} possible solutions! This design problem has two important qualitative properties:

a. Contours of log $\overline{W}$

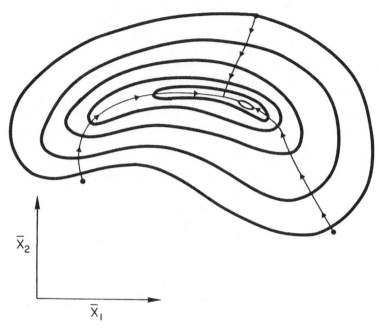

Figure 15.1. This figure shows two adaptive topographies. The axes are the mean genetic value in a population for two characters. The contour lines give contours of equal mean fitness. Populations beginning at different initial states all achieve the same equilibrium state. Figure 15.1*a* shows a simple unimodal adaptive topography. Figure 15.1*b* shows a complex, multimodal topography. Initially similar populations diverge owing only to the influence of selection.

1. *It has a very large number of local optima*. That is, there is a large number of arrangements of circuits with the property that any simple rearrangement increases the number of connections between chips. This means that any search process that simply goes uphill (like our model of genetic evolution) can end up at any one of a very large number of configurations. An unsophisticated optimizing scheme will improve the design only until it reaches one of the many local optima, which one depending upon starting conditions. For example, for the 370 design problem, several runs of a simple hill-climbing algorithm produced between 677 and 730 interconnections. The best design found (using a more sophisticated algorithm) required only 183 connections.

2. *There is a smaller, although still substantial, number of arrangements with close to the optimal number of interconnections*. That is, there are many qualitatively different designs that have close to the best payoff. In the numerical example there are on the order of sqrt(5,000) ≈ 70 such arrangements.

b.

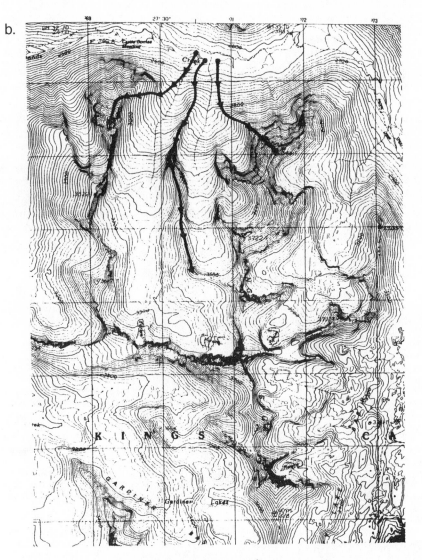

Figure 15.1 (*continued*)

Kirkpatrick et al. (1983) show that two other computer design problems, the arrangement of chips on circuit boards and the routing of wiring among chips, have similar properties. These three computer design problems are not unlike evolutionary "design" problems in biology—the localization of functions in organs, the arrangement of organs in a body, and the routing of the nervous and circulatory networks—that are likely to generate complex adaptive topographies. Moreover, as anyone experienced with the numerical solution of real-world optimization problems will testify, these results are quite typical. To quote from the introduction of a recent textbook on optimization, "many common design problems, from reservoirs to refrigerators, have multiple local optima, as well as false optima, that make conventional [meaning iterative hill-climbing] optimization schemes risky" (Wilde, 1978). Thus, if the analogy is correct, small differences in initial conditions may launch different populations on different evolutionary trajectories, which end with qualitatively different equilibrium phenotypes.

It is important to see that this history-generating property does not depend on the existence of genetic or developmental constraints. At least as defined in Maynard Smith et al. (1985) there are no genetic or developmental constraints in the simple model of selection acting on a complex topography. Every combination of phenotypes can be achieved, and there is no bias in the production of genetic variation. Path dependence results from the facts that different characters interact in a complex way to generate fitness and that the direction of natural selection depends on the shape of the *local* topography.

Of course, developmental constraints could also play a major role in confining lineages to historically determined bauplans, as many biologists have argued (e.g., Seilacher, 1970). Further, complex topographies and developmental constraints may be related. Wagner (1988) hypothesizes, based on a model of multivariate phenotypic evolution, that fitness functions will generally be "malignant" and that developmental constraints act to make phenotypes more responsive to selection. By malignant, Wagner means that the fitness of any one trait is likely to depend on the values of many other traits. For example, larger size may be favored by selection for success in contests for mates but only if many traits of the respiratory, skeletal, and circulatory systems change in concert to support larger size. If phenotype is unconstrained, response to selection will be slow because of the need to change so many independent characters at once, whereas developmental constraints confine the expression of variation to a few axes that can respond rapidly to selection. Thus, the bill is a simple, rather constrained part of the anatomy of birds, yet selection has remodeled bills along the relatively few dimensions available (length, width, depth, curvature) to support an amazing variety of specializations. Developmental constraints may be a solution to the complexity of adaptive topographies, albeit one that limits lineages to elaborating a small set of historically derived basic traits as they respond to new adaptive challenges.

Path dependence can arise from the action of functional processes in a cultural system of inheritance as well. For example, decision-making forces arise when people modify culturally acquired beliefs in the attempt to satisfy some goal. If people within a culture share the same goal, this process will produce an evolutionary trajectory very similar to one produced by natural selection—the rate of change of the distribution of beliefs in a population will depend on the amount of cultural variation and the shape of an analog of the adaptive topography in which fitness is replaced by utility (the extent to which alternative beliefs satisfy the goal) (Boyd and Richerson, 1985, ch. 5). The details of the transmission and selective processes are not crucial, as long as the processes that lead to change can be represented as climbing a complex topography.

It is unclear whether adding genetic constraints will increase or decrease the potential for path dependence. One sort of genetic constraint can be added by allowing significant genetic correlations among characters (Lande, 1986). This assumption means that some mutants are more probable than others. As long as there is some genetic variation in each dimension, the vector of phenotypic means will still go uphill but not necessarily in the steepest direction. The population will come to equilibrium at one of the local peaks, although this might be quite distant from the equilibrium that the population would have reached had there been no genetic correlations (Lande, 1979, 1986). More generally, most genetic architectures do not result in Gaussian distributions of genetic values (Turelli and Barton, 1990), and analyses of two locus models suggest that dynamics resulting from the combination of linkage and selection may create many locally stable equilibria even when the fitness function is unimodal (Karlin and Feldman, 1970). This suggests that adding more genetic realism would increase the potential for path dependence. On the other hand, computer scientists (Holland, 1975; Brady, 1985) have found that optimization algorithms closely modeled on multilocus selection are *less* likely to get stuck on local optima than simple iterative hill-climbing algorithms. The issue of genetic constraints is still open.

The situation in cultural evolution is similar, even if not so well studied. On the one hand, many anthropologists stress the rich structure of culture. To the extent that such structure exists, path dependence is likely to be important. On the other hand, Bandura (1977), a pioneering student of the processes of social learning, argues that there is relatively little complex structuring of socially learned behavior. The many examples of cultural syncretism and diffusion of isolated elements of technology suggest his view ought to be taken seriously. Perhaps complex structure is most important in the symbolic aspects of culture, but symbolic variation may be only weakly constrained by functional considerations (Cohen, 1974). According to Cohen, we have to use purely contingent historical explanations for things such as linguistic variation, while simple functional explanations suffice for economic, political, and social-organizational phenomena.

Long-term nonstationary trends in evolution can result if there is some process that causes populations to shift from one peak to another and if that process acts on a longer time scale than adaptive processes like natural selection. So far we have assumed that populations are large and the environment is unchanging. With these assumptions, populations will usually rapidly reach an adaptive peak and then stay there indefinitely. They will not exhibit the kind of long-run change that we have required for change to be historical. Wright (e.g., 1977) long argued that drift plays an important role in causing populations to shift from peak to peak, and then competition among populations favors the population on the higher peak. Chance variations in gene frequency in small populations could lead to the occasional crossing of adaptive valleys and the movement to higher peaks. Recently, several authors have considered mathematical models of this process (Barton and Charlesworth, 1984; Newman, Cohen, and Kipnis, 1985; Lande, 1986; Crow, Engels, and Denniston, 1990). These studies suggest that the probability that a shift to a new peak will occur during any time period is low; however, when a shift does occur, it occurs very rapidly. If this view is correct, drift should generate a long-run pattern of change in which populations wander haltingly up the adaptive topography from lower local peaks to higher ones. It is also implausible that environments remain constant either in space or in time. As environments change, the shape of the adaptive topography shifts, causing peaks to merge, split, disappear, or temporary ridges to appear, connecting a lower peak to a higher one. Thus, populations will occasionally slide from one peak to another. As long as such events are not too common, environmental change will also lead to long-run change. Such change might appear gradual if there are many small valleys to cross or punctuational if there are a few big ones.

Adding social or ecological realism to the basic adaptive hill-climbing model of evolution probably increases the potential for multiple stable equilibria. In the simple model, an individual's fitness depended only on his phenotype. When there are social or ecological interactions among individuals within a population, individual fitness will depend on the composition of the population as a whole. When this is the case, evolutionary dynamics can no longer be represented in terms of an invariant adaptive topography. However, they may still be characterized by multiple stable equilibria. Moreover, the fact that many quite simple models of frequency dependence have this property suggests that frequency dependence may usually increase the potential for path-dependent historical change.

Models of the evolution of norms provide an interesting example of how frequency dependence can multiply the number of stable equilibria. Hirshleifer and Rasmusen (1989) have analyzed a model in which a group of individuals interact over a period of time. During each interaction, individuals first have the opportunity to cooperate and thereby produce a benefit to the group as a whole but at some cost to themselves; they then have a chance to punish defectors at no cost to themselves. These authors show that strategies in which individuals cooperate, and punish noncooperators and nonpunishers, are stable in the game-theoretic sense. However, they also show that punishment strategies of this kind can stabilize any behavior—cooperation, noncooperation, wearing white socks, or anything else. We (chapter 9) show that the same conclusions apply in an evolutionary model even when punishment is costly. This form of social norm can stabilize virtually any form of behavior as long as the fitness cost of the behavior is small compared to the costs of being punished.

More generally, coordination is an important aspect of several kinds of social interactions (Sugden, 1986). In a pure game of coordination, it does not matter what strategy is used, as long as it is the strategy that is locally common. Driving on the left versus right side of the road is an example. It does not matter which side we use, but it is critical that we agree on one side or the other. This property of arbitrary advantage to the common strategy is shared by many symbolic and communication systems and allows multiple equilibria whenever there are multiple conceivable strategies. In many other common kinds of social interactions, elements of coordination and conflict are mixed. In such games, all individuals are better off if they use the same strategy, even though the relative advantages of using the strategy differ greatly from individual to individual, and some individuals would be much better off if another strategy were common. As long as the coordination aspect of such interactions is strong enough, multiple stable equilibria will exist. Arthur (1990) shows how locational decisions of industrial enterprises could give rise to historical patterns due to coordination effects. It is often advantageous for firms to locate near other firms in the same industry because specialized labor and suppliers have been attracted by preexisting firms. The chance decisions of the first few firms in an emerging industry can establish one as opposed to another area as the Silicon Valley of that industry. More generally, historical patterns can arise in the many situations where there are increasing returns to scale in the production of a given product or technology. Merely because the "qwerty" keyboard is common, it is sensible to adopt it despite its inefficiencies.

Interactions between populations and societies (or elements within societies such as classes) can give rise to multiple stable equilibria. Models of the co-evolution of multiple populations have many of the same properties as fre-quency- and density-dependent selection within populations, although the theory is less well developed (Slatkin and Maynard Smith, 1979). The evolution of one population or society depends upon the properties of others that interact with it, and many different systems of adjusting the relationships between the populations may be possible. For example, Cody (1974:201) noted that com-peting birds replace each other along an altitudinal gradient in California but latitudinally in Chile. Given the rather similar environments of these two places, it is plausible that both systems of competitive replacement are stable and which one occurs is due to accidents of history.

The stratification of human societies into privileged elites and disadvantaged commoners derives from the ability of elites to control high-quality resources or to exploit commoners using strategies that are similar to competitive and predatory strategies in nature. Insko et al. (1983) studied the evolution of social stratification in the social psychology laboratory. They showed that elites could arise in both an experimental condition that mimicked freely chosen trade re-lations and one that mimicked conquest. Elites were approximately as well off in both conditions and, insofar as they controlled things, would have no motivation to change social arrangements. It seems plausible that the diversity of political forms of complex societies could result from many arrangements of relations between constituent interest groups being locally stable. The distinctive differ-ences between the Japanese, American, and Scandinavian strategies for operating technologically advanced societies could well derive from historic differences in social organization that have led to different, stable arrangements between in-terest groups, in spite of similar revolutionary changes in production techniques of the last century or two.

Social or ecological interactions may also give rise to dynamic processes that are sensitive to initial conditions and have no stable equilibria. Lande (1981) analyzed a model of one such process, sexual selection in which females have a heritable preference for mates that is based on a heritable, sex-limited male character. According to his model, when the male character and female pre-ferences are sufficiently correlated genetically, female choice can create a self-reinforcing "runaway" process that causes the mean male character and the mean female preference either to increase or decrease indefinitely, even in the presence of stabilizing selection on the male character. Selection cannot favor female variants that choose fitter males (in the usual sense of fitter) because most females are choosing mates with an exaggerated character. The "sensible" female's sons will be handicapped in the mating game. The direction that evolution takes depends on the details of the initial conditions in Lande's model. His quantitative character will be elaborated in one direction or the other depending on how evolution drifts away from an unstable line of equilibria. Although the interpretation of this model is controversial, it is easy to imagine that the exaggerated characters of polygynous animals like birds of paradise and peacocks result from the runaway process. We (Boyd and Richerson, 1985, ch. 8, 1987; Richerson and

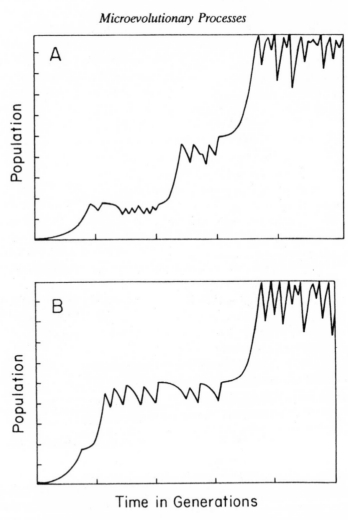

Microevolutionary Processes

Time in Generations

Figure 15.2. Both parts show the trajectories of population growth generated by the same model of social evolution for two slightly different initial population sizes.
In 15.2a the society goes through three distinct phases of growth, while in 15.2b, there are only two.

Boyd, 1989a) have argued that quite similar processes may arise in cultural evolution when individuals are predisposed to imitate some individuals on the basis of culturally heritable characteristics. The use of some character associated with prestige (stylish dress, for example) as an index of whom to imitate has the same potentially unstable runaway dynamic as Lande's model of mate choice sexual selection, and even casual observation suggests that prestige systems do follow contingent historical trajectories. Fashions in clothing, for example, evolve in different directions in different societies, often without much regard for practicality.

Perhaps the most clearly historical patterns of change result when social or ecological interaction leads to "chaotic" dynamics. For example, Day and Walter (1989) have analyzed an extremely interesting model of social evolution in which population growth leads to reduced productivity, social stratification, and eventually to a shift from one subsistence technology to a more productive one. The resulting trajectories of population size are shown in figure 15.2. Population grows, is limited by resource constraints, and eventually technical substitution occurs, allowing population to grow once more. The only difference between figure 15.2a and 15.2b is a very small difference in initial population size. Nonetheless, this seemingly insignificant difference leads to qualitatively different trajectories—one society shows three separate evolutionary stages, and the second only two.

Conclusion

Scientific and historical explanations are not alternatives. Contingent, diverging pathways of evolution and long-term secular trends can result from processes that differ only slightly from those that produce rapid, ahistorical convergence to universal equilibria. Late nineteenth- and early twentieth-century scientists gave up restricting the term "scientific" for deterministic, mechanistic explanations and began to admit "merely" statistical laws into the fundamental corpus of physics (very reluctantly in some cases—recall Einstein's famous complaint about God not playing dice with the universe to express his distaste for quantum mechanics). Similarly, historical explanations cannot be distinguished from other kinds of scientific explanations except that some models (and, presumably, the phenomena they represent) generate trajectories that meet our definition of being historical. These history-generating processes do not depend on exotic forces or immaterial causes that ought to excite a scientist's skepticism; perfectly mundane things will do. There are challenging complexities in historical processes. For example, even well-understood processes will not allow precise predictions of future behavior when change is historical. However, all the tools of conventional scientific methods can be brought to bear on them. For example, it should be possible to use measurement or experiment to determine if a process is in a region of parameter values where chaotic behavior is expected. At the same time, the historian's traditional concern for critically dissecting the contingencies that contribute to each unique historical path is well taken. Process-oriented "scientific" analyses help us understand how history works, and "historical" data are essential to test scientific hypotheses about how populations and societies change.

In the biological and social domains, "science" without "history" leaves many interesting phenomena unexplained, while "history" without "science" cannot produce an explanatory account of the past, only a listing of disconnected facts. The generalizing impulses of science require historical methods, because the phenomena to be understood are genuinely historical and because historical data are essential for developing generalizations about evolutionary processes. In return, generalizations derived from history and by the study of contemporary systems would seem to be essential for an understanding of particular cases. The amount of data available from the past is usually very limited, and the number of possible reconstructions of the past is correspondingly large. Some sort of theory has to be applied to make some sense of the isolated facts. Historians (e.g., Braudel, 1972) and paleontologists (e.g., Valentine, 1973) often cast their nets rather widely in search of help in interpreting the documents and fossils. McNeill (1986) advocates a "scientific," generalization-seeking approach to history much in this spirit. Consider the question of which of the potential history-producing processes we have discussed are most important in explaining the changes in human societies over the last few tens of thousands of years. Generalizing disciplines such as climatology and cultural ecology are certainly relevant to the task in general and to the understanding of how particular societies changed in particular environments (Henry, 1989). At the same time, because these historical societies faced Pleistocene climates and the transition to the Holocene, and because they developed a series of technical, social, and ideological innovations that are the foundation of modern human societies by processes that are not open to direct observation, the historical and archaeological records provide crucial data not available from ahistorical study. To the extent that the processes we have described are important, "science" and "history" cannot be disentangled as separate intellectual enterprises.

Darwinian models of organic and cultural evolution illustrate how little distinction can be made between the two approaches. Such models can produce historical patterns of change by a rather large number of different mechanisms. We have argued that historical change is distinguished by two attributes: the tendency of initially similar systems to diverge and the occurrence of long-term change. Evolutionary models, including those that assume that selection or analogous cultural processes increase adaptiveness in each generation, readily generate multiple stable equilibria. Populations with similar initial conditions may evolve toward separate equilibria. Random genetic drift and analogous cultural processes, coupled with environmental change, may cause populations to shift from one equilibrium to another. It is plausible that peak shifting by populations (or the shifting of peaks due to environmental change) occurs at a slow enough rate to explain long-term secular trends.

Many anthropologists take as their task the explanation of differences among human societies and suggest that most such differences are historical in character. If explanation of such variation is mainly historical, then anthropologists might reasonably ask, what is the point of Darwinian models of cultural change when historical or "contextual" explanations will be much more productive? The reasons are as follows.

First, the premise is often incorrect. Genuine convergences are common and explaining them requires some theory based on common processes of cultural change. Perhaps the most spectacular cultural example is the convergence of social organization in stratified, state-level societies in the Old and New Worlds. For example, Cortez in 1519 found that Aztec society was quite similar to his own in important ways: it contained familiar roles, hereditary nobility, priests, warriors, craftsmen, peasants, and so on. The bureaucracy was organized hierarchically. This convergence is remarkable because the Spanish and Aztec states evolved independently from a hunter-gatherer ancestry. The cultural lineages that resulted in these two states were without known cultural contact for several thousand years before state formation began in either (Wenke, 1980).

Second, Darwinian models can make useful predictions. They can tell us why some forms of behavior or social organization are never observed and others are common. For example, kinship is an extremely common principle of social organization. Contrarily, there would seem to be lots of advantages to a free market in babies—for the individual, it would allow easy adjustment of family size, age composition, sex ratio, and so on, and for society, a division of labor in child rearing would allow better use of human resources. The sociobiological theory of kin selection explains why there are no societies with free trade in infants and why kinship is generally an important feature of social organization. If most of the historic context is taken as given, Darwinian arguments can be very powerful heuristics. This is especially clear for genetic evolution. For example, given haplodiploidy, a theory based directly on the expected equilibrium outcome of natural selection can make surprising and extremely fruitful predictions about patterns of behavior in social insects. Who, for example, would have thought to connect sex ratio among reproductives and "slave making" in ant species? In recent years, similar ideas have been usefully applied to understanding human behavior. For example, Hill, Kaplan, and their colleagues (reviewed in Hill and Kaplan, 1988) have used theory from behavioral ecology to relate patterns of foraging, mate preference, and child care among Ache hunter-gatherers, and Borgerhoff-Mulder (1988) has explained variation in bride price among Kipsigis pastoralists in terms of parameters that predict future female fitness.

Finally, it is useful in and of itself to know that even the most strongly functional Darwinian models can give rise to historical change. The same processes that give rise to convergence in one case can generate differences in another, given only small changes in the structure of the process or in initial conditions. Brandon (1990) argues that "why possibly" explanations are useful in evolutionary biology. By this, he means explanations that tell us how some character *could* have evolved are useful even if we cannot determine whether the explanation is true. The theoretical models in population genetics provide a good example: Hamilton's (1964) kin selection models show how natural selection could give rise to self-sacrificial behavior. However, we usually do not know whether any particular case of altruism arose as a result of kin selection. The lack of *any* "why possibly" explanation would cast doubt on other aspects of our knowledge of how selection shapes behavior.

Understanding how adaptive processes could give rise to historical change is useful for analogous reasons. There is considerable evidence that people's choices about what to believe and what to value are affected by the consequences in material well-being, social status, and so on (e.g., Boyd and Richerson, 1985). This view has a venerable history in anthropology (e.g., Barth, 1981; Harris, 1979), plays a foundation role in economics, and is taken for granted in many historians' explanations for particular sequences of events. If cultural change is affected by consequence-driven individual choice or natural selection, then it follows that there will be a process that will act to modify the distribution of cultural variation in a population in much the same way that natural selection changes genetic variation (Boyd and Richerson, 1985, chs. 4 and 5). The fact that functional processes like natural selection readily lead to history allows one to hold this view without having necessarily to search for external environmental differences to explain the differences among apparently similarly situated human societies.

NOTES

We thank James Griesemer, Matthew H. Nitecki, Eric A. Smith, and two anonymous reviewers for most helpful comments on previous drafts of this chapter.

1. This project is quite different from the better-known, classical studies of cultural evolution developed by Leslie White (1959) and other scholars in anthropology. This work focused descriptively on the large-scale patterns of cultural evolution rather than on the details of the processes by which cultural evolution occurs (Campbell, 1965, 1975). The research tradition White represents derives from the progressivist ideas of Herbert Spencer, rather than from Darwin.

2. The additive genetic value of a particular individual for a particular character is the average value of that character for offspring produced when that individual mates at random with a large number of other individuals in the population. For example, the additive genetic value of a bull for fat content is the average fat content of all its offspring where mates were chosen at random. The distribution of genetic values is Gaussian when the probability that an individual has a given genetic value is given by the normal (or Gaussian) probability distribution. Genetic correlations exist when the distributions of genetic values for different characters are not probabilistically independent. For example, if bulls whose genetic value for size also tend to have a higher genetic value for fat content, then body size and fat content are genetically correlated. Genotype environment correlations arise when individuals with the same genotype develop different phenotypes in different environments.

REFERENCES

Alexander, R. D. 1979. *Darwinism and human affairs*. Seattle: University of Washington Press.

Arthur, W. B. 1990. Positive feedbacks in the economy. *Scientific American* (February): 92–99.

Bandura, A. 1977. *Social learning theory.* Englewood Cliffs, NJ: Prentice Hall.

Barth, F. 1981. *Process and form in social life.* London: Routledge and Kegan Paul.

Barton, N. H., & B. Charlesworth. 1984. Genetic revolutions, founder effects, and speciation. *Annual Review of Ecology and Systematics* 15:133–164.

Borgerhoff-Mulder, M. 1988. Kipsigis bridewealth payments. In: *Human reproductive behavior: A Darwinian perspective,* L. Betzig, & P. Turke,' eds. (pp. 65–82). Cambridge: Cambridge University Press.

Boyd, R., & P. J. Richerson. 1985. *Culture and the evolutionary process.* Chicago: University of Chicago Press.

Boyd, R., & P. J. Richerson. 1987. The evolution of ethnic markers. *Cultural Anthropology* 2:65–79.

Brady, R. M. 1985. Optimization strategies gleaned from biological evolution. *Nature* 317:804–806.

Brandon, R. 1990. *Adaptation and environment.* Princeton, NJ: Princeton University Press.

Braudel, F. 1972. *The Mediterranean and the Mediterranean world in the age of Philip II,* vol. 1. New York: Harper and Row.

Campbell, D. T. 1965. Variation and selective retention in sociocultural evolution. In: *Social change in developing areas: A reinterpretation of evolutionary theory,* H. R. Barringer, G. I. Blanksten, & R. W. Mack, eds. (pp. 19–49). Cambridge: Schenkman.

Campbell, D. T. 1975. On the conflicts between biological and social evolution and between psychology and moral tradition. *American Psychologist* 30:1103–1126.

Cavalli-Sforza, L. L., & M. W. Feldman. 1973. Models for cultural inheritance. I. Group mean and within group variation. *Theoretical Population Biology* 4:42–55.

Cavalli-Sforza, L. L., & M. W. Feldman. 1981. *Cultural transmission and evolution: A quantitative approach.* Princeton, NJ: Princeton University Press.

Cavalli-Sforza, L. L. & M. W. Feldman. 1983. Cultural versus genetic adaptation. *Proceedings of the National Academy of Sciences, USA* 90:4993–4996.

Cloak, F. T., Jr. 1975. Is a cultural ethology possible? *Human Ecology* 3:161–182.

Cloninger, C. R., J. Rice, & T. Reich. 1979. Multifactorial inheritance with cultural transmission and assortative mating. II. A general model of combined polygenic and cultural inheritance. *American Journal of Human Genetics* 31:176–198.

Cody, M. L. 1974. *Competition and the structure of bird communities.* Princeton, NJ: Princeton University Press.

Cohen, A. 1974, *Two dimensional man: An essay on the anthropology of power and symbolism in complex society.* Berkeley: University of California Press.

Collingwood, R. G. 1946. *The idea of history.* Oxford: Clarendon.

Conklin, H. C. 1969. An ethnoecological approach to shifting agriculture. In: *Environment and cultural behavior,* A. P. Vayda, ed. Garden City, NY: Natural History Press.

Crow, J. F., W. R. Engels, & C. Denniston. 1990. Phase three of Wright's shifting balance theory. *Evolution* 44:233–247.

Day, R. H., & X. Walter. 1989. Economic growth in the very long run: On the multiple phase interaction of population, technology, and social infrastructure. In: *Economic complexity: Chaos, sunspots, bubbles, and nonlinearity,* W. A. Barnett, J. Geweke, & K. Schell, eds. Cambridge: Cambridge University Press.

Diamond, J. 1978. The Tasmanians: Longest isolation, simplest technology. *Nature* 273:185–186.

Durham, W. H. 1976. The adaptive significance of cultural behavior. *Human Ecology* 4:89–121.

Eaves, L. J., A. Last, P. A. Young, & N. G. Martin. 1978. Model fitting approaches to the analysis of human behavior. *Heredity* 41:249–320.

Eldredge, N. 1989. *Macroevolutionary dynamics.* New York: McGraw-Hill.

Elton, G. R. 1967. *The practice of history.* London: Methuen.

Futuyma, D. J. 1986. *Evolutionary biology.* 2nd ed. Sunderland, MA: Sinauer.

Hallpike, C. R. 1986. *Principles of social evolution.* Oxford: Clarendon.

Hamilton, W. D. 1964. The genetical evolution of social behavior. *Journal of Theoretical Biology* 7:1–52.

Harris, M. 1979. *Cultural materialism: The struggle for a science of culture.* New York: Random House.

Henry, D. O. 1989. *From foraging to agriculture: The levant at the end of the Ice Age.* Philadelphia: University of Pennsylvania Press.

Hill, K., & H. Kaplan. 1988. Tradeoffs in male and female reproductive strategies among the Ache. In: *Human reproductive behavior: A Darwinian perspective,* L. Betzig, & P. Turke, eds. (pp. 227–306). Cambridge: Cambridge University Press.

Hirshleifer, D., & E. Rasmusen. 1989. Cooperation in a repeated prisoner's dilemma with ostracism. *Journal of Economic Behavior and Organization* 12:87–106.

Holland, H. D. 1984. *The chemical evolution of the atmosphere and oceans.* Princeton, NJ: Princeton University Press.

Holland, J. H. 1975. *Adaptation in natural and artificial systems.* Ann Arbor: University of Michigan Press.

Ingold, T. 1985. *Evolution and social life.* Cambridge: Cambridge University Press.

Insko, C. A., R. Gilmore, S. Drenan, A. Lipsitz, D. Moehle, & J. Thibaut. 1983. Trade versus expropriation in open groups: A comparison of two types of social power. *Journal of Personality and Social Psychology* 44:977–999.

Jackson, J. B. C. 1988. Does ecology matter? *Paleobiology* 14:307–312.

Karlin, S., & M. W. Feldman, 1970. Linkage and selection: Two-locus symmetric viability model. *Theoretical Population Biology* 1:39–71.

Kirkpatrick, S., C. D. Gelatt, & M. P. Vecchi. 1983. Optimization by simulated annealing. *Science* 220:671–680.

Labov, W. 1972. *Sociolinguistic patterns.* Philadelphia: University of Pennsylvania Press.

Lande, R. 1979. Quantitative genetic analysis of multivariate evolution, applied to brain: body size allometry. *Evolution* 33:402–416.

Lande, R. 1981. Models of speciation by sexual selection on polygenic traits. *Proceedings of the National Academy of Sciences, USA* 78:3721–3725.

Lande, R. 1986. The dynamics of peak shifts and the pattern of morphological evolution. *Paleobilogy* 12:343–354.

Levinton, J. 1988. *Genetics, paleontology, and macroevolution.* Cambridge: Cambridge University Press.

Lowe-McConnell, R. H. 1975. *Fish communities in tropical freshwaters.* London: Longmann.

Lumsden, C. J., & E. O. Wilson. 1981. *Genes, mind and culture.* Cambridge, MA: Harvard University Press.

Maynard Smith, J., R. Burian, S. Kauffman, P. Alberch, J. Campbell, B. Goodwin, R. Lande, D. Raup, & L. Wolpert. 1985. Developmental constraints and evolution. *Quarterly Review of Biology* 60:265–287.

McNeill, W. H. 1980. *The human condition: An ecological and historical view.* Princeton, NJ: Princeton University Press.

McNeill, W. H. 1986. A defense of world history. In: *Mythistory and other essays by W. H. McNeill* (pp. 71–95). Chicago: University of Chicago Press.

Mitchell, W. A., & T. J. Valone. 1990. The optimization research program—Studying adaptations by their function. *Quarterly Review of Biology* 65:43–52.

Newman, C. M., J. E. Cohen, & C. Kipnis. 1985. Neo-Darwinian evolution implies punctuated equilibria. *Nature* 315:400–401.

Pulliam, H. R., & C. Dunford. 1980. *Programmed to learn: An essay on the evolution of culture.* New York: Columbia University Press.

Raup, D. M. 1977. Probabilistic models in evolutionary paleobiology. *American Scientist* 65:50–57.

Reed, C. A. 1977. Origins of agriculture: Discussion and some conclusions. In: *Origins of agriculture*, C. A. Reed, ed. (pp. 879–993) The Hague: Mouton.

Richerson, P. J., & R. Boyd. 1989a. A Darwinian theory for the evolution of symbolic cultural traits. In: *The relevance of culture*, M. Freilich, ed. (pp. 124–197). S. Hadley, MA: Bergin and Garvey.

Richerson, P. J., & R. Boyd. 1989b. The role of evolved predispositions in cultural evolution: Or human sociobiology meets Pascal's wager. *Ethology and Sociobiology* 10:195–219.

Rogers, A. R. 1989. Does biology constrain culture? *American Anthropologist* 90: 819–831.

Rogers, E. M. 1983. *The diffusion of innovations.* 3rd ed. New York: Free Press.

Ruyle, E. E. 1973. Genetic and cultural pools: Some suggestions for a unified theory of biocultural evolution. *Human Ecology* 1:201–215.

Seilacher, A. 1970. Arbeitskonzept zur Konstructions-morphologie. *Lethaia* 3:393–396.

Slatkin, M., & J. Maynard Smith. 1979. Models of coevolution. *Quarterly Review of Biology* 54:233–263.

Sugden, R. 1986. *The economics of rights, co-operation and welfare.* Oxford: Basil Blackwell.

Trigger, B. 1978. *Time and traditions.* Edinburgh: Edinburgh University Press.

Turelli, M., & N. Barton. 1990. The dynamics of polygenic characters under selection. *Theoretical Population Biology* 38:138–193.

Valentine, J. W. 1973. *The evolutionary paleoecology of the marine biosphere.* Englewood Cliffs, NJ: Prentice-Hall.

Valentine, J. W., & E. M. Moores. 1972. Global tectonics and the fossil record. *Journal of Geology* 80: 167–184.

Vermeij, G. J. 1987. *Evolution and escalation: An ecological history of life.* Princeton, NJ: Princeton University Press.

Wagner, G. P. 1988. The significance of developmental constraints for phenotypic evolution by natural selection. In: *Population genetics and evolution*, G. de Jong, ed. (pp. 222–229). Berlin: Springer-Verlag.

Wenke, R. 1980. *Patterns in prehistory.* Oxford: Oxford University Press.

Westoby, M. 1989. Australia's high lizard density: A new hypothesis. *Trends in Ecology and Evolution* 4:38.

White, L. A. 1959. *The evolution of culture.* New York: McGraw-Hill.

Wilde, D. J. 1978. *Globally optimal design.* New York: John Wiley.

Wright, S. 1977. *Evolution and the genetics of populations.* vol. 3. *Experimental results and evolutionary deductions.* Chicago: University of Chicago Press.

16   Are Cultural Phylogenies Possible?

With Monique Borgerhoff Mulder and
William H. Durham

Biology and the social sciences share an interest in phylogeny. Biologists know that living species are descended from past species and use the pattern of similarities among living species to reconstruct the history of phylogenetic branching. Social scientists know that the beliefs, values, practices, and artifacts that characterize contemporary societies are descended from past societies, and some social science disciplines (e.g., linguistics and cross-cultural anthropology) have made use of observed similarities to reconstruct cultural histories. Darwin appreciated that his theory of descent, with modification, had many similarities of pattern and process to the already well-developed field of historical linguistics. In many other areas of social science, however, phylogenetic reconstruction has not played a central role.

Phylogenetic reconstruction plays three important roles in biology. First, it provides the basis for the classification. Entities descended from a common ancestor share novel, or derived, characters inherited from that ancestor. Therefore, it is possible to group them into hierarchically organized series of groups—species, genus, family, order, and so on in the biological case.

Second, knowledge of phylogeny often allows inferences about history. The knowledge that humans are more closely related to chimpanzees and gorillas than to orangutans provides evidence that the human lineage arose in Africa. Phylogenetic reconstructions based on the characters of extant species or cultures often allow us to reconstruct the history in the absence of a historical, archaeological, or fossil record. In practice, the history of many biological and cultural groups is so poorly known that only by combining phylogenetic and historical or archaeological information can reliable reconstructions be obtained.

Third, entities descended from a common ancestor share features that may constrain the pathways that more recent evolution has followed. For example, selection for terrestrial locomotion may lead to quadrupedal locomotion in a small monkey that runs along the tops of branches but to bipedal locomotion in a large arboreal ape that swings below branches (Foley, 1987). The latter pattern allows the hand to specialize in manipulative tasks and, on many accounts, is why the ape, but not the monkey lineage, eventually was able to produce a cultural species.

The importance of *descent* is the crux of some of the deepest controversies of all the historical sciences. Some social scientists and biologists (e.g., Boyd and Richerson, 1992; Hallpike, 1986; Sahlins, 1976) have argued that history strongly constrains adaptation and, as a result, strictly limits adaptive interpretations of current behavior. As Francis Galton taught both biologists and social scientists in the nineteenth century, to account for the effects of common ancestry, the study of adaptation or function requires that patterns of descent be known. Our inability to provide appropriate roles for history and function is a chronic source of controversy.

If the analogy is real, an interdisciplinary exchange of concepts and tools could pay great dividends. Social scientists may be particularly interested in the near-revolutionary developments in systematics (Ridley, 1986) and comparative methods (Harvey and Pagel, 1991) developed by evolutionary biologists in the last two decades.

The purpose of this chapter is to examine the role of descent in culture evolution theory. We believe that the critical question is whether human cultures, or parts of them, are isolated from one another to the same degree as biological entities like species and genes. Cultures are frequently characterized by sharp ingroup-outgroup boundaries (LeVine and Campbell, 1972) that may function to limit the flow of ideas from one population to another (Boyd and Richerson, 1987). However, there are also many examples of the diffusion of cultural traits across such boundaries (Rogers, 1983). Are the isolating processes sufficiently strong to provide at least a core of important cultural traits that are sufficiently protected from diffusion so that phylogenetic analysis is possible? If so, concepts and methods from biological systematics can be used to reconstruct the history of cultures. If not, human cultures are more like subspecies or local populations linked by gene flow than like reproductively isolated species. In this case, it may be useful to make separate phylogenies for each subunit of culture that is substantially protected from diffusion, in much the same way that modern molecular procedures are used to reconstruct the phylogeny of subgenomic units, especially individual genes. It may also be that there are no cultural units with sufficient coherence and therefore that phylogenetic methods are useless.

We begin by reviewing the notions of descent used in evolutionary biology. Biologists have been making use of the concept of descent ever since Darwin, and they have developed a sophisticated appreciation for the concept and its problems that may be helpful in the human case. The complexity and diversity of biological systems of inheritance is wondrous to those brought up on the simple Mendelism of 20 years ago (Falk and Jablonka, 1997). Although it is likely that the process of cultural descent with modification is different from the analogous process in organic evolution, we believe that much can be learned from biologists' century of hard work. We then consider data from the social sciences that indicate the extent to which cultures form bounded wholes, analogous to species. Finally, we consider how the descent concepts, partly borrowed from biology, might be used to tackle important questions in the social sciences.

Descent in Organic Evolution

In biology, two different entities exhibit the clear patterns of descent with modification. The most familiar example is the species. The collection of individuals who make up a species during any generation is descended, and perhaps slightly modified, from the collection of individuals who made up the species during the previous generation. A new species is formed by the splitting of an existing species. Then each of the daughter species is descended from the single ancestral species that gave rise to them.

Much the same holds for genes one by one. Because genes result from the copying of DNA, every gene is descended from the gene that provided its template. Modified genes arise from existing genes by mutation, recombination, and gene conversion at a given locus. A genetic locus can give rise to another locus by duplicating itself on the chromosome, after which the daughter locus begins independent evolution. The relationships among genes is not simply the relationships among the species that carry them (although this is often the case). We can keep track of the relationship of genes *within* a single species (e.g., various forms of hemoglobin within human populations). It is also possible to speak of relationships among genes that are inconsistent with relationships among species. For example, genes for globin molecules in vertebrates and certain plants seem to share a more recent common ancestor than the genes in vertebrates and arthropods, as surprising as this seems at first blush (Jeffreys et al., 1983).

Descent relationships are often represented using branching diagrams like that shown in figure 16.1. The diagram conveys the idea that both A and B are descended from an ancestor C. (Systematists use similar branching diagrams called cladograms to represent patterns of similarity without reference to time, or ancestor-descendant relationships; statistical clustering algorithms create treelike dendrograms also without any pretense to representing ancestor-descendant relationships. Tree diagrams are used here to represent phylogeny.) The same diagram is used to represent the relationship among different kinds of things. For biologists A, B, and C may represent species or genes. Social scientists use similar diagrams to express the relationship among languages, or other aspects of culture, often with the explicit intention of representing a phylogeny. What, if anything, do the descents of genes and species have in common? Can these commonalities provide some help in analyzing the descent of cultures, languages, and technologies?

The Descent of Genes

To answer this question, let us begin with the simpler case—the descent relationship among genes. If we ignore for a moment the possibility of recombination,

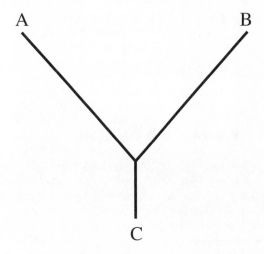

Figure 16.1. A hypothetical phylogeny in which species *A* and *B* are descended from species C.

every gene is a copy of another gene. Of course, that gene was the copy of yet another gene, and so on. Thus, if we pick any two genes, *A* and *B*, we can, in principle, trace back through a series of copies until we find a gene, C, that served as a template for both. We say that genes *A* and *B* are descended from C. If mutations have occurred, *A* or *B* may be different from C and each other. As long as mutations are rare and the gene includes enough bases, then genes that share more derived mutations are more likely to be related. Taxonomists use this fact to reconstruct the branching pattern among genes sampled from living species. Notice that there is nothing in the discussion that specifies that C, *A*, and *B* have to belong to the same (or different) species. The same argument would hold regardless of whether *A* and *B* are genes found within a single species or among distantly related species (e.g., humans and bean plants).

Units with Reticulated Phylogenies

Recombination—the shuffling of chromosomes of the genes along a chromosome and the sequence within a gene—complicates matters because it leads to what cladists call *reticulated phylogenies*. Figure 16.2 shows the lineages of three genes. Recombination has occurred within the gene three times. After each recombination event, each of the daughter genes is a copy of part of each of the two parents. The daughter genes are no longer descended from the parental genes in the same way that they were in the absence of recombination. They are no longer almost exact copies of the parents; rather, they are partial copies of both parents. Further recombination events create yet more complicated patterns of relationship. After some time, every copy of the gene is related to a large number of other genes in some complicated way that utterly obscures descent. Recombination within a gene is rare, but recombination within chromosomes between different genes is quite common. Deep phylogenies can be reconstructed for genes, but only shallow ones for chromosomes.

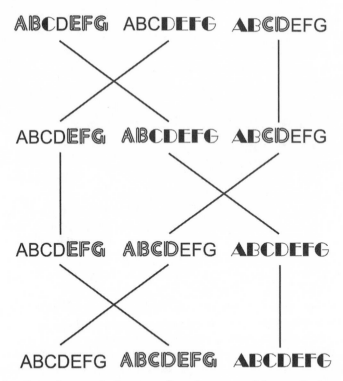

Figure 16.2. Recombination leads to complicated patterns of descent. Each string of letters represents a segment of the chromosome. Each generation each gene is replicated, sometimes with recombination. After four generations, each chromosome is partly descended from all three of the original chromosomes.

Gene flow (migration) among subpopulations of a species has a similar effect. Any given local group will have acquired genes from many different local groups in the past. Even if most subpopulations are created by the subdivision of a single parental population, a relatively small rate of individual-level migration between subpopulations will carry genes evolved in one daughter subpopulation to its sisters. Fairly shortly, descent at the subpopulation level will be impossible to detect. Thus, there is a large range of genetic units ranging in size from roughly small chromosome segments to the subspecies for which phylogenetic analysis is usually impossible.

Some large gene collections, such as mitochondrial genomes, are protected from recombination because they are transmitted asexually. Mitochondrial phylogenies of some depth can be constructed, although they illustrate another process that eliminates phylogenetic information in the long run. Mitochondria are subject to high mutation rates. In a matter of a few million years, every descendant pair of mitochondrial genes will have independently mutated more than once, and the traces of descent will be lost. Conservative genes like the cytochrome genes have slow rates of evolution and can be used to reconstruct phylogenetic relationships reaching back to near the origin of life, but these are exceptional. More typically, deep phylogenetic reconstructions based on less faithful structures are quite controversial even when we can be almost certain that recombination and migration have not confused the picture.

The Descent of Species

Species and higher taxa are the classic focus of phylogenetic analysis in biology. Linnean systematists formalized the common observation that the organic world comes in readily observable clusters. Species and higher taxa seem to be separated by distinctive gaps that do not occur within species or among many other natural objects. Darwin's theory of descent with modification gave a theoretical underpinning to the trees of relationships that Linnaeus had enshrined in a hierarchical classification system, although Darwin had little to say about the species-isolating mechanisms that enforce the gaps between species. His followers have made up for this deficiency; the issue of speciation is a major topic in modern evolutionary biology.

In the basic picture constructed by architects (e.g., like Ernst Mayr) of the midcentury neo-Darwinian synthesis, species are created when a barrier to gene flow evolves to isolate two sets of populations. Once isolated, the evolution of the two new species is independent, and slowly changes accumulate due to natural selection, genetic drift, mutation, and so forth. There may be some evolutionary differentiation within a population due to selection or drift. But interbreeding among populations unites a species, whereas absolute speciating barriers definitively separate them from other species. Over the long run, species become different enough to be classified as new genera, families, orders, and so on, up Linnaeus's hierarchy. In the classic picture, complete isolation and the slow accumulation of differences allow for the reconstruction of relationships of descent by splitting over great time depths.

The basic picture provides a clear causal explanation of the temporal and spatial coherence of species. Advocates of the biological species concept hold that only when this picture applies do we have species, properly speaking. However, several lines of evidence suggest that the absence of gene flow is neither necessary nor sufficient for the existence of coherent species in the sense of lumpy entities that show clear evidence of descent. Species can maintain their coherence without gene flow within the species, and species boundaries may be maintained despite gene flow between species.

Some species have maintained species-typical phenotypes, including the ability to form fertile hybrids despite long periods without any gene flow. For example, the checkerspot butterfly is found in scattered populations throughout California. Members of different populations are very similar morphologically and are all classified as members of the species *Euphydryas editha*. However, careful study has shown that there is virtually no gene flow among widely separated populations (Ehrlich and Raven, 1969). There are also many examples (Levinton, 1988) of cryptic "sibling" species that are long isolated but have evolved no detectable morphological differences. Some taxonomists claim that it is no more difficult to detect species in asexual organisms than it is in sexual organisms (e.g., Mishler and Brandon, 1987), despite the fact that there is no gene flow to unite asexual populations.

Some species persist despite substantial gene flow (Barton and Hewitt, 1989). A hybrid zone can exist between what seem to be good species, and often a few genes have clearly leaked across the boundary from one species to another. It would seem as if such species must either be formerly geographically isolated subspecies that will hybridize away or incipient species that will eventually evolve an isolating barrier. In fact, active hybrid zones between rather distinct species sometimes persist for long periods of time. Selection can apparently maintain the coherence of species both without any help from gene flow and in the face of substantial amounts of it.

Things are not always so neat. In bacteria, genes are frequently transmitted horizontally among lineages (Eberhardt, 1990). Bacterial DNA exists in two distinct forms: most of the DNA is contained in a large chromosome, but about 1 percent is contained in small loops of DNA called *plasmids*. The two forms of DNA are transmitted differently. For the most part, the chromosomal DNA is transmitted vertically. When bacteria divide, the chromosomal DNA is duplicated, and each daughter cell contains a copy. In contrast, plasmid DNA is transmitted horizontally from one bacteria to another during conjugation. Moreover, bacteria that are classified as belonging to different genera or families according to their chromosomal DNA readily conjugate and exchange plasmid DNA. As a result, genes carried on plasmids may jump from one lineage to another quite distant one. It is not certain that the two types of DNA are completely separate. Sometimes plasmid DNA may be incorporated into the chromosome, although if this occurs it is probably quite rare (Eberhardt, 1990). In the case of bacteria, there are really two sets of phylogenies: one for the chromosomal DNA and one for the mitochondrial. Relationships between these phylogenies break down rapidly because of the horizontal transmission of plasmids across chromosomal lineages.

The opposite situation occurs with the lineages of hosts and parasites and predators in many animals and plants. For example, ectoparasites like lice and fleas are often isolated within their hosts, so that host and parasite phylogenies are similar despite no transfer between host and parasite genomes.

The Common Properties of Genes and Species

Genes and species are units at quite different levels of organization. For them, but not units between them on the scale of organization, deep phylogenies can usually be constructed. The reason is a pair of similarities. First, both units are replicated with great fidelity and change slowly due to ongoing evolution. Second, when daughter genes and species change, these changes are not effectively shared with sister lineages by mixing or any other form of communication. For systems with high rates of change, like mitochondrial genomes, deeper descent is obscured because recently evolved differences completely obliterate the ancient similarities that are necessary to detect descent. In the case of units like chromosomes and local populations with high rates of mixing, descent is generally untraceable because descent-derived differences are erased as rapidly as they arise.

Genealogy is by itself not enough to generate much descent. There is a hierarchy of genealogical entities in biology: genes, chromosomes, individuals, populations, species, and communities. These are genealogical entities because they are all descendants of other entities at the same level. In the face of rapid mixing or evolution (or both), genealogy alone cannot preserve detectable patterns of descent, at least not for long. Note that patterns of descent are a matter of timescale. If we are interested in relationships over only a few splittings of daughter entities, these may be detectable in the face of considerable mixing and high rates of evolution. If we want to know relationships traceable many splits ago, the criteria are more demanding.

Reconstructing Cultural Phylogenies

Can we apply these ideas from biology to the analysis of human culture? Darwinian models of cultural evolution hold that culture is information transmitted from individual to individual by imitation, teaching, and other forms of social learning. Various processes cause the pool of cultural variants that characterize a population's change through time.

This view of culture and cultural evolution implies the existence of a hierarchy of genealogical entities analogous to the genealogical hierarchy of organic evolution. We do not know what is the smallest unit of cultural inheritance because we do not know in detail how culture is stored in brains. Nevertheless, scholars have proposed histories of quite small elements: particular words, particular innovations, elements of folk stories, and components of ritual practice. Such small elements are linked together in larger, culturally transmitted entities: systems of morphology, myth, technology, and religion. Such medium-scale units are collected together into "subcultures" and "cultures" that characterize human groups of different scales: kin group, village, ethnic group, nation, and so forth. Cultural subunits sometimes crosscut one another in complex ways, as when religion or occupation crosscuts ethnicity (much like bacterial chromosomes and plasmids).

Four Hypotheses

Reconstructing cultural phylogenies is possible to the extent that there are genealogical entities that have sufficient coherence, relative to the amount of mixing and independent evolution among entities, to create recognizable history. There is a continuum of possible views about what units in the hierarchy of cultural descent satisfy these desiderata. It is useful to identify four regions along this continuum.

Cultures as Species

Cultures are isolated from one another or are tightly integrated. They contain within them powerful sources of isolation (ethnocentric discrimination against strangers) or coherence (such as organizing systems of thought that act as biases against ideas one by one, rather than strangers as whole individuals). Both mechanisms could cause cultures to act as single entities or "individuals" in the course of cultural evolution (see e.g., Marks and Staski, 1988). By one mechanism or another, there is little cross-cultural borrowing of any significance. New cultures are formed completely by the fissioning of populations and subsequent divergence. In this case, whole cultures are analogous to species or mitochondrial genomes. Biological methods of systematics can be applied almost intact, and deep cultural phylogenies are relatively easy to infer for at least the bulk of a people's culture.

Cultures with Hierarchically Integrated Systems

Although cross-cultural borrowing may be frequent for many peripheral components, a conservative "core tradition" in each culture is rarely affected by diffusion from other groups. New core traditions mainly arise by the fissioning of populations and subsequent divergence of daughter cultures. Isolation and integration protect the core from the effects of diffusion, although peripheral elements are much more heavily subject to cross-cultural borrowing. In this case, core traditions are analogous to the bacterial chromosomes and the peripheral components to plasmids. Biological methods of systematics can be modified to deal with cross-cultural borrowing. Reasonably deep core-cultural phylogenies can still be inferred, but this requires disentangling the effects of borrowing by distinguishing core and peripheral elements, and especially by methods to identify elements that "introgressed" into the core.

Cultures as Assemblages of Many Coherent Units

Cultures could be quite ephemeral assemblages of small units, but the latter may have limited mixing and slow evolution. Culture may have no species, but it might have genes, plasmids, and mitochondria. Different domains may have different patterns of inheritance and different evolutionary histories. The components may be fairly large, plasmid, or mitochondrion-like, such as language, or small, solitary memes, such as the idea of using a magnetized needle to point north. Any given culture is an assemblage of many such units acquired from diverse sources. Methods of phylogeny can be applied independently to each domain. The essential problem is to determine the boundaries of the domains and establish that they are stable in time and space.

Cultures as Collections of Ephemeral Entities

There are no observable units of culture that are sufficiently coherent for phylogeny reconstruction to be useful. Observable aspects of culture could be the result of units that are beneath the resolution of current methods to observe. The forms of Acheulean hand axes are so similar that they cannot be used to infer anything about descent among their makers. Perhaps there were really many traditional ways to reach this apparently uniform end result. If we knew the details, we could reconstruct cultural phylogenies of hand ax making. There may be observable differences, but if they are the product of many recombining elements that cannot be observed, there is no information that would allow us to construct a phylogeny of the bits. Alternatively, if cultural evolution is sufficiently rapid, behavior may reflect such recent history that all phylogeny is lost. The "jukebox" culture, in which cultures are rapidly modeled and remodeled to serve current adaptive purposes, would have this effect due to functional convergence rapidly destroying any trace of history.

There are two issues at stake. First, when using the term *descent*, what do we mean? Proponents of the view that whole cultures are like species use *descent* to describe cultural replication of complex coherent groups by the mechanism of group fission or budding, whereas those who believe that only components of culture cohere would use *descent* to describe ancestor–descendant relationships resulting from any pattern of culture preserving the footprints of its history. We shall try to be clear in our own usages, but this is a merely terminological issue to which we devote no further space. Second, what is the world like? This is a much more interesting question, to which we devote the rest of the chapter. At one end of the continuum, all of the elements that make up a culture cohere and resist recombination. Cultures as a whole are analogous to species. At the other end, the observed elements of culture are the result of memes diffused or invented on a timescale too short for phylogenetic reconstruction. What is culture really like?

Mechanisms

Several general mechanisms might cause longevity and coherence in cultural units so that descent can be determined.

Longevity of Historical Traces

As in the case of genes, the phylogenetic process of cultural transmission provides some level of historical continuity. As with genes, the deepest phylogenies are possible when culture changes slowly and is not subject to functional convergence. Slow evolution will occur when people either cannot, or have no reason to, invent new forms. Surprisingly simple bits of culture are often apparently too obscure to reinvent, and all known modern exemplars derive from a single invention. Needham (1988) gave many plausible examples of Chinese technology that subsequently diffused to the rest of Eurasia (e.g., the magnetic compass). Nonetheless, in the long run, functional convergence seems to be the rule for technology. A long tradition in the social sciences, including the classic cultural ecology of Steward (1955) and modern evolutionary anthropology, it trades on the reality of substantial convergent evolution in human cultures. As in the biological case, the best elements for historical analysis are those that are functionally arbitrary and symbolic. Language and other symbolically meaningful, but nonfunctional, variations are often used as indices of descent, much as functionally neutral flower form is used in plant systematics. Flowers are a plant's way of communicating with pollinators, so the analogy with language is real.

The next subsection describes some mechanisms that may prevent mixing between coherent elements. Similar mechanisms may act to slow the rate of evolution if internal innovations or innovators are perceived as strange, either because of a poor internal fit or because they arouse suspicions of heresy or deviance on the part of innovators.

Processes That Give Rise to Coherence

What general processes could give cultural elements an enduring coherence, leaving aside the size of cohering units and their relation to one another? In the symbolic and interpretive anthropology literature, the "glue" has been attributed to the "meaning" that inheres in culture. Meaningful cultural information provides a convincing and compelling weltanschauung for its bearers. Meaningful components help organize and make sense of other parts of the cultural system and natural world. They also legitimize and justify the system in the minds of its bearers. For this reason, meaningful components have variously been called "root paradigms" (Turner, 1977), "ultimate sacred postulates" (Rappaport, 1979), "core principles" (Hallpike, 1986), and the like. Because it is critically important to a people's understanding of the world and its place within it, they often have a special, even sacred, status. The notion of meaning is often linked to the idea of cultural holism. There is no logical reason for this limitation, and the idea may apply to cores or much smaller units. Subcultural units as small as the individual social scientific disciplines, street gangs, and clans often appear to have well-articulated systems of meaning.

The special status of meaningful elements could provide coherence in several ways. First, the internal logic of a coherent block of culture may discriminate against intrusive elements. Diffused elements may be known to individuals, but the mismatch of meanings between whole cultures or subcultures entails that "foreign" values and ideas be misunderstood, disliked, and neglected. The mismatch may be between foreign elements but also between domains within a single culture (e.g., gender marked identities or even sets of subsistence skills).

Second, meaningful culture often involves markers of group identity that are especially salient to the definitions of ingroup and outgroup. Contexts where coherent units of meaning-rich culture are available for acquisition from foreigners are likely to involve marked ritual observances or ceremonies that mobilize ethnocentric sentiments more thoroughly than mundane contacts like trade, in which symbolically less marked elements may diffuse readily. Ethnocentrism can provide an effective isolating barrier to diffusion of cultural elements in theory and apparently in practice (Boyd and Richerson, 1987) at the whole-culture level. Class, caste, gender, occupation, and even hobby groups are symbolically marked within some societies. Within bounded groups, however large they may be, intermarriage, diffusion, and other mixing processes create cultural uniformity, but there are sharp differences among them. This is a form of indirect bias.

Third, to the extent that what coheres in culture is a symbolic system of organizing meanings, rather than the meanings themselves, it is protected from ordinary adaptive evolutionary pressures. In language at least, the symbol system is so rich and flexible that quite novel new meanings can be coded with the existing system; only linguistically trivial changes in lexicons were needed to adapt modern languages to the industrial revolution.

Finally, elements may cohere because certain combinations are adaptive and favored by natural selection or derivative adaptive decision-making rules. Adaptive forces may simply discriminate so strongly against recombinants that coherence is maintained despite massive mixing, as seems to be the case in certain hybrid boundaries in the biological case (Barton and Hewitt, 1989). A related sort of selective "glue" could come from the multiplicity of evolutionarily stable strategies that seem to exist in social systems. Perhaps the stability of coherent features comes from the failure of new or foreign social practice to fit into actual arrangements, rather than from inconsistencies at the cognitive/affective meaning level. The symbolic or ideological level may follow the social, rather than dictate it.

Rushforth and Chisholm (1991) gave a possible example in their discussion of Athapaskan "structures of communicative social interaction." According to these investigators, a core "framework of meaning and moral responsibility" has persisted among Bearlake Athapaskan of northern Canada with "extraordinarily little change" across many generations and hundreds of years (p. 64). Moreover, remarkably similar beliefs and values—urging industriousness, generosity, autonomy, and restraint—have been documented among more than 30 other Athapaskan-speaking peoples across three geographically discontinuous clusters in Canada and Alaska, the Pacific Northwest, and the American Southwest.

A deeply rooted family of social norms such as these might directly underpin social institutions. The norms that underpin social interactions are good candidates to be maintained as a coherent block because they are part of local evolutionarily stable strategies. In game theory, at least, it is easy to imagine locally and evolutionarily stable strategies for complex social institutions that are impossible to change at the margin by either diffusion or within-lineage change because small movements away from current practice are disadvantageous.

Would the multiple evolutionarily stable strategies (ESS) explanation account for the remarkable cultural persistence of Athapaskan norms? Focusing on the Bearlake version, Rushforth and Chisholm (1991) suggested that "the Bearlake interpretive scheme has persisted because of the historically stable composition of the [social interaction] strategies it informs" (p. 119). They argued that Bearlakers pursue goals in daily life that are defined and valued by their interpretive framework of beliefs and values. The interactions that follow generate regular rewards or "payoffs" that encourage individuals to convey certain intentions to others. But the actions that convey these intentions are precisely those defined by the framework. In short, the framework persists as "an unintended consequence of the strategic behavior of individuals operating in their own interests" (p. 121).

Sometimes coherent traditions are "acquired" by imposition by an invading, dominant culture, or assimilation to an attractive one. Even in this case, little admixture from the competing coherent structure of the adopting culture need result from its transfer from one biological population to another, as in the imposition of a common Greco-Roman urban civilization on a host of "barbarian" peoples in ancient Europe and Western Asia. Note that individual people can move readily without disturbing the integrity of the coherent elements, as the assimilation of many immigrant people to at least aspects of Anglo-American culture over the last two centuries testifies. Nevertheless, replication by transfer to a new biological population is arguably normally accompanied by much mixing of old and new, and the fission of one population into two daughters probably conserves coherence more effectively. Similarly, high rates of immigration need not necessarily result in high rates of erosion of coherence, but cultural diffusion does seem likely to be stimulated by immigration in typical cases.

Evidence

The Descent of Cultures as Wholes

Commentators such as Marks and Staski (1988) sometimes imply that they defend this position. According to McNeill (1986), historians such as Toynbee imply a position as extreme as this end of our continuum, although without any specific defense. McNeill's own magisterial *Rise of the West* was written to demonstrate how it was not possible to write a world history without ac-knowledging the exchange of ideas among major culture areas, much less within them. Holistic arguments, ultimately deriving from Wittgensteinian philosophy, once enjoyed great appeal in history and many branches of the social sciences, and echoes remain. For example, in linguistics, de Saussure (1959) is often cited as a proponent of extreme systemicity in language, and even today some linguists espouse this view (Wardhaugh, 1992). The limitations of such arguments have long been recognized by philosophers, and more recently by social scientists. There is such overwhelming evidence for substantial diffusion and rapid evolution in many components of culture that it is unlikely that any tenable empirical defense of a completely holistic cultures-as-species position can be offered.

The Descent of Core Traditions

The hierarchical hypothesis of large-scale cultural coherence rooted in a core tradition is a point along the continuum that warrants closer examination. Like the previous hypothesis it assumes that culture is an ideational system (i.e., it consists of widely shared ideas, values, and beliefs that shape behavior in local human populations; the named cultures of anthropologists). In this model, cultures are viewed as hierarchically integrated systems, each with its own internal gradient of coherence. At one extreme in the gradient are the "core" components of a culture—those ideational phenomena that constitute its basic conceptual and interpretive framework and influence many aspects of social life. At the other are peripheral elements that change rapidly or are widely shared by diffusion. On this hypothesis, the processes of coherence generate one main, central core unit. But this central unit does not equally organize all elements of culture. There may be many other smaller elements that are only lightly or not at all influenced by the core.

Core versus Periphery.   Regardless of whether the core gets its coherence from meaning, protection, diffusion, structured social interaction, or from all these sources, the key assertion of this model is that core components exhibit a remarkable resilience in the course of cultural history. The core "sticks to-gether" as a cohesive bundle even through repeated episodes of culture birth, giving rise to a set of descendant branches that then share the same "tradition." As Vansina (1990) argued based on his case study, such traditions are based upon

> the fundamental continuity of a concrete set of basic cognitive patterns and concepts.... [The] continuity concerns basic choices which, once made, are never again put into question.... These fundamental acqui-sitions then act as a touchstone for proposed innovations, whether from within or without. The tradition accepts, rejects, or molds borrowings to fit. It transforms even its dominant institutions while leaving its principles unquestioned. (p. 258)

Despite these numerous sources of cohesion, the hierarchical hypothesis holds that many "peripheral" components exist that are only loosely tied to the core framework. These diffuse freely and readily, as in the well-studied case of technical innovations (Rogers, 1983). Peripheral components may include ide-ational elements that make sense on their own and can be socially transmitted without a lot of supplementary cultural information. Such components are as-sumed to play little or no organizational role within the broader ideational system, and they must be relatively easy to learn. Such components are expected to be highly "contagious," rather like Dawkins's (1993) viruses of the mind.

New forms will be adopted quickly, simply, and smoothly, particularly if there is some perceived functional advantage and low cost. In this instance, change is quick and easy: different components come and go as independent interchangeable parts. They are likely to spread horizontally among cultures, regardless of whether those cultures are related historically by branching. For this reason, their phylogenies will have the vine-like appearance mentioned earlier. Kroeber (1948) gave a long list of well-known examples (e.g., days of the week, tobacco, printing, paper, gunpowder, etc.). Unlike the descent-of-wholes hypothesis, the hierarchical hypothesis recognizes that cores are not as com-pletely isolated as good biological species. Kroeber's "tree of culture" implies that cultural descent is like a rain forest canopy tree—one whose crown is a tangle of branches (related by birth) and vines (related by diffusion). For some substantial period of time, one can easily distinguish what grows as branches from what grows as vines with more care, even in a thick, old tangle. Eventually, however, over the course of thousands of years, vines will proliferate and come to obscure the branches. At the same time, processes of coherence will integrate elements with separate histories. Old vines will coalesce to form a solid trunk—much like the strangler fig that starts out as a viny parasite of a tree, but gradually forms a solid trunk about its host, which then dies.

The hierarchical model also acknowledges the rapidity of cultural evolution, compared with the biological case. The evidence of a history of common descent will gradually disappear in independent lineages. Barth (1987) gave a detailed account of the rapid evolution of the core tradition of the Mountain Ok of New Guinea due to a mutation-like process. The case is probably unusual because the core traditions are transmitted in rare secret rituals that create high "mutation" rates via forgetting. But even in the absence of diffusion, evidence of common ancestry in sister cultures will degrade on the millennial timescale (compared with hundreds of millions of years, in the case of sister species of mammals). We know from the massive convergence of agricultural technology and state-level social institutions in the pre-Columbian New and Old Worlds that cultural evolution can produce spectacular adaptive change on the timescale of a few thousand years. We can almost be certain that Old–New World similarities were independently derived convergences, but only because we have the evidence of hundreds of cultures on both branches to help distinguish the vines. Notoriously, careless historians who ignore the massively redundant evidence have no trouble "finding" false descent relationships between Old and New World cultures (e.g., Heyerdhal, 1950).

The Practice of Constructing Core-Cultural Phylogenies.   The hierarchical hypothesis is supported to the extent that it can be shown that a large complex of core traits has a common pattern of descent. The core traditions in question must be related through a sequence of population fissionings (allowing for the odd core transfer). The existence of only one deep element, such as language, cannot be used alone to infer the existence of a full core of shared traditions among cultures related by language only. Because language phylogenies can be traced to considerable depth using conservative aspects of vocabulary and phonology, language trees are the usual starting point for attempting to trace out the descent patterns of larger core units. Related traditions can then be used as a basis for reconstructing a fuller culture history, including the "proto-tradition" out of which they evolved (see Aberle, 1984, 1987). Sometimes genetic relatedness of the populations involved provides supplementary evidence, given that full core replication by processes other than fission of a parent culture is unusual. However, if diffusion and rapid evolution swamped all traces of relationship by birth, anthropology could not speak of branches, only vines, and hypothesis 3 would be supported.

The work of Rushforth and Chisholm on Athapaskan similarities illustrates the method. Linguistic evidence indicates that Athapaskans are part of a second wave of Native Americans that arrived from Asia a few thousand years after the migration that contributed most known pre-Columbian populations. At contact, the Athapaskan language family was spoken by people in quite isolated clusters in Canada, California, and the Southwest (the Southwestern group includes the famous Apache and Navajo). According to their analysis, the evidence suggests that a core of meaning related to social behavior coheres with language and that all are "cognate," (i.e., related historically by culture birth; Rushforth and Chisholm, 1991).

First, the authors implied that the pertinent beliefs and values in Athapaskan populations are distinct from those of the surrounding populations belonging to other language groups (although it is also true that the differences are not thoroughly documented in their presentation). Second, similarity by diffusion can be ruled out because of the highly discontinuous geographical clustering of the carrier populations. Third, independent origins are highly improbable (Rushforth and Chisholm, 1991), even if each cluster of populations is taken as a whole.

Rushforth and Chisholm (1991) concluded that the pertinent beliefs and values are all "genetically" related, having "originated in, and developed from, a common, ancestral cultural tradition that existed among Proto-Athapaskan or, perhaps, even among [the ancestral] NaDene peoples" (p. 71). As they put it, "simplicity strongly argues" that "this cultural framework originated once, early in Proto-Athapaskan or NaDene history and has persisted (perhaps with some modifications) in different groups after migrations separated them from contact with each other" (p. 78).

The work of Indo-Europeanists to reconstruct the descent of societies speaking this family of languages is the most ambitious attempt yet made to reconstruct a pattern of descent for a core. According to some Indo-Europeanists like George Dumézil and Marija Gimbutas, the Indo-Europeans are the bearers of a core tradition consisting of language elements, myths, and a distinctive tripartite pattern of social organization that had its origin in a particular culture of steppe horse nomads. Gimbutas's reconstructed "Kurgans" lived about 6,500 years ago between the Black and Caspian Seas. Her Kurgan proposal is widely respected but also widely criticized; a reconstruction of such breadth and depth tests the margins of the hierarchical hypothesis (Mallory, 1989).

Shared core traditions have been proposed for people in a number of different regions of the world, each with time horizons dating back at least a few thousand years. Recently reviewed in Durham (1992), these include the oft-cited case of cultural similarity among Polynesian islanders (see especially Kirch, 1986; Kirch and Green, 1987; see critical review in Terrell, 1986), the Athapaskan (Rushforth and Chisholm, 1991) and Indo-European traditions mentioned earlier (e.g., Gamkrelidze and Ivanov, 1990; Hallpike, 1986; but see Mallory, 1989), Mayans (Vogt, 1964), Tibetans (Durham, 1991), and Tupi speakers among native South Americans (Durham and Nassif, 1991). Although one could always argue that the Polynesian case is exceptional because of the inherent isolation of its populations, plausible examples of enduring shared traditions among cultures related by birth have now been proposed for a diverse array of continental populations as well.

Consider Vansina's (1990) recent comprehensive study of political tradition in equatorial Africa. Through a controlled comparison of some 200 distinct societies in the basin of the Zaire river and its tributaries, Vansina concluded that these "widely differing societies arose out of [a] single ancestral tradition" (p. 191) by way of 3,000–4,000 years of historical transformations. As reconstructed by Vansina, the original ancestral tradition came into the region with the immigration of western Bantu-speaking farmers. They brought with them a single distinct pattern of social organization based on fragile temporary alliances into House (capital H in original), village, and district, and a common ideology and world view to go with it (see Vansina, 1990).

From this common baseline, Vansina (1990) argued, through successive splits, migrations, and expansions, "widely differing societies arose out of the single ancestral tradition by major transformations" (p. 191). The variation included, for example, two kinds of segmentary lineage societies, four kinds of associations, and five kinds of chiefdoms or kingdoms. All the while, "the principles and fundamental options inherited [at birth] from the ancestral tradition remained a gyroscope in the voyage through time: they determined what was perceivable and imaginable as change" (p. 195).

Vansina made it clear that outside influences—"the new habitants, the autochthons [indigenous hunter-gatherers in the region], the non-Bantu, the eastern Bantu farmers with their different legacies—each influenced the development of this ancestral tradition differently from place to place" (p. 69). Yet as he repeatedly showed, change "was not mainly induced by outside influences. In all these cases [for example, in the inner Zaire basin] a chain of reactions fed continuous internal innovations. Outside innovations were accepted only insofar as they made sense in terms of existing structures" (p. 126). Even in regions where external influences played a relatively heavy role, the internal sovereignty of distinct polities meant that "internal dynamics always remain determining" (p. 192). Even with the establishment of Atlantic trade after 1480 and the attendant challenges of slave raiding and more, "the tradition was not defeated. It adapted. It invented new structures. [N]o foreign ideals or basic concepts were accepted and not even much of a dent was made in the aspirations of individuals" (p. 236f). Inherited at birth in each equatorial society, the tradition lived on for hundreds of years more, only to be destroyed by European conquest between 1880 and 1920.

Why Core Homology Matters. Vansina's (1990) study illustrates a key proposition of the hierarchical model. Even in continental areas with high contact between peoples, one can still trace "the historical course of a single tradition" (p. 261). But there is a second important implication as well: reconstructing the histories of peoples without written records requires that one distinguish between homologies (similarities produced by culture birth), analogies (similarities produced by convergence or parallel change), and synologies (similarities produced by diffusion or borrowing). The reason, as Vansina noted, is that the reconstruction of past cultures requires that one "seeks out homologies first" (p. 261). Only by identifying genuine cultural homologies can one establish the nature of the initial ideational system that was later transformed by historical processes. To the extent that hypothesis 2 of the four proves valid, it offers a useful tool that societies with no written records can use to gain access to their own histories.

The Descent of Small Cultural Components

On this hypothesis, there is no central core culture that deserves special attention in phylogenetic analysis. Rather, there are multiple "cores" and sometimes quite small units whose descent can be usefully traced. To characterize a narrow region on the continuum of possible hypotheses, we suppose that even the biggest deeply coherent blocks of culture are fairly small.

Definition.   The components are collections of memes that are transmitted as units with little recombination and slow change, and therefore their phylogenies can be reliably reconstructed to some depth. (As for the hierarchical hypothesis, how much recombination and change are tolerable depends on the timescale—deeper phylogenies require more coherent units and slower rates of evolution.) On this hypothesis, different components diffuse and recombine at a rapid rate, compared with the rates of elements within components, so that core-like complexes of components will have shallower phylogenies than their smaller constituent components.

The processes that provide "glue" for the hierarchical core hypothesis also explain the coherence within these smaller units. The amendments needed are only quantitative. If the scope of integration provided by internal processes is limited, and if ethnocentric barriers to diffusion are weak or shifting in kinds of components is protected, recombination between large blocks of memes will be high, although the same processes may protect many small sets of coherent memes. In practice, the units have to be large enough to have significant internal complexity, or their actual documented history has to be good. Otherwise, the amount of information available for descent reconstruction is limited. Thus, before the advent of modern molecular techniques, the functionally similar genes in various bacteria had a pattern of descent, but the traces of history needed to reconstruct the pattern were absent. When genes can be sequenced, a vastly greater array of data is available by reading the DNA strands directly. Strings of functionally irrelevant, highly improbable similarities and differences in the strands can now be used to construct phylogenies where classical biologists despaired.

Is there any theoretical reason to expect smaller, rather than larger, coherent units in the cultural case? The fact that different cultural variants can be acquired from different people during different parts of the life cycle makes genealogical processes less effective at maintaining coherence than the analogous processes in the case of genetic evolution. We all have many cultural parents, with the attendant potential for independent samples of culture from many sources. At the same time, mixing could be less effective within small units because one can learn some things from one person or a small group of closely related mentors and other things from a quite different set of mentors. This may lead to small, but coherent, subcultures within a larger culture complex. For example, the culture of science is fairly coherent and coexists within the same society as the culture of rock climbers, but people from each of these partial cultures may share the partial culture of the English language. (Of course, to some extent, science, rock climbing, and English are international institutions and provide avenues of communication among the cultures that play host to them.) On this argument, maintaining cultural coherence over large units faces a considerable mechanical obstacle due to the hyperrecombinatorial nature of the cultural transmission system.

If one focuses on one special unit, such as those few features of language that cohere over long timescales, one may indeed find a few correlated units of other types that persist in having a pattern of descent in common with the language features, merely as a matter of chance. From one attempt at deep reconstruction to another, different pseudocore elements will be discovered.

The linguistic characters used by historical linguists (basic lexicon, phonological rules) provide good examples of what is meant by a cultural component. Linguists can reconstruct a phylogeny for a basic lexicon and phonological rules that tells us the pattern of relationships among variants of this character. For example, we know that the basic lexicon and phonological rules that characterize English and German share a more recent ancestor than either does with French. In other words, we believe that we can trace back the sizable complex of memes that underlie the English basic lexicon and phonology through a series of ancestor-descendant pairs to a point where the same people speak a language that has phonological rules and a basic lexicon that also forms the ancestor of German.

Examples of Coherence of Small Units and Recombination among Them.   A clear example of how sets of memes exhibit considerable coherence when borrowed between groups can be seen in the adoption of the "age organization" principle by Bantu peoples in Central and Eastern Africa (LeVine and Sangree, 1962). Age sets are an institution in which children born within a few years of one another are simultaneously initiated into a group of adolescents of nearly the same age (boys and girls into different sets). After initiation, a given age set is a corporate organization that is formally charged with a series of roles in succession (warrior, married man, elder, etc.), with formal graduation from role to role of the whole set.

The Tiriki (an offshoot of the Abaluhyia Bantu), for example, currently have an age organization almost identical to that of their Nilotic neighbors, the Terik, while remaining distinctively Abaluhyia in language and culture. This situation arose as a result of intense political turmoil in the mid-eighteenth century, when the Terik offered asylum to refugee segments of Abaluhyia lineages on condition that their men would become incorporated into the Terik warrior groups. At this time, the Tiriki warriors accepted the full set of initiation rituals for their sons (circumcision and seclusion) and adopted the seven named age-set system. In addition, the grades of warrior, retired warrior, judicial elder, and ritual elder emerged as the principal corporate units of political significance at the local level, and the Nilotic ideology of bravery and prowess in battle became predominant. Indeed, there is some evidence that the Tiriki became a distinct group within the Abaluhyia as a result of their adoption of Terik customs, as is indeed suggested by their name. Interestingly, the practice of female circumcision was viewed with disfavor by the Tiriki, such that they never adopted this trait. In short, this example shows how a number of cultural elements can be borrowed as a package, although not indiscriminately so, and the packages are often smallish.

Linguistics also provides many good examples. Important components of the language spoken by a group of people often have a different evolutionary history from the basic lexicon and phonology of the same language. A substantial fraction of the words in the English lexicon (but not in the basic lexicon) share more recent common ancestors with words in French than with German. This is also true of English syntax, subject-verb-object like French, not subject-object-verb like most Germanic languages. It is even true of aspects of English phonology. For example, English speakers distinguish *veal* and *feel*, apparently as a result of the influence of Norman loan words. Thus, we can identify coherent cultural entities, words, and syntactical and phonological rules that are longer lived than the larger complex called the English language, and whose ancestry can be traced back through independent series of ancestor–descendant relationships. Thomason and Kaufman (1991) provided numerous other examples, including the Ma'a language spoken in northern Tanzania, which, despite classification as a Nilotic language, has a basic lexicon related to Cushitic languages and a grammar related to Bantu languages. (We return to the problems that this example raises for the practice of linguistic classification later.)

Less formal data suggest that important social organizational rules and values are often decoupled rather rapidly from descent, as can be reckoned by the user of a basic lexicon and phonology. In Central and East Africa, for example, cyclical and linear age sets, alternating generation classes, genital mutilation of males and females, warrior organizations, and many other associated practices are common among people whose basic lexicons are categorized as Nilotic, Cushitic, and Bantu. Although it was once thought that these customs were essentially of Cushitic origin, it is now clear from Ehret's (1971) linguistic analyses and voluminous ethnographic sources that different customs associated with the recruitment, function, and ritual validity of age organizations have been repeatedly borrowed between protolinguistic units over the last 5,000 years, reflecting periods of proximity, expansion, and dependence. The resulting situation is one of a thorough intertwining of social organization and language.

In some cases, the distribution of cultural traits appears to represent functional convergences, as in the case of the Tiriki, who adopted age sets and male circumcision in response to the turbulent militaristic conditions of the times. In other cases, there is evidence of a decoupling of apparently nonfunctional details. Thus, the Bantu Gusii conduct male and female genital mutilation but apparently have never organized their men into age sets (LeVine and Sangree, 1962); the Datoga dropped the 5–8 cycling age-set system of their protosouthern Nilotic ancestors for noncycling generation classes (Ehret, 1971). The Bantu Kuria provide a particularly revealing example of this complexity (Tobisson, 1986). Men belong to age sets almost indistinguishable in name from those of the southern Nilotes but are recruited on entirely different principles (father's set membership, rather than circumcision cohort). However, the Kuria have important military units; these are based on circumcision but are organized quite differently from those of the Nilotes and are quite unrelated to the age-set system that among the Kuria bears Nilotic names. The inescapable conclusion to be drawn from these complex observations is that the phylogeny of language and other cultural characters are often distinct.

Religious practices provide many further examples: the spread of the Sun Dance on the Great Plains, the spread of Islam from Western to Central and Eastern Asia and Northern Africa, millenarian movements in Melanesia, and so on. Ethnographic details are sometimes available for such borrowings, and the motives involved do not seem to be such as to enforce much coherence. For example, Sierra Leonean Creoles first adopted freemasonry in the late 1940s. The reason seems to have been that exclusive occupation of elite political roles had long served Creoles with an integrative community symbolic system. When Creoles lost power to the large majority of tribal peoples without a slave background, this symbol system was lost. Freemasonry happened to be an available substitute and quickly became very important (Cohen, 1974). Of course, national and imperial powers sometimes maintain symbolic units over wide areas for impressive periods of time. The Habsburgs' success in defending Catholicism and expelling Protestantism and Islam from their dominions during the life of the Austro-Hungarian Empire is a famous example. However, the need to exercise a large measure of brute force to succeed in such an enterprise is perhaps testimony to the long-run weakness of large-scale coherence.

There also may be rather well-bounded subcultures within a language group (as defined by a basic lexicon), as in the Indian caste system or the class, occupational, and religious subunits of many other state-level agricultural societies. Here, some memes are confined to some subset of the group—the castes, the guild, and so on. These subgroups may be marked by boundaries that are rather impervious to the flow of at least some kinds of memes. This phenomenon reaches its extreme in contemporary societies like the United States, where a diverse array of specialized subcultures of many types exists.

These subgroups may be far more enduring than the "cultures" to which they bear a somewhat temporary allegiance. For example, East Africanists often question the attribution of any time depth to the ethnic units currently residing in the area. This is not simply a consequence of European colonialist policy. Thus, Waller (1986) painted a picture of the nineteenth-century and earlier ephemeral political associations of clans with different linguistic and cultural backgrounds, linked through diverse patterns of intermarriage, trade, expansion, and dependency. These flexible and highly inclusive concepts of group identity are seen as an adaptation to heterogeneous and somewhat unpredictable environmental conditions (i.e., circumstances by no means unique to East Africa). Knauft (1985) told a similar story about the Gebusi and their neighbors, the Bedamini, in the Fly River area of Papua New Guinea. According to this picture, there would be frequent recombination of memes due to temporary association of peoples who exchange memes while in contact.

Comparison of Core and Small Units Hypotheses.   Whether such examples are more representative than those given by supporters of the core hypothesis is an important, but unanswered, question. The little anthropological work done is not capable of answering this question. There are a few studies, but they are indecisive. Jorgensen's (1967, 1980) studies of the Salish and larger-scale analysis of the Indians of western North America are examples of the kind of comprehensive cultural analysis that might deliver. However, his methods are based on measures of overall similarity and difference and do not constitute proper analyses of descent. Biological systematists argue that the only evidence for membership in a given branch of a descent tree is given by characters that are shared by that branch alone but not more ancient or more recent similarities, much less similarities acquired by convergence.

Even in the case of language, "wave" models of linguistic evolution have long contended with "genetic" analyses based on strict criteria of descent (Jorgensen, 1980; Mallory, 1989; Renfrew, 1987). Many features of Indo-European languages seem easier to account for if we assume that the whole family was in contact throughout most of its history and that innovative features tended to diffuse from multiple centers to neighboring languages. Treelike models of relationship can certainly be constructed for data that are substantially influenced by wavelike processes (e.g., with clustering algorithms). Just because a tree diagram explains much of the variation in a set of data, it does not guarantee that the descent hypothesis is correct. It would be quite interesting to see the modern "cladistic" methods of biological systematists formally applied to such cultural descent problems. At least part of the solution to the debate between proponents of hierarchical core and small units hypotheses will rest on the application of sharper methodological tools, and biologists have something to offer.

The Descent of Memes

The boundary of the small units hypothesis toward the small end of the continuum is not well defined. It is also possible that, aside from core vocabulary and phonology, there are few multimeme cultural units that are well protected from diffusion. It could be that each of the cultural things we observe is affected by many memes, that these memes readily diffuse from one socially or linguistically defined group to another, and that memes that affect different cultural components readily recombine. For example, a religious system might be affected by many different memes: beliefs about causation, beliefs about the role of men and women, beliefs about disease, and so on. This system could diffuse from one group to another, and then some of the memes could recombine with other aspects of the culture. Beliefs about the roles of men and women that came with the new religious system might then recombine with preexisting beliefs about subsistence practices, generating new, observable subsistence variants. If we could actually measure the memes that characterize different human groups, this case would be much like the previous one, except we would reconstruct the phylogenies of memes largely instead of whole cultural components.

Descent Analysis: Impossible or Uninteresting?

There are several situations in which descent analysis regarding culture is impossible. If we observe phenotype, and not the mental representations that are stored and transmitted, we cannot directly measure memes. The fact that many memes affect any given observable cultural attribute makes it difficult to trace the path of recombining memes, and reconstructing phylogenies is likely to be impossible. If the actual units to which descent might apply are as small as or smaller than our practically observable units, descent is impossible to trace simply because there is not enough information available to separate common descent from other hypotheses, such as independent origins. A quantitative character subject to blending inheritance is an extreme example.

In some cases, methodological improvements may increase resolution. Comparative ethnographic data with age sets scored as present/absent, or as a quantitative variable on political importance, would not contain enough detail to reconstruct much history in East Africa. A richer data set offers more possibilities, as we have seen.

The existence of coherent cultures will depend on the rate of diffusion and independent evolution. If the rate of diffusion among cultures for most characters is high, then there will be no cultural unit larger than some small atomistic unit of which to track the descent. Between the time that a newly formed group buds off its parent, and the time it creates buds itself, many new traits will have entered the group from outside. If the rate of evolution is high, the trace of history also vanishes. High rates of random evolution, especially simple characters with few observable states, will eventually result in so many random "hits" that descendant characters will have occupied all states fairly recently. Similar simple artistic motifs are found in many cultures, perhaps because artists frequently rediscover and abandon them. Functional convergence presents similar problems. Around the world, tropical horticulturalists often live in small-scale societies that are murderously hostile to their neighbors. This commonality is presumably a by-product of the population densities and level of political organization supportable in wet tropical climates, not due to common ancestry.

Even when descent analysis is possible, it may be uninteresting. The few components that resist diffusion—basic lexicon and so on—will be descended from the grandparental group (defined in terms of basic lexicon), but most components will not be descendants of components in that same grandparental group. Put another way, a culture is nothing more than its most elementary components. Each component may well be traceable back to a grandparental society. But a neighboring society may share particular grandparents for particular traits at random. Phylogenetic analysis could still be conducted for an element-by-element case, and this might be of interest or utility for some special cases. However, one important use of phylogeny is to make manageable the overwhelming complexity of populations and cultures. With no coherence, the analysis of descent could promise nothing in this regard.

Partial Phylogenies and the Study of Adaptation

Good phylogenies are crucial for the proper study of adaptation using the comparative method. Comparative studies attempt to determine the function of various attributes by looking for predicted correlations among societies. For example, Thornhill (1991) hypothesized that inbreeding avoidance rules function to preserve capital in powerful families. To test this hypothesis, she collected data on inbreeding rules and social stratification, predicting (accurately) that the degree of elaboration of rules would positively correlate with the degree of social stratification.

Similar studies utilizing correlations among species are widely used in comparative biology. A key problem in such comparative studies is determining the extent to which different societies (or species) are independent data points. In comparative biology, only independently derived associations are counted as separate data points. Thus, if an innovation arises and then the lineage speciates, preserving the innovation in both daughter species, the daughter species should be counted as a single data point. The first step in the proper exercise of the comparative method is phylogenetic reconstruction (Harvey and Pagel, 1991). In cross-cultural anthropology, this problem is referred to as "Galton's problem." Scholars working in this discipline attempt to select their samples so as to include only unrelated cultures or correct for diffusion by using statistical methods (Burton and White, 1987).

Adaptations acquired by diffusion from other groups are related by descent to the adaptations in those groups. If one analogizes with the practice in biology, such adaptations would not be counted as independent cases because the adaptation in the borrowing group is *not* an innovation. However, to the extent that diffusion represents the goal-driven choices of individuals in the borrowing group (or some other potentially adaptation-producing process), the borrowed trait *is* independent. If it had not been an adaptation, it would not have been adopted. This problem is particularly acute given that the rate of diffusion of new cultural adaptations through biased transmission is likely to be much higher than the rate of innovation. If this is so, most groups will adapt by borrowing, and it is unreasonably conservative to disregard these cases.

The relationship between the Sun Dance and the buffalo-hunting ecology of the Great Plains people illustrates this difficulty. A summer ceremonial called the Sun Dance characterized all the Great Plains buffalo-hunting people. One might hypothesize that such a ceremony is related to the fission-fusion social organization that characterized the buffalo-hunting ecology of those people. But does one count this as one case, or several? It is likely that this ceremony originated with the Crow and diffused to other tribes, so the various versions of the ceremony are not independent inventions. However, each group did adopt the ceremony, perhaps because it served the hypothesized need. Moreover, it could be that, in the absence of diffusion, each group would have independently developed a summer ceremonial but did not because the rate of adaptation by diffusion is faster than independent invention (Oliver, 1962).

On a longer temporal and spatial scale, the problem is also well illustrated by basic technical innovations like agriculture or iron working. The number of independent inventions of these techniques were few indeed—fewer even than the number of language-based descent groups that have subsequently adopted them. It seems absurd to say that we cannot really decide whether iron working is adaptive because all examples of iron-working technology are derived from a single common ancestor in Asia Minor about 3,400 years ago. Regardless of our answer of how many cases of iron working to count for purposes of estimating its adaptive value, it seems clear that language-based descent groups are largely irrelevant to solving this problem. We say "largely irrelevant" because it does seem that an association of an important adaptive innovation with a linguistic unit sometimes lasts long enough to carry the language area great distances, as with iron working and the Bantu expansion in Africa in the last millennium B.C. and the first millennium A.D. (Ehret, 1982); the use of abundant, but low-quality, plant resources and the spread of Numic languages in the American Great Basin (Bettinger and Baumhoff, 1982); and the domestication of the horse, invention of wheeled transport, and spread of Indo-European (Mallory, 1989). Note that such associations tend to persist only for a millennium or so, although the expansion of the innovating group tends to preserve the association.

Conclusion

It seems that, as regards most meme complexes, specific cultures are more like local populations within a species than like species. The whole human species is united by complex flows of ideas from one culture to another. This has always been so, although the geographical isolation of the New World, Australia, and a few other areas from each other and Eurasia may have substantially isolated large blocks of cultures on multimillennial timescales. On smaller time and space scales, other mechanisms of isolation and coherence do generate some patterns of descent that are traceable for a few millennia.

The use of descent analysis for cultural units has a long, but controversial, history. Many authors claim a degree of success in reconstructing the history of descent of fairly large cultural units fairly far into the past. The most interesting outstanding question is the size and timescale of coherent units of culture. Do single cores in an interrelated complex have real histories that reach back five millennia or more? There seems to be no doubt that many small units have descent relationships that can be reliably inferred for this depth, but the upper size/time limit is not well defined by current methods. There is an ill-explored neutral analogy worth further work here. The cladistic revolution in systematic biology has sharpened concepts and built new tools for phylogenetic analysis. Might they be used, despite the problem of high diffusion rates among cultures compared with species, to help advance the resolution of genetic versus wave explanation of culture history?

REFERENCES

Aberle, D. F. 1984. The language family as a field for historical reconstruction. *Journal Of Anthropological Research* 40:129–136.

Aberle, D. F. 1987. Distinguished lecture: What kind of science is anthropology? *American Anthropologist* 89(3):551–566.

Barth, F. 1987. *Cosmologies in the making: A generative approach to cultural variation in inner New Guinea.* Cambridge: Cambridge University Press.

Barton, N. H., & G. M. Hewitt. 1989. Adaptation, speciation, and hybrid zones. *Nature* 341:497–502.

Bettinger, R. L., & M. A. Baumhoff. 1982. The Numic spread, Great Basin cultures in competition. *American Antiquity* 47:485–503.

Boyd, R., & P. J. Richerson. 1987. The evolution of ethnic markers. *Cultural Anthropology* 2:65–79.

Boyd, R., & P. J. Richerson. 1992. How microevolutionary processes give rise to history. In: *History and evolution*, M. H. Nitecki, & D. V. Nitecki, eds. Albany: SUNY.

Burton, M. L., & D. R. White 1987. Cross cultural surveys today. *Annual Review of Anthropology* 16:143–160.

Cohen, A. 1974. *Two-dimensional man.* Berkeley: University of California Press.

Dawkins, R. 1993. *Viruses of the mind.* In: Dennett and his critics: Demystifying the mind, B. Dahlbom, ed. London: Basil Blackwell.

Durham, W. H. 1991. *Coevolution.* Palo Alto, CA: Stanford University Press.

Durham, W. H. 1992. Applications of evolutionary culture theory. *Annual Review of Anthropology* 21:331–356.

Durham, W. H., & R. C. Nassif. 1991. *Managing the competition: A Tupi adaptation in Amazonia.* Presented at UNESCO Conference Food and Nutrition in the Tropical Forest, Paris.

Eberhardt, W. G. 1990. Evolution in bacterial plasmids and levels of selection. *Quarterly Review of Biology* 65:3–22.

Ehret, C. 1971. *Southern Nilotic history: Linguistic approaches to the study of the past.* Evanston, IL: Northwestern University Press.

Ehret, C. 1982. Linguistic inferences about early Bantu history. In: *Archaeological and linguistic reconstruction of African history,* C. Ehret, & M. Posnansky, eds. (pp. 57–65). Berkeley: University of California Press.

Ehrlich, P., & P. Raven. 1969. Differentiation of populations. *Science* 165:1228–1232.

Falk, R., & E. Jablonka. 1997. Inheritance: Transmission and development. In: *Human By Nature: Between Biology and the social sciences.*

Foley, R. 1987. *Another unique species.* London: Longham.

Gamkrelidze, T. V., & V. V. Ivanov. 1990. The early history of Indo-European languages. *Scientific American* 262(3):110–116.

Hallpike, C. 1986. *The principles of social evolution.* Oxford: Clarendon.

Harvey, P. H., & M. Pagel. 1991. *The comparative method in evolutionary biology.* Oxford: Oxford University Press.

Heyerdhal, T. 1950. *Kon-tiki: Across the Pacific by raft.* New York: Garden City Books.

Jeffreys, A. J., S. Harris, P. A. Barrie, D. Wood, A. Blanchetot, & S. Adams. Evolution of gene families: The globin genes. In: *Evolution from molecules to men,* J. S. Bendall, ed. (pp. 175–196). Cambridge: Cambridge University Press.

Jorgensen, J. G. 1967. *Salish language and culture.* Indiana University Language Science Monographs, vol. 3, Bloomington, IN.

Jorgensen, J. G. 1980. *Western Indians: Comparative environments, languages, and cultures of 172 Western American Indian tribes.* San Francisco: W. H. Freeman.

Kirch, P. V. 1986. *The evolution of the Polynesian chiefdoms.* Cambridge, MA: Cambridge University Press.

Kirch, P. V., & R. C. Green. 1987. *History, phylogeny, and evolution in Polynesia. Current Anthropology* 28(4):431–456.

Knauft, B. 1985. *Good company and violence: Sorcery and social Action in a lowland New Guinea society.* Berkeley: University of California Press.

Kroeber, A. L. 1948. *Anthropology.* New York: Harcourt Brace.

LeVine, R. A., & W. H. Sangree. (1962). The diffusion of age-group organization in East Africa: A controlled comparison. *Africa* 32:97–110.

LeVine, R. A., & D. T. Campbell. 1972. *Ethnocentrism: Theories of conflict, ethnic attitudes, and group behavior.* New York: Wiley.

Levinton, J. 1988. *Genetics, paleontology, and macroevolution.* Cambridge: Cambridge University Press.

Mallory, J. P. 1989. *In search of the Indo-Europeans: Language, myth, and archaeology.* London: Thames and Hudson.

Marks, J., & E. Staski. 1988. Individuals and the evolution of biological and cultural systems. *Human Evolution* 3(3):147–161.

McNeill, W. H. 1986. *Mythistory and other essays.* Chicago: University of Chicago Press.

Mishler, B., & R. Brandon. 1987. Individuality, pluralism, and the species concept. *Biology and Philosophy* 2:397–414.

Oliver, S. 1962. Ecology and cultural continuity as contributing factors to social organization of the Plains Indians. *University of California Publications in Archaeology and Ethnology* 48:1–90.

Needham, J. 1988. *Science and civilization in China.* Cambridge: Cambridge University Press.

Rappaport, R. A. 1979. *Ecology, meaning and religion.* Richmond, CA: North Atlantic.

Renfrew, C. 1987. *Archaeology and language: The puzzle of Indo-European origins.* London: J. Cape.

Ridley, M. 1986. *Evolution and classification: The reformation of cladism*. London: Longman.

Rogers, E. M. 1983. *Diffusion of innovations*, 3rd ed. New York: Free Press.

Rushforth, S., & J. S. Chisholm. 1991. *Cultural persistence: Continuity in meaning and moral responsibility among Bearlake Athapaskans*. Tucson: University of Arizona Press.

Sahlins, M. 1976. *Culture and practical reason*. Chicago: University of Chicago Press.

Saussure, F. de. 1959. *Course in general linguistics*. New York: McGraw Hill.

Stebbins, G. L. 1950. *Variation and evolution in plants*. New York: Columbia University Press.

Steward, J. 1955. *Theory of culture change: The methodology of multilinear evolution*. Urbana: University of Illinois Press.

Terrell, J. 1986. *Prehistory of the Pacific islands: A study of variation in language, customs, and human biology*. Cambridge, MA: Cambridge University Press.

Thomason, S. G., & T. Kaufman. 1991. *Language contact, Creolization, and genetic linguistics*. Berkeley: University of California Press.

Thornhill, N. W. 1991. An evolutionary analysis of rules regulating human inbreeding and marriage. *Behavioral and Brain Sciences* 14:247–293.

Tobisson, E. 1986. *Family dynamics among the Kuria: Agropastoralists in Northern Tanzania*. Gothenburg Studies in Social Anthropology no. 9. Gothenberg.

Turner, V. 1977. Process, system and symbol: A new anthropological synthesis. *Daedalus* I:61–80.

Vansina, J. 1990. *Paths in the rainforests: Toward a history of political tradition in equatorial Africa*. Madison: University of Wisconsin Press.

Vogt, E. Z. 1964. The genetic model and Maya cultural development. In: *Desarrollo cultural de los Mayas*, E. Z. Vogt, & A. Ruz L. eds. (pp. 9–48). Mexico City: University Nacional Autonomade Mexico.

Waller, R. (1986). Ecology, migration, and expansion in East Africa. *African Affairs* 85:347–370.

Wardhaugh, R. 1992. *An introduction to sociolinguistics*. 2nd ed. Oxford: Blackwell.

17 Was Agriculture Impossible during the Pleistocene but Mandatory during the Holocene?

A Climate Change Hypothesis

With Robert L. Bettinger

Evolutionary thinkers have long been fascinated by the origin of agriculture. Darwin (1874) declined to speculate on agricultural origins, but twentieth-century scholars were bolder. The Soviet agronomist Nikolai Vavilov, the American geographer Carl O. Sauer, and the British archaeologist V. Gordon Childe wrote influential books and articles on the origin of agriculture in the 1920s and 1930s (see Flannery, 1973, and MacNeish, 1991:4–19, for the intellectual history of the origin of agriculture question). These explorations were necessarily speculative and vague but stimulated interest in the question.

Immediately after World War II, the American archaeologist Robert Braidwood (Braidwood et al., 1983) pioneered the systematic study of agricultural origins. From the known antiquity of village sites in the Near East and from the presence of wild ancestor species of many crops and animal domesticates in the same region, Braidwood inferred that this area was likely a locus of early domestication. He then embarked on an ambitious program of excavation in the foothills of the southern Zagros Mountains using a multidisciplinary team of archaeologists, botanists, zoologists, and earth scientists to extract the maximum useful information from the excavations. The availability of ^{14}C dating gave his team a powerful tool for determining the ages of the sites. Near Eastern sites older than about 15,000 B.P. excavated by Braidwood (Braidwood and Howe, 1960) and others were occupied by hunter-gatherers who put much more emphasis on hunting and unspecialized gathering than on collecting and processing the seeds of especially productive plant resources (Goring-Morris and Belfer-Cohen, 1998; Henry, 1989). Ages are given here as calendar dates before present (B.P.), where present is taken to be 1950, estimated from ^{14}C dates according to Stuiver et al.'s (1998) calibration curves. The Braidwood team showed that about 11,000 years ago, hunter-gatherers were collecting wild seeds, probably the ancestors of wheat and barley, and were hunting the wild ancestors of domestic goats and sheep. At the 9,000 B.P. site of Jarmo, the team excavated an early farming village. Using much the same seed-processing technology as their hunter-gatherer ancestors 2,000 years before, the Jarmo people were settled in permanent villages cultivating early-domesticated varieties of wheat and barley.

Numerous subsequent investigations now provide a reasonably detailed picture of the origins of agriculture in several independent centers and its subsequent diffusion to almost all of the earth suitable for cultivation. These investigations have discovered no region in which agriculture developed earlier or faster than in the Near East, though a North Chinese center of domestication of millet may prove almost as early. Other centers seem to have developed later, or more slowly, or with a different sequence of stages, or all three. The spread of agriculture from centers of origin to more remote areas is well documented for Europe and North America. Ethnography also gives us cases where hunters and gatherers persisted to recent times in areas seemingly highly suitable for agriculture, most notably much of western North America and Australia. Attempts to account for this rather complex pattern are a major focus of archaeology.

Origin of Agriculture as a Natural Experiment in Cultural Evolution

The processes involved in such a complex phenomenon as the origin of agriculture are many and densely entangled. Many authors have given climate change a key explanatory role (e.g., Reed, 1977:882–883). The coevolution of human subsistence strategies and plant and animal domesticates must also play an important role (e.g., Blumler and Byrne, 1991; Rindos, 1984). Hunting-and-gathering subsistence may normally be a superior strategy to incipient agriculture (Cohen and Armelagos, 1984; Harris, 1977), and, if so, some local factor may be necessary to provide the initial impetus to heavier use of relatively low-quality, high-processing-effort plant resources that eventually result in plant domestication. Population pressure is perhaps the most popular candidate (Cohen, 1977). Quite plausibly, the complex details of local history entirely determine the evolutionary sequence leading to the origin and spread of agriculture in every region. Indeed, important advances in our understanding of the origins of agriculture have resulted from pursuit of the historical details of particular cases (Bar-Yosef, 1998; Flannery, 1986).

Nonetheless, we propose that much about the origin of agriculture can be understood in terms of two propositions:

Agriculture was impossible during the last glacial age. During the last glacial age, climates were variable and very dry over large areas. Atmospheric levels of CO_2 were low. Probably most important, last-glacial climates were characterized by high-amplitude fluctuations on timescales of a decade or less to a millennium. Because agricultural subsistence systems are vulnerable to weather extremes, and because the cultural evolution of subsistence systems making heavy, specialized use of plant resources occurs relatively slowly, agriculture could not evolve.

In the long run, agriculture is compulsory in the holocene epoch. In contrast to the Pleistocene climates, stable Holocene climates allowed the evolution of agriculture in vast areas with relatively warm, wet climates, or access to irrigation. Prehistoric populations tended to grow rapidly to the carrying capacity set by the environment and the efficiency of the prevailing subsistence system. Local communities that discover or acquire more intensive subsistence strategies will increase in number and exert competitive pressure on smaller populations with less intensive strategies. Thus, in the Holocene epoch, such intergroup competition generated a competitive ratchet favoring the origin and diffusion of agriculture.[1]

The great variation among local historical sequences in the adoption and diffusion of agriculture in the Holocene provides data to test our hypothesis. In the Near East, agriculture evolved rapidly in the early Holocene and became a center for its diffusion to the rest of western Eurasia. At the opposite extreme, hunting-and-gathering subsistence systems persisted in most of western North America until European settlement, despite many ecological similarities to the Near East. *Thus, each local historical sequence is a natural experiment in the factors that limit the rate of cultural evolution of more intensive subsistence strategies.* For our hypothesis to be correct, the evolution of subsistence systems must be rapid compared to the time cognitively modern humans lived under glacial conditions without developing agriculture, but slow relative to the climate variation that we propose was the main impediment to subsistence intensification in the late Pleistocene epoch. By cultural evolution, we simply mean the change over time in the attitudes, skills, habits, beliefs, and emotions that humans acquire by teaching or imitation. In our view (Bettinger, 1991; Boyd and Richerson, 1985), culture is best studied using Darwinian methods. We classify the causes of cultural change into several "forces." In a very broad sense, we recognize three classes of forces: those due to random effects (the analogs of mutation and drift), natural selection, and decision making (invention, individual learning, biased imitation, and the like). The decision-making forces will tend to accelerate cultural evolution relative to organic evolution, but by how much is a major issue in the explanation of agricultural origins.

Was Agriculture Impossible in the Pleistocene?

The Pleistocene geological epoch was characterized by dramatic glacial advances and retreats. Using a variety of proxy measures of past temperature, rainfall, ice volume, and the like, mostly from cores of ocean sediments, lake sediments, and ice caps, paleoclimatologists have constructed a stunning picture of climate deterioration over the last 14 million years (Bradley, 1999; Cronin, 1999; Lamb, 1977; Partridge, et al., 1995). The Earth's mean temperature dropped several degrees and the amplitude of fluctuations in rainfall and temperature increased. For reasons that are as yet ill understood, glaciers wax and wane in concert with changes in ocean circulation, carbon dioxide, methane and dust content of the atmosphere, and changes in average precipitation and the distribution of precipitation (Broecker, 1995). The resulting pattern of fluctuation in climate is very complex. As the deterioration proceeded, different cyclical patterns of glacial advance and retreat involving all these variables have dominated the pattern. A 21,700-year cycle dominated the early part of the period, a 41,000-year cycle between about 3 and 1 million years ago, and a 95,800-year cycle during the last million years (deMenocal and Bloemendal, 1995). Milankovich's hypothesis that these variations are driven by changes in the earth's orbit, and hence the solar radiation income in the different seasons and latitudes, fits the estimated temperature variation well, although doubts remain (Cronin, 1999: 185–189).

Rapid Climate Variation in the Late Pleistocene

The long timescale climate change associated with the major glacial advances and retreats is not directly relevant to the origins of agriculture because it occurs so slowly compared to the rate at which human populations adapt by cultural evolution. However, the ice ages also have great variance in climate at much shorter timescales. For the last 400,000 years, very high-resolution climate proxy data are available from ice cores taken from the deep ice sheets of Greenland and Antarctica. Resolution of events lasting little more than a decade is possible in Greenland ice 80,000 years old, improving to monthly resolution 3,000 years ago. During the last glacial, the ice core data show that the climate was highly variable on time scales of centuries to millennia (Clark, Alley, and Pollard, 1999; Dansgaard et al., 1993; Ditlevsen, Svensmark, and Johnsen, 1996; GRIP 1993). Figure 17.1 shows data from the GRIP Greenland core. The $\delta^{18}O$ curve is a proxy for temperature; less negative values are warmer. Ca^{2+} is a measure of the amount of dust in the core, which in turn reflects the prevalence of dust-producing arid climates. The last glacial period was arid and extremely variable compared to the Holocene. Sharp millennial-scale excursions occur in estimated temperatures, atmospheric dust, and greenhouse gases. The intense variability of the last glacial carries right down to the limits of the nearly 10-year resolution of the ice core data. The highest resolution records in Greenland ice (and lower latitude records) show that millennial-scale warmings and coolings often began and ended very abruptly and were often punctuated by quite large spikes of relative warmth and cold with durations of a decade or two (e.g., Grafenstein et al., 1999). Figure 17.2 shows Ditlevsen et al.'s (1996) analysis of a Greenland ice core. Not only was the last glacial age much more variable on timescales of a century and a half or more (150-year low-pass filter) but also on much shorter timescales (150-year high-pass filter). Even though diffusion and thinning within the ice core progressively erases high-frequency variation in the core (visible as the narrowing with increasing age of the 150-year high-pass data in figure 17.2), the shift from full glacial conditions about 18,000 years ago to the Holocene interglacial is accompanied by a dramatic reduction in variation on timescales shorter than 150 years. The Holocene (the last relatively warm, ice-free 11,600 years) has been a period of very stable climate, at least by the standards of the last glacial age.[2]

The climate fluctuations recorded in high-latitude ice cores are also recorded at latitudes where agriculture occurs today. Sediments overlain by anoxic water

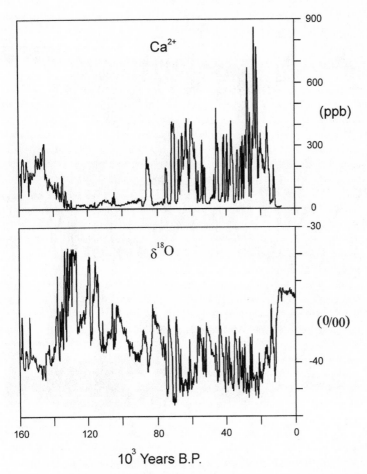

Figure 17.1. Profiles of a temperature index, $\delta^{18}O$, and an index of dust content, Ca^{2+}, from the GRIP Greenland ice core. 200-year means are plotted. The parts of the GRIP profile representing the last interglacial may have been affected by ice flow so their interpretation is uncertain (Johnsen et al., 1997). Note the high-amplitude, high-frequency variation in both the temperature and dust records during the last glacial age. The Holocene epoch is comparatively much less variable. Plotted from original data obtainable at: ftp://ftp.ngdc.noaa.gov/paleo/icecore/greenland/summit/grip/iso-topes/gripd18o.txt and ftp://ftp.ngdc.noaa.gov/paleo/icecore/greenland/summit/grip/chem/ca.txt.

that inhibits sediment mixing by burrowing organisms are a source of low- and mid-latitude data with a resolution rivaling ice cores. Events recorded in North Atlantic sediment cores are closely coupled to those recorded in Greenland ice (Bond et al., 1993), but so are records distant from Greenland. Hendy and Kennett (2000) report on water temperature proxies from sediment cores from the often-anoxic Santa Barbara Basin just offshore of central California. This data shows millennial- and submillennial-scale temperature fluctuations from 60–18

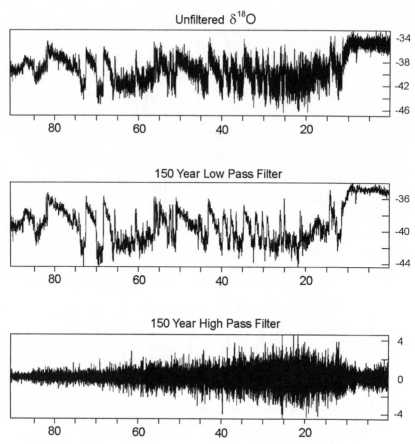

Figure 17.2. High-resolution analysis of the GRIP ice core $\delta^{18}O$ data by Ditlevsen et al. (1996). The low-pass filtered data show that the Holocene epoch is much less variable than the Pleistocene on timescales of 150 years and longer. The high-pass filtered data shows that the Pleistocene was also much more variable on timescales less than 150 years. The high- and low-pass filtering used spectral analytic techniques. These are roughly equivalent to taking a 150-year moving average of the data to construct the low-pass filtered series and subtracting the low-pass filtered series from the original data to obtain the high-pass filtered record. Since layer thinning increasingly affects deeper parts of the core by averaging variation on the smallest scales, the high-pass variance is reduced in the older parts of the core. In spite of this effect, the Pleistocene/Holocene transition is very strongly marked.

thousand years ago with an amplitude of about 8°C, compared to fluctuations of about 2°C in the Holocene epoch. As in the Greenland cores, the millennial-scale events often show very abrupt onsets and terminations and are often punctuated by brief spikes of warmth and cold. Schulz, von Rad, and Erlenkeuser (1998) analyzed organic matter concentrations in sediment cores at oxygen minimum depths from the Arabian Sea deposited over the past 110 thousand years. The variation in organic matter deposited is thought to reflect the strength of upwelling, driven by changes in the strength of the Arabian Sea monsoon. AMS ^{14}C dating of both the Arabian Sea and Santa Barbara cores gives good time control in the upper part of the record, and the climate proxy variation is easily fit to Greenland ice millennial-scale interstadial-stadial oscillations. Allen, Watts, and Huntley (2000) examine the pollen profiles from the laminated sediments of Lago Grande di Monticcio in southern Italy. Changes in the proportion of woody taxa in the core were dominated by large-amplitude changes near the limits of resolution of the data, about a century. The millennial-scale variations in this core also correlate with the Greenland record. Peterson et al. (2000) show that proxies for the tropical Atlantic hydrologic cycle have a strong millennial-scale signal that likewise closely matches the Greenland pattern.

Reports of proxy records apparently showing the ultimate Younger Dryas millennial-scale cold episode, strongly expressed in the North Atlantic records 12,600–11,600 B.P., have been reported from all over the world, including southern German oxygen isotope variations (Grafenstein et al., 1999), organic geochemistry of the Cariaco Basin, Venezuela (Werne et al., 2000), New Zealand pollen (Newnham and Lowe, 2000), and California pollen (West, 2000). The Younger Dryas episode has received disproportionate attention because the time period is easily dated by ^{14}C and is sampled by many lake and mountain glacier cores too short to reach older millennial-scale events. As Cronin (1999:202–221) notes, the Younger Dryas is frequently detected in a diverse array of Northern Hemisphere climate proxies from all latitudes. The main controversy involves data from the Southern Hemisphere, where proxy data often do not show a cold period coinciding with the Younger Dryas, although some records show a similar Antarctic Cold Reversal just antedating the Northern Hemisphere Younger Dryas (Bennett, Haberle, and Lumley, 2000).

Other records provide support for millennial-scale climate fluctuations during the last glacial age that cannot be convincingly correlated with the Greenland ice record. Cronin (1999:221–236) reviews records from the deep tropical Atlantic, Western North America, Florida, China, and New Zealand. Recent notable additions to his catalog include southern Africa (Shi et al., 2000), the American Midwest (Dorale et al., 1998), the Himalayas (Richards, Owen, and Rhodes, 2000), and northeastern Brazil (Behling et al., 2000). Clapperton (2000) gives evidence for millennial-scale glacial advances and retreats from most of the American cordillera—Alaska and western North America through tropical America to the southern Andes.

While the complex feedback processes operating in the atmosphere-biosphere-ocean system are not completely understood (Broecker 1995:241–270), plausible physical mechanisms could have linked temperature fluctuation in both hemispheres. For example, Broecker and Denton (1989) proposed an explanation based upon the effects of glacial meltwater on the deep circulation of the North Atlantic. Today, cold, salty water from the surface of the North Atlantic is the source of about half of the global ocean's deep water. This large outflow of deep water currently must be balanced by an equally enormous inflow of warm surface and intermediate water into the high North Atlantic. If glacial meltwater lowered the salinity of the North Atlantic and interrupted the flow of deep water, the whole coupled atmosphere-ocean circulation system of the world would be perturbed. Broecker and Denton's hypothesis explains how the northern and southern Hemisphere temperature and ice fluctuations could have been in phase even though the direct effects of orbital-scale variation on the two hemispheres are out of phase.

Impacts of Millennial-Scale and Submillennial-Scale Variation on Agriculture

We believe that high-frequency climate and weather variation would have made the evolution of methods for intensive exploitation of plant foods extremely difficult. Holocene weather extremes significantly affect agricultural production (Lamb, 1977). For example, the impact of the Little Ice Age (400–150 B.P.) on European agriculture was quite significant (Grove, 1988). The Little Ice Age is representative of the Holocene millennial-scale variation that is very much more muted than last-glacial events of similar duration. Extreme years during the Little Ice Age caused notable famines and such extremes would have been more exaggerated and more frequent during last glacial times. The United Nations Food and Agriculture Organization's (2000) Global Information and Early Warning System on Food and Agriculture gives a useful qualitative sense for the current impacts of interannual weather variation on food production. Quantitative estimates of current crop losses due to weather variation are difficult to make, but reasonable estimates run 10 percent on a country-wide basis (Gommes, 1999) and perhaps 10–40 percent on a state basis in Mexico, depending upon mean rainfall (Eakin, 2000). Gommes believes that weather problems account for half of all crop losses.

If losses in the Holocene are this high and if high-frequency climate variation in the last glacial age increased at lower latitudes roughly as much as at Greenland, a hypothetical last-glacial farming system would face crippling losses in more years than not. Devastating floods, droughts, windstorms, and other climate extremes, which we experience once a century, might have occurred once a decade. In the tropics, rainfall was highly variable (Broecker, 1996). Few years would be suitable for good growth of any given plant population. Even under relatively benign Holocene conditions agriculturalists and intensive plant collectors have to make use of risk-management strategies to cope with yield variation. Winterhalder and Goland (1997) use optimal foraging analysis to argue that the shift from foraging to agriculture would have required a substantial shift from minimizing risk by sharing to minimizing risk by field dispersal. Some ethnographically known Eastern Woodland societies that mixed farming and hunting, for example, the Huron, seemed not to have made this transition and to have suffered frequent catastrophic food shortages. Storage by intensive plant collectors and farmers is an excellent means of meeting seasonal shortfalls, but is a marginal means of coping with interannual risk, much less multiyear shortfalls (Belovsky, 1987:60).[3]

If Winterhalder and Goland are correct that considerable field dispersal is required to manage Holocene yield risks, it is hard to imagine that further field division would have been successful at coping with much larger amplitude fluctuations that occurred during the last glacial age. We expect that opportunism was the most important strategy for managing the risks associated with plant foods during the last glacial age. Annual plants have dormant seed that spreads their risk of failure over many years, and perennials vary seed output or storage organ size substantially between years as weather dictates. In a highly variable climate, the specialization of exploitation on one or a few especially promising species would be highly unlikely, because "promise" in one year or even for a decade or two would turn to runs of years with little or no success. However, most years would likely be favorable for some species or another, so generalized plant-exploitation systems are compatible with highly variable climates. The acorn-reliant hunter-gatherers of California, for example, used several kinds of oak, gathering less favored species when more favored ones failed (Baumhoff, 1963:table 2). Reliance on acorns demanded this generalized pattern of species diversification because the annual production of individual trees is highly variable from year to year, being correlated within species but independent between species (Koenig et al., 1994). Pleistocene hunter-gatherer systems must have been even more diversified, lacking the kind of commitment to a single resource category (acorns) observed in California.

The evolution of intensive resource-use systems like agriculture is a relatively slow process, as we document. If ecological timescale risks could be managed some way, or if some regions lacked the high-frequency variation detected by the as yet few high-resolution climate proxy records, the evolution of sophisticated intensive strategies would still be handicapped by millennial-scale variation. Plant and animal populations responded to climatic change by dramatically shifting their ranges, but climate change was significant on the timescales shorter than those necessary for range shifts to occur. As a result, last-glacial natural communities must have always been in the process of chaotic reorganization as the climate varied more rapidly than they could reach equilibrium. The pollen record from the Mediterranean and California illustrates how much more dynamic plant communities were during the last glacial age (Allen et al., 1999; Heusser 1995). Pleistocene fossil beetle faunas change even more rapidly than plants because many species, especially generalist predators, change their ranges more rapidly than plants. Hence, they are better indicators of the ecological impacts of the abrupt, large-amplitude climate changes recorded by the physical climate proxies from the last glacial (Coope, 1987).

Could the evolution of intensive plant-exploitation systems have tracked intense millennial- and submillennial-scale variation? Plant food-rich diets take considerable time to develop. Plant foods are generally low in protein and often high in toxins. Some time is required to work out a balanced diet rich in plant foods, for example, by incorporating legumes to replace part of the meat in diets. Whether intensification and agriculture always lead to health declines due to nutritional inadequacy is debatable, but the potential for them to do so absent sometimes-subtle adaptations is clear (Cohen and Armelagos, 1984; Katz, Hediger, and Valleroy, 1974). The seasonal round of activities has to be much modified, and women's customary activities have to be given more prominence relative to men's hunting. Changes in social organization either by evolution in situ or by borrowing tend to be slow (Bettinger and Baumhoff, 1982; North and Thomas, 1973). We doubt that even sophisticated last-glacial hunter-gatherers would have been able to solve the complex nutritional and scheduling problems associated with a plant-rich diet while coping with unpredictable high-amplitude change on timescales shorter than the equilibration time of plant migrations and shorter than actual Holocene trajectories of intensification. In keeping with our argument, the direct archaeological evidence suggests that people began to use intensively the technologies that underpinned agriculture only after about 15,000 B.P. (Bettinger, 2000).

Carbon Dioxide Limitation of Photosynthesis

Plant productivity was also limited by lower atmospheric CO_2 during the last glacial. The CO_2 content of the atmosphere was about 190 ppm during the last glacial age, compared to about 250 ppm at the beginning of the Holocene (figure 17.3). Photosynthesis on earth is CO_2-limited over this range of variation (Cowling and Sykes, 1999; Sage, 1995). Beerling and Woodward (1993; see also Beerling et al., 1993) have shown that fossil leaves from the last glacial age have higher stomatal density, a feature that allows higher rates of gas exchange needed to acquire CO_2 under more limiting conditions. This higher stomatal conductance also causes higher transpiration water losses per unit CO_2 fixed, exacerbating the aridity characteristic of glacial times. Beerling (1999) estimates the total organic carbon stored on land as a result of photosynthesis during the Last Glacial Maximum using a spatially disaggregated terrestrial plant production model coupled to two different global climate models to provide the environmental forcing for plant growth. The model results differ substantially, one indicating a 33 percent lower, and the other a 60 percent lower, terrestrial carbon store at the Last Glacial Maximum compared to the Holocene. Mass-balance calculations based on stable isotope geochemistry also indicate a qualitatively large drop, but uncertainties regarding terrestrial $\delta^{13}C$ lead to a similarly large range of estimates. Low mean productivity, along with greater variance in productivity, would have greatly decreased the attractiveness of plant resources during the last glacial age.

Lower average rainfall and carbon dioxide during the last glacial age reduced the area of the earth's surface suitable for agriculture (Beerling, 1999). Diamond (1997) argues that the rate of cultural evolution is more rapid when innovations in local areas can be shared by diffusion. Thus, a reduction in the area suitable for agriculture and the isolation of suitable areas from one another will have a tendency to reduce the rate of intensification and make the evolution of agriculture less likely in any given unit of time. Since the slowest observed rates of intensification in the Holocene epoch failed to result in agriculture until the European invasions of the last few hundred years, a sufficient slowing of the rate of evolution of subsistence could conceivably in itself explain the failure of agriculture to emerge before the Holocene. A slower rate of cultural evolution would also tend to prevent the rapid adaptation of intensive strategies during any favorable locales or periods that might have existed during the last glacial.

On present evidence we cannot determine whether aridity, low CO_2 levels, millennial-scale climate variability, or submillennial-scale weather variation was

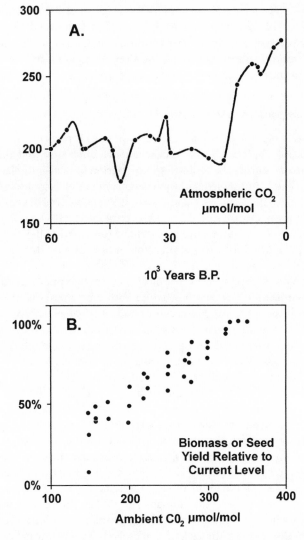

Figure 17.3. Panel A shows the curve of atmospheric CO_2 as estimated from gas bubbles trapped in Antarctic glacial ice. Data from Barnola et al. (1987). Panel B summarizes responses of several plant species to experimental atmospheres containing various levels of CO_2. Based on data summarized by Sage (1995).

the main culprit in preventing the evolution of agriculture. Low CO_2 and climate variation would handicap the evolution of dependence on plant foods everywhere and were surely more significant than behavioral or technological obstacles. Hominids evolved as plant-using omnivores (Milton, 2000), and the basic technology for plant exploitation existed at least 10 thousand years before the Holocene (Bar-Yosef, 1998). At least in favorable localities, appreciable use seems to have been made of plant foods, including large-seeded grasses, well back into the Pleistocene (Kisley, Nadel, and Carmi, 1992). Significantly, we believe, the use of such technology over spans of last-glacial time that were sufficient for successive waves of intensification of subsistence in the Holocene led to only minor subsistence intensification, compared to the Mesolithic, Neolithic, and their ever-more-intensive successors.

Subsistence Responses to Amelioration

As the climate ameliorated, hunter-gatherers in several parts of the world began to exploit locally abundant plant resources more efficiently, but only, current evidence suggests, during the Bølling-Allerød period of near-interglacial warmth and stability. The Natufian sequence in the Levant is the best-studied and so far earliest example (e.g., Bar-Yosef and Valla, 1991). One last siege of glacial climate, the Younger Dryas from 12,900 B.P. until $\approx$ 11,600 B.P., reversed these trends during the Late Natufian (e.g., Goring-Morris and Belfer-Cohen, 1998). The Younger Dryas climate was appreciably more variable than the preceding Allerød-Bølling and the succeeding Holocene (Grafenstein et al., 1999; Mayewski et al., 1993). The 10 abrupt, short, warm-cold cycles that punctuate the Younger Dryas ice record were perhaps felt as dramatic climate shifts all around the world. After 11,600 B.P., the Holocene period of relatively warm, wet, stable, CO_2-rich environments began. Subsistence intensification and eventually agriculture followed. Thus, while not perfectly instantaneous, the shift from glacial to Holocene climates was a very large change and took place much more rapidly than cultural evolution could track.

Might we not expect agriculture to have emerged in the last interglacial 130,000 years ago or even during one of the even older interglacials? No archaeological evidence has come to light suggesting the presence of technologies that might be expected to accompany forays into intensive plant collecting or agriculture at this time. Anatomically modern humans may have appeared in Africa as early as 130,000 years ago (Klein 1999: ch. 7), but they were not behaviorally modern. Humans of the last interglacial were uniformly archaic in behavior. Very likely, then, the humans of the last interglacial were neither cognitively nor culturally capable of evolving agricultural subsistence. However, climate might also explain the lack of marked subsistence intensification during previous interglacials. Ice cores from the thick Antarctic ice cap at Vostok show that each of the last four interglacials over the last 420,000 years was characterized by a short, sharp peak of warmth, rather than the 11,600-year-long stable plateau of the Holocene (Petit et al., 1999). Further, the GRIP ice core suggests the last interglacial (130,000–80,000 B.P.) was more variable than the Holocene, although its lack of agreement with a nearby replicate core for this time period makes this interpretation tenuous (Johnsen et al., 1997). On the other hand, the atmospheric concentration of CO_2 was higher in the three previous interglacials than during the Holocene and was stable at high levels for about 20,000 years following the warm peak during the last interglacial. The highly continental Vostok site unfortunately does not record the same high-frequency variation in the climate as most other proxy climate records, even those in the southern hemisphere (Steig et al., 1998). Some northern hemisphere marine and terrestrial records suggest that the last interglacial was highly variable, while other data suggest a Holocene-length period of stable climates ca. 127,000–117,000 B.P. (Frogley, Tzedakis, and Heaton, 1999). Better data on the high-frequency part of the Pleistocene beyond the reach of the Greenland ice cores is needed to test hypotheses about events antedating the latest Pleistocene. Long marine cores from areas of rapid sediment accumulation are beginning to reveal the millennial-scale record from previous glacial-interglacial cycles (McManus, Oppo, and Cullen, 1999). At least the last five glacials have millennial-scale variations much like the last glacial. The degree of fluctuations during previous interglacials is still not clear, but at least some proxy data suggest that the Holocene has been less variable than earlier interglacials (Poli, Thunell, and Rio, 2000).

During the Holocene, Was Agriculture Compulsory in the Long Run?

Once a more productive subsistence system is possible, it will, over the long run, replace the less-productive subsistence system that preceded it. The reason is simple: all else being equal, any group that can use a tract of land more efficiently will be able to evict residents that use it less efficiently (Boserup, 1981; Sahlins and Service, 1960:75–87). More productive uses support higher population densities, or more wealth per capita, or both. An agricultural frontier will tend to expand at the expense of hunter-gatherers as rising population densities on the farming side of the frontier motivate pioneers to invest in acquiring land from less-efficient users. Farmers may offer hunter-gatherers an attractive purchase price, a compelling idea about how to become richer through farming, or a dismal choice of flight, submission, or military defense at long odds against a more numerous foe. Early farmers (and other intensifiers more generally) are also liable to target opportunistically high-ranked game and plant resources essential to their less-intensive neighbors, exerting scramble competitive pressure on them even in the absence of aggressive measures. Thus, subsistence improvement generates a *competitive ratchet* as successively more land-efficient subsistence systems lead to population growth and labor intensification. Locally, hunter-gatherers may win some battles (e.g., in the Great Basin; Madsen, 1994), but in the long run the more intensive strategies will win wherever environments are suitable for their deployment.

The archaeology supports this argument (Bettinger, 2000). Societies in all regions of the world undergo a very similar pattern of subsistence efficiency increase and population increase in the Holocene, albeit at very different rates. Holocene hunter-gatherers developed local equilibria that, while sometimes lasting for thousands of years, were almost always replaced by more intensive equilibria.

Alternative Hypotheses Are Weak

Aside from other forms of the climate-change hypotheses described, archaeologists have proposed three prominent hypotheses—climate stress, population growth, and cultural evolution—to explain the *timing* of agricultural origins. They were formulated before the nature of the Pleistocene-Holocene transition was understood but are still the hypotheses most widely entertained by archaeologists (MacNeish, 1991). None of the three provides a close fit with the empirical evidence or to theory.

Climate Stress Was First Too Common, Then Too Rare

Childe (1951) proposed that terminal Pleistocene desiccation stressed forager populations and led to agriculture. Wright (1977) argued that Holocene climate amelioration brought pre-adapted plants into the Fertile Crescent areas where agriculture first evolved. Bar-Yosef (1998) and Moore and Hillman (1992) argue that Late Natufian sedentary hunter-gatherers probably undertook the first experiments in cultivation under the pressure of the Younger Dryas climate deterioration. Natufian peoples lived in settled villages and exploited the wild ancestors of wheat and barley beginning in the Allerød-Bølling warm period (14,500–12,900 B.P.) (Henry, 1989) and then reverted to mobile hunting-and-gathering during the sharp, short Younger Dryas climate deterioration (12,600–11,600 B.P.), the last of the high-amplitude fluctuations that were characteristic of the last glacial (Bar-Yosef and Meadow, 1995; Goring-Morris and Belfer-Cohen, 1997). Post-Natufian cultures began to domesticate the same species as warm and stable conditions returned after the Younger Dryas, around 11,600 B.P. Unfortunately, a flat spot in the ^{14}C/calendar-year calibration curve makes precise dating difficult for the most critical several hundred years centered on 11,600 B.P. (Fiedel, 1999). As a component of an explanation of a local sequence of change, such hypotheses may well be correct. Yet they beg the question of why the 15 or so similar deteriorations and ameliorations of the last glacial age did not anywhere lead to agriculture or why most of the later origins of agriculture occurred in the absence of Younger Dryas-scale deteriorations. Note also that, in principle, populations can adjust downward to lower carrying capacities through famine mortality even more quickly than they can grow up to higher ones. Such hypotheses cannot, we believe, explain the longer time- and larger spatial-scale problem of the absence of agriculture in the Pleistocene and its multiple origins and rapid spreads in the Holocene.

The details of subsistence responses to the Younger Dryas in the areas of early origins of agriculture will eventually produce a sharp test of the variability hypothesis. We suggest that the late Natufian de-intensification in response to the Younger Dryas was a retreat from the trend leading to agriculture and was unlikely to have produced the first steps toward domestication. More likely, the late Natufian preserved remnants of earlier, more intensive Natufian technology and social organization that served to start the Levantine transition to agriculture at an unusually advanced stage after the Younger Dryas ended. Events in the Younger Dryas time period also provide an opportunity to investigate the effects of CO_2 concentration partly independently of climate variability. The rise in CO_2 concentration in the atmosphere began two to three millennia before temperatures began to rise and continued to increase steadily through the Younger Dryas (Sowers and Bender, 1995). The Younger Dryas period de-intensification of the Natufian suggests an independent effect of millennial or submillennial variability.

Population Growth Has the Wrong Timescale

Cohen's (1977) influential book argued that slowly accumulating global-scale population pressure was responsible for the eventual origins of agriculture beginning at the 11,600 B.P. time horizon. He imagines, quite plausibly, that subsistence innovation is driven by increases in population density, but, implausibly we believe, that a long, slow buildup of population gradually drove people to intensify subsistence systems to relieve shortages caused by population growth, eventually triggering a move to domesticates. Looked at one way, population pressure is just the population growth part of the competitive ratchet. However, this argument fails to explain why pre-agricultural hunter-gatherer intensification and the transition to agriculture began in numerous locations after 11,600 years ago (Hayden, 1995). Assuming that humans were essentially modern by the Upper Paleolithic, they would have had 30,000 years to build up a population necessary to generate pressures for intensification. Given any reasonable estimate of the human intrinsic rate of natural increase under hunting-and-gathering conditions (somewhat less than 1% yr^{-1} to 3% yr^{-1}), populations substantially below carrying capacity will double in a century or less, as we will see in the models that follow.

A Basic Model of Population Pressure

Since the population explanation for agriculture and other adaptive changes[4] connected with increased subsistence efficiency remains very popular among archaeologists, we take the time here to examine its weakness formally. The logistic equation is one simple, widely used model of the population growth. The rate of change of population density, N, is given by:

$$\frac{dN}{dt} = rN\left(1 - \frac{N}{K}\right) \tag{1}$$

where r is the "intrinsic rate of natural increase"—the rate of growth of population density when there is no scarcity—and K is the "carrying capacity," the equilibrium population density when population growth is halted by density-dependent checks. In the logistic equation, the level of population pressure is given by the ratio N/K. When this ratio is equal to zero, the population grows at its maximum rate; there is no population pressure. When the ratio is one, density dependence prevents any population growth at all. It is easy to solve this equation and calculate the length of time necessary to achieve any level of population pressure, $\pi = N/K$.

$$T(\pi) = -\frac{1}{r}\ln\left(\left(\frac{\pi}{1 - \pi}\right)\left(\frac{1 - \pi_0}{\pi_0}\right)\right) \tag{2}$$

where π_0 is the initial level of population pressure. Let us very conservatively assume that the initial population density is only 1 percent of what could be sustained with the use of simple agriculture and that the maximum rate of increase of human populations unconstrained by resource limitation is 1 percent per year. Under these assumptions, the population will reach 99 percent of the maximum population pressure (i.e., $\pi = .99$) in only about 920 years. Serendipitous inventions (e.g., the bow and arrow) that increase carrying capacity do not fundamentally alter this result. For example, only the rare single invention is likely to so much as double carrying capacity. If such an invention spreads within a population that is near its previous carrying capacity, it will still face half the maximum population pressure and thus significant incentive for further innovation. At an r of 1 percent, such an innovating population will again reach 99 percent of the maximum population pressure in 459 years.

One might think that this result is an artifact of the very simple model of population growth. However, it is easy to add much realism to the model without any change of the basic result. In Appendix 1 we show that a more realistic version of the logistic equation actually leads to even more rapid growth of population pressure.

Allowing for Dispersal

Once, after listening to one of us propound this argument, a skeptical archaeologist replied, "But you've got to fill up all of Asia, first." This understandable intuitive response betrays a deep misunderstanding of the timescales of exponential growth. Suppose that the initial population of anatomically modern humans was only about 10^4 and that the carrying capacity for hunter-gatherers is very optimistically 1 person per square kilometer. Given that the land area of the Old World is roughly $10^8 \, \text{km}^2$, $\pi_0 = 10^4/10^8 = 10^{-4}$. Then using equation 2 and again assuming $r = .01$, Eurasia will be filled to 99 percent of carrying capacity in about 1,400 years. The difference between increasing population pressure by a factor of 100 and by a factor of 10,000 is only about 500 years!

Moreover, this calculation seriously *over*estimates the amount of time that will pass before any segment of an expanding Eurasian population will experience population pressure because populations will approach carrying capacity locally long before the entire continent is filled with people. R. A. Fisher (1937) analyzed the following partial differential equation that captures the interaction between population growth and dispersal in space:

$$\frac{\partial N(x)}{\partial t} = \underbrace{rN(x)\left(1 - \frac{N(x)}{K}\right)}_{\text{population growth}} + \underbrace{d^2 \frac{\partial^2 N(x)}{\partial x^2}}_{\text{dispersal}} \tag{3}$$

Here $N(x)$ is the population density at a point x in a one-dimensional environment. Equation (3) says that the rate of change of population density in a particular place is equal to the population growth there plus the net effect of random, density-independent dispersal into and out of the region. The parameter d measures the rate of dispersal and is equal to the standard deviation of the distribution of individual dispersal distances. In an environment that is large compared to d, a small population rapidly grows to near carrying capacity at its

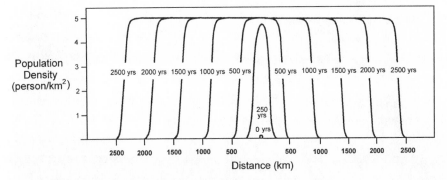

Figure 17.4. A numerical simulation of Fisher's equation showing that after an initial period, population spreads at a constant rate so that at any point in space population pressure increases to its maximum in less than 500 years for reasonable parameter values. (Redrawn from Ammerman and Cavalli-Sforza, 1984).

initial location, and then, as shown in figure 17.4 (redrawn from Ammerman and Cavalli-Sforza, 1984), begins to spread in a wave-like fashion across the environment at a constant rate. Thus, at any given point in space, populations move from the absence of population pressure to high population pressure as the wave passes over that point. Figure 17.4 shows the pattern of spread for $r = .01$ and $d \approx 30$. With these quite conservative values, it takes less than 200 years for the wave front to pass from low population pressure to high population pressure. More realistic models that allow for density-dependent migration also yield a constant, wave-like advance of population (Murray, 1989), and although the rates vary, we believe that the same qualitative conclusion will hold.

The Dynamics of Innovation

So far we have assumed that the carrying capacity of the environment is fixed (save where it is increased by fortuitous inventions). However, we know that people respond to scarcity caused by population pressure by intensifying production, for example, by shifting from less labor-intensive to more labor-intensive foraging, or by innovations that increase the efficiency of subsistence (Boserup, 1981). Since innovation increases carrying capacity, intuition suggests that it might therefore delay the onset of population pressure. However, as the model in Appendix 2 shows, this intuition, too, is faulty.

Figure 17.5 shows the results of the model in Appendix 2. A small population initially grows rapidly. As population pressure builds, population growth rate slows to a steady state in which population pressure is constant, and just enough innovation occurs to compensate for population growth. For plausible parameter values the second phase of population growth steady state is reached in less than a thousand years. Interestingly, increasing the intrinsic rate of innovation or the innovation threshold reduces the waiting time until population pressure is important. Innovation allows greater population increases over the long run, but it does not change the timescales on which population pressure

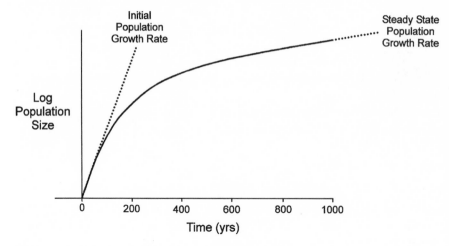

Figure 17.5. This plots the logarithm of population size as a function of time for the model described in Appendix 2. Initially, when there is little population pressure, population grows at a high rate. As the population grows, per capita income decreases, and people intensify. Eventually the population growth rate approaches a constant value at which the growth of intensification balances growth in population. For reasonable parameters ($a = 0.005$, $r = 0.02$, $y_m = 1$, $y_s = 0.1$, $y_i = 0.2$, initial population size 1 percent of initial carrying capacity), it takes less than 500 years to shift from the initial low population pressure mode of growth to the final high population pressure mode of growth.

occurs. The most important factor on timescales of a millennium or greater (if not a century or greater, given realistic starting populations) is the rate of intensification by innovation, not population growth.

This picture of the interaction of demography and innovation leads to predictions quite different from those of scholars like Cohen (1977). For example, we do not expect to see any systematic evidence of increased population pressure immediately prior to major innovations, an expectation consistent with the record (Hayden, 1995). If people are motivated to innovate whenever population pressure rises above an innovation threshold, and if, in the absence of successful innovation, populations adjust relatively quickly to changes in K by growth or contraction, then evidence of extraordinary stress—for example, skeletal evidence of malnutrition—is likely only when rapid environmental deterioration exceeds a population's capacity to respond via a combination of downward population adjustment and innovation.[5] Thus, for parameter values that seem anywhere near reasonable to us, population growth on millennial timescales will be limited by rates of improvement in subsistence efficiency, not by the potential of populations to grow, just as Malthus argued. Populations can behave in non-Malthusian ways only under extreme assumptions about population dynamics and rates of intensification, such as the modern world in which the rate of innovation, but also the rate of population growth, is very high.

Of course, in a time as variable as the Pleistocene epoch, populations may well have spent considerable time both far above and far below instantaneous carrying capacity. If agricultural technologies were quick and easy to develop, the population-pressure argument would lead us to expect Pleistocene populations to shift in and out of agriculture and other intensive strategies as they find themselves in subsistence crises due to environmental deterioration or in periods of plenty due to amelioration. Most likely, minor intensifications and de-intensifications were standard operating procedure in the Pleistocene. However, the time needed to progress much toward plant-rich strategies was greater than the fluctuating climate allowed, especially given CO_2- and aridity-limited plant production.

Cultural Evolution Has the Wrong Timescale

The timing of the origin of agriculture might possibly be explained entirely by the rate of intensification by innovation. For example, Braidwood (1960) argued that it took some time for humans to acquire enough familiarity with plant resources to use them as a primary source of calories, and that this "settling in" process limited the rate at which agriculture evolved. This proposal may explain the post-Pleistocene timing of the development of agriculture. However, if we interpret his argument to be that the settling-in process began with the evolution of behaviorally modern humans, the timescale is wrong again. There is no evidence that people were making significant progress at all toward agriculture for 30,000 years, and Braidwood's excavations at Jarmo show that some 4,000 years was enough to go from an unintensive hunting-and-gathering subsistence system to settled village agriculture in a fast case. Ten thousand years in the Holocene was ultimately sufficient for the development of plant-intensive gathering technologies or agriculture everywhere except in the coldest, plant-poor environments.

The Pattern of Intensification across Cases Implicates Climate Change

We have argued that Malthusian processes lead to population pressure much more quickly than assumed by such writers as Cohen (1977) and that the rate of cultural "settling in" and intensification is faster than Braidwood (1960) imagined, but not fast enough to intensify more than a small distance toward agriculture in the highly variable environments of the Pleistocene. Thus, our hypothesis that the abrupt transition from glacial to Holocene climates caused the origin of agriculture requires that Holocene rates of intensification be neither too slow nor too fast.

Agriculture Was Independently Evolved about 10 Times

The sample of origins is large enough to support some generalizations about the processes involved. Table 17.1 gives a rough time line for the origin of agriculture in seven fairly well-understood centers of domestication, two more controversial centers, three areas that acquired agriculture by diffusion, and two areas that were without agriculture until European conquest.[6] The list of independent centers is complete as far as current evidence goes, and while new

Table 17.1. Dates before present in calendar years of achievement of plant-intensive hunting and gathering and agriculture in different regions, mainly after Smith (1995)

Region	Intensive foraging	Agriculture
Centers of domestication		
Near East: Bar-Yosef and Meadow, 1995	15,000	11,500
North China: An, 1991; Elston et al., 1997	11,600	> 9,000
South China: An, 1991	12,000?	8,000
Sub-Saharan Africa: Klein, 1993	9,000	4,500
Southcentral Andes: Smith, 1995	7,000	5,250
Central Mexico: Smith, 1995	7,000	5,750
Eastern United States: Smith, 1995	6,000	5,250
Controversial centers		
Highland New Guinea: Golson, 1977	?	9,000?
Amazonia: Pearsall, 1995	13,000?	9,000?
Acquisition by diffusion		
Northwestern Europe	12,500	7,000
Southwestern U.S.: Cordell, 1984; Doelle, 1999	6,000	3,500
Japan: Aikens and Akazawa, 1996; Crawford, 1992	10,500	3,000
Never acquired agriculture		
California: Bettinger, 2000	4,000	n/a
Australia: Hiscock, 1994; Smith, 1987	3,500	n/a

centers are not unexpected, it is unlikely that the present list will double. Numerous areas acquired agriculture by diffusion (societies acquire most of their technological innovations by diffusion, not independent invention), so the three areas in table 17.1 are but a small sample. The number of nonarctic areas without agriculture at European contact is small and the two listed, western North America and Australia, are the largest and best known.

Two lines of evidence indicate that the seven centers of domestication are independent. First, the domesticates taken up in each center are distinctive, and no evidence of domesticates from other centers turns up early in any of the sequences. For example, in the eastern North American center a sunflower, a goosefoot, marsh elder, an indigenous squash, and other local plants were taken into cultivation around 6,000 B.P. Mesoamerican maize subsequently appeared here around 2,000 B.P. but remained a minor domesticate until around 1,100 B.P., when it suddenly crowded out several traditional cultivars (Smith, 1989). Second, archaeology suggests that none of the centers had agricultural neighbors at the time that their initial domestications were undertaken. The two problematic centers, New Guinea and Lowland South America, present difficult archaeological problems (Smith, 1995). Sites are hard to find and organic remains are rarely preserved. The New Guinea evidence consists of apparently human-constructed ditches that might have been used in controlling water for taro cultivation. The absence of documented living sites associated with these features makes their interpretation quite difficult. The lowland South American evidence consists of starch grains embedded in pottery fragments and phytoliths, microscopic silicious structural constituents of plant cell walls. The large size of some early starch grains and phytoliths convinces some archaeologists that root crops were brought under cultivation in the Amazon Basin at very early dates.

The timing of initiation of agriculture varies quite widely. The Near Eastern Neolithic is the earliest so far attested. In northern, and possibly southern, China, however, agriculture probably followed within a thousand years of the beginning of the Holocene, even though the best-documented, clearly agricultural complexes are still considerably later (An, 1991; Crawford, 1992; Lu, 1999). Agriculture may prove to be as early in northern China as in the Near East, since the earliest dated sites, which extend back to 8,500 B.P., represent advanced agricultural systems that must have taken some time to develop. Excavations in northern China north of the earliest dated agricultural sites document a technological change around 11,600 B.P., signaling a shift toward intensive plant and animal procurement that may have set this process in motion (Elston et al., 1997).

The exact sequence of events also varies quite widely. For example, in the Near East, sedentism preceded agriculture, at least in the Levantine Natufian sequence, but in Mesoamerica crops seem to have been added to a hunting-and-gathering system that was dispersed and long remained rather mobile (MacNeish, 1991:27–29). For example, squash seems to have been cultivated around 10,000 B.P. in Mesoamerica, some 4,000 years before corn and bean domestication began to lead to the origin of a fully agricultural subsistence system (Smith, 1997). Some mainly hunting-and-gathering societies seem to have incorporated small amounts of domesticated plant foods into their subsistence system without this leading to full-scale agriculture for a very long time. Perhaps American domesticates were long used to provide specialized resources or to increase food security marginally (Richard Redding, personal communication) and initially raised human carrying capacities relatively little, thus operating the competitive ratchet quite slowly. According to MacNeish, the path forward through the whole intensification sequence varied considerably from case to case.

A Late Intensification of Plant Gathering Precedes Agriculture

In all known cases, the independent centers of domestication show a late sequence of intensification beginning with a shift from a hunter-gatherer subsistence system based upon low-cost resources using minimal technological aids to a system based upon the procurement and processing of high-cost resources, including small game and especially plant seeds or other labor-intensive plant resources, using an increasing range of chipped and ground stone tools (Hayden, 1995). The reasons for this shift are the subject of much work among archaeologists (Bettinger, 2000). The shifts at least accelerate and become widespread only in the latest Pleistocene or Holocene. However, a distinct tendency toward intensification is often suggested for the Upper Paleolithic more generally. Stiner et al. and commentators (2000) note that Upper Paleolithic peoples often made considerable use of small mammals and birds in contrast to earlier populations.

These species have much lower body fat than large animals, and excessive consumption causes ammonia buildup in the body due to limitations on the rate of urea synthesis ("rabbit starvation"; Cordain et al., 2000). Consequently, any significant reliance on low-fat small animals implies corresponding compensation with plant calories, and at least a few Upper Paleolithic sites, such as the Ohalo II settlement on the Sea of Galilee (Kislev et al., 1992), show considerable use of plant materials in Pleistocene diets. Large-seeded annual species like wild barley were no doubt attractive resources in the Pleistocene when present in abundance and would have been used opportunistically during the last glacial age. If our hypothesis is correct, in the last glacial age no one attractive species like wild barley would have been consistently abundant (or perhaps productive enough) for a long enough span of time in the same location to have been successfully targeted by an evolving strategy of intensification, even if their less intensive exploitation was common. The broad spectrum of *species*, including small game and plants, reflected in these cases is not per se evidence of intensification (specialized use of more costly but more productive resources using more labor and dedicated technology), as is sometimes argued (Flannery, 1971). In most hunter-gatherer systems, marginal diet cost and diet richness (number of species used) are essentially independent (Bettinger, 1994:46–47), and prey size is far less important in determining prey cost than either mode or context of capture (Bettinger, 1993:51–52; Bettinger and Baumhoff, 1983:832; Madsen and Schmitt, 1998). For all these reasons, quantitative features of subsistence *technology* are a better index of Pleistocene resource intensification than species used. We believe that the dramatic increase in the quantity and range of small chipped stone and groundstone tools only after 15,000 B.P. signals the beginning of the pattern of intensification that led to agriculture.

Early intensification of plant resource use would have tended to generate the same competitive ratchet as the later forms of intensification. Hunter-gatherers who subsidize hunting with plant-derived calories can maintain higher population densities and thus will tend to deplete big game to levels that cannot sustain hunting specialists (Winterhalder and Lu, 1997). Upper Paleolithic people appear to be fully modern in their behavioral capacities (Klein, 1999). Important changes in subsistence technology did occur during the Upper Paleolithic, for example, the development of the atlatl. Nevertheless, modern abilities and the operation of the competitive ratchet drove Upper Paleolithic populations only a relatively small distance down the path to the kind of heavy reliance on plant resources that in turn set the stage for domestication.

Braidwood's reasoning that pioneering agriculturalists would have gained their intimate familiarity with proto-domesticates first as gatherers is logical and supported by the archaeology. Once the climate ameliorated, the rate of intensification accelerated immediately in the case of the Near East. In other regions changes right at the Pleistocene-Holocene transition were modest to invisible (Straus et al., 1996). The full working out of agrarian subsistence systems took thousands of years. Indeed, modern breeding programs illustrate that we are still working out the possibilities inherent in agricultural subsistence systems.

The cases where Holocene intensification of plant gathering did not lead directly to agriculture are as interesting as the cases where it did. The Jomon of

Japan represents one extreme (Imamura, 1996). Widespread use of simple pottery, a marker of well-developed agricultural subsistence in western Asia, was very early in the Jomon, contemporary with the latest Pleistocene Natufian in the Near East. By 11,000 years B.P., the Jomon people lived in settled villages, depended substantially upon plant foods, and used massive amounts of pottery. However, the Jomon domesticated no plants until rather late in the sequence. Seeds of weedy grasses are found throughout, but only in later phases (after about 3,000 B.P.) do the first unambiguous domesticates occur, and these make up only a small portion of the seeds in archaeological contexts (Crawford, 1997). Sophisticated agriculture came to Japan with imported rice from the mainland only about 2,500 B.P. Interestingly, acorns were a major item of Jomon subsistence. The people of California were another group of sedentary hunter-gatherers that depended heavily on acorns. However, in California the transition to high plant dependence began much later than in the Jomon (Wohlgemuth, 1996). Milling-stones for grinding small seeds became important after 4,500 B.P., although seeds were of relatively minor importance overall. After 2,800 B.P. acorns processed with mortars and pestles became an important subsistence component and small seeds faded in comparative importance. In the latest period, after 1,200 B.P. quantities of small seeds were increasingly added back into the subsistence mix alongside acorns in a plant-dominated diet. Other peoples with a late onset of intensification include the Australians. The totality of cases tells us that any stage of the intensification sequence can be stretched or compressed by several thousand years but reversals are rare (Harris, 1996; Price and Gebauer, 1995). Farming did give way to hunting-and-gathering in the southern and eastern Great Basin of North America after a brief extension of farming into the region around 1,000 B.P. (Lindsay, 1986). A similar reversal occurred in southern Sweden between 2,400 and 1,800 B.P. (Zvelebil, 1996). Horticultural Polynesian populations returned substantially to foraging for a few centuries while population densities built up on reaching the previously uninhabited archipelagos of Hawaii and New Zealand (Kirch, 1984). Had intensification on plant resources been possible during the last glacial age, even the slowest Holocene rates of intensification were rapid enough to produce highly visible archaeological evidence on the 10 millennium timescale, one third or less time than Upper Paleolithic peoples lived under glacial climates.

More Intensive Technologies Tend to Spread

One successful and durable agricultural origin in the last glacial age on any sizeable land mass would have been sufficient to produce a highly visible archaeological record, to judge from events in the Holocene epoch. Once well-established agricultural systems existed in the Holocene, they expanded at the expense of hunting-and-gathering neighbors at appreciable rates (Bellwood, 1996). Ammerman and Cavalli-Sforza (1984) summarize the movement of agriculture from the Near East to Europe, North Africa, and Asia. The spread into Europe is best documented. Agriculture reached the Atlantic seaboard about 6,000 B.P. or about 4,000 years after its origins in the Near East. The regularity of the spread, and the degree to which it was largely a cultural diffusion process as opposed to a population dispersion as well, are matters of debate.

Cavalli-Sforza, Minozzi, and Piazza, (1994:296–299) argue that demic expansion by western Asians was an important process with the front of genes moving at about half the rate of agriculture. They imagine that pioneering agricultural populations moved into territories occupied by hunter-gatherers and intermarried with the preexisting population. The then-mixed population in turn sent agricultural pioneers still deeper into Europe. They also suppose that the rate of spread was fairly steady, though clearly frontiers between hunter-gatherers and agriculturalists stabilized in some places (Denmark, Spain) for relatively prolonged periods. Zvelebil (1996) emphasizes the complexity and durability of frontiers between farmers and hunter-gatherers and the likelihood that in many places the diffusion of both genes and ideas about cultivation was a prolonged process of exchange across a comparatively stable ethnic and economic frontier. Further archaeological and paleogenetic investigations will no doubt gradually resolve these debates. Clearly, the spread process is at least somewhat heterogeneous.

Other examples of the diffusion of agriculture are relatively well documented. For example, maize domestication is dated to about 6,200 B.P. in Central Mexico, spreading to what is now the southwestern United States (New Mexico) by about 4,000 B.P. (Matson, 1999; Smith, 1995). In this case, the frontier of maize agriculture stabilized for a long time, only reaching the area now in eastern United States at the comparatively late date noted. Maize failed entirely to diffuse westward into the Mediterranean parts of California even though peoples growing it in the more arid parts of its range in the Southwest used irrigation techniques that have eventually worked in California with modest modifications to cope with dry-season irrigation. As with the origin process, the rate of spread of agriculture exhibits an interesting degree of variation.

Changes in the Cultural Evolutionary System?

A possible alternative to our hypothesis would be that a substantial modernization of the cultural system occurred coincidently at the end of the Pleistocene epoch and that this resulted in a general acceleration of rates of cultural evolution, including subsistence intensification. The modernization of culture capacities leading up to the Upper Paleolithic transition was presumably such an event, as were later inventions like literacy (Donald, 1991; Klein, 1999: ch.7). We are not aware of any proposals for major changes in the intrinsic rate of cultural evolution coincident with the Pleistocene-Holocene boundary. Students of the evolution of subsistence intensification and social complexity in the Holocene have suggested a series of plausible processes that will probably turn out to be at least part of the explanation for why the trend to intensification has taken such diverse forms in different regions (table 17.2). This list of diversifying and rate-limiting processes does not include any that should have operated more stringently on Upper Paleolithic, as opposed to Mesolithic and Neolithic, societies, climate effects aside. Holocene rates of intensification do have the right timescale to be drastically affected by millennial- and submillennial-scale variation that is rapid with respect to observed rates of cultural evolution in the Holocene.

Table 17.2. Processes that may retard the rate of cultural evolution and create local optima that halt evolution for prolonged periods

Process	Authors (examples)
Geography: Eurasia, having the largest land mass, has more local populations to exchange innovations by diffusion, hence the fastest Holocene rate of subsistence intensification.	Diamond, 1997
Minor climate change: The late medieval onset of the Little Ice Age caused the extinction of the Greenland Norse colony. Agriculture at marginal altitudes in places like the Andes seems to respond to Holocene climatic fluctuation.	Kent, 1987; Kleivan, 1984
Preadapted plants: The Mediterranean Old World is unusually well endowed with large-seeded grasses susceptible to domestication pressures. American domesticates, especially maize, may outcross too much to respond quickly to selection.	Blumler, 1992; Blumler and Byrne, 1991; Diamond, 1997; Hillman and Davies, 1990
Diseases: Density-dependent epidemic diseases may evolve that slow or stop the population growth, pending the evolution of resistance, that would otherwise drive the competitive ratchet. Local diseases that attack foreigners may protect otherwise-vulnerable systems.	Cavalli-Sforza, et al. 1994; Crosby, 1986; Gifford-Gonzalez, 2000; McNeill, 1976
New technological complexes evolve slowly: Nutritional adequacy in plant-rich diets requires discovering cooking techniques, acquiring balancing domesticates, developing the potential of animal domesticates, and the like.	Katz et al., 1974
New social institutions evolve slowly: Social institutions are generally deeply involved in subsistence but are also liable to be regulated by norms that make adaptive evolution away from current local optima difficult.	Bettinger, 1999; Bettinger and Baumhoff, 1982; North and Thomas, 1973; Richerson and Boyd, 1999
Ideology may play a role: The evolution of fads, fashions, and belief systems may act to drive cultural evolution in nonutilitarian directions that sometimes carry them to new adaptive slopes.	Weber, 1930

If climate variation did not limit intensification during the last glacial age to vanishingly slow rates compared to the Holocene epoch, the failure of intensive systems to evolve during the tens of millennia anatomically and culturally modern humans lived as sophisticated hunter-gatherers before the Holocene is a considerable mystery.

Conclusion

The large, rapid change in environment at the Pleistocene-Holocene transition set off the trend of subsistence intensification of which modern industrial innovations are just the latest examples. If our hypothesis is correct, the reduction in climate variability, increase in CO_2 content of the atmosphere, and increases in rainfall rather abruptly changed the earth from a regime where agriculture was impossible everywhere to one where it was possible in many places. Since groups that use efficient, plant-rich subsistence systems will normally outcompete groups that make less efficient use of land, the Holocene has been characterized by a persistent, but regionally highly variable, tendency toward subsistence intensification. The diversity of trajectories taken by the various regional human subpopulations since $\approx 11,600$ B.P. are natural experiments that will help us elucidate the factors that control the tempo of cultural evolution and that generate historical contingency against the steady, convergent adaptive pressure toward ever more intense production systems. A long list of processes (table 17.2) interacted to regulate the nearly unidirectional trajectory of subsistence intensification, population growth, and institutional change that the world's societies have followed in the Holocene. Notably, even the slowest evolving regions generated quite appreciable and archaeologically visible intensification, demanding some explanation for why similar trajectories are absent in the Pleistocene.

Those who are familiar with the Pleistocene epoch often remark that the Holocene is just the "present interglacial." The return of climate variation on the scale that characterized the last glacial age is quite likely if current ideas about the Milankovich driving forces of the Pleistocene are correct. Sustaining agriculture under conditions of much higher amplitude, high-frequency environmental variation than farmers currently cope with would be a considerable technical challenge. At the very best, lower CO_2 concentrations and lower average precipitation suggest that world average agricultural output would fall considerably.

Current anthropogenic global warming via greenhouse gases might at least temporarily prevent any return to glacial conditions. However, we understand the feedbacks regulating the climate system too poorly to have any confidence in such an effect. Current increases in CO_2 threaten to elevate world temperatures to levels that in past interglacials apparently triggered a large feedback effect producing a relatively rapid decline toward glacial conditions (Petit et al., 1999). The Arctic Ocean ice pack is currently thinning very rapidly (Kerr, 1999). A dark, open Arctic Ocean would dramatically increase the summer heat income at high northern latitudes and have large, difficult-to-guess impacts on the Earth's climate system. No one can yet estimate the risks we are taking of a rapid return to colder, drier, more variable environment with less CO_2 or evaluate exactly the threat such conditions imply for the continuation of agricultural production. Nevertheless, the intrinsic instability of the Pleistocene climate system, and the degree to which agriculture is likely dependent upon the Holocene stable period, should give one pause (Broecker, 1997).

APPENDIX 1: More Realistic Population Dynamics

The logistic equation assumes that an increment to population density has the same effect on population pressure at low densities as at high densities. We know that this assumption is not correct in all cases. For example, hunters pursuing herd animals may generate much population pressure at low human population densities because killing only a small fraction of the herd makes the many survivors difficult to hunt. On the other hand, subsistence farmers spreading into a uniform fertile plain may feel little population pressure until all farmland is occupied. If returns to additional labor on shrinking farms then drop steeply, most population pressure will be felt at densities near K. To allow for such effects, ecologists often utilize a generalized logistic equation:

$$\frac{dN}{dt} = rN\left[1 - \left(\frac{N}{K}\right)^{\theta}\right]$$ (A1.1)

Population pressure is now given by the term $(N/K)^{\theta}$. If $\theta > 1$, population pressure does not increase until densities approach carrying capacity, as is usually the case for species like humans that have flexible behavior and considerable mobility, and thus can mitigate the effects of increasing population density over some range of densities. It seems intuitive that this would increase the length of time necessary to reach a given level of population pressure. However, this intuition is wrong. The generalized logistic can be used to derive a differential equation for $\pi = (N/K)^{\theta}$:

$$\frac{d\pi}{dt} = \frac{\theta}{N}\left(\frac{N}{K}\right)^{\theta}\frac{dN}{dt}$$
$$= r\theta\left(\frac{N}{K}\right)^{\theta}\left(1 - \left(\frac{N}{K}\right)^{\theta}\right)$$ (A1.2)
$$= \theta\pi(1 - \pi)$$

Thus, the differential equation for population pressure is always the ordinary logistic equation in which $K = 1$ and r is multiplied by θ. This means that when $\theta > 1$, it takes *less* time to reach a given amount of population pressure than would be the case if $\theta = 1$. Reduced population pressure at low densities leads to more rapid initial population growth. Population growth is close to exponential longer and this more than compensates for the fact that higher densities have to be reached to achieve the same level of population pressure.

APPENDIX 2: The Dynamics of Innovation

Consider a population of size N in which the per capita income of the population is given by:

$$y = \frac{y_m I}{I + N}$$ (A2.1)

where y_m is the maximum per capita income, and I is a variable that represents the productivity of subsistence technology. Thus, per capita income declines as population size increases, but for a given population size, greater productivity raises per capita income. As in the previous models, we assume that as population pressure, now measured as falling per capita income, increases, population growth decreases. In particular, assume:

$$\frac{dN}{dt} = \rho N(y - y_s) \tag{A2.2}$$

where y_s is the per capita income necessary for subsistence. If per capita incomes are above this value, population increases; if per capita income falls below y_s, population shrinks. If I is fixed, this equation is another generalization of the logistic equation. In an initially empty environment, population initially grows at a rate $\rho(y_m - y_s)$, but then slows and reaches an equilibrium population size:

$$\frac{I(y_m - y_s)}{y_s} \tag{A2.3}$$

To allow for intensification we assume that people innovate whenever their per capita income falls below a threshold value y_i. Thus:

$$\frac{dI}{dt} = aI(y_i - y) \tag{A2.4}$$

When per capita income is less than the threshold value y_i, people innovate, increasing the carrying capacity and therefore decreasing population pressure. When per capita income is greater than the threshold, they "de-innovate." This may seem odd at first, but such abandonment of more efficient technology has been observed occasionally. The maximum rate at which innovation can occur is governed by the parameter a.

If a small pioneer population enters an empty habitat, it experiences two distinct phases of expansion (figure 17.5). Initially, per capita income is near the maximum, and population grows at the maximum rate. As population density increases, per capita income drops below y_i, and the population begins to innovate, eventually reaching a steady state value:

$$\hat{y} = \frac{\rho y_s + a y_i}{\rho + a} \tag{A2.5}$$

The steady state per capita income is above the minimum for subsistence but below the threshold at which people experience population pressure and begin to innovate. At this steady state population growth continues at a constant rate,

$$\hat{\rho} = \frac{a(y_i - y_s)}{\rho + a} \tag{A2.6}$$

that is proportional to the rate of growth of subsistence efficiency.

NOTES

We thank Joe Andrew, Ofer Bar-Yosef, Richard Redding, Bruce Winterhalder, and three anonymous referees for unusually constructive criticism of the manuscript. Thanks to Scott Elias for insights pertaining to Pleistocene seasonality and to Peter Ditlevsen for providing figure 17.2. Peter Lindert's invitation to give a seminar led to the first draft of this chapter. Thanks to Francisco Gil-White for assistance with the Spanish abstract in the original article.

1. We define "efficiency" as the productivity per unit area of land exploited for subsistence. Efficiency of subsistence is favored by strategies that move subsistence

down the food chain, especially to high-productivity plant resources. "Intensification" we define as the use of human labor to add productive lower-ranked resources to the diet or the use of technological innovations to increase the rank of more productive resources. Typically both strategies are employed simultaneously. Since increases in efficiency are achieved by either labor or technical intensification and since increases in efficiency usually also lead to population growth, we use the term "intensification" loosely for the interlinked processes of labor and technical intensification and population growth. We define "agriculture" as dependence upon domesticated crops and animals for subsistence. We mark the origin of agriculture as the first horizon in which plant remains having anatomical markers of domestication are found, or are likely on other grounds to be found in the future. Fully agricultural subsistence systems in the sense of a dominance of domesticated species in the diet typically postdate the origin of agriculture by a millennium or more.

2. It has also been argued that Pleistocene climates were less seasonally variable than during the Holocene, but this idea has scant empirical support (Miracle and O'Brien, 1998). Elias (1999) has used fossil beetle faunas to estimate July and January temperatures in Holocene and Pleistocene deposits. These data suggest that the Pleistocene was *more* seasonal than the Holocene. However, beetle estimates of January temperatures are not very reliable because beetles in temperate and arctic climates overwinter in a dormant state so that their distributions are rather insensitive to winter as opposed to summer temperatures. Plant distributions are similarly affected. No current method of estimating winter temperatures in the Pleistocene is reliable.

3. Agronomists and climatologists have recently become interested in the impacts of climate change and climate variability in the context of CO_2-indcued global warming (Bazzaz and Sombroek, 1996; Downing, Olsthoorn, and Tal, 1999; Kane and Yohe, 2000; Reilly and Schimmelpfennig, 2000; Rosensweig and Hillel, 1998; Schneider, Easterling, and Mearns, 2000). Global climate models suggest that global warming may increase short timescale climate variation as well as creating a steep trend. To some degree, these conditions mimic the millennial and submillennial scale variations in the Pleistocene, and, as crop-and-weather models and empirical data improve, more definitive assessments of impact of last glacial conditions on plant-based subsistence strategies will become possible.

4. By "adaptive," we mean behaviors that, by comparison with available alternatives, have the largest population mean fitness.

5. Some human populations might have curtailed birth rates in order to preserve higher incomes at any given level of intensification. In a sense, such populations have just redefined K to be a lower value that permits higher incomes by employing what Malthus called the "preventative checks" on population growth. The rest of the analysis then applies with K measured in suitably emic terms. Cultural differences in the value of intensification threshold or K (Coale, 1986) will make evidence of stress more likely in populations where the effective carrying capacity is closer to the ultimate subsistence carrying capacity than in populations that reduce growth rates by preventative checks that keep population well below absolute subsistence limits. The perceived costs of population control, given that the main mechanism in non-modern societies was infanticide and sexual abstinence, may mean that most populations intensified labor inputs at any given level of technology efficiency to near subsistence limits (Hayden, 1981). In either event, population pressure will tend to stay constant to the extent that rates of population growth and intensification are successful in adjusting subsistence to current conditions. Normally population growth and decline are quite rapid processes relative to rates of innovation and will keep

average population size quite close to K. Short-term departures from K caused by short-term environmental shocks and windfalls should be the commonest reasons to see especially stressed or unstressed populations. If the rate of innovation is more rapid than exponential population growth for any significant time period, then per capita incomes can rise under a regime of very rapid population growth, as in the last few centuries. This regime, if it had occurred in the past, should be quite visible in the historical and archaeological record because it so rapidly leads to large populations and large-scale creation of durable artifacts. Alternatively, population growth may have been limited in past populations by the analog of the modern demographic transition. Thus, hunter-gatherers might have resisted the adoption of plant-based intensification because they viewed the life style associated with plant collecting or planting as a decrement to their incomes. However, resisting intensifications that increase human densities makes such groups vulnerable to competitive displacement by the intensifiers unless the greater wealth of the population limiters allows them to successfully defend their resource-rich territories. On the evidence of the fairly rapid rate of spread of intensified strategies once invented, such defense is seldom successful (e.g., Ammerman and Cavalli-Sforza, 1984; Bettinger and Baumhoff, 1982).

6. The dates in table 17.1 reflect considerable recent revision stemming from accelerator mass spectrometry ^{14}C dating, which permits the use of very small carbon samples and can be applied directly to carbonized seeds and other plant parts showing morphological changes associated with domestication. Isolated seeds tend to work their way deep into archaeological deposits, and dates based on associated large carbon samples (usually charcoal) often gave anomalously early dates.

REFERENCES

Aikens, C. M., & T. Akazawa. 1996. The Pleistocene-Holocene transition in Japan and adjacent Northeast Asia: Climate and biotic change, broad-spectrum diet, pottery, and sedentism. In: *Humans at the end of the Ice Age*, L. G. Straus, B. V. Eriksen, J. M. Erlandson, & D. R. Yesner, eds. (pp. 215–227). New York: Plenum.

Allen, J. R., U. Brandt, A. Brauer, H.-W. Hubberten, B. Huntley, J. Keller, M. Kraml, A. Mackensen, J. Mingram, J. F. Negendank, N. R. Nowaczyk, H. Oberhansli, W. A. Watts, S. Wulf, & B. Zolitschka. 1999. Rapid environmental changes in Southern Europe during the last glacial period. *Nature* 400:740–743.

Allen, J. R. M., W. A. Watts, & B. Huntley. 2000. Weichselian palynostratigraphy, palaeovegetation and palaeoenvironment: The record from logo grande di Monticcio, Southern Italy. *Quaternary International* 73/74:91–110.

Ammerman, A. J., & L. L. Cavalli-Sforza. 1984. *The Neolithic transition and the genetics of populations in Europe*. Princeton, NJ: Princeton University Press.

An, Z. 1991. Radiocarbon dating and the prehistoric archaeology of China. *World Archaeology* 23:193–200.

Bar-Yosef, O. 1998. The Natufian culture in the Levant, threshold to the origins of agriculture. *Evolutionary Anthropology* 6:159–177.

Bar-Yosef, O., & R. H. Meadow. 1995. The origins of agriculture in the Near East. In: *Last hunters, first farmers: New perspectives on the prehistoric transition to agriculture*, T. D. Price, & B. Gebauer, eds. (pp. 39–94). Santa Fe: School of American Research Press.

Bar-Yosef, O., & F. R. Valla. 1991. *The Natufian culture in the Levant*. International Monographs in Prehistory, Archaeological Series 1, Ann Arbor.

Barnola, J. M., D. Raynaud, Y. S. Korotkevich, & C. Loris. 1987. Vostok ice core provides 160,000-year record of atmospheric CO_2. *Nature* 329:408–414.

Baumhoff, M. A. 1963. Ecological determinants of aboriginal California populations. *University of California Publications in American Archaeology and Ethnology* 49(2):155–236.

Bazzaz, F., & W. Sombroek. 1996. *Global climate change and agricultural production.* Chichester, England: Wiley.

Beerling, D. J. 1999. New estimates of carbon transfer to terrestrial ecosystems between the last glacial maximum and the Holocene. *Terra Nova* 11:162–167.

Beerling, D. J., W. G. Chaloner, B. Huntley, J. A. Peason, & M. J. Tooley. 1993. Stomatal density responds to the glacial cycle of environmental change. *Proceedings of the Royal Society of London B* 251:133–138.

Beerling, D. J., & F. I. Woodward. 1993. Ecophysiological responses of plants to the global environmental change since the last glacial maximum. *New Phytologist* 125:641–648.

Behling, H., H. W. Arz, J. Pätzold, & G. Wefer. 2000. Late quaternary vegetation and climate dynamics in Northeastern Brazil, inferences from marine core geoB 3104–1. *Quaternary Science Reviews* 19:981–994.

Bellwood, P. 1996. The origins and spread of agriculture in the Indo-Pacific region: Gradualism and diffusion or revolution and colonization. In: *The origins and spread of agriculture and pastoralism in Eurasia*, D. R. Harris, ed. (pp. 465–498). Washington, DC: Smithsonian Institution Press.

Bennett, K. D., S. G. Haberle, & S. H. Lumley. 2000. The last glacial-Holocene transition in Southern Chile. *Science* 290:325–328.

Belovsky, G. E. 1987. Hunter-gatherer foraging: A linear programming approach. *Journal of Anthropological Archaeology* 6:29–76.

Bettinger, R. L. 1991. *Hunter-gatherers: Archaeological and evolutionary theory.* New York: Plenum.

Bettinger, R. L. 1993. Doing Great Basin archaeology recently: Coping with variability. *Journal of Archaeological Research* 1:43–66.

Bettinger, R. L. 1994. How, when, and why Numic spread. In: *Across the West: Human population movement and the expansion of the Numa*, D. B. Madsen, & D. R. Rhode, eds. (pp. 44–55). Salt Lake City: University of Utah Press.

Bettinger, R. L. 1999. From traveler to processor: Regional trajectories of hunter-gatherer sedentism in the Inyo-Mono region, California. In: *Settlement pattern studies in the Americas: Fifty years since Viru*, B. R. Billman, & G. M. Feinman, eds. (pp. 39–55). Washington, DC: Smithsonian Institution Press.

Bettinger, R. L. 2000. Holocene hunter-gatherers. In: *Archaeology at the millennium: A sourcebook*, G. M. Feinman, & T. D. Price, eds. (pp. 137–195). New York: Plenum.

Bettinger, R. L., & M. A. Baumhoff. 1982. The Numic spread: Great Basin cultures in competition. *American Antiquity* 47:485–503.

Bettinger, R. L., & M. A. Baumhoff. 1983. Return rates and intensity of resource use in Numic and prenumic adaptive strategies. *American Antiquity* 48:830–834.

Blumler, M. A.1992. *Seed weight and environment in Mediterranean-type grasslands in California and Israel*, Ph.D. Thesis, University of California, Berkeley.

Blumler, M. A., & R. Byrne (A. Belfer-Cohen, R. M. Bird, V. L. Bohrer, B. F. Byrd, R. C. Dunnell, G. Hillman, A. T. M. Moore, D. I. Olszewski, R. W. Redding, & T. J. Riley, commentators).

Blumler, M. A., & R. Byrne. 1991. The ecological genetics of domestication and the origins of agriculture. *Current Anthropology* 32:23–53.

Bond, G., W. Broecker, S. Johnsen, J. McManus, L. Labeyrie, J. Jouzel, & G. Bonani. 1993. Correlations between climate records from North Atlantic sediments and Greenland ice. *Nature* 365(6442):143–147.

Boserup, E. 1981. *Population and technological change: A study of long-term trends.* Chicago: University of Chicago Press.

Boyd, R., & P. J. Richerson. 1985. *Culture and the evolutionary process.* Chicago: University of Chicago Press.

Bradley, R. S. 1999. *Paleoclimatology: Reconstructing climates of the quaternary.* 2nd ed. San Diego: Academic Press.

Braidwood, R. J. 1960. The agricultural revolution. *Scientific American* 203:130–148.

Braidwood, L., R. Braidwood, B. Howe, C. Reed, & P. J. Watson, eds. 1983. *Prehistoric archaeology along the Zagros Flanks.* Studies in Ancient Oriental Civilization 105. Chicago: University of Chicago Oriental Institute.

Braidwood, R. J., & B. Howe. 1960. *Prehistoric investigations in Iraqi Kurdistan.* Studies in Ancient Oriental Civilization 31. Chicago: University of Chicago Oriental Institute.

Broecker, W. S. 1995. *The glacial world according to Wally.* Palisades, NY: Lamont-Doherty Earth Observatory of Columbia University.

Broecker, W. S. 1996. Glacial climate in the tropics. *Science* 272:1902–1903.

Broecker, W. S. 1997. Thermohaline circulation, the Achilles heel of our climate system: Will man-made CO_2 upset the current balance? *Science* 178:1582–1588.

Broecker, W. S., & G. H. Denton. 1989. The role of ocean-atmosphere reorganizations in glacial cycles. *Geochimica et Cosmochimica Acta* 53:2465–2501.

Cavalli-Sforza, L. L., P. Menozzi, & A. Piazza. 1994. *The history and geography of human genes.* Princeton, NJ: Princeton University Press.

Childe, V. G. 1951. *Man makes himself.* London: Watts.

Clapperton, C. 2000. Interhemispheric synchroneity of marine isotope stage 2 glacier fluctuations along the American cordilleras transect. *Journal of Quaternary Science* 15:435–468.

Clark, P. U., R. B. Alley, & D. Pollard. 1999. Northern hemisphere ice-sheet influences on global climate change. *Science* 286:1104–1111.

Coale, A. J. 1986. The decline of fertility in Europe since the eighteenth century as a chapter in human demographic history. In: *The decline in fertility in Europe,* A. J. Coale, & S. C. Watkins, eds. (pp. 1–30). Princeton, NJ: Princeton University Press.

Cohen, M. N. 1977. *The food crisis in prehistory: Overpopulation and the origins of agriculture.* New Haven, CT: Yale University Press.

Cohen, M. N., & G. J. Armelagos. 1984. *Paleopathology at the origins of agriculture.* Orlando, FL: Academic Press.

Coope, G. R. 1987. The response of late Quaternary insect communities to sudden climatic changes. In: *Organization of communities past and present,* J. H. R. Gee, & P. S. Giller, eds. (pp. 421–438). Oxford: Blackwell.

Cordain, L., J. B. Miller, S. B. Eaton, N. Mann, S. H. A. Holt, & J. D. Speth. 2000. Plant-animal subsistence ratios and macronutrient energy estimations in worldwide hunter-gatherer diets. *American Journal of Clinical Nutrition* 71:682–692.

Cordell, L. S. 1984. *Prehistory of the Southwest.* San Francisco: Academic Press.

Cowling, S. A., & M. T. Sykes. 1999. Physiological significance of low atmospheric CO_2 for plant-climate interactions. *Quaternary Research* 52:237–242.

Crawford, G. W. 1992. Prehistoric plant domestication in East Asia. In: *The origins of agriculture,* C. W. Cowan, & P. J. Watson, eds. (pp. 7–38). Washington, DC: Smithsonian Institution Press.

Crawford, G. W. 1997. Anthropogenesis in prehistoric northeastern Japan. In: *People, plants, and landscapes: Studies in paleoethnobotany*, K. J. Gremillion, ed. (pp. 86–103). Tuscaloosa: University of Alabama Press.

Cronin, T. M. 1999. *Principles of paleoclimatology*. New York: Columbia University Press.

Crosby, A. W. 1986. *Ecological imperialism: The biological expansion of Europe*, 900–1900. Cambridge: Cambridge University Press.

Dansgaard, W., S. J. Johnsen, H. B. Clausen, D. Dahl-Jensen, N. S. Gundestrup, C. U. Hammer, C. S. Hvidberg, J. P. Steffensen, A. E. Sveinbjönsdottir, J. Jousel, & G. Bond. 1993. Evidence for general instability of past climate from a 250-kyr ice-core record. *Nature* 364:218–220.

Darwin, C. 1902 [1874]. *The descent of man and selection and selection in relation to sex*. New York: American Home Library.

deMenocal, P. B., & J. Bloemendal. 1995. Plio-Pleistocene climatic variability in subtropical Africa and the paleoenvironment of hominid evolution: A combined data-model approach. In: *Paleoclimate and evolution with emphasis on human origins*, E. S. Vrba, G. H. Denton, T. C. Partridge, & L. H. Burckle, eds. (pp. 262–288). New Haven, CT: Yale University Press.

Diamond, J. 1997. *Guns, germs, and steel*. New York: Norton.

Ditlevsen, P. D., H. Svensmark, & S. Johnsen. 1996. Contrasting atmospheric and climate dynamics of the last-glacial and Holocene periods. *Nature* 379:810–812.

Doelle, W. H. 1999. Early maize in the greater southwest. *Archaeology Southwest* 13: 1–9.

Donald, M. 1991. *Origins of the modern mind*. Cambridge, MA: Harvard University Press.

Dorale, J. A., R. L. Edwards, E. Ito, & L. A. González. 1998. Climate and vegetation history of the midcontinent from 75 to 25 ka: A speleothem record from Crevice Cave, Missouri, USA. *Science* 282:1871–1874.

Downing, T. E., A. J. Olsthoorn, & R. S. J. Tol. 1999. *Climate, change and risk*. London: Routledge.

Durham, W. H. 1991. *Coevolution: Genes, culture, and human diversity*. Stanford: Stanford University Press.

Eakin, H. 2000. Smallholder maize production and climatic risk: A case study from Mexico. *Climatic Change* 45:19–36.

Elias, S. A. 1999. Mid-Wisconsin seasonal temperatures reconstructed from fossil beetle assemblages in Eastern North America: Comparisons with other proxy records from the Northern Hemisphere. *Journal of Quaternary Science* 14:255–262.

Elston, R. G., C. Xu, D. B. Madsen, K. Zhong, R. L. Bettinger, J. Li, P. J. Brantingham, H. Wang, & J. Yu. 1997. New dates For the North China Mesolithic. *Antiquity* 71:985–993.

Fiedel, S. J. 1999. Older than we thought: Implications of corrected dates for Paleoindians. *American Antiquity* 64:95–115.

Fisher, R. A. 1937. The wave of advance of advantageous genes. *Annals of Eugenics (London)* 7:355–369.

Flannery, K. V. 1971. The origins and ecological effects of early domestication in Iran and the Near East. In: *The domestication and exploitation of plants and animals*, S. Struever, ed. (pp. 50–79). American Museum Sourcebooks in Anthropology. Garden City, NY: Natural History Press.

Flannery, K. V. 1973. The origins of agriculture. *Annual Review of Anthropology* 2: 271–310.

Flannery, K. V. 1986. *Guilá Naquitz: Archaic foraging and early agriculture in Oaxaca, Mexico*. Orlando, FL: Academic Press.

Frogley, M. R., P. C., Tzedakis, & T. H. E. Heaton. 1999. Climate variability in Northwestern Greece during the last interglacial. *Science* 285:886–1889.

Gifford-Gonzales, D. 2000. Animal disease challenges to the emergence of pastoralism in sub-Saharan Africa. *African Archaeological Review* 17:95–139.

Golson, J. 1977. No room at the top: Agricultural intensification in the New Guinea Highlands. In: *Sunda and Sahul: Prehistoric studies in Southeast Asia, Melanesia, and Australia*, J. Allen, J. Golson, & R. Jones, eds. (pp. 601–638). London: Academic Press.

Gommes, R. 1999. Production variability and losses. United Nations Food and Agriculture Organization, www.fao.org/sd/eidirect/agroclim/losses.htm.

Goring-Morris, N., & A. Belfer-Cohen. 1998. The articulation of cultural processes and late quaternary environmental change in Cisjordan. *Paléorient* 23:71–93.

Grafenstein, U. von, H. Erlenkeuser, A. Brauer, J. Jouzel, & S. J. Johnsen. 1999. A mid-European decadal isotope-climate record from 15,500 to 5000 years B.P. *Science* 284:1654–1657.

GRIP (Greenland Ice-core Project Members). 1993. Climate instability during the last interglacial period recorded In the GRIP ice core. *Nature* 364:203–207.

Grove, J. M. 1988. *The Little Ice Age*. London: Methuen.

Harris, D. R., ed. 1996. *The origins and spread of agriculture and pastoralism in Eurasia*. London: University College London Press.

Harris, D. R. 1977. Alternative pathways toward agriculture. In: *Origins of agriculture*, C. A. Reed, ed. (pp. 179–243). The Hague: Mouton.

Hayden, B. 1981. Research and development in the Stone Age: Technological transitions among hunter-gatherers. *Current Anthropology* 22:519–548.

Hayden, B. 1995. A new overview of domestication. In: *Last hunters, first farmers: New perspectives on the prehistoric transition to agriculture*, T. D. Price, & B. Gebauer, eds. (pp. 273–299). Santa Fe: School of American Research Press.

Hendy, I. L., & J. P. Kennett. 2000. Dansgaard-Oeschger cycles and the California current system: Planktonic foraminiferal response to rapid climate change in Santa Barbara Basin, ocean drilling program hole 893A. *Paleoceanography* 15:30–42.

Henry, D. O. 1989. *From foraging to agriculture: The Levant at the end of the Ice Age*. Philadelphia: University of Pennsylvania Press.

Heusser, L. E. 1995. Pollen stratigraphy and paleoecologic interpretation of the 160-K.Y. record from the Santa Barbara Basin, hole 893A. *Proceedings of the Ocean Drilling Program, Scientific Results* 146:265–277.

Hillman, G. C., & M. S. Davies. 1990. Measured domestication rates in wild wheats and barley under primitive cultivation, and their archaeological implications. *Journal of World Prehistory* 4:157–222.

Hiscock, P. 1994. Technological response to risk in Holocene Australia. *Journal of World Prehistory* 8:267–292.

Imamura, K. 1996. Jomon and Yoyoi: The transition to agriculture in Japanese prehistory. In: *The origins and spread of agriculture and pastoralism in Eurasia*, D. R. Harris, ed. (pp. 442–464). London: University College London Press.

Johnsen, S. J., H. B. Clausen, W. Dansgaard, N. S. Gundestrup, C. U. Hammer, U. Andersen, K. K. Andersen, C. S. Hvidberg, D. Dahl-Jensen, J. P. Steffensen, H. Shoji, A. E. Sveinbjornsdottir, J. White, J. Jouzel, & D. Fisher. 1997. The [18]O record along the Greenland ice core project deep ice core and the problem of possible Eemian climatic instability. *Journal of Geophysical Research* C 102:26, 397–26, 410.

Kane, S., & G. Yohe. 2000. Societal adaptation to climate variability and change: An introduction. *Climatic Change* 45:1–4.

Katz, S. H., M. L. Hediger, & L. A. Valleroy. 1974. Traditional maize processing techniques in the New World. *Science* 184:765–773.

Kent, J. D. 1987. Periodic aridity and prehispanic Titicaca Basin settlement patterns. In: *Arid land use strategies and risk management in the Andes,* D. L. Browman, ed. (pp. 297–314). Boulder: Westview.

Kerr, R. A. 1999. Will the Arctic Ocean lose all its ice? *Nature* 286:128.

Kirch, P. V. 1984. *The evolution of Polynesian chiefdoms.* Cambridge: Cambridge University Press.

Kislev, M. E., D. Nadel, & I. Carmi. 1992. Epipaleolithic (19,000 BP) cereal and fruit diet at Ohalo II, Sea of Galilee, Israel. *Review of Palaeobotany and Palynology* 73:161–166.

Klein, R. G. 1993. Hunter-gatherers and farmers in Africa. In: *People of the Stone Age: Hunter-gatherers and early farmers,* G. Burenhult, ed. (pp. 39–47, 50–55). San Francisco: Harper.

Klein, R. G. 1999. *The human career: Human biological and cultural origins.* Chicago: University of Chicago Press.

Kleivan, I. 1984. History of Norse Greenland. In: *Handbook of North American Indians 5: Arctic,* D. Damas, ed. (pp. 549–555). Washington, DC: Smithsonian Institution.

Koenig, W. D., R. L. Mumme, W. J. Carmen, & M. T. Stanback. 1994. Acorn production by oaks in Central Coastal California: Variation within and among years. *Ecology* 75:99–109.

Lamb, H. H. 1977. *Climatic history and the future.* Princeton, NJ: Princeton University Press.

Lindsay, L. W. 1986. Fremont fragmentation. In: *Anthropology of the desert West: Essays in honor of Jesse D. Jennings,* C. J. Condie, & D. D. Fowler, eds. (pp. 229–252). Anthropological Papers No. 110. Salt Lake City: University of Utah.

Lu, T. L. D. 1999. *The transition from foraging to farming and the origin of agriculture in China.* BAR International Series #774. Oxford: British Archaeological Reports.

MacNeish, R. S. 1991. *The origins of agriculture and settled life.* Norman: University of Oklahoma Press.

Madsen, D. B. 1994. Mesa Verde and Sleeping Ute Mountain: The geographical and chronological dimensions of Numic expansion, In: *Across the West: Human population movement and the expansion of the Numa,* D. B. Madsen, & D. Rhode, eds. (pp. 24–31). Salt Lake City: University of Utah Press.

Madsen, D. B., & D. N. Schmitt. 1998. Mass collecting and the diet breadth model: A Great Basin example. *Journal of Archaeological Science* 25:445–455.

Matson, R. G. 1999. The spread of maize to the Colorado Plateau. *Archaeology Southwest* 13:10–11.

Mayewski, P. A., L. D. Meeker, S. Whitlow, M. S. Twickler, M. C. Morrison, R. B. Alley, P. Bloomfield, & K. Taylor. 1993. The atmosphere during the Younger Dryas. *Science* 261:195–200.

McManus, J. F., D. W. Oppo, & J. L. Culler. 1999. A 0.5-million-year record of millennial-scale climate variability in the North Atlantic. *Science* 283:971–975.

McNeill, W. H. 1976. *Plagues and peoples.* Garden City, NY: Anchor.

Milton, K. 2000. Hunter-gatherer diets—A different perspective. *American Journal of Clinical Nutrition* 71(3):665–667.

Miracle, P. T., & C. J. O'Brien. 1998. Seasonality of resource use and site occupation at Badanj, Bosnia-Herzegovinia: Subsistence stress in an increasingly seasonal environment? In: *Seasonality and sedentism: Archaeological perspectives from old and new world sites*, T. R. Rocek, & O. Bar-Yosef, eds. (pp. 41–74). Peabody Museum of Archaeology and Ethnology Bulletin 6. Cambridge, MA: Harvard University.

Moore, A. M. T., & G. C. Hillman. 1992. The Pleistocene to the Holocene transition and human economy in Southwest Asia: The impact of the Younger Dryas. *American Antiquity* 57:482–495.

Murray, J. D. 1989. *Mathematical biology*. Berlin: Springer-Verlag.

Newnham, R. M., & D. J. Lowe. 2000. Fine-resolution pollen record of late-glacial climate reversal from New Zealand. *Geology* 28:759–762.

North, D. C., & R. P. Thomas. 1973. *The rise of the Western World: A new economic history*. Cambridge: Cambridge University Press.

Partridge, T. C., G. C. Bond, C. J. H. Hartnady, P. B. deMenocal, & W. F. Ruddiman. 1995. Climatic effects of late Neogene tectonism and vulcanism. In: *Paleoclimate and evolution with emphasis on human origins*, E. S. Vrba, G. H. Denton, T. C. Partridge, & L. H. Burckle, eds. (pp. 8–23). New Haven, CT: Yale University Press.

Pearsall, D. M. 1995. Domestication and agriculture in the New World tropics. In: *Last hunters, first farmers: New perspectives on the prehistoric transition to agriculture*, T. D. Price, & B. Gebauer, eds. (pp. 157–192). Santa Fe: School of American Research Press.

Petersen, L. C., G. H. Haug, K. A. Haugen, & U. Röhl. 2000. Rapid changes in the hydrologic cycle of the tropical Atlantic during the last glacial. *Science* 290: 1947–1951.

Petit, J. R., J. Jouzel, D. Raynaud, N. I. Barkov, J.-M. Barnola, J. Basile, M. Bender, J. Cappellaz, M. Davis, G. Delaygue, M. Delmotte, V. M. Kotlyakov, M. Legrand, V. Y. Lipenkov, C. Lorius, L. Pépin, C. Ritz, E. Saltzman, & M. Stievenard. 1999. Climate and atmospheric history of the past 420,000 years from the Vostok ice core, Antarctica. *Nature* 399:429–436.

Poli, M. S., R. C. Thunell, & D. Rio. 2000. Millennial-scale changes in North Atlantic deep water circulation during marine isotope stages 11 and 12: Linkage to Antarctic climate. *Geology* 28:807–810.

Price, T. D., & B. Gebauer. 1995. *Last hunters, first farmers: New perspectives on the prehistoric transition to agriculture*. Santa Fe: School of American Research Press.

Reed, C. A. 1977. *The origins of agriculture*. The Hague: Mouton.

Reilly, J., & D. Schimmelpfennig. 2000. Irreversability, uncertainty, and learning: Portraits of adaptation to long-term climate change. *Climatic Change* 45:253–278.

Richards, B. W., L. A. Owen, & E. J. Rhodes. 2000. Timing of Late Quaternary glaciations in the Himalayas of Northern Pakistan. *Journal of Quaternary Science* 15:283–297.

Richerson, R. J., & R. Boyd. 1999. The evolutionary dynamics of a crude super organism *Human Nature* 10:253–289.

Rindos, D. 1984. *The origins of agriculture: An evolutionary perspective*. London: Academic Press.

Rosenzweig, C., & D. Hillel. 1998. *Climate change and the global harvest*. New York: Oxford University Press.

Sage, R. F. 1995. Was low atmospheric CO_2 during the Pleistocene a limiting factor for the origin of agriculture? *Global Change Biology* 1:93–106.

Sahlins, M. D., & E. R. Service. 1960. *Evolution and culture*. Ann Arbor: University of Michigan Press.

Schneider, S. H., W. E. Easterling, & L. O. Mearns. 2000. Adaptation: Sensitivity to natural variability, agent assumptions and dynamic climate changes. *Climatic Change* 45:203–221.

Schulz, H., E. von Rad, & H. Erlenkeuser. 1998. Correlation between Arabian Sea and Greenland climate oscillations of the past 110,000 years. *Nature* 393:54–57.

Shi, N., L. M. Dupont, H-J. Beug, & R. Schneider. 2000. Correlation between vegetation in Southwestern Africa and oceanic upwelling in the past 21,000 years. *Quaternary Research* 54:72–80.

Smith, B. D. 1989. Origins of agriculture in Eastern North America. *Science* 246:1566–1571.

Smith, B. D. 1995. *The emergence of agriculture*. New York: Scientific American Library.

Smith, B. D. 1997. Initial domestication of *Curcurbita pepo* in the Americas 10,000 years ago. *Science* 276:932–934.

Smith, M. A. 1987. Antiquity of seedgrinding in arid Australia. *Archaeology in Oceania* 21:29–39.

Sowers, T., & M. Bender. 1995. Climate records covering the last deglaciation. *Science* 269:210–214.

Steig, E. J., E. J. Brook, J. W. C. White, C. M. Sucher, M. L. Bender, S. J. Lehman, D. L. Morse, E. D. Waddington, & G. D. Glow. 1998. Synchronous climate changes in Antarctica and the North Atlantic. *Science* 282:92–95.

Stiner, M. C., N. D. Munro, & T. A. Surovell (G. Bar-Oz, T. Dayan, N. F. Bicho, A. Deitti, J-P. Brugal, E. Carbonell, K. V. Flannery, S. Newton, & A. Pike-Tay, commentators). 2000. The Tortoise and the hare: Small-game use, the broad-spectrum revolution, and Palolithic demography. *Current Anthropology* 41:39–73.

Straus, L. G., B. V. Eriksen, J. M. Erlandson, & D. R. Yesner. 1996. *Humans at the end of the Ice Age: The archaeology of the Pleistocene-Holocene transition*. New York: Plenum.

Stuiver, M., P. J. Reimer, E. Bard, J. W. Beck, G. S. Burr, K. A. Hughen, B. Kromer, G. McCormack, J. V. D. Plicht, & M. Spurl. 1998. INTCAL98 radiocarbon age calibration, 24,000–0 cal BP. *Radiocarbon* 40:1041–1083.

United Nations Food and Agriculture Organization. 2000. Global Information and Early Warning System for Food and Agriculture. http://www.fao.org/giews/.

Weber, M. 1930. *The protestant ethic and the spirit of capitalism*, translated by Talcott Parsons. London: Allen Unwin.

Werne, J. P., D. J. Hollander, T. W. Lyons, & L. C. Peterson. 2000. Climate-induced variations in productivity and planktonic ecosystem structure from the Younger Dryas to Holocene in the Cariaco Basin, Venezuela. *Paleoceanography* 15:19–29.

West, G. J. 2000. Pollen analysis of late Pleistocene-Holocene sediments from core CL-73-5, Clear Lake, Lake County, California: A terrestrial record of California's cismontane vegetation and climate change inclusive of the Younger Dryas event. In: *Proceedings of the Seventeeth Annual Pacific Climate Workshop, May 22–25, 2000*, G. J. West, & L. D. Buffaloe, eds. (pp. 91–106). Interagency Ecological Program for the San Francisco Estuary Technical Report 67. Sacramento: California Department of Water Resources.

Winterhalder, B., & C. Goland. 1997. An evolutionary ecology perspective on diet choice, risk, and plant domestication. In: *People, plants, and landscapes: Studies in paleoethnobotany*, K. J. Gremillion, ed. (pp. 123–160). Tuscaloosa: University of Alabama Press.

Winterhalder, B., & F. Lu. 1997. A forager-resource population ecology model and implications for indigenous conservation. *Conservation Biology* 11:1354–1364.

Wohlgemuth, E. 1996. Resource intensification In prehistoric central California: Evidence from archaeobotanical data. *Journal of California and Great Basin Archaeology* 18:81–103.

Wright, H. E., Jr. 1977. Environmental change and the origin of agriculture in the Old and New Worlds. In: *Origins of agriculture*, C. A. Reed, ed. (pp. 281–318). The Hague: Mouton.

Zvelebil, M. 1996. The agricultural frontier and the transition to farming in the Circum-Baltic region. In: *The origins and spread of agriculture and pastoralism in Eurasia*, D. R. Harris, ed. (pp. 323–345). Washington, DC: Smithsonian Institution Press.

PART 5

Links to Other Disciplines

Biology is an immense enterprise whose purview ranges from the physics of enzyme catalysis to the role of gene expression in cell differentiation to the evolutionary origins of flight to the global carbon cycle. Nonetheless, biology is a single discipline that is taught as a coherent, integrated subject to first-year university students. By contrast, each social science has its own independent introductory course, one that usually makes little reference to other disciplines. The rigid division of human sciences into disciplines has always seemed quite odd to us. In the great scheme of things, humans surely present a smaller range of phenomena than all the rest of biology. One reason why biology remains a unified discipline is that the science has a small set of unifying problems at its core. Physics and chemistry underpin everything. Genetics, cell metabolism, ecology, and evolution are relevant to all organisms, and physiology is common to all multicellular life. A good basic biology course will show how these integrating subdisciplines relate to one another. Practicing biologists often discover that they need to know something of each of these integrating subjects in their professional careers. How can the human sciences possibly be very different?

We have no clear idea of why the human sciences have evolved so differently from biology. Our mentor Donald T. Campbell took an interest in such matters (Campbell, 1969) and supposed that the social sciences would become much more interdisciplinary than they in fact have. In this part, we argue that evolutionary theory, specifically the theory of cultural evolution, stands ready to play much the same role that organic evolution does in biology. The basic argument is very simple. What is the most dramatic feature of human life? Certainly one candidate is its dramatic variation in time and space.

No other species changes its behavior so rapidly, and none occupies such a wide range of environments using such a wide range of economic strategies. Evolutionary processes produce this diversity; every culture has descended from some immediate ancestor, ultimately tracing back to a common African ancestor. Every discipline in the human sciences is centrally concerned with cultural evolution and cultural diversity, whether called by these names or not. Anthropologists have made the study of cultural diversity their specialty. Historians study cultural change in all its forms. For economists, the evolution (or growth) of economies is a central theme. Political scientists study opinion, policy, and constitutional change; sociologists, institutional change. Cultural evolutionists have something to say about some central topics, such as the explanation of human cooperation and social institutions (parts 2 and 3 contain examples). Should human scientists care to emphasize unifying problems, cultural evolution can share a portion of the burden.

Chapter 18 shows economists how a theory of cultural evolution quite naturally complements the rational choice theory that is basic to their discipline. Rational choice theory is one of the other candidates to be a major unifying element in the human sciences. Yet rational choice theory famously lacks psychological realism (Simon, 1959) and lacks an explicit temporal dimension (Nelson and Winter, 1982). Here we derive the basic Darwinian theory of cultural evolution from Bayesian assumptions applicable to the standard rational actor. The behaviors of others are merely a form of proxy information about the world, a resource to be tapped in deciding how to behave oneself. In a world where gaining information tends to be costly, imitating what others do is an excellent strategy under a wide variety of circumstances. An inheritance system provides time-tested information. Using your parents' beliefs or those of others as Bayesian priors is highly adaptive. Doing so allows an individual to concentrate scarce resources on updating decent priors rather than on starting with less information-rich priors, such as those furnished by a generic human nature. Adding these bits of psychological realism yields a theory of cultural evolution within which boundedly rational actors play a fundamental role. The theory also accounts for important human oddities such as our extraordinary cooperation and our susceptibility to certain types of maladaptations. Neat as one of Adam Smith's pins we thought, and still think, though the manuscript was rejected by the *American Economic Review* after protracted adventures with editors and reviewers. We suspect the baleful influence of the lack of a synthetic first-year course is at work here. Subjects not legitimated in that course, which purports to encompass all someone needs to know, are deeply suspect, and culture is generally absent from Econ 1. At the same time, an economic anthropologist who taught us a lot about the science of culture knew little of what is taught in Econ 1.

Chapter 19 is directed at those in the social sciences unfamiliar with a style of deploying mathematical models that is second nature to economists, evolutionary biologists, engineers, and others. Much science in many disciplines consists of a toolkit of very simple mathematical models. To many not familiar with the subtle art of the simple model, such formal exercises have two seemingly deadly flaws. First, they are not easy to follow. The modern style of mathematical analysis uses a very compact notation that facilitates algebra but is quite hard to read. Even the initiated reader might take days to deeply understand even a rather elementary model. The untrained are nearly helpless. Second, motivation to follow the math is often wanting because the model is so cartoonishly simple relative to the real world being analyzed. Critics often level the charge "reductionism" with what they take to be devastating effect. The modeler's reply is that these two criticisms actually point in opposite directions and sum to nothing. True, the model *is* quite simple relative to reality, but *even so*, the analysis is difficult. The real lesson is that complex phenomena like culture require a humble approach. We have to bite off tiny bits of reality to analyze and build up a more global knowledge step by patient step. Experimentalists know the same lesson. To achieve virtues of experimental control of variables, you have to examine only one or a few variables at a time. Similarly, observational studies must examine a relatively few dimensions if any explanatory power is to result. Simple models, simple experiments, and simple observational programs are the best the human mind can do in the face of the awesome complexity of nature. The alternatives to simple models are either complex models or verbal descriptions and analysis. Complex models are sometimes useful for their predictive power, but they have the vice of being difficult or impossible to *understand*. The heuristic value of simple models in schooling our intuition about natural processes is exceedingly important, even when their predictive power is limited. (The predictive power of complex models is no better; they often sacrifice much transparency for little improvement in predictive power.) Verbal reasoning is exceedingly important because the human mind seems to be a verbal organ. However, words alone can be snares and delusions. Unaided verbal reasoning can be unreliable—words are polysemic, and the phenomena of the world have quantitative dimensions poorly captured by the qualitative concepts of natural language. The lesson, we think, is that all serious students of human behavior need to know enough math to at least appreciate the contributions simple mathematical models make to the understanding of complex phenomena. The idea that social scientists need less math than biologists or other natural scientists is completely mistaken.

Chapter 20 deals with the vexatious concept of memes. On the one hand, we have great sympathy with the views of the "universal" Darwinists like Daniel Dennett, Robert Aunger, and Susan Blackmore, who, following Richard Dawkins, employ the term to stress the analogies between genes and culture. On the other hand, we have several worries. One is academic punctilio. When Dawkins (1976) coined the term *meme*, he quite frankly admitted that he had done no scholarship in the social sciences. Fair enough in the context of a trade book, but, in fact, another pioneering universal Darwinist, Donald Campbell (1965, 1975), had done significant work on cultural evolution by 1976. Lucca Cavalli-Sforza and Marc Feldman (1973) had already published their pioneering formal models of cultural evolution. The second, more substantive problem is that the analogy between genes and culture is not very deep. The two are similar in that important information is transmitted between individuals. Both systems create patterns of heritable variation, which in turn implies that the population-level properties of both systems are important. Population-level properties require broadly Darwinian methods for analysis. But this just about exhausts the similarities. The list of differences is much larger. Culture is not based on direct replication but upon teaching and imitation. The transmission of culture is temporally extended. It is not necessarily particulate. Psychological processes have a direct impact on what is transmitted and remembered. These psychological effects can produce complex adaptations in the absence of natural selection. Users of the meme concept seem to us to believe that it does more work than it really does. Third, most users of the meme concept follow Dawkins in being rather incurious about the existing scholarship on the

nature of cultural transmission. A large amount of data already exists on how culture works as an inheritance system and as an evolutionary system. Linguists are perhaps the most advanced students of memes (e.g., Bloom, 2000). Building upon such existing scholarship is surely the most effective way to make progress. Other domains of culture—social organization, technology, folk science—may be governed by rather different principles. The job of synthesizing what we already know and drawing lessons for future work is left undone to the extent that we think that the analogy with genes is a sufficient foundation for a science of culture. It isn't.

We believe that the Darwinian theory of cultural evolution will make contributions across the broad sweep of problems in the human sciences, but the project is one of introducing *additional* useful tools and unifying concepts rather than an imperial ambition to *replace* great swaths of existing theory or methods.

REFERENCES

Bloom, P. 2000. *How children learn the meanings of words*. Cambridge, MA: MIT Press.

Campbell, D. T. 1965. Variation and selective retention in socio-cultural evolution. In: *Social change in developing areas: A Reinterpretation of evolutionary theory*, H. R. Barringer, G. I. Blanksten, & R. W. Mack, eds. (pp. 19–49). Cambridge, MA: Schenkman.

Campbell, D. T. 1969. Ethnocentrism of disciplines and the fish-scale model of omniscience. In: *Interdisciplinary relationships in the social sciences*, M. Sherif, & C. W. Sherif, eds. (pp. 328–348). Chicago: Aldine.

Campbell, D. T. 1975. On the conflicts between biological and social evolution and between psychology and moral tradition. *American Psychologist* 30(12):1103–1126.

Cavalli-Sforza, L. L., & M. W. Feldman. 1973. Models for cultural inheritance: I. Group mean and within group variation. *Theoretical Population Biology* 4:42–55.

Dawkins, R. 1976. *The selfish gene*. Oxford: Oxford University Press.

Nelson, R. R., & S. G. Winter. 1982. *An evolutionary theory of economic change*. Cambridge, MA: Belknap Press of Harvard University Press.

Simon, H. A. 1959. Theories of decision-making in economics and behvioral science. *American Economic Review* 49:253–283.

18   Rationality, Imitation, and Tradition

When the quality of information is poor, people often rely on tradition in making economic decisions. What is the best retail markup percentage? When should one refinance one's home? What is the right safety factor in designing a building? Retailers, homeowners, and engineers typically make such decisions using traditionally acquired rules-of-thumb. This tactic has both advantages and disadvantages. It can be useful because solving problems from scratch is difficult and costly. On the other hand, the uncritical adoption of traditional solutions to problems can lead people to acquire outmoded or even completely unfounded beliefs. Peasants sometimes resist beneficial innovations proffered by development agencies and retain traditional agricultural practices; many contemporary Americans maintain the unfounded belief that there are innate differences between the members of different ethnic groups.

The fact that tradition is sometimes reliable and other times misleading creates an interesting problem for economists. Traditions often work; when they do, they are useful because they reduce the costs of acquiring information and lower the possibility of making errors. However, if everyone were to depend exclusively on traditional rules, what would cause traditional rules to be modified in response to changes in the environment, and what would initially cause useful and reliable behaviors to become traditions?

Conventional economic theory is not helpful in answering this question (Conlisk, 1980). Economists have adopted the Bayesian theory of rational choice as the natural extension of the utility-maximizing view of human behavior when there is uncertainty and use it as a positive theory to predict people's behavior in a wide variety of contexts (Hirshleifer and Riley, 1978). Within the context of this theory, a person's beliefs about the world are represented as a subjective probability distribution. Once this distribution is specified, the theory tells us how rational people should behave and how they should modify their beliefs in accord with their experience. The theory does not tell us why people initially come to have the beliefs that they do but simply takes them as given.

The role of traditional knowledge has been discussed by some economists, but the processes that lead to sensible traditions seem to have been largely ignored. Hayek (1978) believes that limited knowledge and cognitive abilities force people to rely on traditional beliefs and values and argues that traditions are sensible because groups with favorable traditions survive longer and attract more members. Proponents of evolutionary models of firms (Alchian, 1950; Nelson and Winter, 1982) assume that beliefs, values, and other determinants of firm behavior are transmitted within firms and that these beliefs are shaped by the natural selection of firms. The only formal theoretical treatment of tradition seems to be the interesting article of Conlisk (1980) in which the individuals who optimize compete with individuals who acquire their behavior by imitation. If optimization is costly, Conlisk shows that imitation can persist in the population.

In this chapter, we introduce tradition into conventional theory by assuming that people acquire their initial subjective probabilities by imitating their parents, relatives, teachers, business associates, and friends, but otherwise behave as classical Bayesian rationalists. Several lines of empirical evidence support the assumption that people acquire their beliefs about the world by imitation and similar processes. Psychologists have shown that children readily acquire behavioral traits from moral beliefs to rules of grammar by imitating adult models (Bandura, 1977; Rosenthal and Zimmerman, 1978). Data collected on familial resemblances show high parent-offspring correlations for a wide variety of cognitive traits (I.Q.; Scarr and Weinberg, 1976), behaviors (child abuse, alcoholism; Smith, 1975), and indicators of beliefs (religious and political-party affiliation; Fuller and Thompson, 1960). A wealth of anthropological data suggests that human groups possess considerable cultural inertia; members of groups with different cultural histories behave quite differently even when living in similar environments (e.g., Edgerton, 1971). There is also evidence that individuals acquire new beliefs by imitation when they enter organizations such as business firms (Van Maanen and Schein, 1979) and that this process causes distinct cultures to develop in different organizations. (This body of evidence is reviewed in more detail in Boyd and Richerson, 1985:38–60.)

The assumption that people acquire their beliefs by imitation leads to models that keep track of the processes that change the frequency of alternative beliefs in a population of decision makers. To understand why a particular person acquires a particular set of beliefs, we must know to what kinds of behavior naive individuals are exposed. This in turn will depend on the distribution of beliefs (and thus behaviors) that exist in the population. A person in a village in which many people have adopted modern farming practices is more likely to acquire the beliefs that underlie such practices than a person exposed only to traditional lifeways. To predict the distribution of beliefs in the population at some future time, we must know the present distribution of beliefs and account for all of the processes that change that distribution through time. Here we present several such models of cultural change. For a more extensive exposition of our views, see Boyd and Richerson (1985), and for related work, see Pulliam and Dunford (1980), Cavalli-Sforza and Feldman (1981), Lumsden and Wilson (1981), and Rogers (1989).

These models are different from Conlisk's in two important ways: (1) Conlisk regards imitation as an alternative to optimization; individuals are either imitators or optimizers. We assume that imitation is a precondition for optimization; everyone must acquire beliefs about the world before they can optimize. (2) Conlisk simply posits dynamical relations between variables that describe a whole population of decision makers; we are more concerned to show how the details of individual imitation and decision-making processes lead to the dynamics of the distribution of beliefs in a population through time. As we shall see, the optimal behavior in these models is usually for individuals to mix imitation and individual decision making, depending on how the temporal dynamics work out.

We think that there are three lessons to be drawn from our theory of traditions: first, there are plausible circumstances in which it is optimal to depend nearly completely on tradition at equilibrium. Second, there are plausible genetic and cultural mechanisms that could cause people to achieve this equilibrium. Third, when people do depend largely on tradition, processes other than individual choice may have important effects on why people behave the way they do. We will begin by modeling a reference case in which people acquire their initial subjective probabilities by imitation and then modify them in accordance with their own experience in a uniform and constant environment. This model indicates that when beliefs are transmitted culturally, greater reliance on tradition always leads to higher expected utility. We will then add environmental variability to the model. When the optimal behavior varies because individuals encounter different environments, there is an optimal level of dependence on tradition. If there is a substantial chance that individuals and the people that they imitate experience the same environment, and if the information available to update priors is poor, it can be an evolutionary equilibrium to rely almost completely on tradition. In the simplest model, a population of such individuals will, on the average, behave almost as if they were perfect-information optimizers. However, in such a population other processes, which can lead to both beneficial (but poorly understood) beliefs or deleterious superstitions, may also be important. Finally, we will argue that there are cultural processes that may cause people to be characterized by an optimal reliance on tradition.

The Basic Model

In the first and simplest model there are only two processes that affect the distribution of beliefs in a population of decision makers. First, individuals use available information to update their subjective probability distributions. Second, the frequency of different beliefs is changed by the transmission of these beliefs to another generation. The model has three parts: a description of how single individuals modify their beliefs in light of their experience (a process we refer to as "individual learning"), a consideration of how individual learning affects the distribution of beliefs in a population of individuals, and a mechanism for passing one generation's beliefs to the next.

Consider the following very simple decision problem. An individual decision maker has the following utility function:

$$u(y,z) = -u_0(z-y)^2 \tag{1}$$

where z is a decision variable under his control, y is a variable that represents the state of the world, and u_0 is a constant. While the quadratic form of this utility function is unconventional in the theory of the consumer, it is a mathematically convenient representation of the usual view of individual choice. To see this, consider the following example: suppose that the decision maker is a young professional just beginning his or her career and that z represents the amount of time devoted to career advancement. The remainder of the young professional's time, t, is devoted to family and recreation. Then t and z are arguments of a personal "production function," which gives amounts of various "commodities," for example, income and marital happiness, produced for each combination of t and z. The consumption of these commodities in turn generates utility. By using the constraint that total time is fixed and assuming that the young professional's personal production and utility functions have the appropriate convexity properties, one could derive a unimodal function giving utility as a function of z. The optimum value of this function, y, would depend on the properties of the personal production function, which in turn will depend on the state of the world. For example, the relationship between time devoted to work and income might depend on what kind of firm the young professional has entered. While the utility function so derived is unlikely to be exactly quadratic, this functional form is a reasonable caricature of a more general unimodal function. In fact, one could think of it as the first two terms of a Taylor's series expansion of an arbitrary utility function in the neighborhood of the optimum. Because we have not specified how commodities map onto utilities, this model can represent any degree of risk preference.

The individual does not know the value of y with certainty, but his or her beliefs about the likelihood that y takes on various values conform to a normal probability distribution with mean $\hat{y}$ and variance L. Note that y is not a random variable; in a given environment there is an optimum amount of time devoted to career. The probability distribution describes the decision maker's subjective beliefs about what value of z is optimum.

Before making his or her choice, the decision maker has the opportunity to review a certain amount of evidence about the state of the world. For example, by observing the effects of time devoted to work on career advancement and home life, the young professional could get an estimate of the optimal amount of time to devote to work. Because our young professional's initial rate of advancement and domestic satisfaction might depend on a variety of factors other than the amount of time devoted to work, this estimate will be imperfect. Suppose that this evidence can be quantified by the variable x. The decision maker believes (correctly) that the value of x is normally distributed with mean y

and variance V_e. After using this evidence and Bayes's law, the decision maker's updated subjective probability distribution is normal with mean $\hat{y}'$ where

$$\hat{y}' = \frac{V_e\hat{y} + Lx}{V_e + L} \tag{2}$$

To simplify the development here, assume that the decision maker does not update the variance of his or her subjective probability distribution.

The decision maker uses the updated distribution to calculate his or her expected utility as a function of z:

$$E\{u(z,y)|\hat{y}',x\} = -u_0[(z - \hat{y}')^2 + L] \tag{3}$$

and, thus, the value of z that maximizes his or her expected utility, z^* is the following:

$$z^* = \hat{y}' \tag{4}$$

That is, the optimal behavior is the individual's posterior estimate of the most likely state of the environment.

Now, suppose that there is a large population of decision makers. The individuals who make up this population differ in only two respects: (1) they have different prior beliefs about the most likely state of the world, and (2) they are exposed to different evidence about the state of the world. To formalize the first assumption, we assume that the frequency distribution of $\hat{y}$ *in the population* before the subjective probability distributions have been updated, $Q_t(\hat{y})$, is normal with mean M_t and variance B_t. Notice that this is a description of the population, not a probability density. To formalize the second assumption, we assume that the value of x experienced by each different individual is an independent random variable with the density $p(y)$, which has a mean equal to the true state of the world, y, and variance V_e. Otherwise, all individuals are identical; in particular, they all have the same utility function and their subjective probability distribution is characterized by the same value of L.

Let us now consider how the use of Bayes's law by individuals to modify their beliefs changes the frequency distribution of $\hat{y}$ in the population. The distribution of $\hat{y}$ in the population of decision makers after updating, Q'_t, is as follows:

$$Q'_t(\hat{y}) = \iint h(\hat{y}|\hat{y}',x)Q_t(\hat{y}')p(x)d\hat{y}'dx \tag{5}$$

where $h(\hat{y}|\hat{y}',x)$ is the conditional density of an individual's belief after updating, given that the individual had beliefs characterized by $\hat{y}'$ before updating and observed x. Then $Q'_t(\hat{y})$ is normal with this mean:

$$M'_t = \frac{M_t V_e + yL}{V_e + L} \tag{6}$$

and variance:

$$B'_t = \frac{B_t V_e^2 + V_e L^2}{(V_e + L)^2} \tag{7}$$

Thus, after updating, the mean value of $\hat{y}$ moves closer to the correct value, y; the variance may either increase or decrease depending on the magnitudes of B_t, V_e, and L.

So far, we have followed the usual practice of taking the decision maker's initial subjective probabilities as given. We are now in a position to consider the effect of the transmission of these beliefs to another "generation" of decision makers by imitation. For example, suppose that the young professionals advance in their firm and are eventually replaced by a new cohort of entry-level professionals, who form a new population of decision makers and face the same decision problem that their predecessors faced. Initially the individuals in this second "generation" are naive; they have no beliefs of any kind about how much time should be devoted to work. However, each naive individual has been able to observe n models of behavior of the previous generation of professionals. Based on the behavior of their models, naive individuals are able to infer what each model believes about how much time should be devoted to one's profession. Then each of the naive individuals adopts the mean of the n inferred values of $\hat{y}$ that characterize their models as the mean of their own subjective probability distributions. We assume that the variance, L, remains constant at the same value as in the previous generation.

With these assumptions the distribution of $\hat{y}$ in the population just before updating in generation, $t + 1, Q_{t+1}(\hat{y})$, is normal with mean, $M_{t+1} = M'_t$, and variance, $B_{t+1} = (1/n)B'_t$. Because the distribution of $\hat{y}$ remains normal, the state of the population of decision makers at any time can be specified by the mean and variance of $\hat{y}$. If the environment remains constant, the values of the mean and variance in the population will eventually reach a unique stable equilibrium, $\hat{M}$ and $\hat{B}$, where

$$\hat{M} = y \tag{8}$$

and

$$\hat{B} = \frac{V_e L^2}{n(V_e + L)^2 - V_e^2} \tag{9}$$

Equations (6) and (8) say that the effect of the repeated application of Bayesian inference and accurate imitation on the mean value of $\hat{y}$ is unambiguous: the average of the best guesses about the state of the environment in the population converges monotonically to the actual state of the environment. According to (7) and (9), however, the variance of $\hat{y}$ is affected by competing processes. New variation is introduced each generation by errors in individual learning; this process acts to increase $\hat{B}$. On average, however, inference causes beliefs about the environment to become more accurate, and this decreases $\hat{B}$. Finally, if $n > 1$, imitation itself acts to decrease the variance of $\hat{B}$ in the population.

The Evolutionary Stable Amount of Tradition

The relative importance of tradition and individual learning is determined by the relative magnitudes of the width of each individual's initial prior probability distribution (L) and the quality of the information available to individuals (V_e). If L is small compared to V_e, young professionals' work habits will be mostly determined by the beliefs that they acquire by imitation. If L is large, the information that individuals gather for themselves will be more important.

In this section we determine the evolutionary stable, or ESS, value of L. To do this, we find the value of L that when common in a population has higher expected utility than slightly different values of L. One way to justify the ESS approach is to assume that L is a genetically variable character and that utility is monotonically related to fitness. The ESS value of L is the value that prevents the rare genotypes from invading under the influence of natural selection. Some models of cultural transmission have very similar properties to genetic ones, and for our immediate purposes, we can think of L as evolving under the influence of either process. Clearly, cultural and genetic transmission also differ in important ways, for example, in the timescale over which they are relevant. Variations in reliance on tradition among contemporary societies likely require a cultural explanation, while a genetic model would be appropriate for studying the evolution of humans from apes. The penultimate section of the chapter will address several explicitly cultural mechanisms that can lead to the ESS.

Consider a population in which most individuals have a learning rule characterized by the parameter value, L, and that has reached the associated equilibrium values $\hat{M}$ and $\hat{B}$. The expected utility of an individual whose learning rule is characterized by parameter L' is the following:

$$E\{u(\hat{y},x)\} = - U_0 \frac{V_e^2}{(V_e + L)^2}[(y=\hat{M})^2 + \hat{B}] + \frac{L^2 V_e}{(V_e + L)^2} \qquad (10)$$

One can show that this expression for expected utility is concave with a global maximum at the value of L, $L^\dagger$,

$$L^\dagger = (y - \hat{M})^2 + \hat{B} \qquad (11)$$

The term $(y - \hat{M})^2 + \hat{B}$ measures the closeness of the population's beliefs about the state of the world to its actual state; V_e measures the accuracy of the information gained by each individual through his own experience. Relation (11) (together with [1]) says individuals should rely on imitation in proportion to the accuracy of the distribution of beliefs. If $(y - \hat{M})^2 + \hat{B}$ is large compared to V_e, individuals should rely mainly on their own experience; if $(y - \hat{M})^2 + \hat{B}$ is small compared to V_e, then it is optimal to depend mainly on imitation. This expression does not depend on the assumption that the population is in equilibrium nor that the environment is constant.

Now, suppose that natural selection, or an analogous cultural process, favors L, which increases expected utility. Then because $\hat{B}$ is a function of L, the population will eventually reach an ESS value of L, L^*, such that $L^* = \hat{B}(L^*)$. Using the expression for $\hat{B}$ given in equation (9), one can show that the ESS amount of imitation is $L^* = 0$. At equilibrium, individuals will depend completely on tradition and totally disregard the evidence presented by one's own experience.

This result has an intuitive explanation. At equilibrium, the relative merit of tradition and learning depends on the relative "noisiness" of the two sources of information. Learning has two effects on the variance in the population. On average, learning causes individual's estimates of y to move toward the correct value and thus acts to reduce the variation in the population. However, errors made during learning increase the variation of the population. Once the population reaches equilibrium in a constant environment, the net effect of learning is to maintain erroneous beliefs in the population. Decreasing L always decreases $\hat{B}$. Thus, any process that acts to change L so as to increase expected utility will reduce L until experience plays no role in determining individual beliefs.

Heterogeneous Environments

There are good reasons to doubt the robustness of the conclusion of the previous section. So far, we have assumed that (1) every member of the population experienced the same state of the world, (2) the state of the world did not vary from generation to generation, and (3) all individuals had the same utility function. Relaxing any one of these assumptions reduces the usefulness of tradition. For example, consider a heterogeneous environment in which different individuals experience different states of the world, but in which there is some chance that individuals in one environment draw models from other environments. In a given environment, people's beliefs will tend toward the optimum in that environment, but drawing models from diverse environments will reduce the likelihood that an individual acquires beliefs that are appropriate to its own environment. The models in this section show that a substantial reliance on tradition may still be evolutionarily stable in a heterogeneous environment or in a population in which utility functions vary. We have shown elsewhere that this conclusion also holds true in an environment that changes through time (Boyd and Richerson, 1983, 1985: ch. 4).

The essential feature of a heterogeneous environment is that different individuals in the population experience different states of the world, formalized in terms of the value of y. Such variation might arise for many reasons. For example, different young professionals might work in different firms, practice different professions, or live in different regions. We will model heterogeneous environments by assuming that the probability that an individual in the population experiences the environment specified by the value y is given by a normal density function, $f(y)$, with mean 0 and variance H. Setting the mean to 0 can be done without loss of generality since it sets only the origin from which different environments are measured. The variance, H, is a measure of the amount of environmental variation.

Suppose that in the environment characterized by the value y, the frequency of individuals with a subjective probability distribution characterized by a mean $\hat{y}$ before updating is normal with mean $M_t(y)$ and variance $B_t(y)$. Then the mean and variance after updating in that environment are given by equations (6) and (7) with the appropriate value of y. Further, suppose that there is a probability $1 - m$ that given models experience the same environment that their naive imitators will experience and a probability, m, that models are drawn at random from the population as a whole. Thus, for example, some of a particular young professional's models might be drawn from another firm in which more (or less) dedication is required to succeed. This model also applies to a population of individuals who live in a uniform environment but whose utility functions have different optima.

With these assumptions, one can derive recursions for the mean and variance of the distribution of prior beliefs in each environment. One can show that the equilibrium mean in habitat y is shown here:

$$\hat{M}(y) = \frac{(1-m)yL}{mV_e + L} \tag{12}$$

Equation (12) says that in a heterogeneous environment on average individuals have incorrect beliefs about their environment. The mean value of $\hat{y}$ in any environment y results from the balance of two forces. The Bayesian learning process tends to move the mean toward the correct value for that environment, but the exposure to models drawn from other environments moves the mean toward the mean for the entire population, 0. To find the equilibrium variance, we proceed exactly as in the previous section.

By averaging the expressions for the equilibrium mean and variance over all habitats, and using the expression for the ESS value of L given by equation (11), one can calculate L^* in a heterogeneous environment. The results of this calculation are shown in figure 18.1, which plots the relative importance of imitation in determining behavior, $V_e/(L^* + V_e)$, as a function of V_e for several

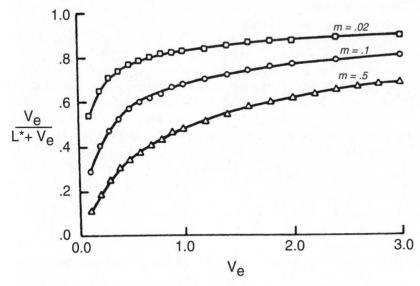

Figure 18.1. Plot of the fractional importance of tradition in determining behavior when the propensity to rely on tradition is at its equilibrium value, $V_e/(L^* + V_e)$, as a function of the quality of information available to individuals (V_e) assuming a heterogeneous environment, $n = 1$ and $H = 1.0$. Increasing values of m represent increasing amounts of mixing of models among different environments.

values of m. This figure indicates that the equilibrium optimum amount of imitation increases as the quality of the information available through individual experience declines and as the probability that models are drawn from foreign environments decreases.

These results make sense. The amount of imitation favored by evolutionary processes depends on the relative quality of two sources of information, the information available to individuals through their own experience and through observing the behavior of their models. As V_e increases, the quality of the information available to individuals through experience declines. As m decreases, the probability that an individual's models will exhibit behavior that is appropriate in the local environment increases. Thus, both increasing V_e and decreasing m cause the equilibrium value of L to increase.

These results suggest that the conclusions of the first section are not entirely misleading. When the amount of mixing between environments is not too large and information is of low quality, individuals achieve the highest expected utility by relying mainly on tradition. We think that this combination of circumstances is not uncommon. The world is complicated and poorly understood and the effects of many decisions are experienced over the course of a lifetime. In deciding how much time to devote to their families, young professionals must estimate not only the immediate effect on their careers and homelives but also the long-run effects on the development of their children's adolescent behavior. In such cases the information available to individuals may be very poor indeed, and it is plausible that they are best off relying almost entirely on traditional beliefs. Also notice that figure 18.1 is a worst case for tradition because it assumes that there is only one model $(n = 1)$. As n increases, the equilibrium variance within environments decreases, and, therefore, tradition is relatively more reliable.

It is important to note that even when the amount of individual learning is small, it plays an important role in the evolutionary dynamics of the population. Some individual learning is necessary if traditional beliefs are to remain utilitarian in local environments in the face of imitation of experienced individuals from other environments. However, a relatively small amount of individual learning is sufficient to keep traditional behaviors on average reasonably near utilitarian optima, so long as mixing between heterogeneous environments is not too great.

Biased Imitation

To this point, we have assumed that individuals adopt a simple unbiased average of the beliefs of the models to which they are exposed. This may not be the most sensible procedure. It would seem better to preferentially imitate models whose behavior has been successful. Young professionals might imitate models who are particularly accomplished in their work and content in their private lives. More generally, naive individuals may imitate prosperous models, contented models, prestigious models, or devout models. By doing this, naive individuals will be more likely to acquire beliefs that lead to prosperity, devotion, contentment, or prestige. In this section we show how this form of biased cultural transmission can increase the frequency of correct beliefs in a population, even when individuals do not understand the causal connection between beliefs and their consequences.

Suppose that instead of simply averaging the beliefs of their models, naive individuals weight models according to their utility-models achieving higher utility having a greater influence on a naive individual's initial belief than individuals with lower utility. There are many plausible observable correlates of utility, such as level of consumption. It seems likely that by imitating individuals with higher levels of consumption, naive individuals might increase their chances of acquiring beliefs that lead to higher utility. In particular, suppose that the initial value of $\hat{y}$ acquired by a naive individual exposed to models with the utilities $u_1, \ldots, u_n$, and beliefs $\hat{y}_1, \ldots, \hat{y}_n$, is this expression:

$$\hat{y} = \frac{\sum_{i=1}^{n} \hat{y}_i (1 + bu_i)}{\sum_{i=1}^{n} (1 + bu_i)} \tag{13}$$

where b is a positive constant small enough that terms of order b^2 can be ignored.

With this assumption, it can be shown (Boyd and Richerson, 1985) the mean in the population after transmission is shown here:

$$M_{t+1} = M'_t + (1 - 1/n) B'_t E \{\text{Reg}[\hat{y}, u(\hat{y})]\} \tag{14}$$

where $E\{\text{Reg}(\hat{y}, u(\hat{y}))\}$ is the regression of utility on $\hat{y}$ averaged over all possible sets of models. According to equation (14), the change in the mean due to biased transmission depends on two factors: the amount of variability within sets of models $[(1 - 1/n) B'_t]$ and the extent to which beliefs about the world are predictably related to utility $[E\{\text{Reg}[\hat{y}, u(\hat{y})]\}]$. Variability within sets of models is important because biased transmission is a culling process that works because some models are more attractive than others. If all models are identical, biased transmission can have no effect. The regression of utility on $\hat{y}$ is a measure of the average effect of a change in an individual's beliefs on his or her utility. If it is positive, individuals with larger values of $\hat{y}$ will have higher utility and, therefore, be more likely to be imitated. This will cause the mean value of $\hat{y}$ in the population to increase. Both the sign and the magnitude of $E\{\text{Reg}[\hat{y}, u(\hat{y})]\}$ depend on the distribution of $\hat{y}$ in the population. If M_t is less than the optimum value (y), larger values of $\hat{y}$ will on average lead to higher utility, and the regression will be positive. The reverse will occur if $M_t < y$. This means that biased transmission will leave the mean unchanged only if it is at the optimal value.

Biased transmission is of interest because it can explain the existence of "folk wisdom," beneficial but poorly understood customs. The preferential imitation of successful people will tend to increase beliefs and practices that lead to success; there is no need for individuals to understand the causal connection between traditional practice and success, even on the part of the individuals who invent the practices.

Natural Selection

So far we have assumed that the probability that a naive individual is exposed to models who are characterized by given beliefs (i.e., a given value of $\hat{y}$) is equal to the frequency of that kind of individual in the previous generation. There is good reason to suppose that this assumption is often violated. For example, the probability that young professionals are advanced in their firm is likely to depend on how much time they devote to work. Underachievers are likely to be fired and overachievers to be promoted. Thus, models who are available for imitation within a firm may represent a biased sample of the original population. More generally, if the behaviors that are shaped by the beliefs acquired by imitation are important, they may affect many aspects of individuals' lives: whom they meet, how long they live, how many children they have, or whether they get tenure. All of these factors could affect the probability that an individual becomes available as a model for others. This means that individuals characterized by some values of $\hat{y}$ will end up being more likely to be imitated than individuals with other values. All other things being equal, it is intuitive that this process, which we will term "natural selection" because of its close resemblance to the biological process, will increase the frequency of the variants most likely to "survive" to enter the pool of models. For a more extensive discussion of the natural selection of culturally transmitted behaviors, see Boyd and Richerson (1985:173–203).

To formalize this idea, we suppose that the probability that an individual who chooses behavior z becomes available as a model, $W(z)$, is the following:

$$W(z) = \exp\{ - (z - w)^2 / 2K\} \tag{15}$$

where w is behavior that maximizes the probability of being in the model pool and $1/K$ is a measure of the intensity of the selection process. Note w need not equal y; for example, individuals who devote more than the utility maximizing amount of time to their work may be more likely to be promoted within the firm.

Using (15) one can show that the mean value of $\hat{y}$ in the population of models (after selection), M_t'', in this equation:

$$M_t'' = \frac{M_t'K + wB_t'}{B_t' + K} \tag{16}$$

Thus, selection moves the mean value of $\hat{y}$ in the population toward the value that maximizes the probability of entering the pool of models, w. One can also show that it reduces the variance of $\hat{y}$ in the population. The strength of both these effects is proportional to the variance in $\hat{y}$ in the population and the intensity of the selection process.

Natural selection is important because it explains how a reliance on tradition can lead to erroneous or deleterious beliefs. Many social and economic processes affect the kinds of individuals available as models. Some of these processes act on the level of the individual, as in the case of the young professional. Others affect whole firms or institutions. For example, firms composed of overachievers may be more likely to survive and expand than firms composed of utility maximizers. When culturally acquired beliefs are important in determining people's behavior, these selective processes will affect what kinds of people are available for imitation and therefore what beliefs will characterize the population. Since there is no reason to believe that such selective processes always favor utility maximizing behavior, selection may cause the most common beliefs in a population to be deleterious. Nonetheless, if information is imperfect and costly to acquire, it may still be sensible to rely on tradition; a modest systematic error may be preferable to a larger random error.

As an aside, we could also interpret the case of a naive manager being socialized by overachievers as the acquisition of a new utility function by considering that preferences are transmitted by tradition and modified by evolutionary processes such as selection. Such a model would allow a more general account of the relationship between learning and tradition than the Bayesian framework used here permits in order to reflect other models of the decision-making process (e.g., Nelson and Winter's, 1982, evolving "routines"). To enlarge on these problems is, however, outside the scope of this chapter. Here we want to emphasize that the standard, and normatively appropriate, Bayesian model is incomplete without a theory of tradition.

Cultural Mechanisms Leading to the ESS Amount of Imitation

So far we have assumed that natural selection acting on genetic variation or an analogous cultural process causes the value of L to change in the direction of increasing expected utility. In this section we consider such cultural processes in more detail. Suppose that the relative dependence on tradition versus one's own experience itself is a culturally transmitted trait. Then each of the three mechanisms we have just studied can, under the right circumstances, act like natural selection to change L in the direction that increases expected utility.

First, however, it is important to clarify why, within the context of the model outlined so far, it is not possible for individuals to choose directly the appropriate value of L. An essential assumption of this chapter is that the information available to individuals is limited; they know the results of their own direct experience and the observable behavior of the individuals whom they had available to imitate, but they do not know the optimum behavior, y. From equation (11), the optimal amount of imitation is given by the term $\hat{B} + (\hat{M} - y)^2$. Individuals can estimate $\hat{B}$ and $\hat{M}$ from their sample of models, and under some circumstances this information might be sensibly used to modify L. They cannot choose the optimum value of L, however, because that value depends on how close the mean belief in the population is to the optimum, y.

How do people acquire their attitudes toward tradition? Assume that people acquire their value of L by imitation during an earlier episode of social learning. With this assumption, any of the processes that change the frequency of a culturally transmitted trait could affect the evolution of the mean value of L in the population:

1. *Ordinary learning.* Individuals might acquire an initial value of L by imitation or teaching and then modify it in accordance with their experience. For example, during enculturation, individuals must acquire many different beliefs and behaviors. They might experiment

with different values of L during early episodes of learning, re-
taining the value that seems to yield the best results. This process
would change the mean value of L among members of the population in the direction that increased average utility.

2. *Biased transmission.* Suppose that available models are variable, some
of them relying on tradition to a greater degree than others. Moreover, suppose that naive individuals can observe some behavior of
their models that serves as a useful index of the model's utility. Then
if naive individuals are predisposed to imitate successful models, the
mean value of L in the population will move toward the optimum.
Notice that this can be true even if, as we have assumed, individuals
have no understanding of why certain beliefs lead to higher utilities.

3. *Natural selection.* Once again assume that individuals vary in their
attitudes toward tradition. Individuals with different values of L
will, on average, behave differently. If an individual's behavior
affects the probability that he or she becomes a model, natural
selection will change the mean value of L in the direction that
increases the chance of acquiring behaviors that make an individual
likely to become a model. To the extent that there is a correlation
between the utility associated with a behavior and the probability
that an individual with the same behavior will become a model,
natural selection would modify L in a utility maximizing direction.

To see how these processes might work, consider how attitudes toward tradition
might change as a society undergoes industrialization. It is often thought that in
pre-industrial agricultural societies people rely heavily on tradition. If one supposes that in such societies information is costly, then their reliance on tradition
is sensible according to our model. Now, suppose that during industrialization,
technical and institutional change makes information less costly. According to
the model, people would be better off if they relied more on their own experience and less on tradition. This might come about by any of the three processes
mentioned. To some extent, individuals might have been able to infer from their
own experience that a lower reliance on tradition improved their lot. More
plausibly, during industrialization people with a tendency to rely more on their
own experience and less on traditional beliefs might more readily acquire non-
traditional skills that lead to wealth and other kinds of observable markers of
success. If successful individuals are more likely to be imitated, biased transmission would decrease average reliance on tradition. Or less traditional individuals might simply be more successful at becoming teachers, managers, and
bureaucrats in modernizing societies. The natural selection mechanism could
have favored a reduced dependence on tradition through differential achievement of roles that are important in socialization.

Invoking processes that affect earlier episodes of imitation to understand the
nature of a subsequent episode clearly creates a problem of explanatory regress.
Each of the three processes mentioned depends on some aspect of the imitation
process, which then must be explained. In the case of ordinary learning, individuals must have some way of weighting the importance of the value of L that

they acquired by imitation against the value that their experience indicates is best. Do they rely on their experience or on imitation? In the case of biased transmission, individuals must have some criteria of success—do they imitate wealthy individuals? Content individuals? Even natural selection will differ in its effects depending on whom naive individuals are prone to imitate. Are they disproportionately affected by their parents, or are other individuals important?

Ultimately, these are questions about human nature. The answers must be sought in the long-run processes that govern the interactions of cultural and genetic evolution in our species. This topic has been discussed at length by us (Boyd and Richerson, 1985) and others (Pulliam and Dunford, 1980; Lumsden and Wilson, 1981; Durham, 1978). Our work supports two generalizations that are relevant here:

1. If there is genetic variation that affects the tendency of people to imitate, natural selection will tend to modify this tendency so that it maximizes genetic fitness. Thus, to the extent that people prefer fitness-enhancing outcomes, selection would increase average utility.
2. There are a variety of conditions in which the fitness-maximizing values of L are near 1. Thus, it is plausible that even the earliest episodes of imitation are not directly subject to genetic influences.

Discussion

The economic theory of rational choice under uncertainty is incomplete because it is silent about the source of people's initial beliefs about the world. People are not immortal; sometime between birth and adulthood they acquire a set of beliefs about the world. Because rational behavior, including the rational response to new information, depends on the nature of an individual's prior beliefs, virtually any behavior can be rational, and therefore explicable, given some set of prior beliefs. A peasant's initial resistance to a beneficial innovation is explicable if one supposes that he believes that traditional ways are superior to modern ones. His ultimate rejection of modern practices may also be rational if his beliefs are described by "tight" priors.

In this chapter we have extended the economic theory of choice under uncertainty by assuming that individuals acquire their initial subjective probability distribution by imitation. In particular, we supposed that each naive individual observes the behavior of a number of experienced models sampled from a larger population, induces the belief that led to the observed behavior, and then adopts an average of those beliefs as his own initial beliefs. Then to understand why people acquire the initial beliefs that they do, we must understand why the population is characterized by a particular distribution of beliefs. This means that models that allow for imitation must account for all of the processes that will arise from individual learning and decision making, while others result from social and economic processes that have different effects on people with different beliefs.

This amendment to economic theory is not proposed as a behavioral alternative to the usual assumption that people are rational optimizers. Whether they are optimizers or not, mortal individuals must acquire their initial beliefs from others. It well may be that the particular model of imitation we have chosen is incorrect, that Bayesian optimizing is a poor model of how humans make choices, or that genetic inheritance is important in determining people's behavioral predispositions. In any case, we believe that a complete theory of human behavior would have a similar structure to the models outlined here; it would keep track of the dynamics of a population of decision makers by accounting for the processes that change the distribution of beliefs or other predispositions in the population. Some of these processes will result from people's attempts at improving their lot, while others will result from what happens to them because they hold the beliefs that they do.

There are two lessons that can be drawn from the models presented here: first, they suggest that a strong reliance on tradition may indeed be sensible. At equilibrium, individuals may rely almost entirely on traditional knowledge and ignore any other information that may be available to them. When (1) the quality of information available to individuals is low and improving it is costly, (2) there is a good chance that the individuals' models experienced the same environment that they experience. Traditional solutions to problems may be much closer to the optimal behavior, on the average, than the solutions that individuals could devise on their own.

The theory also suggests, however, that when traditions are substantially more important in determining people's beliefs than their own experience, a variety of processes other than individual learning may affect the commonness of different beliefs. When tradition is important, it acts like a system of inheritance to create heritable variation within and among groups. Processes like biased transmission and natural selection can then affect the frequency of different beliefs by making it more likely that some beliefs will be transmitted from one generation to the next. When the effect of individual experience is small, it is plausible that such processes may have an important effect on the way that people behave.

Some of these processes, such as biased transmission, may increase the frequency of utility-enhancing behaviors. This fact is of interest because it may explain "folk wisdom," that is, the fact that people hold beneficial traditions that they do not understand. The most striking examples of folk wisdom come from anthropological research. For example, in many parts of the New World native peoples treated maize as a strong base to produce foods such as hominy or masa as part of their traditional cuisine. Katz, Hediger, and Valleroy (1974) have shown that such treatment makes more of the amino acid lysine available (lysine is the least plentiful amino acid in maize). They have also shown that there was a strong negative correlation between the use of alkali treatment and the availability of protein from sources other than maize. Given that many factors influence nutrition, and that only small, uncontrolled samples were available, it is difficult to see how individuals in these cultures could have detected the effect of the treatment. Indeed, although Africans have been using maize as a staple for a few centuries, alkali cooking has not yet developed there. It seems more likely that it could spread because eating treated maize made people more successful or more likely to survive and, therefore, more likely to be imitated. Folk wisdom also plays a role in economic thinking. Hayek (1978) argues that traditional beliefs and institutional arrangements reflect wisdom beyond the ken of any individual, and he bases many political and economic prescriptions on this view. Similarly, proponents of an evolutionary view of the firm (e.g., Alchian, 1950; Nelson and Winter, 1982) argue that inherited decision rules that determine a firm's response to market conditions may be sensible in ways that nobody in the firm understands.

However, for other processes that affect the frequency of alternative beliefs in a population, such as natural selection, there is no guarantee that utility-maximizing behaviors will be favored. This may explain the existence of behavior that seems paradoxical under the usual assumption of individual rationality. In our example of natural selection on behaviors transmitted in the workplace, people could come to work harder than they would desire. Such behaviors could remain in a population because on average the traditions transmitted within a firm are more useful than alternative behaviors individuals could acquire by their own efforts. In other words, a reliance on tradition causes individuals to trade systematically suboptimal behaviors transmitted within the firm for the randomly suboptimal ones that can be discovered by individual effort. Elsewhere we show that processes other than natural selection can have this general effect (Boyd and Richerson, 1985).

Finally, models of the kind described here may also be useful in clarifying the relationship between human evolution and contemporary human behavior. Hirshleifer (1977) has argued that one of the attractive features of sociobiological theory is that it provides an independent way to derive utility functions; namely, human preferences have been shaped by natural selection so that, at least in the context of a hunter-gatherer society, they enhanced genetic fitness. While we are sympathetic to this general approach, we have argued (Boyd and Richerson, 1985) that many human preferences are difficult to explain on this basis. For example, many contemporary professionals seem to sacrifice genetic fitness by delaying marriage, reducing family size, and limiting time devoted to child care in order to gain professional success. Such behavior is explicable, however, if one imagines that individuals who value professional accomplishment for its own sake are more likely to rise to positions of influence than those with more "sociobiological" values. To take another example, humans cooperate in large groups of unrelated individuals to provide public goods (such as victory in warfare) in a way that seems difficult to reconcile with individual fitness maximization. In the work cited, we have shown how some forms of cultural transmission, permitting selection on culture at the level of groups, can arise from attempt to use traditions to enhance the ends of genetic fitness. To take advantage of the economies of information acquisition that tradition offers requires a measure of blind trust of traditional wisdom. Such weak rational control on tradition by its users may be sensible but at the same time allows culture to respond to blind evolutionary processes unique to the cultural system of inheritance. These processes may ultimately have important effects on what individuals prefer as well as on what they believe.

NOTE

We thank Robert Brandon, John Conlisk, Jack Hirshleifer, Richard Nelson, Eric A. Smith, John Staddon, Robert Seyfarth, Joan Silk, Michael Wade, and John Wiley for providing comments on an earlier version of this chapter; we also thank John Gillespie and Ron Pullman for crucial insights about modeling environmental variation and learning, respectively. As tradition dictates, we stipulate that any errors are our own.

REFERENCES

Alchian, A. A. 1950. Uncertainty, evolution and economic theory. *American Economic Review* 58:219–238.

Bandura, A. 1977. *Social learning theory.* Englewood Cliffs, NJ: Prentice Hall.

Boyd, R., & P. J. Richerson. 1983. The cultural transmission of acquired variation: Effect on genetic fitness. *Journal of Theoretical Biology* 100:567–596.

Boyd, R., & P. J. Richerson. 1985. *Culture and the evolutionary process.* Chicago: University of Chicago Press.

Cavalli-Sforza, L. L., & M. W. Feldman. 1981. *Cultural transmission and evolution.* Princeton, NJ: Princeton University Press.

Conlisk, J. 1980. Costly optimizers versus cheap imitators. *Journal of Economic Behavior and Organization* 1:275–293.

Durham, W. H. 1978. Toward a coevolutionary theory of human biology and culture. In: *The sociobiology debate,* A. Caplan, ed. New York: Harper & Row.

Edgerton, R. B. (with W. Goldschmidt). 1971. *The individual in cultural adaptation: A study of four East African Peoples.* Berkeley: University of California Press.

Fuller, J. L., & R. W. Thompson. 1960. *Behavior genetics.* New York: Wiley.

Hayek, F. A. 1978. *The three sources of human values.* L. T. Hobhouse Memorial Trust Lecture 44. London: London School of Economics and Political Science.

Hirshleifer, J. 1977. Economics from a biological viewpoint. *Journal of Law and Economics.* 20:1–52.

Hirshleifer, J., & J. Riley. 1979. The analytics of uncertainty and information: An expository survey. *Journal of Economic Literature* 17:1375–1421.

Katz, S., M. Hediger, & L. Valleroy. 1974. Traditional maize processing techniques in the new world. *Science* 184:765–773.

Lumsden, C., & E. O. Wilson. 1981. *Genes, mind, and culture.* Cambridge, MA: Harvard University Press.

Nelson, R. R., & S. G. Winter. 1982. *An evolutionary theory of economic change.* Cambridge, MA: Harvard University Press.

Pulliam, R., & C. Dunford. 1980. *Programmed to learn.* New York: Columbia University Press.

Rogers, A. 1989. Does biology constrain culture. *American Anthropologist* 90:819–831.

Rosenthal, T., & B. Zimmerman. 1978. *Social learning and cognition.* New York: Academic Press.

Scarr, S., & R. A. Weinberg. 1976. I.Q. performance of black children adopted by white families. *American Psychology* 31:726–739.

Smith, S. S. 1975. *The battered child syndrome.* London: Butterworths.

Van Maanen, J., & E. H. Schein. 1979. Toward a theory of organizational socialization. *Research on Organization and Behavior* 1:209–263.

19   Simple Models of Complex Phenomena

The Case of Cultural Evolution

A great deal of the progress in evolutionary biology has resulted from the deployment of relatively simple theoretical models. Staddon's, Smith's, and Maynard Smith's contributions illustrate this point. Despite their success, simple models have been subjected to a steady stream of criticism. The complexity of real social and biological phenomena is compared to the toylike quality of the simple models used to analyze them and their users charged with unwarranted reductionism or plain simplemindedness.

This critique is intuitively appealing—complex phenomena would seem to require complex theories to understand them—but misleading. In this chapter we argue that the study of complex, diverse phenomena like organic evolution requires complex, multilevel theories but that such theories are best built from toolkits made up of a diverse collection of simple models. Because individual models in the toolkit are designed to provide insight into only selected aspects of the more complex whole, they are necessarily incomplete. Nevertheless, students of complex phenomena aim for a reasonably complete theory by studying many related simple models. The neo-Darwinian theory of evolution provides a good example: fitness-optimizing models, one and multiple locus genetic models, and quantitative genetic models all emphasize certain details of the evolutionary process at the expense of others. While any given model is simple, the theory as a whole is much more comprehensive than any one of them.

Our argument is not very original; the conscious use of the strategy of using simple models to study complex phenomena goes back at least as far as Weber's (1949) use of "ideal types" to study human societies. Good modern expositions include those by Levins (1966, 1968), Liebenstein (1976), Wimsatt (1980), and Quinn and Dunham (1983). If we can contribute anything useful to the case for simple models, it is because our work has involved extending standard evolutionary theory to a particularly troublesome complexity, cultural inheritance of humans (and in rudimentary form, of some other organisms). This work makes a variety of uses of starkly simple evolutionary models, including models based on the assumption of fitness optimization. Yet one of our concerns has been to determine the conditions under which fitness optimization models will fail to account for human behavior. Perhaps we have acquired a self-conscious awareness of some of the tactical details of the simple-model strategy that will be of some use to others.

The Complexity and Diversity of Evolutionary Processes

Evolutionary processes are both extremely complex and extremely diverse. On this count, those who are skeptical of simple models are certainly on solid ground. Every evolving population has a complex history in which many processes have contributed to its evolution, including perhaps drift, migration, mutation, and many other things besides selection. Further, each of these processes can be broken down into a series of interacting subprocesses, each encompassing many varieties. Take selection. There is selection on genes with large effects, selection on quantitative characters, selection on correlated characters and pleiotropic genes, frequency- and density-dependent selection, selection on sex-limited and sex-linked characters, sexual selection of a couple of kinds, and so on. Aside from viruses, all organisms have an intimidatingly large number of interacting genes and phenotypic characters. Environments vary in space and time with large effects on migration and selection. Age, sex, and social organization structure populations and affect their response to evolutionary processes. Developmental processes are complex, although poorly understood, and perhaps affect evolution in fundamentally important ways. Organisms affect their environments as they evolve. In the case of cultural evolution, additional complexities are introduced. We must understand the details of how individuals acquire and modify attitudes and beliefs, how different attitudes and beliefs interact with genes and environment to produce behavior, and how behavior and environment interact to produce consequences for individual lives. Obviously, the study of evolutionary processes must somehow cope with this complexity.

Evolutionary processes are diverse because different populations are quite different from one another in terms of their biology and the environments to which they are and have been exposed. Discoveries about the concatenation of processes affecting the evolution of one population or species do not necessarily say very much about those in others. In the case of cultural evolution, the details of the cultural transmission process vary appreciably from culture to culture. In some, fathers are more important in childhood socialization; in others, less. Modern societies depend on formal teachers; in traditional societies members of the extended family are often important, and so on. Our models of cultural evolution suggest that such structural differences can be quite important to understanding what cultural traits might evolve.

Culture and the Evolutionary Process

In this section, in order to provide a body of detailed examples for use in the later sections, we shall sketch some theoretical results from our own work on the complexities in the evolutionary process caused by culture. Other kinds of complexities of the evolutionary process could be used instead, but we know this one best.

In the last few years, a number of scholars have attempted to understand the processes of cultural evolution in Darwinian terms. Social scientists (Campbell, 1965, 1975; Cloak, 1975; Durham, 1976; Ruyle, 1973) have argued that the analogy between genetic and cultural transmission is the best basis for a general theory of culture. Several biologists have considered how culturally transmitted behavior fits into the framework of neo-Darwinism (Pulliam and Dunford, 1980; Lumsden and Wilson, 1981; Boyd and Richerson, 1983a,b). Other biologists and psychologists have used the formal similarities between genetic and cultural transmission to develop theory describing the dynamics of cultural transmission (Cavalli-Sforza and Feldman, 1973, 1981; Cloninger, Rice, and Reich, 1979; Eaves et al., 1978).

The idea that unifies all this work is that social learning or cultural transmission can be modeled as a system of inheritance; to understand the macroscopic patterns of cultural change we must understand the microscopic processes that increase the frequency of some culturally transmitted variants and reduce the frequency of others. Put another way, to understand cultural evolution we must account for all of the processes by which cultural variation is transmitted and modified. This is the essence of the Darwinian approach to evolution. We (Boyd and Richerson, 1985) have been particularly interested in the question of the origin of cultural transmission. Under what circumstances might selection on genes favor the existence of a second system of inheritance based on the principle of the inheritance of acquired variation?

Cultural and genetic transmission are similar in some respects. For example, the skills and dispositions transmitted during enculturation of children by parents create patterns of behavior that are very difficult to distinguish empirically from patterns resulting from genetic influences.

In other respects, cultural and genetic transmission differ sharply. First, culture is transmitted by an individual observing the behavior of others or by the naive being taught by the experienced. This means that behavior modified by trial-and-error learning can subsequently be transmitted; culture is a system for the inheritance of acquired variation. Second, patterns of cultural transmission are quite different from patterns of genetic transmission. Models other than biological parents are often imitated, including peers, grandparents, and so forth. The cultural analogues of generation length and the mating system are different from, and more variable than, the genetic case. Finally, the naive individual acquiring an item of culture is a more or less active decision-making participant in the transmission process. To some extent, we choose what traits we learn from others, but a zygote cannot choose its genes.

The goal of the Darwinian approach to cultural evolution is to understand cultural change in terms of the forces that act on cultural variation as individuals acquire cultural traits, use the acquired information to guide behavior, and act as models for others. What processes increase or decrease the proportion of people in a society who hold particular ideas about how to behave? We thus seek to understand the cultural analogues of the forces of natural selection, mutation, and drift that drive genetic evolution. These are divisible into three classes: random forces, decision-making forces, and natural selection operating directly on cultural variation.

The random forces are the cultural analogues of mutation and drift in genetic transmission. Intuitively, it seems likely that random errors, individual idiosyncracies, and chance transmission play a role in behavior and social learning. For example, linguists have documented a good deal of individual variation in speech, some of which is probably random individual variation (Labov, 1972). Similarly, small populations might well lose rare skills or knowledge by chance, for example, due to the premature death of the only individuals who acquired them (Diamond, 1978).

Decision-making forces result when naive individuals evaluate alternative behavioral variants and preferentially adopt some variants relative to others. Naive individuals may be exposed to a variety of models and preferentially imitate some rather than others. We call this force biased transmission. Alternatively, individuals may modify existing behaviors or invent new ones by individual learning. If the modified behavior is then transmitted, the resulting force is much like the guided, nonrandom variation of classical "Lamarckian" transmission.

The decision-making forces are derived forces (Campbell, 1965). Decisions require rules for making them, and ultimately the rules must derive from the action of other forces. These decision-making rules may be acquired during an earlier episode of cultural transmission, or they may be genetically transmitted traits that control the neurological machinery for acquisition and retention of cultural traits. The latter possibility is the basis of the various sociobiological hypotheses about cultural evolution (Alexander, 1979; Lumsden and Wilson, 1981). The authors of these hypotheses, among others, argue that the course of cultural evolution is determined by natural selection operating indirectly on cultural variation via the decision-making forces.

Natural selection may also operate directly on cultural variation. Selection is an extremely general evolutionary process (Campbell, 1965). Darwin was able to formulate a clear statement of natural selection in the absence of a correct understanding of genetic inheritance because it is a force that will operate on any system of inheritance with a few key properties. There must be heritable variation, the variants must affect phenotype, and the phenotypic differences must affect individuals' chances of transmitting the variants they carry. That variants are transmitted by imitation rather than sexual or asexual reproduction does not affect the basic argument, nor does the possibility that some of the variants were originally acquired under the guidance of individual decisions. Darwin had no problem in imagining that random variation, acquired variation, and natural selection all acted together as forces in organic evolution. In the case of cultural evolution, we see none either.

We have attempted to construct a series of models that represent all of the processes sketched in the previous section. One interesting general result is that the processes of cultural evolution can easily lead to the evolution of behaviors that reduce Darwinian fitness, especially when nonparental individuals are important in cultural transmission. In the simplest model we have analyzed (Richerson and Boyd, 1984) natural selection acting on cultural variation transmitted by a parent and a "teacher" may cause the trait favoring transmission via teachers to go to fixation at a cost in terms of the number of children produced by parents. Some Darwinian students of humans (Alexander, 1979; Lumsden and Wilson, 1981; Durham, 1976) argue that such effects are unlikely to be important because a system of cultural inheritance with such properties would not be favored by selection on genes. Selection, the argument would run, ought to have acted to prevent such distorted cultural adaptations by either (1) the creation of decision-making forces that counteract the effect of selection on nonparentally transmitted cultural variation or (2) preventing nonparental individuals from becoming important in cultural transmission.

We believe this argument is incomplete because it ignores the fact that individual decision making may be costly compared to social learning. If the costs of using individual decision-making processes are high, selection may not favor decision-making forces that would completely compensate for the maladaptive effects of nonparental transmission. Similarly, if nonparental patterns of cultural transmission offer advantages to individuals of economy in information acquisition, selection on the genes that underlie a capacity for asymmetric transmission may be favored.

For example, nonparental individuals may be more useful models than parents because they may be more skilled or knowledgeable than parents. The effort in decision making required to discriminate exactly among the adaptive skills and maladaptive inclinations of teachers and other nonparental models may require extensive, costly, empirical checks of each element of the teacher's behavior. In contrast, the use of relatively simple, low-cost decision-making rules to bias the choice of models or which of their behaviors to imitate may substantially increase a naive person's skills at a tolerable cost of imitating some maladaptive behaviors. We have analyzed the evolutionary consequences of a variety of simple bias rules. These models suggest that nonparental transmission may often be adaptive despite the cost of selection, especially in spatially variable environments (Boyd and Richerson, 1982, 1985: chs. 7 and 8). In essence, humans may accept the cost of imitating maladaptive cultural traits because the alternatives are a high frequency of random errors or extreme decision-making costs. Even when a cultural system of inheritance optimizes genetic fitness when averaged over all the traits it transmits, many traits taken individually may be quite far from those that would optimize fitness.

Even more extreme violations of the genetic fitness–optimizing model are conceivable. For example, if rules of mate choice are transmitted culturally, human genes might be "domesticated" to serve cultural functions. On the other hand, perhaps the critics of these models are correct, and the abstract possibilities demonstrated by such models are empirically unimportant. The essential point is that, like many bits of genetic realism, adding culture to the evolutionary process might make a qualitative difference in the behavior we expect to observe compared to that expected from the simple fitness optimizing caricature of evolution.

Why Families of Simple Models

Disadvantages of Complex Models

In the face of the complexity of evolutionary processes, the appropriate strategy may seem obvious: to be useful, models must be realistic; they should incorporate all factors that scientists studying the phenomena know to be important. This reasoning is certainly plausible, and many scientists, particularly in economics (e.g., Hudson and Jorgenson, 1974) and ecology (Watt, 1968), have constructed such models, despite their complexity. On this view, simple models are primitive, things to be replaced as our sophistication about evolution grows.

Nevertheless, theorists in such disciplines as evolutionary biology and economics stubbornly continue to use simple models even though improvements in empirical knowledge, analytical mathematics, and computing now enable them to create extremely elaborate models if they care to do so. Theorists of this persuasion eschew more detailed models because (1) they are hard to understand, (2) they are difficult to analyze, and (3) they are often no more useful for prediction than simple models. Let us now consider each of these points in turn.

Complex, detailed models are usually extremely difficult to understand. As more realism is added, the myriad interactions within the model become almost as opaque as the real world we wish to understand. When a set of not-so-complex parts is linked into an interacting complex, it is often impossible to understand why the results behave as they do. To substitute an ill-understood model of the world for the ill-understood world is not progress. In the end, the only way to understand how such a model works is to abstract pieces from it or study simplified cases where its behavior is more transparent. Even when complex models are useful, they are so because we understand how they work in terms of simple models abstracted from them.

Costly, complex models are most likely to be scientifically justified when phenomena are complex but not diverse. It is worth studying the complexities of atoms in great detail because there are only a few kinds, and they all obey the same basic laws. The generality of such laws makes them worth knowing even if the task is difficult. The equivalent sophistication in a model of the evolution of a given society or species is possible, perhaps, but unlikely to be justified on scientific grounds because of limited generalizability to other species or societies.

The analysis of complex models is also expensive and time consuming. The complexity of a recursion model is roughly measured by the number of independent variables that must be kept track of from generation to generation. It usually is not possible to analyze nonlinear recursions involving more than a handful of variables without resorting to numerical techniques. Until the advent of digital computers, obtaining numerical solutions was impractical. Since then, however, there have been many attempts to make computer simulation models of complex social and biological processes. These projects have generally been quite costly. As the number of variables in a model increases, the number of interactions between variables increases even faster. This means that even with the fastest computers, it is not practical to explore the sensitivity of a model to changes in assumptions about very many of its constituent interactions. Considerations of economy of effort in scientific practice dictate that we should be satisfied with much simpler models than we could build in principle.

Complex, realistic models are sometimes employed when prediction rather than understanding is the main goal. Numerical weather prediction models and economic forecasting models come to mind. In both cases the gains in *understanding* of atmospheric and economic phenomena are mostly attributable to the constituent simple submodels of particular processes that are individually not much good for prediction. The marginal increase in understanding relative to cost in the large predictive models is so small that only their practical application justifies their expense; scientific discovery would be better served by more attention to the simpler models. As Dupré (1987) observes, explanation differs from prediction in being easier to achieve (leaving aside statistical models that make no pretentions to explanation). We would argue in addition that explanation or understanding is scientifically far more fundamental than prediction. This is most clearly evident in examples such as the simple deterministic models of economic and population processes that can exhibit chaotic behavior (Day, 1982; May, 1976). If these models prove to apply in the real world, they will guarantee that only short-range predictions are possible with less than perfect specification of initial conditions, but they also give a quite satisfactory explanation of why this is so. The problem is well understood in the context of a purely physical problem, weather prediction (Smagorinsky, 1969).

Detailed models of complex social or biological systems are often not much more useful for prediction than are simple models. Detailed models usually require very large amounts of data to determine the various parameter values in the model. Such data are rarely available. Moreover, small inaccuracies or errors in the formulation of the model can produce quite erroneous predictions. The temptation is to "tune" the model, making small changes, perhaps well within the error of available data, so that the model produces reasonable answers. When this is done, any predictive power that the model might have is due more to statistical fitting than to the fact that it accurately represents actual causal processes. It is easy to make large sacrifices of understanding for small gains in predictive power. Contrarily, although evolutionary processes are inherently complex and diverse, models with a few variables may capture enough of the really important processes in a given case or class of cases both to explain and to predict with tolerable accuracy.

The Utility of Simple Models

In the face of these difficulties, the most useful strategy will usually be to build a variety of simple models that can be completely understood but that still capture the important properties of the processes of interest. Liebenstein (1976: ch. 2) calls such simple models "sample theories." Students of complex and diverse subject matters develop a large body of models from which "samples" can be drawn for the purpose at hand. Useful sample theories result from attempts to satisfy two competing desiderata: they should be simple enough to be clearly and completely grasped, and at the same time they should reflect how real processes actually do work, at least to some approximation. A systematically constructed population of sample theories and combinations of them constitutes the theory of how the whole complex process works.

The synthetic theory of evolution provides a good example. Each of the basic processes (e.g., selection, mutation, drift) is represented by a large variety of simple models, some specific to a particular population, and others quite general. These models are combined in different ways to represent interesting phenomena (e.g., sexual selection, speciation). This whole family of models, together with a knowledge of which models are appropriate for what kinds of situations, constitutes the theoretical system of population biology.

A theoretical system so constituted from simple sample models is a complicated and diverse collection of knowledge; it cannot be legitimately labeled simpleminded. Still, every tactical deployment of models to study a question of interest will be quite simple compared to the phenomena that they are intended to represent. The sample models are caricatures. If they are well designed, they are like good caricatures, capturing a few essential features of the problem in a recognizable but stylized manner and with no attempt to represent features not of immediate interest.

Wimsatt (1980, 1981) provides good general discussions of tactical considerations in the deployment of simple models. To Wimsatt, all sample models of evolutionary phenomena should be viewed as "heuristics" rather than universally applicable laws. This terminology has the virtue of emphasizing that all sample models have defects. They usefully apply only over a limited range of phenomena, and even over the range where they are useful they are almost certain to have biases. Even the very best scientific heuristic (or sample model) will fail and possibly mislead if pushed too far or in the wrong direction. It is in attention to details of the use of simple sample theories that these problems are minimized and the maximum understanding gained. The user attempts to discover "robust" results, conclusions that are at least qualitatively correct, at least for some range of situations, despite the complexity and diversity of the phenomena they attempt to describe.

Note that simple models can often be tested for their scientific content via their predictions even when the situation is too complicated to make practical predictions. Experimental or statistical controls often make it possible to expose the variation due to the processes modeled, against the background of "noise" due to other ones, thus allowing a ceteris paribus prediction for purposes of empirical testing. Simple models, in other words, are the formal theoretical parallel of the experimental and comparative methods so widely used in biology and the social sciences.

Generalized Sample Theories

Generalized sample theories are an important subset of the simple sample theories used to understand complex, diverse problems. They are designed to capture the qualitative properties of the whole class of processes that they are used to represent, while more specialized ones are used for closer approximations to narrower classes of cases. Generalized sample theories are useful because we do not seem to be able to construct models of social and biological phenomena that are general, realistic, and precisely predictive (Levins, 1966, 1968). That is, evolutionary biologists and social scientists have not been able to satisfy the epistemological norm derived from the physical sciences that holds that theory be in the form of universal laws that can be tested by the detailed predictions they make about the phenomena considered by the law. This failure is probably a consequence of the complexity and diversity of living things. Basic theoretical constructs like natural selection are not universal laws like gravitation; rather, they are taxonomic entities, general classes of similar processes that nonetheless have a good deal of diversity within the class. A theoretical construct designed to represent the general properties of the class of processes labeled natural selection must sacrifice many of the details of particular examples of selection. On the other hand, a model tailored to the details of a particular case is unlikely to have much relevance beyond that case. Further, the most precise predictions may be obtained by statistical models that sacrifice realism and hence are useless as explanatory devices.

One might agree with the case for a diverse toolkit of simple models but still doubt the utility of *generalized* sample theories. Fitness-maximizing calculations are often used as a simple caricature of how selection ought to work most of the time in most organisms to produce adaptations. Does such a generalized sample theory have any serious scientific purpose? Some might argue that their qualitative kind of understanding is, at best, useful for giving nonspecialists a simplified overview of complicated topics and that real scientific progress still occurs entirely in the construction of specialized sample theories that actually predict. A sterner critic might characterize the attempt to construct generalized models as loose speculation that actually inhibits the real work of discovering predictable relationships in particular systems.

These kinds of objections implicitly assume that it is possible to do science without any kind of general model. All scientists have mental models of the world. The part of the model that deals with their disciplinary specialty is more detailed than the parts that represent related areas of science. Many aspects of a scientist's mental model are likely to be vague and never expressed. The real choice is between an intuitive, perhaps covert, general theory and an explicit, often mathematical, one.

It seems to us that generalized sample models such as fitness-optimizing models do play an important role. Well chosen to represent the stripped-down essence of a much larger set of more specialized models, generalized sample theories serve important functions in scientists' cognitive organization of complex-diverse subject matters and in communication between specialists. For example, we are concerned with the details of how cultural transmission occurs, a subject studied by psychologists (Boyd and Richerson, 1985: ch. 3). Social learning theorists have made many, but not all, of the kinds of measurements that are necessary for specifying good sample theories of cultural transmission. Crucial unknowns include the mechanisms by which variation and covariation are maintained in cultural traits. These properties have important implications for the process of cultural evolution because the selection and bias forces depend on the maintenance of variation for their effectiveness. These deficiencies of social learning theory are not at all apparent in the absence of a theory linking the psychology of enculturation with the macroscopic phenomena of social institutions and long-run outcomes. It seems unlikely that a sensible psychologist would be motivated to make the arduous and costly experiments necessary to determine such processes without a general theoretical argument justifying their importance. This is an example of a common situation: constructing models that make such links, even if they are simple caricatures, often shows that processes with small, relatively hard to measure, effects can produce major results.

The relationship between a generalized sample theory and empirical test or prediction is a subtle one. To insist upon empirical science in the style of physics is to insist upon the impossible. However, to give up on empirical tests and prediction would be to abandon science and retreat to speculative philosophy. Generalized sample theories normally make only limited qualitative predictions. The logistic model of population growth is a good elementary example. At best, it is an accurate model only of microbial growth in the laboratory. However, it captures something of the biology of population growth in more complex cases. Moreover, its simplicity makes it a handy general model to incorporate into models that must also represent other processes such as selection, and intra- and interspecific competition. If some sample theory is consistently at variance with the data, then it must be modified. The accumulation of these kinds of modifications can eventually alter general theory, either by compelling the abandonment of some sample models or by systematizing knowledge about the variation of processes. In extreme cases, major discoveries in some of the components of a general theory can compel the reorganization of the entire edifice, as exemplified by the impact of Mendelian genetics on Darwinian theory in biology. No one nowadays would think of using Karl Pearson's models of the inheritance of acquired variation as a sample theory of genetic inheritance, although they might have some specialized uses in the study of cultural evolution.

A generalized model is useful so long as its predictions are qualitatively correct, roughly conforming to the majority of cases. It is helpful if the inevitable limits of the model are understood. It is not necessarily an embarrassment if more than one alternative formulation of a general theory, built from different sample models, is more or less equally correct. In this case, the comparison of theories that are empirically equivalent makes clearer what is at stake in scientific controversies and may suggest empirical and theoretical steps toward a resolution.

Some Remarks on the Strategy of Building Simple Models

One of the main points of the preceding discussion is that the analysis of evolutionary problems using simple models depends very much on the appropriate choice of those models. How does one go about making such choices? Evolutionary biologists and social scientists use a variety of methods to accomplish this task that, we believe, can be collected under three main headings, corresponding to idealized analytical steps: (1) the choice of problem, (2) the modularization of analysis, and (3) the construction of synthetic hypotheses that we shall call "plausibility arguments."

Choice of Problem

When one uses simple models to understand complex and diverse problems, the choice of the problem to be analyzed exerts a strong influence on the kinds of simplifications one chooses. The idea is to simplify most drastically those aspects that are not centrally related to the problem at hand in order to retain the maximum feasible detail in the features of most direct interest. In the case of our models of cultural evolution, we have been concerned with the evolution of cultural organisms from acultural ancestors. This required us to represent the processes of ordinary organic evolution in most of our modeling efforts. Still, we were also interested in trying to develop preliminary general models of the important structural features and forces that affect cultural evolution. Given this choice of problem, it seemed advisable to use very simple models of genetic processes to represent the evolution of genetic capacities for culture in order that the models of cultural transmission could be made a bit more elaborate. Thus, we frequently asked what parameter value of a model controlling the propensity to acquire culture in a certain way would cause fitness to be optimized. Those models that included specific genetics used only the simplest haploid, one locus, or quantitative models of genetic transmission.

Models emphasizing cultural detail at the expense of genetic detail accept the risk that some particular complexity of the human genetic system plays a direct role in the coevolution of genes and culture. For example, if genes affecting the behavior toward relatives are transmitted on the Y chromosome, as Hartung (1976) suggested, the models we constructed might turn out to be seriously misleading. The opposite risk, however, seemed more serious to us in the context of the problem; in models that are too complex, the important details of culture itself might be obscured or lost. Several commentators (Maynard Smith and Warren, 1982; Boyd and Richerson, 1983b; Kitcher, 1985) have remarked that the analysis that led Lumsden and Wilson (1981) to their "thousand year rule" is dubious because key properties of culture disappear as a result of simplifying assumptions. The general formulation of their model is conceptually satisfactory, but its complexity appears to have dictated misleading simplifications in the interests of successful analysis.

Modularization of Analysis

Most interesting evolutionary problems involve the interaction of evolutionary processes and a particular pattern of genetic transmission and gene expression. For example, the interaction of selection and mutation at a diploid locus is a classic problem of the synthetic theory. The sample models of the parts of this problem are less interesting than the combination of them in a model that can help us understand how the two basic forces interact with genetically inherited variation. Similar problems are of interest in cultural evolution. How does learning, acting as an evolutionary force because learned variants can be imitated, interact with selection, both selection on the cultural variants and on the underlying senses of reward and punishment that guide learning? Such combinations of processes inevitably make for relatively complex models. To make any headway, relatively difficult mathematical and experimental procedures have to be introduced, and many simplifying assumptions have to be made. Difficult choices between analytical tractability, comprehensibility, generality, and realism have to be made. Is a fitness optimization representation of the genetic process a reasonable simplification, or can some additional genetic realism be usefully retained in the context of the problem?

The answers to such questions are sought by breaking the problem down first into its constituent sample models and then reassembling them step by step into more complex combinations. This tactic is obvious but easily misunderstood and misused. In the long run, the simple models strategy leads to large families of well-understood sample models, some of which will be relatively complex, specialized, and difficult to understand. Also, relatively complex combinations of models are often useful. However, such relatively complicated models depend on a thorough understanding of the simplest models of each family and of the constituent submodels of compound models. The possibility for artifactual results increases with the complexity of the analysis unless one can be reasonably confident that the constituent sample models are empirically reasonable and mathematically well behaved. It is relatively much easier to conduct experiments and detailed mathematical analysis on processes when they are isolated than when they are imbedded in a complex system. In population biology, both history and pedagogic practice suggest that one must begin with an understanding of the elementary constituents of the theory.

While building models of complex processes composed of simpler modules may be second nature to evolutionary biologists, in our experience it sometimes confuses social scientists who read the present body of theory in cultural evolution. The modularization of complex problems seems reductionistic; even after the parts are reassembled it seems to some readers as if the models are attempting to deduce the properties of wholes from properties of parts. The tactical "reductionism" used to understand a problem does not imply that the interaction of parts might not produce irreducible effects. For example, some models of culture built using this tactic suggest that group selection might be especially likely under some plausible forms of cultural transmission (Boyd and Richerson, 1985: ch. 7).

Sometimes, evolutionary biologists (and social scientists who use similar methods, such as economists) contribute to the confusion by failing to distinguish between the heuristic use of tactical reductionism from a real belief that some particular simple model is a true description of a complex process. Indeed, the relative ease with which interesting, even approximately correct, results can be obtained for intrinsically rather complex processes with simple models can lead the unwary to conclude that successful tactical reduction implies the adequacy of a philosophical reductionist stance. Those who are so tempted should consult Wimsatt's work. Most users of simple models know better. For example, Dawkins (1982), a prototypical genetic reductionist by some accounts (Sober, 1984), begins his discussion (pp. 1–2) by asking the reader to take his idea of selfish genes with extended phenotypes as a heuristic model. Later (by p. 7), Dawkins does express the hope that it may prove more fundamental than a mere heuristic, but the distinction between the two interpretations is clear, and the reader is left the choice.

The development of a formal theory of cultural evolution is in its infancy, and attention has properly concentrated on quite elementary models. This means that the theory to date appears quite reductionistic. For example, most models consider only one cultural trait. On the one hand, an overenthusiast might claim that these models are relatively successful in explaining human behavior and hence that human cultures really can be atomized into traits. On the other hand, a critic might complain that they are completely bankrupt because they do not take account of the fact that cultural traits must interact in complex ways. The fact is that such preliminary models are silent about what complexities might flow from the interaction of multiple traits. That is a difficult question in its own right, but one whose analysis must be deferred until we understand the simpler theoretical elements we might use in such an analysis.

The thorough study of simple models includes pressing them to their extreme limits. This is especially useful at the second step of development, where simple models of basic processes are combined into a candidate generalized model of an interesting question. There are two related purposes in this exercise.

First, it is helpful to have all the implications of a given simple model exposed for comparative purposes, if nothing else. A well-understood simple sample theory serves as a useful point of comparison for the results of more complex alternatives, even when some conclusions are utterly ridiculous.

Second, models do not usually just fail; they fail for particular reasons that are often very informative. Just what kinds of modifications are required to make the initially ridiculous results more nearly reasonable? For example, the failures of the logistic model of population growth suggest the amendments needed to make better models. In the case of culture, models that include only faithful cultural transmission suggest that culture is generally inferior to genes as a mode of inheritance (Cavalli-Sforza and Feldman, 1983). If the evolution of culture in the hominid line was favored by natural selection, there must be more to the story than just the acquisition of behavior by imitation. We have suggested that the ability of culture to couple individual learning to a transmission mechanism, thus to generate a system for the inheritance of acquired variation, could cause capacities for culture to evolve (Boyd and Richerson, 1983a, 1985: ch. 4). However, this analysis also fails because it suggests that the advantages of culture are quite general, and hence that many organisms ought to have "Lamarckian" systems of inheritance. This failure in turn suggests that there are other costs to the inheritance of acquired variation that must be accounted for.

In both of these respects, human sociobiology has made a major contribution by showing what must be true if the genetic fitness optimizing model generally holds when behavioral variation is proximally transmitted by culture. For example, Alexander (1979; see also Flinn and Alexander, 1982) argues that decision-making forces are powerful enough to constrain cultural variation to maximize fitness in most circumstances. Important qualitative predictions flow from this argument. If strong, accurate decision making is possible, then humans need not depend on relatively passive imitation; they can easily invent or choose those behaviors appropriate to the environments they find themselves in. If so, culture will behave more like ordinary mechanisms of phenotypic flexibility than like an inheritance system. Empirically, behavioral variation will be largely explicable, even in the short run, in terms of environmental variation rather than the variation in what traits are available for imitation. This argument also implies that costs of making decisions are low relative to any economies that might result from imitation. In our judgment (Boyd and Richerson, 1985: ch. 5), theory and the available data suggest that Alexander's argument is incorrect in general, although it may well be roughly correct for those traits for which accurate decision making is easy. Regardless of whether we or Alexander ultimately prove more nearly correct, his contribution is substantial; work on the complexities of culture is much aided by having the implications of the simplest genetic fitness-maximizing model incorporating culture cogently developed.

The exhaustive analysis of many sample models in various combinations is also the main means of seeking robust results (Wimsatt, 1981). One way to gain confidence in simple models is to build several models embodying different characterizations of the problem of interest and different simplifying assumptions. If the results of a model are robust, the same qualitative results ought to obtain for a whole family of related models in which the supposedly extraneous details differ. The fact that genetic and game theoretic models of altruism usually lead to similar conclusions reassures us that general results like Hamilton's $k = 1/r$ rule are robust. Similarly, as more complex considerations are introduced into the family of models, simple model results can be considered robust only if it seems that the qualitative conclusion holds for some reasonable range of plausible conditions. Thus, quantitative genetic (Boyd and Richerson, 1982) and multiple-locus models (Uyenoyama and Feldman, 1980) suggest that Hamilton's rule is approximately correct when a variety of complications is introduced. Complications substantially affect the exact form of the rule, but do preserve the qualitative result that kin cooperation can evolve and the propensity to cooperate should be a function of relatedness under most circumstances that seem empirically reasonable. Nevertheless, it is slow and difficult work to make reasonably certain that particular results can be treated as robust (Wimsatt, 1980).

In the case of cultural evolution, we make the tentative claim that the costly information argument is a robust result. In all of the models we have constructed of the novel structural properties of culture and the evolutionary forces that result from them, it seems that optimizing the genetic fitness of a capacity for culture generally leads to a situation in which many individual cultural traits can easily evolve to values quite distant from those that would maximize fitness, so long as decision making is costly. These results do not depend on whether cultural traits are imagined to be discrete characters or continuous quantitative variables, for example. The tentativeness of the claim must be emphasized because the whole corpus of models of cultural evolution is still so small.

Plausibility Arguments

We believe that "plausibility argument" is a useful term for a scientific construct that plays much the same role in the study of complex, diverse phenomena that mutually exclusive hypotheses are supposed to play in the investigation of simpler subject matters. A plausibility argument is a hypothetical explanation having three features in common with a traditional hypothesis: (1) a claim of deductive soundness, of in-principle logical sufficiency to explain a body of data; (2) sufficient support from the existing body of empirical data to suggest that it might actually be able to explain a body of data as well as or better than competing plausibility arguments; and (3) a program of research that might distinguish between the claims of competing plausibility arguments. The differences are that competing plausibility arguments (1) are seldom mutually exclusive, (2) can seldom be rejected by a single sharp experimental test (or small set of them), and (3) often end up being revised, limited in their generality or domain of applicability, or combined with competing arguments rather than being rejected. In other words, competing plausibility arguments are based on the claims that a different set of submodels is needed to achieve a given degree of realism and generality, that different parameter values of common submodels are required, or that a given model is correct as far as it goes, but applies with less generality, realism, or predictive power than its proponents claim. Most frequently, the empirical program suggested by competing plausibility arguments is an arduous series of measurements of the relative strengths of several known processes in a wide range of organisms.

The reason for these differences is that quantitative questions are at the crux of debates about evolutionary processes. For example: how strong is selection among individuals relative to selection among groups? Theoretical analysis suggests that selection among groups must be commonplace, and laboratory experiments (Wade, 1977) demonstrate that it could have important effects. However, it is not at all clear whether selection among groups is important in nature. Sex ratio provides another example. Clear examples of sex ratio distortion exist (Hamilton, 1967), and theory suggests that it should be favored under a wide variety of ecological conditions (Charnov, 1982). Yet this process seems to be relatively rare—at least weak enough to neglect in most cases. Even if we are willing to be content with qualitative knowledge of complex processes, the term "qualitative" must be taken in the sense of rough estimates of quantitative variables, not in the sense of simple acceptance or rejection of mutually exclusive hypotheses. This feature of evolutionary problems is the basis for Quinn and Dunham's (1983) rejection of Popperian falsification as a proper epistemological model in ecology and evolution (see also Rapoport's, 1967, claim that many scientific paradoxes have been resolved when the polar positions were shown to be only opposite ends of a continuum).

Human sociobiology provides a good example of a plausibility argument. The basic premise of human sociobiology is that fitness-optimizing models drawn from evolutionary biology can be used to understand human behavior. Many social scientists have objected to this enterprise on the grounds that evolutionary theory does not account for the existence of culture. As we have already noted, Alexander (1979), Lumsden and Wilson (1981), Durham (1976), and others have defended the fitness-optimizing approach not by denying the importance of culture but by proposing various means by which decision-making forces could evolve under the guidance of selection to constrain cultural evolution so as generally to produce fitness-optimizing behavior. These authors have supported their plausibility argument by constructing an array of simple models that predict the details of human behavior in various circumstances—for example, patterns of adoption, unilineal descent, and child abuse—and compared the results of these simple models with empirical data.

The sociobiological explanations of human behavior and those derived from explicit models of cultural evolution provide an example of competing plausibility arguments. As Flinn and Alexander (1982) argue, there is wide agreement among Darwinian students of the problem of human evolution that culture is important and that the processes of cultural evolution may sometimes fail to keep cultural variation "on track" of genetic fitness (e.g., Alexander, 1979:142). Disagreements revolve around the relative strength of decision-making forces compared to natural selection on cultural variation, the degree to which cultural transmission acts like an inheritance system rather than an ordinary mechanism for phenotypic flexibility, the importance of nonparental transmission, and so forth. For example, we have argued that decision making is frequently costly and that this allows culture a certain autonomy, while Durham (1976) argues that cultural evolution will be constrained to produce behaviors that approximately maximize fitness most of the time.

We think that the clearest way to address the controversial questions raised by competing plausibility arguments is to try to formulate models with parameters such that for some values of the critical parameters the results approximate one of the polar positions in such debates, while for others the model approximates the other position. If the parameters that produce these contrasting results capture some real features of the processes of cultural and genetic coevolution, it may be possible to understand at least what is at stake in the controversy. In the models we have constructed, several parameters control the extent to which a typical cultural trait will be at the fitness optimum. If decisions about what cultural behaviors to adopt or invent can be made easily and accurately, and the rules that guide choices are ultimately transmitted genetically and subject to selection, culture will be very strongly constrained to maximize genetic fitness. Similarly, if important cultural traits are transmitted mostly from biological parents to offspring, cultural variation will act much like an extra chromosome of a biochemically odd kind. Even if decision-making forces are weak, selection on cultural variation will favor individual (inclusive) reproductive success, subject only to the same kinds of qualifications that obtain for a genetic locus. This result seems to approximate Durham's (1976) argument. As decision-making costs and nonparental transmission are allowed to become more important, cultural evolution becomes less directly constrained by selection on genes that control culture and it is possible to approximate positions like the group-functionalism of many social scientists and the afunctional position of Sahlins (1976).

As primitive as our own models are in this regard (see also Pulliam and Dunford, 1980; Werren and Pulliam, 1981; Pulliam, 1982, 1983), we think they are a promising step. The costs of decision making and the extent to which important items of culture are transmitted by nonparental individuals are empirical issues that can be resolved. Indeed, data already exist on these points (Boyd and Richerson, 1985: chs. 3 and 5). It would be overenthusiastic to claim that any of the controversial questions surrounding the application of Darwinism to human culture are resolved, but we do believe that the modest body of formal theory so far developed, and empirical argument derived from the theory, has clarified the issues to the extent that rapid progress is now possible.

A well-developed plausibility argument differs sharply from another common type of argument that we call a programmatic claim. Most generally, a programmatic claim advocates a plan of research for addressing some outstanding problem without, however, attempting to construct a full plausibility argument. Programmatic claims can be exceedingly useful; the development of a Darwinian theory of culture was greatly stimulated by mostly programmatic essays such as those by Campbell (1965), Ruyle (1973), and Cloak (1975). However, they are useful only insofar as they indicate the possibility of, or need for, new plausibility arguments. An attack on an existing, often widely accepted, plausibility argument on the grounds that the plausibility argument is incomplete is a kind of programmatic claim. Critiques of human sociobiology are commonly of this type. Burden-of-proof claims are another variant. For example, sociobiologists often seem to imply that the general success of adaptive reasoning in biology means that the existence of any prima facie plausible adaptive interpretation of human behavior is a sufficient counter to anything but a perfect case for a nonadaptive explanation.

Programmatic attacks and burden-of-proof claims can be positively harmful when taken, by themselves, as sufficient substitutes for a sound plausibility argument. We have argued that theory about complex-diverse phenomena is necessarily made up of simple models that omit many details of the phenomena under study. It is very easy to criticize theory of this kind on the grounds that it is incomplete (or defend it on the grounds that it one day will be much more complete). Such criticism and defense is not really very useful because all such models are incomplete in many ways and may be flawed because of it. What is required is a plausibility argument that shows that some factor that is omitted could be sufficiently important to require inclusion in the theory of the phenomenon under consideration, or a plausible case that it really can be neglected for most purposes. Thus, for example, it is not enough to attack a purportedly general plausibility argument with a few special cases, for it is (or ought to be) stipulated that generalized models are always likely to account more or less poorly for many special cases. In contrast, the success of genetic fitness-maximizing theory in biology cannot be used to defend that generalized model in the face of plausible arguments that cultural evolution is a divergent special case.

It seems to us that until very recently, "nature-nurture" debates have been badly confused because plausibility arguments have often been taken to have been successfully countered by programmatic claims. It has proved relatively easy to construct reasonable and increasingly sophisticated Darwinian plausibility arguments about human behavior from the prevailing general theory. It is also relatively easy to spot the programmatic flaws in such arguments; conventional Darwinian models do not allow for human culture. The problem is that programmatic objections have not been taken to imply a promise to deliver a full plausibility claim. Rather, they have been taken as a kind of declaration of independence of the social sciences from biology. Having shown that the biological theory is in principle incomplete, the conclusion is drawn that it can safely be ignored. Sahlins's (1976) objections to human sociobiology seem to us to have been as much in this tradition as Tarde's (1903:xxi–xxii) very early one. Both arguments ignore that Darwinian plausibility arguments ordinarily contain a serious rationale for accepting their claims despite the unique aspects of the human species. Certainly this is the case with contemporary human sociobiology and explains why it has attracted support by social scientists like van den Berghe (1979, 1981), who cannot be accused of simpleminded hereditarianism.

The Importance of Scientific Pluralism

Jared Diamond (personal communication) has drawn the following useful lesson from his experience as both a physiologist and a community ecologist: in physiology, controversial issues are ordinarily settled quickly by definitive experiments. As a result, debate over contending hypotheses is quite restrained and polite. One or the other contending claim is almost certain to turn out wrong in short order, and any grandiose pronouncements, ad hominem attacks, or similar departures from polite scientific discourse can be held against the loser. As long as scientists know that they can easily be proven wrong by a few critical experiments in the next few years, they will refrain from such departures. In ecology, major controversies last much longer because the issues are more complex and testing contending plausibility arguments is a long-drawn-out affair. The result is that individual claimants are often unlikely to be proven cleanly right or wrong, at least during their own lifetimes. Rhetorical excesses thus cannot be clearly proven as such by the failure of the programmatic claim or plausibility argument to which they are attached, and consequently the motivation to avoid them is reduced.

Perhaps differences between these two disciplines can be understood in terms of Campbell's (1979) general discussion of scientific honesty. According to Campbell, scientists are more honest in their occupational behavior than other professionals, but not because they are morally superior as individuals. Rather, they are careful to present honest work because other scientists are very discriminating consumers. Scientists frequently replicate crucial experiments and can gain prestige by detecting errors. In a controversy, many members of the community will act as relatively unbiased judges of the acceptability of contending hypotheses because their own work depends on using the correct result—say, to make a more accurate measurement instrument. Such acceptors have an interest in the resolution of the controversy but not a vested interest in any particular outcome. It seems likely that this mechanism will work much more effectively when controversial issues are resolved quickly, and consumer/acceptors can confidently use secure results in their own work. In the case of evolutionary and ecological problems, ambiguity lasts longer, and consumers may be forced to choose among plausibility arguments, thus coming to have a vested interest in the controversy. The extensive empirical program of the complex-diverse disciplines reduces the incentive to replicate individual experiments directly because they make so small a contribution to the total program.

Campbell (1969, 1986) contributed an insightful analysis of another potentially serious problem in the study of complex-diverse subject matters: the social complexity of the sciences that study them. Specialization is obviously demanded by complexity and diversity. But there is no guarantee that disciplines will not evolve what Campbell characterized as parochial "tribal" norms and customs that impede scientific progress. His argument is illustrated with reference to the arbitrary disciplinary boundaries, schools within disciplines, and the resulting "ethnocentrism" within the social sciences. Our impression is that the scientific endeavor becomes more prone to "ethnocentrism" as problems become more complex and diverse; certainly evolutionary biology, despite the unifying value of Darwinism, is not immune. As the enforcement of the universalistic norms of scientific discourse weaken, very human motives, such as a desire for collegial relations within one's discipline, a tendency to find that one's extrascientific ideology can be squared one way or another with one's science, career considerations, and a need to economize on information, can easily lead the social structure of science in directions that reduce its collective ability to solve complex-diverse problems. The mental effort of keeping multiple, partly conflicting, plausibility arguments in mind, the ambiguous relationship of these to ideas and norms derived from other roles, and the need to have some knowledge of several unfamiliar disciplines might be psychological motivations that encourage the formation of independent disciplines and schools with little communication between them. Nevertheless, it seems inescapable that complex-diverse subjects demand free communication between specialists and a wide tolerance for the pursuit of temporarily divergent plausibility claims.

Deriving norms from this diagnosis is by no means straightforward. Perhaps new disciplines and new ideas need a measure of isolation, which the development of ethnocentric and sectarian attitudes affords (Campbell, 1985; Beatty, 1987). On the other hand, unchecked, this process can result in a declaration of independence for a mature discipline, such as Sahlins offers for anthropology, which may be wholly harmful. There may be an optimal amount of disciplinary and research program "ethnocentrism" for maximizing scientific progress at any given time.

Nonetheless, we think that the following two norms would, if adopted, improve scientific debate surrounding complex, diverse subjects.

Ad hominem attacks on particular positions and the use of self-serving programmatic claims should be viewed as tacky. Given the deep importance of human behavior to humans, the weakness of the consumer/acceptor mechanism for regulating academic discourse, and the fact of the evolution of "ethnocentric" norms within disciplines, it is utopian to expect that the temptation to behave in such ways will always be resisted, particularly by those who are legitimately pursuing a position. Widespread agreement that such behavior is moderately offensive is a practical norm perhaps and might help to further productive debate over real issues.

Scientists should be encouraged to take a sophisticated attitude toward empirical testing of plausibility arguments (Quinn and Dunham, 1983; Diamond, 1986). Folk Popperism among scientists has had the very desirable result of reducing the amount of theory-free descriptive empiricism in many complex-diverse disciplines, but it has had the undesirable effect of encouraging a search for simple mutually exclusive hypotheses that can be accepted or rejected by single experiments. By our argument, very few important problems in evolutionary biology or the social sciences can be resolved in this way. Rather, individual empirical investigations should be viewed as weighing marginally for or against plausibility arguments. Often, empirical studies may themselves discover or suggest new plausibility arguments or reconcile old ones.

Conclusion

We confess to being somewhat puzzled by the debate between the "adaptationists" and their critics. We suspect that most evolutionary biologists and philosophers of biology on both sides of the dispute would pretty much agree with the defense of the simple models strategy presented here. To reject the strategy of building evolutionary theory from collections of simple models is to embrace a kind of scientific nihilism in which there is no hope of achieving an understanding of how evolution works. On the other hand, there is reason to treat any given model skeptically. As Kitcher (1987) notes, his criticisms of optimality arguments are not meant as "forlorn skepticism," but rather as helpful "in pinpointing strategies for improving hypotheses about selective pressures and functional significance" (p. 99). Kitcher quite properly and quite explicitly calls attention to the fact that because diversity and complexity are real, the tactics of seeking understanding via simple models is something that must be done with care. No one ought to disagree.

Unfortunately, the critics of "adaptationism" are not always as sophisticated as this; they sometimes seem to want to benefit rhetorically from a programmatic critique that implies scientific nihilism without having to face the real (and extremely unpleasant) consequences of actually adopting it. It may be possible to defend the proposition that the complexity and diversity of evolutionary phenomena make any scientific understanding of evolutionary processes impossible. Or, even if we can obtain a satisfactory understanding of particular cases of evolution, any attempt at a general, unified theory may be impossible. Some critics of adaptationism seem to invoke these arguments against adaptationism without fully embracing them. The problem is that alternatives to adaptationism must face the same problem of diversity and complexity that Darwinians use the simple model strategy to finesse. The critics, when they come to construct plausibility arguments, will also have to use relatively simple models that are vulnerable to the same attack. If there is a vulgar sociobiology, there is also a vulgar criticism of sociobiology. Perhaps because we have devoted a considerable effort to building a plausibility argument for the novel and sometimes maladaptive role of culture in human evolution, we are very sensitive to the strength of the sociobiologists' plausibility arguments and the weakness of most of the objections to them.

In our opinion, human sociobiology has been a successful research program because it has made rather good use of the simple models strategy. Its practitioners have taken care to construct sound plausibility arguments and, in the spirit of scientific pluralism, to use the work of social scientists. As pursuers of a somewhat narrow range of plausibility arguments, their work is not above criticism in detail or in general. As befits pursuers, they have usefully driven the fitness-optimizing postulate to extremes that are not likely to be ultimately warranted. Less usefully, they have used a burden-of-proof claim to attempt to insulate sociobiology from counterarguments. On the other hand, the attacks on sociobiology are a good source of negative object lessons. The criticism of human sociobiology has far too frequently depended on mere programmatic claims (often invalid ones at that, as when sociobiologists are said to ignore the importance of culture and to depend on genetic variation to explain human differences). These claims are generally accompanied by dubious burden-of-proof arguments. Some critics also show little sense of the importance of scientific pluralism.

NOTE

We thank D. T. Campbell, J. M. Diamond, J. M. Emlen, G. Macey, A. Rosenberg, E. A. Smith, J. Staddon, & S. Vail for comments on drafts of this chapter. We also benefited from conversations with J. Quinn and J. Griesemer.

REFERENCES

Alexander, R. D. 1979. *Darwinism and human affairs*. Seattle: University of Washington Press.

Beatty, J., 1987. Natural selection and the null hypothesis. In: *The Latest on the Best*, J. Dupré, ed. (pp. 53–75). Cambridge, MA: MIT Press.

Boyd, R., & P. J. Richerson. 1982. Cultural transmission and the evolution of cooperative behavior. *Human Ecology* 10:325–352.

Boyd, R., & P. J. Richerson. 1983a. The cultural transmission of acquired variation: *Journal of Theoretical Biology* 100:567–596.

Boyd, R., & P. J. Richerson. 1983b. Why is culture adaptive? *Quarterly Review of Biology* 58:209–214.

Boyd, R., & P. J. Richerson. 1985. *Culture and the evolutionary process*. Chicago: University of Chicago Press.

Campbell, D. T. 1965. Variation and selective retention in sociocultural evolution. In: *Social change in developing areas: A reinterpretation of evolutionary theory*, H. R. Barringer, G. I. Blanksten, & R. W. Mack, eds. (pp. 19–49). Cambridge: Schenkman.

Campbell, D. T. 1969. Ethnocentrism of disciplines and the fish-scale model of omniscience. In: *Interdisciplinary relationships in the social sciences*, M. & C. W. Sherif, eds. (pp. 328–348). Chicago: Aldine.

Campbell, D. T. 1975. On the conflicts between biological and social evolution and *American Psychologist* 30:1103–1126.

Campbell, D. T. 1979. A tribal model of the social system vehicle carrying scientific knowledge. *Knowledge: Creation, Diffusion, Utilization* 1:181–201.

Campbell, D. T. 1985. Science policy from a naturalistic sociological epistemology. In: *PSA 1984*, vol. 2, P. D. Asquith, & P. Kitcher, eds. (pp. 14–29). East Lansing, MI: Philosophy of Science Association.

Campbell, D. T. 1986. Science's social system of validity enhancing collective belief change and the problems of the social sciences. In: *Metatheory in the social sciences: Pluralisms and subjectivities*, D. W. Fisk, & R. A. Shweder, eds. (pp. 86–105). Chicago: University of Chicago Press.

Cavalli-Sforza, L. L., & M. W. Feldman. 1973. Cultural versus biological inheritance: Phenotypic transmission from parents to children (a theory of the effect of parental phenotypes on children's phenotypes). *Journal of Human Genetics* 25:618–637.

Cavalli-Sforza, L. L., & M. W. Feldman. 1981. *Transmission and evolution: A quantitative approach*. Princeton, NJ: Princeton University Press.

Cavalli-Sforza, L. L., & M. W. Feldman. 1983. Cultural versus genetic adaptation. *Proceedings of the National Academy of Sciences (USA)* 80:4993–4996.

Charnov, E. 1982. *A theory of sex allocation*. Princeton, NJ: Princeton University Press.

Cloak, F. T., Jr. 1975. Is a cultural ethology possible? *Human Ecology* 3:161–182.

Cloninger, C. R., J. Rice, & T. Reich. 1979. Multifactorial inheritance with cultural transmission and assortative mating. II. A general model of combined polygenic and cultural inheritance. *American Journal of Human Genetics* 31:176–198.

Dawkins, R. 1982. *The extended phenotype: The gene as the unit of selection*. San Francisco: Freeman.

Day, R. H. 1982. Irregular growth cycles. *American Economic Review* 72:406–414.

Diamond, J. 1978. The Tasmanians: The longest isolation, the simplest technology. *Nature* 273:185–186.

Diamond, J. 1986. Overview: Laboratory experiments, field experiments, and natural experiments. In: *Community ecology*. J. Diamond, & T. J. Case, eds. (pp. 3–22). New York: Harper and Row.

Dupré, J. 1987. Introduction. In: *The Latest on the Best*, J. Dupré, ed. (pp. 1–24). Cambridge, MA: MIT Press.

Durham, W. H. 1976. The adaptive significance of cultural behavior. *Human Ecology* 4:89–121.

Eaves, L. J., K. A. Last, P. A. Young, & N. G. Martin. 1978. Model-fitting approaches to the analysis of human behavior. *Heredity* 41:249–320.

Flinn, M. V., & R. D. Alexander. 1982. Culture theory: The developing synthesis from biology. *Human Ecology* 10:383–400.

Hamilton, W. D. 1967. Extraordinary sex ratios. *Science* 156:477–488.

Hartung, J. 1976. On natural selection and the inheritance of wealth. *Current Anthropology* 17:607–622.

Hudson, E. A., & D. W. Jorgenson. 1974. *The Long term interindustry transactions model: A simulation model for energy and economic analysis*. Washington, DC: Federal Preparedness Agency, General Services Administration.

Kitcher, P. 1985. *Vaulting ambition: Sociobiology and the quest for human nature*. Cambridge, MA: MIT Press.

Kitcher, P. 1987. Why not the best? In: *The Latest on the Best*, J. Dupré, ed. (pp. 77–102). Cambridge, MA: MIT Press.

Labov, W. 1972. *Sociolinguistic patterns*. Philadelphia: University of Pennsylvania Press.

Levins, R. 1966. The strategy of model building in population biology. *American Scientist* 54:421–431.

Levins, R. 1968. *Evolution in changing environments: Some theoretical explorations*. Princeton, NJ: Princeton University Press.

Liebenstein, H. 1976. *Beyond economic man: A new foundation for microeconomics*. Cambridge, MA: Harvard University Press.

Lumsden, C., & E. O. Wilson. 1981. *Genes, mind and culture*. Cambridge, MA: Harvard University Press.

May, R. M. 1976. Simple mathematical models with very complicated dynamics. *Nature* 261:459–467.

Maynard Smith, J., & N. Warren. 1982. Models of cultural and genetic change. *Evolution* 36:620–627.

Pulliam, H. R. 1982. A social learning model of conflict and cooperation in human societies. *Human Ecology* 10:353–363.

Pulliam, H. R. 1983. On the theory of gene-culture co-evolution in a variable environment. In: *Animal cognition and behavior*, R. L. Mellgren, ed. (pp. 427–443). Amsterdam: North-Holland.

Pulliam, H. R., & C. Dunford. 1980. *Programmed to learn: An essay on the evolution of culture*. New York: Columbia University Press.

Quinn, J. F., & A. E. Dunham. 1983. On hypothesis testing in ecology and evolution. *American Naturalist* 122:602–617.

Rapoport, A. 1967. Escape from paradox. *Sci. Am.* 217 (July):50–56.

Richerson, P. J., & R. Boyd. 1984. Natural selection and culture. *BioScience* 34:430–434.

Ruyle, E. E. 1973. Genetic and cultural pools: Some suggestions for a unified theory of biocultural evolution. *Human Ecology* 1:201–215.

Sahlins, M. 1976. *Culture and practical reason*. Chicago: University of Chicago Press.

Smagorinsky, J. 1969. Problems and promises of deterministic extended range forecasting. *Bulletin of the American Meteorological Society* 50:286–311.

Sober, E. 1984. *The nature of selection: Evolutionary theory in philosophical focus*. Cambridge, MA: MIT Press.

Tarde, G. 1903/1962. *The laws of imitation*. Gloucester, MA: Peter Smith.

Uyenoyama, M., & M. W. Feldman. 1980. Theories of kin and group selection: A population genetics perspective. *Theoretical Population Biology* 17:380–414.

van den Berghe, P. L. 1979. *Human family systems*. New York: Elsevier.

van den Berghe, P. L. 1981. *The ethnic phenomenon*. New York: Elsevier.

Wade, M. J. 1977. An experimental study of group selection. *Evolution* 31:134–153.

Watt, K. E. F. 1968. *Ecology and resource management*. New York: McGraw-Hill.

Weber, M. 1949. *The methodology of the social sciences*. Glencoe, IL: Free Press.

Werren, J. H., & H. R. Pulliam. 1981. An intergenerational model of the cultural evolution of helping behavior. *Human Ecology* 9:465–483.

Wimsatt, W. C. 1980. Reductionistic research strategies and their biases in the units of selection controversy. In: *Scientific discovery*, II: *Case studies*, T. Nickles, ed. (pp. 213–259). Dordrecht: D. Reidel.

Wimsatt, W. C. 1981. Units of selection and the structure of the multi-level genome. In: *PSA-1980*, vol. 2, R. Giere, & P. Asquith, eds. (pp. 121–183). East Lansing, MI: The Philosophy of Science Assn.

20   Memes

Universal Acid or a Better Mousetrap?

Among the many vivid metaphors in *Darwin's Dangerous Idea*, one stands out. The understanding of how cumulative natural selection gives rise to adaptations is, Daniel Dennett says, like a "universal acid"—an idea so powerful and corrosive of conventional wisdom that it dissolves all attempts to contain it within biology. Like most good ideas, this one is very simple: once replicators (material objects that are faithfully copied) come to exist, some will replicate more rapidly than others, leading to adaptation by natural selection. The great power of the idea is that the resulting adaptations can be understood by asking what leads to efficient, rapid replication. Given that ideas seem to replicate, it is natural that Dawkins (1976, 1982), Dennett (1995), and others have explored the possibility of using this idea to explain cultural evolution.

Natural selection was not Darwin's only powerful, far-reaching idea. Ernst Mayr (1982) has argued that what he calls "population thinking" was also among Darwin's foundational contributions to biology. Before Darwin, species were thought to be essential, unchanging types, like geometric figures and chemical elements. Darwin saw that species were populations of organisms that carried a variable pool of inherited information through time. To understand the evolution of species, biologists had to account for the processes that changed the nature of that inherited information. Darwin thought that the most important processes were natural selection, sexual selection, and the "inherited effects of use and disuse." We now know that the last process is not important in organic evolution—unlike Darwin, modern biologists do not believe that the sons of blacksmiths inherit their father's mighty biceps. Nowadays biologists think many processes that Darwin never dreamed of are important, including segregation, recombination, gene conversion, and meiotic drive. Nonetheless, modern biology is fundamentally Darwinian because its explanations of evolution are rooted in population thinking. If Darwin were to be resurrected tomorrow through some miracle of cloning, we think he would be quite happy with his legacy.

In this chapter we want to convince you that population thinking, not natural selection, is the key to conceptualizing culture in terms of material causes. This argument is based on three well-established facts:

1. *There is persistent cultural variation among human groups.* Any explanation of human behavior must account for how this variation arises and how it is maintained.
2. *Culture is information stored in human brains.* Every human culture contains vast amounts of information. Important components of this information are stored in human brains.
3. *Culture is derived.* The psychological mechanisms that allow culture to be transmitted arose in the course of hominid evolution. Culture is not simply a by-product of intelligence and social life.

Much of culture is information stored in human brains—information that got into those brains by various mechanisms of social learning. It follows that to explain the distribution of information stored in the brains of the members of the current generation, any coherent theory will have to account for the cultural information in the brains of the previous generation. The theory will also have to explain how this information, together with genes and environmental contingencies, caused the present generation to acquire the cultural information that it did. Unfortunately, we do not understand how this process works. It may be that cultural information stored in brains takes the form of discrete memes that are replicated faithfully in each subsequent generation, or it may not. This is an empirical question that at present is unanswered, and we will see that other models are possible. In every case, the Darwinian population approach will illuminate the process by which the cultural information that is stored in a population of brains is transformed from one generation to the next.

We also want to convince you that population thinking can play an important, constructive role in the human sciences. The fact that population thinking is logically necessary for a natural, causal, theory of culture does not necessarily mean that such a theory will be useful. Thus, we know that human culture must be consistent with quantum mechanics, but it is unlikely that such a connection will help us understand, say, ethnic conflict. However, we think Darwinian models of culture *are* useful for two reasons. First, they serve to connect the rich models of behavior based on individual action developed in economics, psychology, and evolutionary biology with the data and insights of the cultural sciences, anthropology, archaeology, and sociology. In doing so, we think that they can help shed light on important unsolved problems in the social sciences. Second, population thinking is useful because it offers a way to build a mathematical theory of human behavior that captures the important role of culture in human affairs. Population thinking is not a universal acid that will dissolve existing social sciences. But it is a better mousetrap, providing useful new tools that can help solve outstanding problems in the human sciences.

Culture Is Heritable at the Group Level

One of the striking facts about the human species is that there are important, persistent differences between human groups that are created by culturally transmitted ideas, not genetic differences, or differences in the physical or biotic environment. Sonya Salomon's (1992) research on immigrant communities in the United States shows how cultural differences can give rise to different behaviors in the same environment. One of Salomon's studies focused on two farming communities in southern Illinois. "Freiburg" (a pseudonym), is inhabited by the descendants of German-Catholic immigrants who arrived in the area during the 1840s. "Libertyville" (also a pseudonym) was settled by people from other parts of the United States—mainly Kentucky, Ohio, and Indiana—when the railroad arrived in 1870. These two communities are only about 20 miles apart and have been carefully matched for similar soil types.

The people in these two communities have different values about family, property, and farm practice, and these differences seem consistent with their ethnic origins. The farmers of Freiburg tend to value farming as a way of life, and they want at least one son or daughter to continue as a farmer. In Freiburg, wills specify that the farm will go to a child who will farm the land and use farm proceeds to buy out any nonfarming siblings. Parents put considerable pressure on children to become farmers. They place little importance on education, knowing that advanced education often results in young people not returning to the farm. Salomon argues that these "yeoman" values are similar to those observed among peasant farmers in Europe and elsewhere. In contrast, the "Yankee" farmers of Libertyville regard their farms as profit-making businesses. They buy or rent land depending on economic conditions, and if the price is right, they sell. Many Yankee farmers would prefer their children to continue farming, but they see it as an individual decision. Some families help their children enter farming, but many do not, and they generally place a strong value on higher education.

The difference in values between Freiburg and Libertyville lead to measurable differences in farm practices despite the proximity of the two towns and the similarity of their soils. Farms are substantially larger in Libertyville—the mean size of farm operations in Libertyville is 518 acres compared to 276 acres in Freiburg. The Libertyville farms are larger because Yankee farmers rent more land. They rent more land because Yankees demand a higher income to stay in farming. Yeomen, who so value farming for its own sake, are content with lower incomes and fear the risks of debt-financed expansion.

The two communities also show striking differences in farm operations. In Libertyville, as in most of southern Illinois, farmers specialize in grain production. It is the primary source of income for 77 percent of the farmers in Libertyville. In Freiburg, many people mix grain production with dairying or livestock raising, activities that are almost absent in Libertyville. Because animal husbandry is labor-intensive, it allows Germans to accommodate their larger families on their more limited acreage. Yankee farmers decided against dairying and stock raising because grain farming is more profitable and less work.

The fact that culturally distinctive human groups behave differently in the same environment implies that culture is heritable, at least at the group level. Many beliefs and values that are common in a group at one point in time are also common among the descendants of the same group. Any theory of how culture works must be consistent with this fact. It must explain why the German farmers of Freiburg hold different beliefs about life and land than their Yankee neighbors almost 150 years after leaving Europe.

Culture Is Information in Stored Human Brains

Every human culture contains an enormous amount of information. Consider how much information must be transmitted to maintain a particular distinctive spoken language. A lexicon requires something like 10,000 associations between words and their meanings. Grammar entails a complex set of rules regulating morphosyntax, and although it is unclear the extent to which these rules arise from innate, genetically transmitted structures, it is clear that the rules that underlie the grammatical differences that separate English and Chinese are culturally transmitted. Subsistence techniques also entail large amounts of information. For example, Blurton-Jones and Konner (1976) showed that the !Kung San have a very detailed knowledge of the natural history of the Kalahari—so detailed, in fact, that the researchers were unable to judge the accuracy of much of !Kung knowledge because in some aspects it exceeded Western biology. As anyone who has ever tried to make a decent stone tool can attest, the manufacture of even the simplest tool requires lots of knowledge; more complex technologies require even more. Imagine the instruction manual for constructing a seaworthy kayak from materials available on the North Slope of Alaska. The institutions that regulate social interactions incorporate still more information. Property rights, religious custom, roles, and obligations all require a considerable amount of detailed information.

The vast store of information that exists in every culture cannot simply float in the air. It must be encoded in some material object. In societies without widespread literacy, the most important objects in the environment capable of storing this information are human brains and human genes. It is undoubtedly true that some cultural information is stored in artifacts. It may well be that the designs that are used to decorate pots are stored on the pots themselves and that when young potters learn how to make pots they use old pots, not old potters, as models. In the same way, the architecture of the church may help store information about the rituals performed within. Without writing, however, the ability of artifacts to store culture is quite limited. First, many artifacts are very difficult to reverse-engineer. The young potter cannot learn how to select clay and temper or how to fire a pot by studying existing ones. Second, much cultural information is semantic knowledge—how can an artifact store the notion that Kalahari porcupines are monogamous? Or the rules that govern bride-price transactions?

It is also clear that much cultural information is not stored in human genes. In one sense this is obvious. The evidence is very clear that very little cultural variation results from genetic differences. We know that genetic differences do not explain why some people speak Chinese and others English, or why the !Kung know a lot more about the biology of porcupines than most readers of this chapter.

However, there is a subtle and much more plausible way that genes could store cultural information. It could be that most human culture is innate, genetically transmitted information that is evoked by environmental cues. Pascal Boyer (1994) argues that much of religious belief has this character. For example, the Fang, a group Boyer studied in Cameroon, have elaborate beliefs about ghosts. For the Fang, ghosts are malevolent beings that want to harm the living; they are invisible and can pass through solid objects, and so on. Boyer argues that most of what the Fang believe about ghosts is not culturally transmitted; rather, it is based on the innate, epistemological assumptions that underlie all cognition. Once a young Fang child learns that ghosts are sentient beings, she does not need to learn that ghosts can see or that they have beliefs and desires—these components are provided by cognitive machinery that reliably develops in every environment. According to this view, cultural differences arise because different environmental cues evoke different innate information. A friend of ours believes in angels instead of ghosts because he grew up in an environment in which people talked about angels. However, most of what he knows about angels comes from the same cognitive machinery that gives rise to Fang beliefs about ghosts, and the information that controls the development of this machinery is stored in the genome.

This picture of culture is a useful antidote to the simplistic view that culture is simply poured from one head into another. Evolutionary psychologists are surely right that every form of learning, including social learning, requires an information-rich innate psychology and that much of the adaptive complexity we see in cultures around the world stems from this information. However, it is a big mistake to ignore transmitted cultural information. The single most important adaptive feature of culture is that it allows the gradual, cumulative assembly of adaptations over many generations—adaptations that no single individual could invent on his own. Cumulative adaptation cannot be based solely on innate, genetically encoded information.

Consider the evolution of a relatively simple form of technology, the mariners' magnetic compass (Needham, 1978). First, Chinese geomancers noticed the peculiar tendency of small magnetite objects to orient in the earth's magnetic field, an effect that they used for purposes of divination. Then, Chinese mariners learned that magnetized needles could be floated on water to indicate direction at sea. Next, over several centuries Chinese seamen developed a dry compass mounted on a vertical pin-bearing, like a modern toy compass. Europeans acquired this type of compass in the late medieval period. European seamen then developed the fixed card compass that allowed a helmsman to steer an accurate course by aligning the bow mark with the appropriate compass point. Compass makers later learned to adjust iron balls near the compass to zero out the magnetic influence from the ship and to gimbal the compass and fill it with liquid to damp the motion imparted to the card by the roll and pitch of the ship. Even such a relatively simple tool was the product of at least seven or eight innovations separated in time by centuries and in space by the breadth of Eurasia.

This sort of adaptation occurs only because novel information can accumulate in human populations, be stored in human brains, and be transmitted through time by teaching and imitation.

Evolutionary psychologists argue that our psychology is built of complex, information-rich, evolved modules that are adapted for the hunting and gathering life that we pursued until the origins of agriculture a few thousand years ago. On this argument, humans can easily and naturally do the things we are really adapted to do like learn a language or understand the feelings of others. Inventing complex modern artefacts like the compass is hard, but what about skills necessary for hunting and gathering? Couldn't we learn these as easily as we learn language? Doesn't our brain contain the information necessary to follow hunting and gathering ways? Our ancestors lived as hunter-gatherers of some kind for the last 2 or 3 million years. If we had to do so, couldn't we reinvent that stuff, just as Fang children invent the properties of their ghosts, or children can invent a grammar?

Good questions, but we think the answer is almost certainly "Are you nuts?!" Consider the following thought experiment. Suppose you are stranded in some not-too-extreme desert environment, not the Empty Quarter or the Atacama, but the desert between Sonoita, Mexico, and Yuma, Arizona. Your task is to survive and raise your kids without modern technology. You will be given the resources to survive a few months to get your feet on the ground before we take away your last tin of food and your last steel tool—a little time to see what comes naturally. Will you make it?

We don't think so. The stretch between Sonoita and Yuma is known as El Camino del Diabolo, "the Devil's Road." It was one leg of the main overland route from Old Mexico to California until the coming of railroads. For more than a century it was used by Spanish, Mexican, and American travelers. To get that far, every traveler had to already be an experienced frontiers-person, and no doubt most were hardbitten, desert-wise, and well equipped with familiar technology. It was the best of several bad routes and was comparatively well known and well marked. Still, it was an infamous leg of the journey, and many travelers ended up in the hasty graves that litter the route.

Now, consider that the Camino del Diabolo was also the home to Papago Indians who, with a few pounds of wood, stone, and bone equipment, an impressive amount of hard-won knowledge, and a well-adapted system of social institutions, lived and raised their children in the very same desert that killed so many pioneers. If our task was to survive in this desert without our accustomed industrial technology, we would certainly trade a few hours of tutoring by a traditional Papago for any number of months trying to summon an innate knowledge of the desert.

Culture Is Derived

Simple forms of social learning, often termed "protoculture," occur in many other species of animals. In a review of the social transmission of foraging behavior, Levebre and Palameta (1988) give 97 examples of protocultural variation in foraging behavior in animals as diverse as baboons, sparrows, lizards, and fish. Much of the evidence for protoculture in other animals consists of observations of different behavior by populations of the same species living in similar environments. For example, chimpanzees in the Mahale Mountains of Tanzania often adopt a unique grooming posture in which both partners extend one arm over their heads, clasp hands, and then groom one another's exposed arm pits. These grooming hand-clasps occur often and are performed by all members of the group. Chimpanzees at Gombe, who live less than 100 kilometers away in a similar type of habitat, often groom but never perform this behavior. Sometimes scientists have observed the spread of a novel behavior. One famous example comes from Japan where a group of Japanese macaques, whose range included a sandy beach, were provisioned with sweet potatoes. A young female macaque accidentally dropped her sweet potato into the sea as she was trying to rub the sand off it. She must have liked the result, as she began to carry all of her potatoes to the sea to wash them. Other monkeys followed suit. However, it took other members of the group quite some time to acquire the behavior and many monkeys never washed their potatoes. Finally, some evidence for protoculture in other animals comes from experiments that demonstrate that behavior is socially transmitted. The most famous case is the transmission of song dialects in birds like the white-crowned sparrow.

There is little evidence, however, of cumulatively evolved cultural traditions in other species. With a few exceptions, social learning leads to the spread of behaviors that individuals could have learned on their own. For example, food preferences are socially transmitted in rats. Young rats acquire a preference for a food when they smell the food on the pelage of other rats (Galef, 1988). This process can cause the preference for a new food to spread within a population. It can also lead to behavioral differences among populations living in the same environment, because current foraging behavior depends on a history of social learning. However, it does not lead to the cumulative evolution of complex new behaviors that no individual rat could learn on its own. Thus, in other animals it is quite plausible that most of the detailed information that creates protocultural differences is stored and transmitted genetically.

Circumstantial evidence suggests that the ability to acquire novel behaviors by observation is essential for cumulative cultural change. Students of animal social learning distinguish *observational learning*, which occurs when younger animals observe the behavior of older animals and learn how to perform a novel behavior by watching them, from a number of other mechanisms of social transmission, which also lead to behavioral continuity without observational learning (Galef, 1988; Visalberghi and Fragazy, 1990; Whiten and Ham, 1992). One such mechanism, *local enhancement*, occurs when the activity of older animals increases the chance that younger animals will learn the behavior on their own. Imagine a young monkey acquiring its food preferences as it follows its mother around. Even if the young monkey never pays any attention to what its mother eats, she will lead it to locations where some foods are common and others rare, and the young monkey may learn to eat much the same foods as mom.

Local enhancement and observational learning are similar in that they can both lead to persistent behavioral differences among populations, but only observational learning allows *cumulative* cultural change (Tomasello, Kruger, and Ratner, 1993). To see why, consider the cultural transmission of stone tool use. Suppose that occasionally early hominids learned to strike rocks together to make useful flakes. Their companions, who spent time near them, would be exposed to the same kinds of conditions, and some of them might learn to make flakes too, entirely on their own. This behavior could be preserved by local enhancement because groups in which tools were used would spend more time in proximity to the appropriate raw materials. However, that would be as far as tool-making would go. Even if an especially talented individual found a way to improve the flakes, this innovation would not spread to other members of the group because each individual learned the behavior anew, without any detailed guidance from innovators who have improved on the common technique. Local enhancement is limited by the learning capabilities of individuals and the fact that each new learner must start from scratch. With observational learning, on the other hand, innovations can be incorporated into others' behavioral repertoires if younger individuals are able to acquire the improved behavior by observational learning. To the extent that observers can use the behavior of models as a starting point, observational learning can lead to the cumulative evolution of behaviors that no single individual could invent on its own.

Adaptation by cumulative cultural evolution is apparently not a by-product of intelligence and social life. Capuchin monkeys are among the world's cleverest creatures. They resemble apes in having quite large brains for their size. In nature, they perform many complex behaviors, and in captivity they can be taught extremely demanding tasks. Capuchins live in social groups and have ample opportunity to observe the behavior of other individuals of their own species. Yet good laboratory evidence indicates that these monkeys make little or no use of observational learning (Visalberghi and Fragazy, 1990). Observational learning is not simply a by-product of intelligence and the opportunity to observe conspecifics. Rather, it seems to require special psychological mechanisms (Bandura, 1986). This conclusion suggests that the psychological mechanisms that enable humans to learn by observation are adaptations that have been shaped by natural selection in the human lineage because culture is beneficial.

Cultural Evolution Is Darwinian

Now, let us consider what these facts imply for a theory of culture. Consider a population of individuals who are culturally interconnected; they speak dialects of a single language, use similar technology, share relatively similar beliefs about the world, and have similar moral values. People in this population think and behave differently from other peoples, in part, because they have different culturally transmitted information stored in their brains. Next consider the descendants of this population, say 100 years later. The culture of the descendant population will be similar in many ways to that of their predecessors. Their language will be similar, and they may often use similar technology, have similar beliefs about the world, and subscribe to a similar moral system. The fact that culture depends on behavior stored in the brains of this population requires us to account for how the information that generates these similarities was transmitted from the brains in the first population to the brains in the second.

Of course, there will also be differences between the two populations, some small, some great. Some of these differences will arise because some behaviors are more common in the second population—for example, perhaps what was previously a rare usage or form of pronunciation has become common. Other differences will arise because genuinely new behavior is present, either as a result of borrowing from neighboring populations or due to genuine innovation. Thus, a complete theory would also have to account for why some forms of cultural information spread, and why some forms have diminished, and how innovation occurs.

Cumulative cultural change requires observational learning. People observe the behavior of others, and (somehow) acquire the information necessary to produce a reasonable facsimile of the same behavior. In any given time period, each person observes only a sample of the people who make up his population. A very small child is exposed mainly to the people in her family, older children are exposed to peers and teachers, and adults to yet a wider range of people. We will refer to this group of people as an individual's "cultural sample." For most of human history cultural samples were small, but nowadays they may be immense. On the other hand, for some elements of culture many people may be disproportionately influenced by a single charismatic leader or acknowledged expert.

The fact that cultures often persist over time with little change means that the commonness of a behavior in an individual's cultural sample must have a positive effect on the probability that the individual ultimately acquires the cultural information that generates that behavior. Such a tendency could arise in several different ways: if observational learning takes the form of approximately unbiased copying, then common behaviors will be more frequent in cultural samples, and therefore will be more likely to be copied. It could also be that the psychology of observational learning itself predisposes people to acquire more common behaviors. Finally, it could be that rare behaviors are typically disadvantageous and less likely to be retained as a result of individual learning and experimentation, or even by natural selection against them.

It follows that cultural change is a population process. The argument proceeds in several steps:

- To understand how a person behaves, we have to know the nature of the information stored in her brain
- To understand why people have the beliefs that they do, we must know what kinds of behaviors characterized their cultural sample
- To predict the distribution of cultural samples that exists, one must know the cultural composition of the population
- Therefore, to understand how people behave, we must understand why the population has the cultural composition that it does

Similarities between descendant and ancestral populations arise because the necessary information has been transmitted from individual to individual through time without significant change. Differences occur because some variants have become more common, others have become more rare, and some completely new variants have been introduced. Thus, to account for both continuity and change we need to understand the population processes by which ideas are transmitted through time.

Culturally Transmitted Skills and Beliefs May Not Be Replicators

In *The Extended Phenotype*, Richard Dawkins (1982) argues that the cumulative evolution of complex adaptations requires what he calls replicators, things in the physical world that produce copies of themselves and have the following three additional properties:

1. *Fidelity*. The copying must be sufficiently accurate that even after a long chain of copies the replicator remains almost unchanged.
2. *Fecundity*. At least some varieties of the replicator must be capable of generating more than one copy of themselves.
3. *Longevity*. Replicators must survive long enough to affect their own rate of replication.

Replicators give rise to cumulative adaptive evolution because replicators are targets of natural selection. Genes are replicators—they are copied with astounding accuracy, they can spread rapidly, and they persist throughout the lifetime of an organism, directing its machinery of life. Dawkins thinks that beliefs and ideas are also replicators. On the face of it, this is an apt analogy. Beliefs and ideas can be copied from one mind to another, spreading through a population, controlling the behavior of people who hold them.

But there are reasons to doubt that beliefs and skills are replicators, at least in the same sense that genes are. Unlike genes, ideas are not copied and transmitted intact from one brain to another. Instead, the information in one brain generates some behavior; somebody else observes this behavior and then (somehow) creates the information necessary to generate very similar behavior. The problem is that there is no guarantee that the information in the second brain is the same as the first. For any phenotypic performance, there are potentially an infinite number of rules that would generate that performance. Information will be transmitted from brain to brain only if most people induce a unique rule from a given phenotypic performance. While this may often be the case, it is also plausible that genetic, cultural, or developmental differences among people may cause them to infer different beliefs from the same overt behavior. To the extent that these differences shape future cultural change, the replicator model captures only part of cultural evolution.

The generativist model of phonological change illustrates the problem. According to the generativist school of linguistics, individual pronunciation is governed by a complex set of rules that takes as input the desired sequence of words and produces as output the sequence of sounds that will be produced (Bynon, 1977). Generativists also believe that, as adults, people can modify their pronunciation only by adding new rules that act at the *end* of the chain of existing rules. Children, on the other hand, are not constrained by the rules used to generate adult speech. Instead, they induce the simplest set of grammatical rules that will account for the performances they hear, and these may be quite different than the rules used by adult speakers. Although the new rules produce the same performance, they can have a different structure and, therefore, allow further changes by rule addition that would not have been possible under the old rules.

The following example (from Bynon, 1977) illustrates this phenomenon. In some dialects of English, people pronounce words that begin with *wh* using what linguists call an "unvoiced" sound while they pronounce words beginning with *w* using a voiced sound. (Unvoiced sounds are produced with the glottis open, resulting in a breathy sound, whereas voiced sounds are produced with the glottis closed, causing a resonant tone.) People who speak such dialects must have mental representations of the two sounds and rules to assign them to appropriate words. Now suppose that people who speak such a dialect come into contact with other people who only use the voiced *w* sound. Further suppose that this second group of people is more prestigious, and accordingly people in the first group modify their speech so that they too use only voiced *w*s. According to the generativists, they will accomplish this change by adding a new rule that says "voice all unvoiced *w*s." So, Larry wants to say *Whether it is better to endure*. The part of his brain that takes care of such things looks up the mental representations for each of the words, including *whether*, which has an unvoiced *w* (because that is the way Larry learned to speak as a child). Then after any other processing for stress or tone, the new rule changes the unvoiced *w* in *whether* to a voiced *w*. Children learning language in the next generation never hear an unvoiced *w*, and, according to generativists, they adopt the same underlying representation for *whether* and *weather*. Thus, even though there is no difference in the phenotypic performance among parents and children, children do not acquire the same mental representation as their parents. This difference may be important because it will affect further changes. For example, it might make it less likely that the two sounds would split again in the future. The adult version of the rule still has a latent distinction between the voiced and unvoiced pronunciation that could serve as the basis for renewing the distinction, whereas, if the generativists are correct, the latent distinction is unavailable to child learners who hear only one usage.

Replicators Are Not Necessary for Cumulative Adaptive Evolution

We also doubt that replicators are necessary for the cumulative evolution of complex features. Here is an example of a transmission system that does just that. When you speak, the kind of sounds that come out of your mouth depends on the geometry of your vocal tract. For example, the consonant *p* in *spit* is created by momentarily bringing your lips together with the glottis open. Narrowing the glottis converts this consonant to *b* as in *bib*. Leaving the glottis open and slightly opening the lips produces *pf*, as in the German word *apfel* (apple). Linguists have shown that even within a single speech community individuals vary in the exact geometry of the vocal tract used to produce any given word.

Thus, it seems plausible that individuals vary in the culturally acquired rule about how to arrange the inside of the mouth when they are speaking any particular word. Languages vary in the sounds used and this variation can be very long-lived. For example, in dialects spoken in the northwest of Germany, *p* is substituted for *pf* in *apfel* and many similar words. This difference arose about AD 500 and has persisted ever since (Bynon, 1977).

So how are different rules governing speech production transmitted from generation to generation? Consider two models.

First, suppose that each child learning language is exposed to the speech of a number of adults. These adults vary in the way that they produce the *pf* sound in *apfel*. Each child figures out how she would need to position her tongue to produce the same *pf* sound as each adult model, and then she adopts *one* of these as her own rule. Here, a mental rule that governs speech production is transmitted from one individual to another. The mental rule is a replicator; it clearly has fidelity. It has longevity because it potentially persists for generations, and it would have fecundity if the rule was more attractive than competing rules. And because it is a replicator, it can evolve.

Now consider a second model. As before, children are exposed to the speech of a number of adults who vary in the way that they pronounce *pf*. Each child unconsciously computes the average of all the pronunciations that he hears and adopts the tongue position that produces this average. Here, mental rules are not transferred from one brain to another. The child may adopt a rule that is unlike any of the rules in the brains of his models. The rules in particular brains do not replicate because no rule is copied faithfully. The phonological system can nonetheless evolve in a quite Darwinian way. More attractive forms of pronunciation can increase if they have a disproportionate effect on the average. Rules affecting different aspects of pronunciation can recombine and thus lead to the cumulative evolution of complex phonological rules. It is true that the act of averaging will tend to decrease the amount of variation in the population each generation. However, phenotypic performances will vary as a result of age, social context, vocal tract anatomy, and so on. Learners will often misperceive an utterance. These sorts of errors in transmission will keep pumping variation into a population as averaging bleeds it away. In fact, averaging might be necessary to prevent high noise levels from injecting too much variation into the population (see Cavalli-Sforza and Feldman, 1981; Boyd and Richerson, 1985).

There are still other possibilities that differ even more radically from the replicator model. For example, a propensity to imitate the common type in the population can be coupled with high rates of individual learning to create a model in which there is little heritable variation at the individual level, but substantial heritability of group differences (Henrich and Boyd, 1998). In such a model the cumulative evolution of adaptive complexity can occur, and occur rapidly, through selective processes that act at the group level (Boyd and Richerson, 1990, 2002). Similarly, in recent models of the evolution of social institutions (Young, 1998), there is no cultural transmission at the individual level. Although individuals simply acquire the best response to their social environment by trial-and-error learning, the structure of social interactions creates persistent, heritable variation at the group level.

We do not understand in detail how culture is stored and transmitted, so we do not know whether culturally transmitted ideas and beliefs are replicators or not. If the application of Darwinian thinking to understanding cultural change depended on the existence of replicators, we would be in trouble. Fortunately, culture need not be closely analogous to genes. Ideas must be gene-like to the extent that they are somehow capable of carrying the cultural information necessary to give rise to the cumulative evolution of complex cultural patterns that differentiate human groups. They exhibit the essential Darwinian properties of fidelity, fecundity, and longevity, but, as the example of phonemes shows, this can be accomplished by a most ungene-like, replicatorless process of error-prone phenotypic imitation. All that is really required is that culture constitutes a system maintaining heritable variation.

Darwinian Models Are Useful

Science on the frontier often has an anarchic, nervy flavor because it must deal with multiple uncertainties. Of course, we would be better off knowing exactly what memes are. Papering over the uncertainties of how culture is stored and transmitted no doubt leads to errors and conceals areas of fruitful inquiry. But as the psychologists explore one part of the frontier, the evolutionists should probe others. Studying the population properties of cultural information has lots of implications for human cognitive psychology, and vice versa. For example, when a child has the chance to copy the behavior of several different people, does she choose a single model for a given, discrete cultural attribute? Or does she average, or in some other way combine, the attributes of alternative models? The minute you try to build a population model of culture, you see that this question is crucial. However, despite conducting thousands of experiments on social learning, psychologists apparently have never thought to answer this question. Just as at a four-way stop, it makes no sense for everyone to wait for everyone else. Watch what the other drivers are doing, certainly, but go whenever the road ahead is clear.

Many social scientists have reacted to the advent of Darwinian models of culture with palpable distaste (e.g., Hallpike, 1986), while others have embraced these ideas with enthusiasm (e.g., Runciman, 1998). Much of this variation can be explained by people's feelings about the current Balkanization of the social sciences. The world of social science is divided into self-sufficient "ethnies" like anthropology and economics that are content to follow the questions and presuppositions that govern their discipline. The inhabitants of this world regard other disciplines with a mixture of fear and contempt and take little interest in what they have to say about questions of mutual interest. Clearly, this is not a satisfactory state of affairs.

We believe that Darwinian models can help rectify this problem. Disciplines such as economics, psychology, and evolutionary biology take the individual as the fundamental unit of analysis. These disciplines differ about how to model the individual and his psychology, but because they have the same fundamental structure, there has been much substantive interaction between them. Nowadays, many economists and psychologists work closely together, and a rich new body of work, often called "behavioral economics," has rapidly become mature enough to be applied to important practical problems such as the effect of retirement accounts on national savings rates. In the same way, economists and evolutionary biologists have found it relatively easy to work together on evolutionary models of social behavior, a rapidly growing field in both disciplines.

Other disciplines like cultural anthropology and sociology emphasize the role of culture and social institutions in shaping behavior, and researchers in sociology, anthropology, and history find interaction with each other relatively comfortable. Bridging the gap between the individual and cultural disciplines has proved much more difficult. Darwinian models are useful precisely because they incorporate both points of view within a single theoretical framework in which individuals and culture are articulated in a way that captures some, if not all, of the properties that their respective specialists claim for them. In population-based models, culture and social institutions arise from the interaction of individuals whose psychology has been shaped by their social milieu. As a bonus, Darwinian models come with tools to investigate the population-wide, long-term consequences of the interactions between individuals and their culture and social institutions.

To see how useful population-based models can be, consider the problem of human cooperation. There is no coherent explanation for the vast scale of cooperation in contemporary human societies, or why the scale of cooperation has increased many 1000-fold over the last 10,000 years. Models in economics and evolutionary biology predict that cooperation should be limited to small groups of relatives and reciprocators. Many theories in anthropology simply assume (often implicitly) that cooperative societies are possible and that culturally transmitted beliefs and social institutions serve the interest of social groups, but no attempt is made to reconcile this assumption with the fact that people are at least partly self-interested. Darwinian models provide one cogent mechanism to explain human cooperation by identifying the conditions under which groups will come to vary culturally and predicting when such variation will lead to the spread of culturally transmitted beliefs that support large-scale cooperation (Soltis, Boyd, and Richerson, 1995). In such models, the effect of different culturally transmitted beliefs on group prestige and group survival shapes the kinds of beliefs that survive and spread. These group-level effects in turn influence what people want and what they believe and, therefore, their behavior. Other recent work on the evolution of institutions (Young, 1998; Richerson and Boyd, 2002) makes us optimistic that Darwinian models may have widespread utility.

Population thinking is also useful because it offers a way to build mathematical theory of human behavior that captures the important role of culture in human affairs. Mathematical theory has the great advantage of allowing conclusions to be reliably deduced from assumptions. Experience in economics and evolutionary biology also suggests that it leads to a kind of clear understanding that is difficult to achieve with verbal reasoning alone. Of course there is also a cost—mathematical theory is necessarily based on simplified models. However, the combination of mathematical and verbal reasoning is superior to either alone.

Memes are not a universal acid, but population thinking is a better mousetrap. Population modeling of culture offers social science useful conceptual tools and handy mathematical machinery that will help solve important, long-standing problems. It is not a substitute for rational actor models, or careful historical analysis. But it is an invaluable complement to these forms of analysis that will enrich the social sciences.

REFERENCES

Bandura, A. 1986. *Social foundations of thought and action: A social cognitive theory.* Englewood Cliffs, NJ: Prentice-Hall.

Blurton-Jones, N., & M. J. Konner. 1976. !Kung knowledge of animal behavior. In: *Kalahari hunter-gatherers: Studies of the !Kung San and their neighbors,* R. Lee, & I. DeVore, eds. Cambridge, MA: Harvard University Press.

Boyd, R., & P. J. Richerson. 1985. *Culture and the evolutionary process.* Chicago: University of Chicago Press.

Boyd, R., & P. J. Richerson. 1990. Group selection among alternative evolutionarily stable strategies. *Journal of Theoretical Biology* 145:331–342.

Boyd, R., & P. J. Richerson. 2002. Group beneficial norms spread rapidly in a structured population. *Journal of Theoretical Biology* 215:287–296.

Boyer, P. 1994. *The naturalness of religious ideas: A cognitive theory of religion.* Berkeley: University of California Press.

Bynon, T. 1977. *Historical linguistics.* Cambridge: Cambridge University Press.

Cavalli-Sforza, L. L., & M. Feldman. 1981. *Cultural transmission and evolution.* Princeton, NJ: Princeton University Press.

Dawkins, R. 1976. *The selfish gene.* Oxford: Oxford University Press.

Dawkins, R. 1982. *The extended phenotype.* Oxford: Oxford University Press.

Dennett, D. 1995. *Darwin's dangerous idea.* London: Penguin.

Galef, B. G. 1988. Imitation in animals: History, definitions, and interpretation of data from the psychological laboratory. In *Social learning, psychological and biological perspectives,* T. Zentall, & B. G. Galef, Jr., eds. (pp. 3–29). Hillsdale, NJ: Lawrence Erlbaum.

Hallpike, C. R. 1986. *The principles of social evolution.* Oxford: Clarendon Press.

Henrich, J., & R. Boyd. 1998. The evolution of conformist transmission and the emergence of between-group differences. *Evolution and Human Behavior* 19:215–242.

Lefebvre, L., & B. Palameta. 1988. Mechanisms, ecology, and population diffusion of socially-learned, food-finding behavior in feral pigeons. In: *Social learning, psychological and biological perspectives,* T. Zetall, & B. G. Galef, Jr., eds. (pp. 141–165). Hillsdale, NJ: Lawrence Erlbaum.

Mayr, E. 1982. *The growth of biological thought.* Cambridge, MA: Harvard University Press.

Needham, J. 1978. *The shorter science and civilisation in China* (vol. 1). Cambridge: Cambridge University Press.

Richerson, P. J., & R. Boyd. 1999. Complex societies: The evolutionary origins of a crude superorganism. *Human Nature* 10:253–289.

Runciman, W. G. 1998. Greek hoplites, warrior culture, and indirect bias. *Journal of the Royal Anthropological Institute* 4:731–751.

Salomon, S. 1992. *Prairie patrimony: Family, farming, and community in the Midwest.* Chapel Hill: University of North Carolina Press.

Soltis, J., R. Boyd, & P. J. Richerson. 1995. Can group functional behaviors evolve by cultural group selection? An empirical test. *Current Anthropology* 36:473–494.

Tomasello, M., A. C. Kruger, & H. H. Ratner. 1993. Cultural learning. *Behavioral and Brain Sciences* 16:495–552.

Visalberghi, E., & D. M. Fragazy. 1990. Do monkeys ape? In: *Language and intelligence in monkeys and apes*, S. Parker, & K. Gibson, eds. (pp. 247–273). Cambridge: Cambridge University Press.

Whiten, A., & R. Ham. 1992. On the nature and evolution of imitation in the animal kingdom: A reappraisal of a century of research. In: *Advances in the study of behavior*, vol. 21, P. J. B. Slater, J. S. Rosenblatt, C. Beer, & M. Milkinski, eds. (pp. 239–283). Academic Press, New York.

Young, H. P. 1998. *Individual strategy and social structure: An evolutionary theory of institutions*. Princeton, NJ: Princeton University Press.

Author Index

Subject Index